T0202945

Fundamentals of Spacecraft Attitude Determination and Control

SPACE TECHNOLOGY LIBRARY
Published jointly by Microcosm Press and Springer

The Space Technology Library Editorial Board

Managing Editor: **James R. Wertz** (Microcosm, Inc., El Segundo, CA)

Editorial Board: **Roland Doré** (Professor and Director International Space University, Strasbourg)
Tom Logsdon (Senior member of Technical Staff, Space Division, Rockwell International)
F. Landis Markley (NASA, Goddard Space Flight Center)
Robert G. Melton (Professor of Aerospace Engineering, Pennsylvania State University)
Keiken Ninomiya (Professor, Institute of Space & Astronautical Science)
Jehangir J. Pocha (Letchworth, Herts.)
Rex W. Ridenoure (CEO and Co-founder at Ecliptic Enterprises Corporation)
Gael Squibb (Jet Propulsion Laboratory, California Institute of Technology)
Martin Sweeting (Professor of Satellite Engineering, University of Surrey)
David A. Vallado (Senior Research Astrodynamicist, CSSI/AGI)
Richard Van Allen (Vice President and Director, Space Systems Division, Microcosm, Inc.)

For further volumes:
http://www.springer.com/series/6575

F. Landis Markley • John L. Crassidis

Fundamentals of Spacecraft Attitude Determination and Control

Springer

F. Landis Markley
Attitude Control Systems
Engineering Branch
NASA Goddard Space Flight Center
Greenbelt, MD, USA

John L. Crassidis
Mechanical and Aerospace Engineering
University at Buffalo
State University of New York
Amherst, NY, USA

ISBN 978-1-4939-5569-5 ISBN 978-1-4939-0802-8 (eBook)
DOI 10.1007/978-1-4939-0802-8
Springer New York Heidelberg Dordrecht London

© Springer Science+Business Media New York 2014
Softcover reprint of the hardcover 1st edition 2014
This work is subject to copyright. All rights are reserved by the Publisher, whether the whole or part of the material is concerned, specifically the rights of translation, reprinting, reuse of illustrations, recitation, broadcasting, reproduction on microfilms or in any other physical way, and transmission or information storage and retrieval, electronic adaptation, computer software, or by similar or dissimilar methodology now known or hereafter developed. Exempted from this legal reservation are brief excerpts in connection with reviews or scholarly analysis or material supplied specifically for the purpose of being entered and executed on a computer system, for exclusive use by the purchaser of the work. Duplication of this publication or parts thereof is permitted only under the provisions of the Copyright Law of the Publisher's location, in its current version, and permission for use must always be obtained from Springer. Permissions for use may be obtained through RightsLink at the Copyright Clearance Center. Violations are liable to prosecution under the respective Copyright Law.
The use of general descriptive names, registered names, trademarks, service marks, etc. in this publication does not imply, even in the absence of a specific statement, that such names are exempt from the relevant protective laws and regulations and therefore free for general use.
While the advice and information in this book are believed to be true and accurate at the date of publication, neither the authors nor the editors nor the publisher can accept any legal responsibility for any errors or omissions that may be made. The publisher makes no warranty, express or implied, with respect to the material contained herein.

Springer grants the United States government royalty-free permission to reproduce all or part of this work and to authorize others to do so for United States government purposes.

Printed on acid-free paper

Springer is part of Springer Science+Business Media (www.springer.com)

To Gail, Michelle, and Isaac.
To Pam and Lucas,
and in memory of Lucas G.J. Crassidis.

Preface

This text provides the fundamental concepts and mathematical basis for spacecraft attitude determination and control. It is intended to serve as both a textbook for undergraduate and graduate students and a reference guide for practicing professionals. A primary motivation of this text is to develop the theory of attitude determination from first principles to practical algorithms, because very few of the existing texts on spacecraft control treat spacecraft attitude determination in depth. We emphasize specific applications; so the reader can understand how the derived theory is applied to actual orbiting spacecraft. We also highlight some simplified analytical expressions that can serve as first cut analyses for more detailed studies and are especially important in the initial design phase of a spacecraft attitude determination and control system. For example, Sect. 6.3 shows a simple single-axis analysis that is widely used by spacecraft engineers to design actual attitude estimation hardware and software configurations and to predict their performance. Analyses of this type can also help to determine whether anomalous behavior encountered by an orbiting spacecraft is due to a hardware problem, a control system design error, or a programming error, and to find a path to resolving the anomaly. In writing this book the authors hope to make a significant contribution toward expediting the process most newcomers must go through in assimilating and applying attitude determination and control theory.

Chapter 1 provides an introduction to the concepts presented in the text. It serves to motivate the reader by giving a historical review of the subject matter, including discussions of several actual missions that cover a broad spectrum of attitude determination and control designs. It also provides examples of how an inadequate theoretical analysis can cause a failure to meet mission objectives or, even worse, loss of a mission. Some examples of creative strategies to recover from potential mission failures are also presented. Chapter 2 begins with a review of linear algebra and moves on to the basic ideas of spacecraft attitude studies: reference frames, transformations between reference frames, and alternative representations of these transformations. Chapter 3 provides detailed derivations of attitude kinematics and dynamics, including a treatment of the torques acting on spacecraft. Chapter 4 provides the mathematical models behind the most common types of spacecraft sensors and actuators. Actual hardware specifications of these sensors and actuators are not given because the rapid development of this technology would quickly render them obsolete. Chapter 5 is a detailed treatment of attitude determination methods that do not depend on a retained memory of past observations. The most famous method in this category, the solution of Wahba's problem, is introduced, accompanied by several algorithms for solving it and a rigorous statistical analysis of its estimation errors. Chapter 6 covers attitude determination methods that mitigate the effects of sensor errors by incorporating dynamic models into a filtering process to retain a memory of past observations. Here the focus is on Kalman filtering, including both calibration and mission mode filters. Chapter 7 shows the fundamentals of attitude control, including some recent theoretical advancements on the effects of noise on the control system design.

Chapter 8 gives a list of quaternion identities, many of which have not appeared in open literature. Chapter 9 presents the explicit equations for the attitude matrices and kinematic equations for all 12 Euler angle representations of the attitude. Chapter 10 gives an overview of orbital dynamics to provide the background required to understand several topics in the text, such as the local-vertical/local-horizontal frame in Sect. 2.6.4. Chapter 11 provides a summary of environment models, which is crucial to understanding their effect on spacecraft translational and rotational motion. Chapter 12 reviews the theoretical basis of general control and estimation in enough detail to refresh the reader's memory of the concepts that underlie the applications in the text.

This book is the product of many years of experience possessed by the authors working on actual spacecraft attitude determination and control designs for numerous missions. Several actual mission examples are presented throughout the text to help the reader bridge the gap between theory and practice. For example, Sect. 5.8 presents an attitude determination algorithm that is employed onboard the Tropical Rainfall Measuring Mission (TRMM) as of this writing. Most of the mission examples in this text provide representative examples of typical mission mode designs. However, the authors wish to also show how the theory presented in the text can be applied to nonstandard mission modes derived from unique requirements. Section 7.7 presents the specific example of the range of mission modes actually used in the attitude determination and control system of the Solar, Anomalous, and Magnetospheric Particle Explorer (SAMPEX), each mode responding to its own unique challenges. Section 6.4.5 presents experimental magnetometer calibration results from the Transition Region and Coronal Explorer (TRACE) to give the reader a taste of how algorithm designs perform with real data. All of the examples are based on a "ground up" approach, i.e., starting with the fundamentals and leading towards an actual onboard algorithm.

As stated previously, this text can be used for an undergraduate course in spacecraft dynamics and controls. The second author has taught a senior level course in this area for many years, which is split into two parts: (1) orbital dynamics and (2) attitude kinematics and dynamics. Chapter 10 provides the material that is used for the orbital dynamics portion of the course. Chapters 2 and 3 provide the necessary material for the attitude kinematics and dynamics portion of the course.

Material for a graduate level course can include a brief review of orbital dynamics from Chapter 10 followed by the environmental models shown in Chapter 11. Chapters 5 and 6 can provide the bulk of a graduate course covering attitude determination and estimation. Chapter 7 can also be used in the graduate course to provide an introduction to attitude control. The authors believe that the entire book can serve as a reference or refresher for practitioners, which provides the primary motivation for including Chap. 4 in the text.

To encourage student learning we have incorporated both analytical and computer-based problems at the end of each chapter. This promotes working problems from first principles. General computer software and coded scripts have deliberately not been included with this text. Instead, a website with computer programs for all the examples shown in the text can be accessed by the reader (see Appendix). Although computer routines can provide some insights into the subject, we feel that they may hinder rigorous theoretical studies that are required to properly comprehend the material. Therefore, we strongly encourage students to program their own computer routines, using the codes provided from the website for verification purposes only.

We are indebted to numerous colleagues and students for contributions to various aspects of this work. In particular, we wish to express our gratitude to Mark Psiaki and especially to John Junkins for encouraging us to write this book. Many students have provided excellent insights and recommendations to enhance the pedagogical value, as well as developing new problems that are used as exercises. Although there are far too many students to name individually here, our sincere appreciation goes out to them. We do wish to acknowledge the significant contributions on the subject matter to the following individuals: Christopher Nebelecky for providing the section on the disturbing forces in Sect. 10.3, Agamemnon Crassidis for providing inputs on the sliding mode control in Sect. 12.2.3, John Downing for suggesting the dodecahedral wheel configuration in Sect. 4.8.3, and Sam Placanica for providing Fig. 4.16. Our heartfelt thanks to the following individuals for proofreading the text: Michael Andrle, J. Russell Carpenter, Yang Cheng, Joanna Hinks, Alice Liu, Adonis Pimienta-Peñalver, Steve Queen, Matthias Schmid, and Matthew Whittaker. We also wish to thank the following individuals for their many discussions and insights throughout the development of this book: K. Terry Alfriend, Roberto Alonso, Penina Axelrad, Itzhack Bar-Itzhack, Frank Bauer, J. Russell Carpenter, Yang Cheng, Daniel Choukroun, Neil Dennehy, Adam Fosbury, Michael Griffin, Christopher Hall, Joanna Hinks, Henry Hoffman, Kathleen Howell, Johnny Hurtado, Moriba Jah, Jer-Nan Juang, N. Jeremy Kasdin, Ken Lebsock, E. Glenn Lightsey, Richard Linares, Michael Lisano, Alice Liu, Manoranjan Majji, D. Joseph Mook, Daniele Mortari, Jim O'Donnell, Yaakov Oshman, Mark Pittelkau, Steve Queen, Reid Reynolds, Hanspeter Schaub, Conrad Schiff, Malcolm Shuster, Tarun Singh, Puneet Singla, Eric Stoneking, Sergei Tanygin, Julie Thienel, Panagiotis Tsiotras, James Turner, S. Rao Vadali, James Wertz, Bong Wie, and Renato Zanetti. Also, many thanks are due to several people at Springer, including our editor Maury Solomon, our assistant editor Nora Rawn, and Harry (J.J.) Blom. Finally, our deepest and most sincere appreciation must be expressed to our families for their patience and understanding throughout the years while we prepared this text. This text was produced using $\LaTeX\,2_{\varepsilon}$(thanks Yaakov and HP!). Any corrections are welcome via email to *Landis.Markley@nasa.gov* or *johnc@buffalo.edu*.

Greenbelt, MD, USA F. Landis Markley
Amherst, NY, USA John L. Crassidis

Contents

Chapter 1
Introduction

Spacecraft attitude determination and control covers the entire range of techniques for determining the orientation of a spacecraft and then controlling it so that the spacecraft points in some desired direction. The attitude estimation and attitude control problems are coupled, but they can be considered separately to some extent. The *separation theorem* for linear systems shown in Sect. 12.3.9 tells us that the control system can be designed without considering the estimator and vice versa. Specifically, the feedback gains in the control system can be chosen assuming that the system's state is perfectly known. No general separation theorem exists for nonlinear systems, including spacecraft attitude control systems, but the pointing requirements for most space missions have been satisfied by designing the attitude determination and control systems separately.

It is sometimes useful to distinguish between attitude determination and attitude estimation, although this distinction is often blurred. Attitude determination in this strict sense refers to memoryless approaches that determine the attitude point-by-point in time, quite often without taking the statistical properties of the attitude measurements into account. Attitude estimation, on the other hand, refers to approaches with memory, i.e. those that use a dynamic model of the spacecraft's motion in a filter that retains information from a series of measurements taken over time.

Malcolm Shuster made the incisive observation that attitude estimation is the youngest of the four quadrants of astronautics shown in Table 1.1, which is adapted from his work [33]. The table shows the attitude estimation quadrant to be actually empty before the launch of Sputnik, the first artificial satellite, in 1957. As Shuster pointed out,"There were, apparently no eighteenth- or nineteenth-century contributors to attitude estimation of even modest calibre," because "there was simply no great problem in Attitude Estimation waiting to be solved." The field of attitude estimation actually remained in a very underdeveloped state even into the

F.L. Markley and J.L. Crassidis, *Fundamentals of Spacecraft Attitude Determination and Control*, Space Technology Library 33, DOI 10.1007/978-1-4939-0802-8_1,
© Springer Science+Business Media New York 2014

Table 1.1 Founders of astronautics

	Dynamics	Estimation
Orbit	Newton (1642–1727)	Kepler (1571–1630
	Lagrange (1736–1813)	Lagrange (1736–1813)
	Hamilton (1805–1865)	Gauss (1777–1855)
	Einstein (1874–1955)	
Attitude	Euler (1707–1783)	(gone fishin')
	Cayley (1821–1895)	

late 1970s, although hundreds of spacecraft missions existed by that time, starting with Sputnik and including the Apollo missions from the late 1960s to the early 1970s.

A natural question is: "With all these missions why didn't attitude estimation become important until the late 1970s?" The answer does not necessarily lie in the absence of mathematical developments, as one might expect. Harold Black had developed the *algebraic method* for the point-by-point determination of a spacecraft's attitude from a set of two vector observations in 1964 [4, 24]. Shuster later renamed this algorithm TRIAD [35], the name stemming from an IBM Federal Systems Division internal report called "Tri-Axial Attitude Determination System." One year after Black's breakthrough, Grace Wahba published her famous attitude determination problem involving any number of vector observations [40]; but the first solutions of this problem never found practical application [13]. The first practical algorithm was Paul Davenport's celebrated q method, which solved for the quaternion parameterizing the attitude [24]. Unfortunately, this algorithm required performing an eigenvalue/eigenvector decomposition of a 4×4 matrix. It was used to support NASA's High Energy Astronomy Observatories (HEAO 1–3) [10], but was not practical for most missions on even the mainframe computers of the 1970s [34]. Shuster developed his QUaternion ESTimator (QUEST), which did not require an eigenvalue/eigenvector decomposition, to meet the greater throughput demands of the Magsat mission, launched in late 1979 [35]. Many other practical solutions to Wahba's problem have been developed since that time, but QUEST was the first algorithm suited to onboard computer processors and is still the most widely employed [27].

James Farrell published the first known paper using a Kalman filter for attitude estimation in 1970 [12], but a constant-gain filter had been proposed earlier [21], and Kalman filters for attitude estimation had appeared previously in contractor reports [11] and conference reports [30, 39]. Kalman filters were employed during the late 1970s and early 1980s to both filter noisy measurements and estimate for the attitude and gyro biases. The spacecraft dynamics and measurement models are nonlinear, though, so the filter is actually a quasi-linear extended Kalman filter (EKF, see Sect. 12.3.7.2), which does not possess the guaranteed convergence that the linear Kalman filter has. The EKF provides higher accuracy attitude estimates than point-by-point methods, but it is more expensive computationally, and the threat of divergence made attitude control engineers initially slow to adopt it. Farrell used the Euler angle attitude parameterization [12], but implementations using the quaternion parameterization had gained prominence by the mid 1970s [29]. A survey paper on quaternion-based attitude EKFs appeared in 1982, bringing more attention to that approach [23]. However, a conventional Kalman filter using the quaternion parameterization results in a covariance matrix that is nearly singular.[1] This could cause problems, especially with single-precision computers, but they were overcome by using a local unconstrained three-dimensional parameterization for the attitude error and the quaternion for the global parameterization of the attitude.

Reference [42] provides an excellent summary of the state of the art in 1978, but most of the attitude estimation processing described in this reference was performed by large mainframe computers on the ground. Algorithms for performing attitude determination and estimation existed well before that time, but space-hardened computer processors with the power required to execute them onboard spacecraft were not available. This raises the question: "How did spacecraft control their attitude without having attitude estimates?" Some early spacecraft, such as Sputnik and the Echo 1 and 2 satellites (mylar balloons 30 and 40 m in diameter), were spheres with no pointing requirements at all. Other spacecraft used passive stabilization methods: spacecraft spin or a momentum wheel providing a constant angular momentum bias and/or gravity-gradient stabilization (see Chap. 3). Infrequent control torques to be applied by active mechanisms were computed on the ground and telemetered to the spacecraft. Other spacecraft with active control systems derived their control commands directly from sensor data, using simple analog circuitry.

Most modern-day spacecraft missions have specific pointing requirements, otherwise known as *pointing modes*. Some examples include Earth pointing, inertial pointing, and Sun pointing. Communications and broadcast satellites are a special case of Earth-pointing spacecraft. They are typically placed in geostationary orbits along the equator, with an orbital period equal to the Earth's rotational period (i.e. one sidereal day), so they appear from the ground to be stationary in the sky. Their angular momentum vectors are perpendicular to the equator and their attitude rotation rates are also the same as the Earth's rate, keeping their antennas constantly pointed at the Earth. This explains why a satellite TV dish can be pointed to a fixed location for reception purposes. An example of an inertially pointing spacecraft is the Hubble Space Telescope (HST), which can point its main mirror to a fixed location in the sky for several hours to collect enough faint light to provide a good image. An example of a Sun pointing spacecraft is the Solar & Heliospheric Observatory (SOHO) which is used to study the Sun from its deep core to the outer corona and the solar wind. Sometimes the pointing mode changes during the spacecraft mission. This may be due to changing requirements or failure

[1]Reference [23] states that the covariance matrix is exactly singular, but this would only result from a linear constraint, while the quaternion norm constraint is quadratic. Itzhack Bar-Itzhack and Jeremy Kasdin pointed this out to the authors in personal discussions.

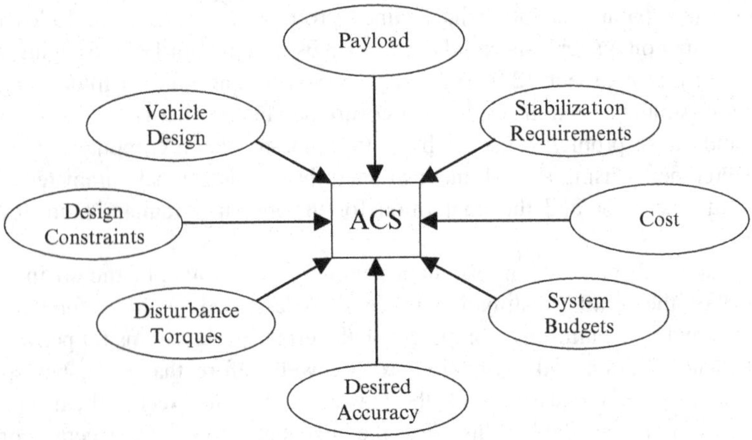

Fig. 1.1 Factors affecting ACS selection

of some hardware component. The Solar, Anomalous and Magnetospheric Particle Explorer (SAMPEX) discussed in Sect. 7.7 provides an example of changing pointing algorithms to meet changing pointing requirements.

A spacecraft's overall *pointing error* is a function of both the attitude determination/estimation errors and control errors, because the control system takes its inputs from the attitude determination/estimation system. As an example consider the non-realistic case where the control errors are exactly zero. If the attitude determination errors are 2°, then the overall pointing errors are also 2° even though the control is perfect. Therefore, when considering the overall pointing design requirements both the control errors and attitude determination/estimation errors must be taken into account. Pointing requirements can include requirements for both holding a desired orientation and for maneuvering the spacecraft from one orientation to another. The desired orientation may be a static resting point or a time varying trajectory. The trajectory itself may be predefined specifically, such as pointing towards a particular location on the ground from low-Earth orbit (LEO), or not specifically defined, such as a general spin that scans the universe. The Wilkinson Microwave Anisotropy Probe (WMAP) spacecraft spent 9 years performing a repetitive "spirograph" scan of the celestial sphere, as described in Sect. 7.3.

Reference [36] provides an excellent overview of the factors that affect an attitude control system (ACS) configuration, as shown in Fig. 1.1. Some of these factors are obvious, such as the payload, cost and desired accuracy, but others take much longer to quantify. More often than not the stabilization requirements may involve multiple pointing modes of the spacecraft. For example, the *mission mode* is the primary mode to complete the mission objectives. In addition to this mode a *safehold mode* is usually required to cope with an anomaly. An example of this mode is a Sun pointing mode that holds the solar arrays to face the Sun and removes power from all non-critical subsystems, minimizing power demands

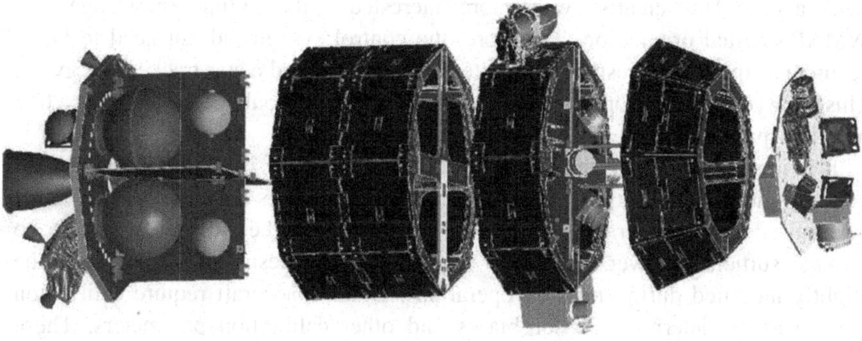

Fig. 1.2 LADEE modular common spacecraft bus architecture

until the anomaly is resolved. The spacecraft's inertia properties and the expected disturbance torques must be accounted for during the ACS design process. The configuration of the vehicle may impose design constraints on the ACS in the form of system budgets for weight, power and volume. This leads to various definitions for the configuration. For example, a CubeSat has a volume of exactly one liter and has a mass of no more than 1.33 kg. Current designs focus on commercial off-the-shelf (COTS) technology, which is a federal acquisition regulation term defining a non-developmental item of supply that is both commercial and sold in substantial quantities in the commercial marketplace. The Lunar Atmosphere and Dust Environment Explorer (LADEE) satellite uses a modular common spacecraft bus architecture along with COTS components, as shown in Fig. 1.2, in order to significantly reduce costs. These types of designs will certainly be prevalent for many future missions.

Although reducing costs is important, it is equally important that a proper ACS achieves its stated objective throughout the lifetime of the mission. This begins with a complete understanding of the theory behind an ACS. Most often the particular mission objectives drive the requirements. Engineers are usually given the objectives and then they must develop a proper ACS that meets these objectives, keeping all the factors shown in Fig. 1.1 in mind. But sometimes engineers are able to drive the objectives. An example of this is an experience that this book's authors had for the contingency mode for the Tropical Rainfall Measuring Mission (TRMM) spacecraft. The scientists asked the authors "Roughly what attitude accuracy do you think can be achieved with three-axis magnetometer and Sun sensors?" After a simple analysis we responded that 0.7° should be doable. The very next day the official requirement from the scientists for the contingency mode was stated to be 0.7°! The authors and other engineers then performed a detailed analysis to prove that this requirement could be met [6]. Other times one system is more important than another system. This is the case for the Wilkinson Microwave Anisotropy Probe (WMAP) spacecraft [25], which is used to make fundamental cosmological measurements. The ACS objectives are to provide a complete scan of the universe twice a year. The scientists were more interested in the attitude knowledge than WMAP's actual orientation. Therefore, the control system did not need to be as accurate as the attitude estimation system, which involved a star tracker and gyros. This allowed for a less sophisticated control design than would be required for a fine pointing spacecraft.

It is important to note that other requirements may be imposed on the overall pointing mode. For example, nearly every spacecraft uses solar arrays for power generation, and the arrays must be pointed towards the Sun well enough to provide sufficient power. This may require that the desired pointing mode be slightly modified during mission operations. Other spacecraft require calibration maneuvers to determine sensor biases and other calibration parameters. These attitude maneuvers typically require that the spacecraft be capable of controlling all three of its axes, which is often broadly termed *three-axis stabilization*.

The history of attitude determination/estimation has been well documented, but the history of attitude control is less known. This is in part due to the fact that this field of study was classified Secret in the early days of spacecraft mission designs. Robert Roberson [32] claims that its birth year is 1952, with the first publicly published works by him and Vladimir Beletskii, who was in the USSR, appearing in the mid 1950s. Early designs were based on linearized equations of motion to study the behavior of the control system. Active control relies on only a handful of available actuators: (1) momentum exchange devices such as reaction wheels, (2) thrusters and (3) magnetic torquers. As early as 1961 magnetic torquers had been proposed both for momentum dumping and active pointing control. According to Roberson, the first Orbiting Solar Observatory (OSO 1) was the first known controlled spacecraft. This used a *dual-spin* design, with a rotating *wheel* to provide gyroscopic stability, and a *sail* driven electrically against the wheel's rotation to point at the Sun. Although modern-day actuators have not changed from their initial purpose, technology has advanced greatly since the early days to provide much greater efficiency and tighter pointing performance. For example, in order to take images of distant, faint objects, HST must point extremely stably and accurately. The telescope is able to lock onto a target without deviating more than 7/1000th of an arcsecond, or the width of a human hair seen at a distance of about 1.5 km. Fine guidance sensors looking through HST's main optics are used to provide the attitude knowledge while reaction wheels provide the necessary control. These reaction wheels must be very accurately balanced and isolated so that vibration is as small as possible; otherwise a jitter effect would ensue causing image blurring.

Spacecraft attitude control methods fall into two broad classes: passive control and active control. Passive control methods use the natural spacecraft dynamics to satisfy their pointing requirements. The simplest of these is spin stabilization, which depends on the fact that the direction of a rigid spinning body's angular momentum is relatively immune to disturbances (see Sect. 3.3.3). The Pioneer series of spacecraft provides a good example of early spinners. Pioneer 10 and 11 each carried a sensor called the Imaging Photopolarimeter (IPP), as shown in Fig. 1.3, which measured the strength of sunlight scattered from the clouds of Jupiter [31]. It then converted this information into digital format of different shades of red

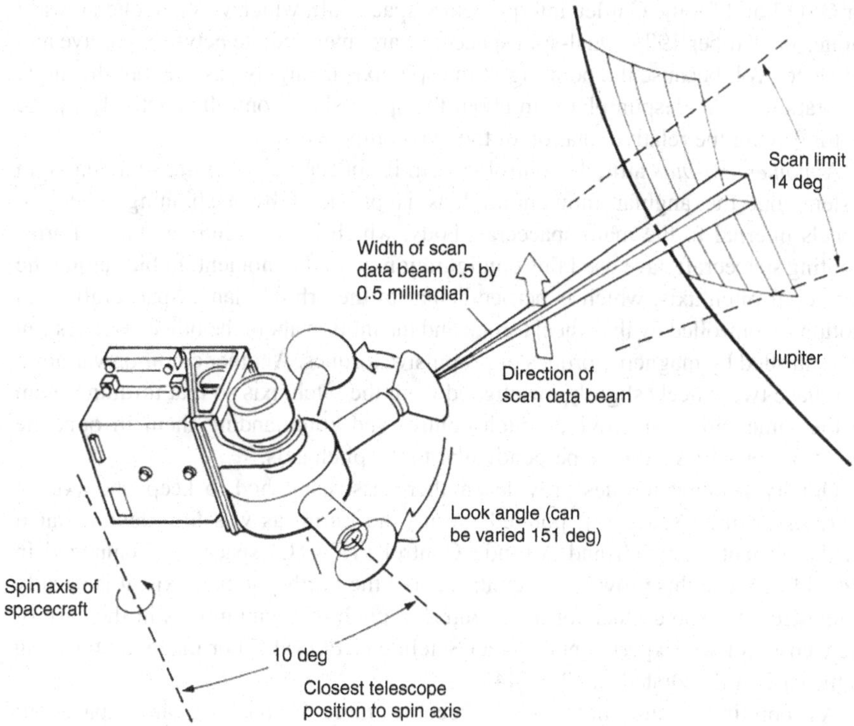

Fig. 1.3 Pioneer imaging photopolarimeter

and blue that made up each image of Jupiter. The telescope axis was swept about the spacecraft spin axis through 360° in the look angle as the spacecraft rotated at its nominal spin rate. It could also be positioned in discrete 0.5 mrad steps of cone angle over a range from 29° from the earthward direction (limited by the spacecraft antenna structure) to 170° from earthward. This allowed two-dimensional mapping of the celestial sphere in any one of three distinct operational modes. The total angle coverage per spin rotation depended on the spacecraft data transmission rate; during encounter it was limited to 70° and 14° (28° at low sample rate) for photopolarimetry and imaging, respectively. Any swath of the 360° rotation could be covered by selecting the starting angle. In the faint light mode the entire roll was covered.

As shown by the example of Pioneer, one advantage of spinning spacecraft is that the spin can be used by ACS and payload sensors to scan targets. A disadvantage, of course, is that a spinning platform is very inconvenient for sensors or antennas that need to point at fixed targets. Dual-spin spacecraft have both a spinning component and a *despun platform* that is not rotating. The dual-spin configuration has mainly been used for communications or broadcast satellites, where the despun platform contains an antenna pointed at a receiving station on the Earth, but it was also used for OSO 1 and for the Galileo interplanetary spacecraft, which was launched toward Jupiter in October 1989. Dual-spin spacecraft are intermediate between passive and active control, because the pointing of the spin axis is largely passive, but the angle of rotation of the despun platform about the spin axis is controlled actively by the motor driving the relative rotation of the two components.

A *momentum-bias* attitude control system is conceptually similar to a dual-spin system, but the angular momentum bias is provided by a spinning wheel or wheels internal to the main spacecraft body, which is not spinning. Many Earth-pointing spacecraft have used this configuration, with the momentum bias along the spacecraft pitch axis, which is perpendicular to the orbital plane. Spacecraft pitch motion is controlled by the wheel torque and the motion about the other two axes can be controlled by magnetic torques or propulsive torques. Another common variation is to have two wheels slightly misaligned from the pitch axis so that torquing them in the same direction provides pitch control and commanding them in opposite directions provides a torque perpendicular to the pitch axis.

Gravity-gradient torques provide another passive method to keep one axis of an Earth-orbiting spacecraft pointed toward the Earth, as was first demonstrated on the Transit Research and Attitude Control (TRAAC) spacecraft, launched in 1962 [45]. Since this provides no control about the Earth-pointing axis, it is usually combined with some other form of control, a pitch momentum bias in the case of the Geostationary Experimental Ocean Satellite (GEOS) [42] or magnetic torquing in the case of the Ørsted satellite [44].

As pointing requirements have become more demanding and spacecraft computers have become more capable, spacecraft ACS designs have largely moved to three-axis stabilization. Most actively-controlled spacecraft use thrusters, reaction wheels or, less frequently, control moment gyros (CMGs) as their primary controllers. Thrusters by their nature use expendable propellants, limiting the life of a mission. Reaction wheels or CMGs, on the other hand, require only electrical power that can be supplied by solar arrays. They also are well suited for repointing and absorbing periodic disturbance torques, which cause no long-term change in their stored angular momentum. Secular disturbances, on the other hand, lead to momentum buildup in the wheels, which must be unloaded (or "dumped") using external torques provided by attitude thrusters or magnetic torquers. Magnetic torquers are preferable because they do not use expendables and are less disruptive to the spacecraft's primary function, but they are generally only employed in LEO.

The Geostationary Operational Environmental Satellites (GOES), which supply the weather pictures seen on televised weather programs, provide an excellent example of the evolution of attitude control system designs. The first three of these satellites, GOES 1–3,[2] were spin-stabilized. Weather images were obtained by a Visible Infrared Spin Scan Radiometer (VISSR) which used the satellite spin to scan in the East-West direction, and a mirror stepped once per spin period to

---

[2]NASA satellites are given letter designations before launch and numerical designations after they attain orbit, so the first three GOES satellites were also known as GOES A-C.

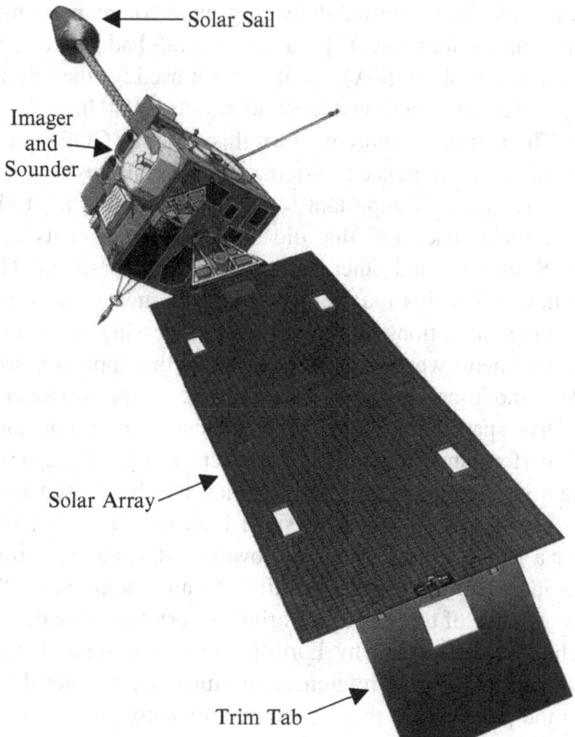

Fig. 1.4 Geostationary operational environmental satellite I–M

scan North to South. GOES 4–7 were dual-spin satellites, with high-gain antennas on the despun platform to increase the speed of data transmission to the Earth. One drawback of these designs was that the VISSR spent only a very small fraction of its spin period actually scanning the Earth. In order to increase the time spent actually gathering weather imagery, the GOES 8–12 satellites, depicted in Fig. 1.4, were designed as momentum-biased three-axis controlled spacecraft, with the scan provided by the science instruments [37]. These were the imager, which senses infrared and visible energy from the Earth's surface and atmosphere, and the sounder, which provides data for vertical atmospheric temperature and moisture profiles, surface and cloud top temperature, and ozone distribution. The main disturbance at geosynchronous altitudes is the torque caused by solar radiation pressure on solar arrays, so geosynchronous spacecraft usually have two solar arrays on opposite sides to balance the torque. The imager and sounder need a cold field of view for their radiators, so GOES could not have balanced solar arrays. A small solar sail on an extended boom was provided to balance the solar torque, but the combination of imperfect symmetry, center of mass shifts, and changing surface optical properties resulted in a seasonally-varying net torque. The resulting accumulated momentum was unloaded by thrusters or by magnetic torquing, the latter with mixed success at geosynchronous altitude. A trim tab at the end of the solar array was adjusted daily by ground command to minimize the need for active momentum dumping [17]. Each spacecraft had a set of gyros, the Digital Integration Rate Assembly (DIRA), but it was not used for the mission mode ACS because the gyros had an operational lifetime of only 2,000 h.

The pointing requirement for this series of GOES satellites was 12 μrad relative to an Earth referenced system, which translates to roughly 4 km knowledge on the ground. It is important to note that this was the tightest pointing requirement ever for a spacecraft that did not use data from its main instruments to aid the ACS, as HST and other fine-pointing spacecraft do. The original ACS designers believed that this requirement could be met using a horizon sensor only and by making corrections through ground processing of the data [20]. Unfortunately, the requirements were not met on orbit with this approach, so a NASA team investigated ACS modifications needed to improve the performance of the next generation of GOES spacecraft [8, 26]. The present authors concluded that the desired level of performance required star trackers and gyros capable of providing continuous data in a stellar-inertial ACS, which had been used for Earth-pointing spacecraft in near-Earth orbit at least since Landsat 4 in 1982, but never to our knowledge for a geostationary satellite. However, star trackers provide an attitude relative to an inertial frame, and converting the attitude to an Earth reference frame requires knowledge of the spacecraft orbit,[3] so orbit errors will contribute to pointing errors. This is the reason why Earth-pointing spacecraft have historically used horizon sensors as pointing references. Fortunately, it is not difficult to reduce orbit errors to the point where they do not significantly degrade pointing performance, so the stellar-inertial design could meet the pointing requirements.

Reference [26] had been presented at the 1991 AIAA Guidance, Navigation, and Control Conference, and it was well known to the spacecraft manufacturers interested in building the next generation of GOES spacecraft. Thus it is no surprise that later GOES satellites starting with GOES 13 have used many of the improvements suggested in that paper, including a stellar-inertial ACS mode [38]. Increased cooling demands of the imager and sounder precluded the use of a solar sail on these spacecraft, so they must perform daily momentum dumps, but the improved performance of the ACS allows them to continue imaging during the dumps and to meet all their performance requirements [15].

The design of the TRMM ACS involved a similar decision between an Earth-referenced concept and a stellar-inertial concept. TRMM was developed at Goddard Space Flight Center concurrently with the Rossi X-Ray Timing Explorer (RXTE), and an effort was made to use common hardware and software wherever feasible. RXTE was an astronomical mission, and its stringent pointing stability requirement of 30 arcsec led it to incorporate a stellar-inertial ACS [3]. The TRMM engineers decided not to use a stellar-inertial system, because their much less stringent attitude knowledge requirement of 0.18° per axis could be met by using pitch and roll measurements from an Earth Sensor Assembly (ESA), and obtaining yaw

[3]This is discussed in Sects. 2.6.4 and 5.7.

knowledge from gyros updated twice each orbit by Sun sensor measurements. Although this decision was controversial, it was supported by the first author of this book on the principle that an Earth-pointing mission should use an Earth-referenced ACS. It appeared to be an unfortunate decision in early 1994, after development of the TRMM ACS was well underway, when ACS engineers became aware of a progressive degradation of similar ESAs on the Defense Meteorological Satellite Program (DMSP) spacecraft. They considered adding a star tracker to TRMM to protect against this possibility, but avoided the severe monetary and schedule penalties that this would entail by developing a contingency mode using gyros and magnetometers, based on results that had been obtained in ground-based attitude determination of the Upper Atmosphere Research Satellite (UARS) using flight gyro and magnetometer data [18].

Implementation of TRMM's contingency mode required an onboard filter to reduce the effects of sensor noise and magnetic field modeling errors. The common hardware procurement with RXTE spacecraft provided gyros that were much more accurate than required by the ESA-based TRMM ACS but provided the dynamic memory needed by the filter. It was not clear that the onboard processor could handle the computational load of a full Kalman filter in addition to the existing ACS software, so the authors of this book conducted a study to develop algorithms that could provide near-optimal attitude estimates with a lower computational load [6]. Our investigation showed that one eigenvalue of the covariance matrix of attitude errors was much larger than the other two, and its eigenvector was within 2.5° of the Sun vector. This reflects the fact that the more accurate Sun sensor cannot reduce the attitude error along the Sun line, which must be estimated using the less-accurate magnetometer. A simplified expression for the covariance using this analysis provided several alternate algorithms that obtained nearly the same accuracy as the Kalman filter at a fraction of the computational load [6].

It was determined, however, that the already-validated RXTE Kalman filter, modified to use magnetometer and Sun sensor data in place of star tracker data, would fit in the TRMM computer, so this alternative was chosen. In August of 2001 TRMM's orbit altitude was raised from 350 to 402 km to save propellant used to compensate for atmospheric drag. The ESA had performed well at the lower altitudes for which it was designed, but it ceased providing valid data above 380 km, and the contingency mode Kalman filter was enabled. It has since provided attitude accuracies of approximately 0.2°, about the same as the ESA had provided and much better than the allotted attitude knowledge accuracy for the contingency mode of 0.7° per axis [1].

In addition to difficulties in meeting requirements, as on GOES 8–12, it is not uncommon to experience an outright failure. Some of these are not recoverable, but ingenious workarounds can often be found. The Lewis spacecraft provides an example of an unrecoverable failure. The Lewis design engineers adopted a Sun-pointing ACS safehold mode that had been designed for the Total Ozone Mapping Spectrometer-Earth Probe (TOMS-EP) spacecraft. Unfortunately, the spacecraft axis that the Lewis safehold mode pointed at the Sun was an unstable axis of intermediate inertia (see Sect. 3.3.3), unlike the stable Sun-pointing axis on

TOMS-EP. This instability should have revealed itself in simulations performed during the design phase, but the simulations assumed perfectly aligned and balanced thrusters. When Lewis entered safehold mode on orbit, a small thruster imbalance caused the spacecraft to spin up around the unstable intermediate axis. This rotation could not be sensed by the Sun sensors, and the spin momentum started to transfer into the controlled principal axis causing the thrusters to fire excessively in an attempt to maintain control. But the ACS processor was programmed to shut down the control system if excessive firings occurred. The spin momentum then was transferred to the principal axis, which rotated the spacecraft 90°, causing the solar arrays to be pointed nearly edge-on to the Sun, resulting in a fatal loss of power from which Lewis could not recover.

The Solar Maximum Mission (SMM) provides an example of a recovery from a potentially fatal malfunction. Launched in February 1970, SMM was the first of the Multimission Modular Spacecraft (MMS) with independent modules designed to be replaceable on orbit [9, 16]. SMM's Modular Attitude Control System (MACS) functioned flawlessly for the first nine months after launch, but then fuse failures permanently disabled three of its four reaction wheels, which were the primary ACS actuators. With only one wheel operational, the experiment axis would drift away from the Sun and severe thermal and power problems would develop. Engineers at Goddard Space Flight Center developed an algorithm to spin-stabilize SMM, which was uplinked to the onboard computer less than six weeks after the loss of the first wheel [19]. A successful spinup was achieved, leaving the spacecraft rotating at about 1 deg/s and pointing near the Sun in a thermal- and power-safe state. Three of the seven observatory experiments continued to operate while spinning to return solar constant measurements as well as data on gamma ray and x-ray emissions from the Sun. SMM remained in the spin-stabilized mode until it was captured in 1984 by Space Shuttle *Challenger* and its MACS module was replaced in the first repair of an orbiting spacecraft.

Another rescue from a potentially fatal malfunction was performed on the International Ultraviolet Explorer (IUE), which was launched in January 1978 into a geosynchronous, but not geostationary, orbit. Because of its high altitude, IUE could not be serviced on orbit. It was equipped with six gyros, of which three were needed for the ACS. Three of the gyros had failed by July 1982, so a replacement two-gyro control algorithm was developed in 1982–1983 to permit continued use of the observatory in the anticipated loss of a fourth gyro [14]. This algorithm was ready for operation by Spring 1983, well before the fourth gyro failed on August 17, 1985. IUE continued science operations until it was decommissioned for non-ACS reasons in September 1996, 18 years and 9 months after launch, although it was built to last for 3 years with a goal of 5 years. A single-gyro control algorithm for IUE was developed and fully tested on orbit, but it was never needed. Their rescues of SMM, IUE, and several other spacecraft earned the engineers in Henry Hoffman's Guidance and Control Branch at Goddard Space Flight Center the appellation "Satellite Saviors" [22].

HST, deployed by Space Shuttle *Discovery* in April 1990, is the only telescope designed to be serviced on orbit by astronauts. Like IUE, HST was launched with six primary high-accuracy single-degree-of-freedom gyros for redundancy. In fact, the gyros are of the same basic design as IUE's, but they were improved to avoid the IUE's failure mode. Unfortunately, they experienced a different failure mode. If four of the six primary gyros were lost, vehicle health and safety could still be maintained by using a set of backup lower-accuracy gyros until the primary gyros could be brought on line, after replacement if necessary. The backup gyros have a limited operational life, however, so they should not be relied on for long-term control. Therefore, when two of HST's primary gyros failed during the first year of HST operation, it was felt prudent to develop a zero-gyro sunpoint (ZGSP) safehold mode [28]. Development of this safehold mode algorithm, which uses magnetometer and coarse Sun sensor data, went from concept to flight code in three months, with uplink to HST on January 20, 1992, and it maintained HST in a power and thermally safe state for 38 consecutive days between the failure of a fourth gyro in November 1999 and a December servicing mission to replace all six gyros.

Our final example of an ACS design using minimal hardware is provided by the Solar Radiation and Climate Experiment (SORCE) spacecraft [2]. The SORCE engineers developed a gyroless ACS design out of necessity after much of the ACS flight software had already been coded, when problems arose with their gyro procurement and no alternate gyros could be procured quickly that met spacecraft interface and resource requirements. Facing a very tight schedule, the ACS team designed and implemented a simple science mode that took advantage of the availability of two star trackers and a fine Sun sensor to derive accurate rates, and a safehold mode deriving rates from magnetometer and coarse Sun sensor measurements. The key system feature enabling the simplicity of these modes was that the science instrument was designed to point at the Sun and therefore could not be damaged by Sun exposure. Reference [5] provides an overview of gyroless pointing modes on several scientific spacecraft, including SORCE.

A sound theoretical foundation is the first and foremost step to achieve a good ACS design. The Lewis issue could have been easily avoided by a simple analysis that would have showed that the safehold mode had a spin about its intermediate axis. Understanding how orbit errors couple with attitude errors in the GOES star tracker design is another example of using theory to aid in the ACS analysis. The TRMM contingency filter, described above, also provides an illustrative case. A classic example is the roll/yaw coupling effect used to control communications satellites (see Sect. 3.3.8), which typically employ a horizon sensor for the ACS. A horizon sensor can provide only roll and pitch information, leaving the yaw rotation angle about the nadir vector unknown and the attitude not fully observable. However, roll/yaw coupling causes an error in yaw to become an error in roll a quarter orbit later, so yaw is controllable even though it is not observable. The yaw pointing accuracy is typically about an order of magnitude worse than roll and pitch, but this is of no concern for most communications satellites because their antennas are still pointed toward the Earth to meet their requirements.

Since the first publications on attitude control this area has been extremely active, producing a vast stream of theoretically-based publications that continues to this day. It is impossible to even begin to scratch the surface of the extensive developments in this area. Rather, we shall highlight some key aspects of the results shown in classic publications. The attitude control problem can be essentially broken down into two parts: (1) attitude regulation and (2) attitude tracking. Regulation drives the attitude toward a fixed orientation and keeps it there. Note that this is exactly the same as an inertial pointing mode. Tracking involves continuously re-orientating the attitude to follow some predefined attitude motion profile. This falls under the category of three-axis stabilization. Control techniques fall under both open-loop and closed-loop approaches, as is further discussed in Sect. 7.1.

As with estimation the parameterization of choice for control purposes is the quaternion. Bong Wie and Peter Barba produced one of the earliest works using this parameterization for the regularization problem in 1985 [43]. This work derives simple feedback laws that are essentially equivalent to classical proportional-derivative controllers and shows them to be stable through a Lyapunov analysis. Note that Wie and Barba did not prove asymptotic stability, but fortunately their control laws do lead to an asymptotically stable closed-loop system as shown in Sect. 7.2. The extension of this work to the tracking case is derived by John Wen and Kenneth Kreutz-Delgado [41]. Control theory has continued to evolve since the time of these classic papers, including robust control approaches, control methods that take into account actuation constraints, controllers that can work when an actuator fails (such as the two-wheel control problem), as well as many others that can be found in the open literature.

Many of the techniques described in this text have been extensively employed for many years, but the field continues to grow. Advances in attitude estimation include a number of nonlinear filtering algorithms developed since the survey paper of 1982 [23], many of which are critically reviewed in a more recent survey paper [7]. The most important theoretical tool that has significantly advanced the mathematics underlying attitude control is the use of Lyapunov-based methods to design controllers that guarantee stability for general nonlinear systems.[4] Although the potential advantages of the new control and estimation algorithms are undeniable, it is wise to apply the old adage "If it ain't broke don't fix it" to the simple but effective control laws and the standard extended Kalman filter, which have proved their worth on a multitude of spacecraft missions. Ultimately, enhanced confidence in the new approaches, coupled with the more stringent pointing requirements of future missions, will bring about more widespread use of these approaches, or of others as yet undiscovered, for spacecraft attitude determination and control.

[4] A review of Lyapunov stability is presented in Sect. 12.2.2.

References

1. Andrews, S.F., Bilanow, S.: Recent flight results of the TRMM Kalman filter. In: AIAA Guidance, Navigation and Control Conference. Monterey (2002). AIAA 2002-5047
2. Baird, G.C., Groszkiewicz, J.E.: Gyroless attitude control for the SORCE spacecraft. In: Chapel, J.D., Culp, R.D. (eds.) Guidance and Control 2004, Advances in the Astronautical Sciences, vol. 118, pp. 593–605. Univelt, San Diego (2004). AAS 04-076
3. Bauer, F.H., Femiano, M.D., Mosier, G.E.: Attitude control system conceptual design for the X-Ray Timing Explorer. In: AIAA Guidance, Navigation and Control Conference. Hilton Head (1992). AIAA 92-4334-CP
4. Black, H.D.: A passive system for determining the attitude of a satellite. AIAA J. **2**(7), 1350–1351 (1964)
5. Bruno, D., Class, B.F., Rovner, D., Baird, G., Groszkiewicz, J., Lebsock, K., Kruk, J.: A comparison of gyroless implementations on scientific spacecraft. In: AIAA Guidance, Navigation and Control Conference. Chicago (2009). AIAA 2009-5942
6. Crassidis, J.L., Andrews, S.F., Markley, F.L., Ha, K.: Contingency designs for attitude determination of TRMM. In: Proceedings of the Flight Mechanics/Estimation Theory Symposium, pp. 419–433. NASA-Goddard Space Flight Center, Greenbelt (1995)
7. Crassidis, J.L., Markley, F.L., Cheng, Y.: Survey of nonlinear attitude estimation methods. J. Guid. Contr. Dynam. **30**(1), 12–28 (2007)
8. Crassidis, J.L., Markley, F.L., Kyle, A.M., Kull, K.: Attitude determination improvements for GOES. In: Proceedings of the Flight Mechanics/Estimation Theory Symposium, pp. 151–165. NASA-Goddard Space Flight Center, Greenbelt (1996)
9. Falkenhayn Jr., E.: Multimission Modular Spacecraft (MMS). In: Proceedings of the AIAA Programs and Technologies Conference. Houston (1988). AIAA 88-3513
10. Fallon III, L., Harrop, I.H., Sturch, C.R.: Ground attitude determination and gyro calibration procedures for the HEAO missions. In: 17th Aerospace Sciences Meeting, AIAA 79-0397. New Orleans (1979)
11. Farrell, J.L.: Attitude determination by Kalman filtering. Contractor Report NASA-CR-598, NASA Goddard Space Flight Center, Washington, DC (1964)
12. Farrell, J.L.: Attitude determination by Kalman filtering. Automatica **6**, 419–430 (1970)
13. Farrell, J.L., Stuelpnagel, J.C.: A least-squares estimate of satellite attitude. SIAM Rev. **7**(3), 384–386 (1966)
14. Femiano, M.D.: Inflight redesign of the IUE attitude control system. In: 2nd Aerospace Maintenance Conference. San Antonio (1986). AIAA 86-1193
15. Gibbs, B.P., Uetrecht, D.S., Carr, J.L., Sayal, C.: Analysis of GOES-13 orbit and attitude determination. In: SpaceOps Conference. Heidelberg (2008). AIAA 2008-3222
16. Guha, A.K.: Solar Maximum Mission–a system overview. In: Proceedings of the AIAA 19th Aerospace Sciences Meeting. St. Louis (1981). AIAA 81-0200
17. Harvie III, E., Rowe, J., Tsui, Y.K.J.: Performance analysis of the GOES trim tab solar pressure torque angular momentum control. In: Proceedings of SPIE, GOES-8 and Beyond, vol. 2812. Denver (1996)
18. Hashmall, J.A., Liu, A.K., Rokni, M.: Accurate spacecraft attitudes from magnetometer data. In: CNES Spacecraft Dynamics (Colloque 95), pp. 169–179. Cépaduès-Éditions, Toulouse (1995)
19. Hoffman, H.C., Donohue, J.H., Flatley, T.W.: SMM attitude control recovery. In: Proceedings of the AIAA Guidance and Control Conference. Albuquerque (1981). AIAA 81-1760
20. Kamel, A.A., Bhat, M.K.P., Gamble, D., Scholtz, J.H.: GOES I–M image motion compensation system. In: 1992 AIAA Aerospace Design Conference. Irvine (1992). AIAA 92-1107
21. Kau, S., Kumar, K.S.P., Granley, G.B.: Attitude determination via nonlinear filtering. IEEE Trans. Aero. Electron. Syst. **AES-5**(6), 906–911 (1969)
22. Kuznik, F.: Satellite saviors. Air Space Smithsonian **6**(3), 66–70 (1991)

23. Lefferts, E.J., Markley, F.L., Shuster, M.D.: Kalman filtering for spacecraft attitude estimation. J. Guid. Contr. Dynam. **5**(5), 417–429 (1982)
24. Lerner, G.M.: Three-axis attitude determination. In: Wertz, J.R. (ed.) Spacecraft Attitude Determination and Control, chap. 12. Kluwer Academic, Dordrecht (1978)
25. Markley, F.L., Andrews, S.F., O'Donnell Jr., J.R., Ward, D.K.: Attitude control system of the Wilkinson Microwave Anisotropy Probe. J. Guid. Contr. Dynam. **28**(3), 385–397 (2005)
26. Markley, F.L., Bauer, F.H., Deily, J.J., Femiano, M.D.: Attitude control system conceptual design for Geostationary Operational Environmental Satellite spacecraft series. J. Guid. Contr. Dynam. **18**(2), 247–255 (1995)
27. Markley, F.L., Mortari, D.: Quaternion attitude estimation using vector observations. J. Astronaut. Sci. **48**(2/3), 359–380 (2000)
28. Markley, F.L., Nelson, J.D.: Zero-gyro safemode controller for the Hubble Space Telescope. J. Guid. Contr. Dynam. **17**(4), 815–822 (1994)
29. Murrell, J.W.: Precision attitude determination for multimission spacecraft. In: Proceedings of the AIAA Guidance, Navigation, and Control Conference, pp. 70–87. Palo Alto (1978)
30. Pauling, D.C., Jackson, D.B., Brown, C.D.: SPARS algorithms and simulation results. In: Proceedings of the Symposium on Spacecraft Attitude Estimation, vol. 1, pp. 293–317. Aerospace Corporation, El Segundo (1969)
31. Pellicori, S.F., Russell, E.E., Watts, L.A.: Pioneer imaging photopolarimeter optical system. Appl. Optic **12**(6), 1246–1258 (1973)
32. Roberson, R.E.: Two decades of spacecraft attitude control. J. Guid. Contr. **2**(1), 3–8 (1979)
33. Shuster, M.D.: In my estimation. J. Astronaut. Sci. **54**(3 & 4), 273–297 (2006)
34. Shuster, M.D.: The quest for better attitudes. J. Astronaut. Sci. **54**(3/4), 657–683 (2006)
35. Shuster, M.D., Oh, S.D.: Attitude determination from vector observations. J. Guid. Contr. **4**(1), 70–77 (1981)
36. Siahpush, A., Gleave, J.: A brief survey of attitude control systems for small satellites using momentum concepts. In: Proceedings of the 2nd AIAA/USU Conference on Small Satellites. Logan (1988). SSC88-IV-24
37. Space Systems Loral: GOES I–M data book (1996). http://goes.gsfc.nasa.gov/text/goes.databook.html (accessed Jan. 27, 2014)
38. The Boeing Company: GOES N data book (2005). http://goes.gsfc.nasa.gov/text/goes.databooknop.html (accessed Jan. 27, 2014)
39. Toda, N.F., Heiss, J.L., Schlee, F.H.: SPARS: The system, algorithms, and test results. In: Proceedings of the Symposium on Spacecraft Attitude Estimation, vol. 1, pp. 361–370. Aerospace Corporation, El Segundo (1969)
40. Wahba, G.: A least-squares estimate of satellite attitude. SIAM Rev. **7**(3), 409 (1965)
41. Wen, J.T.Y., Kreutz-Delgado, K.: The attitude control problem. IEEE Trans. Automat. Contr. **36**(10), 1148–1162 (2002)
42. Wertz, J.R. (ed.): Spacecraft Attitude Determination and Control. Kluwer Academic, Dordrecht (1978)
43. Wie, B., Barba, P.M.: Quaternion feedback for spacecraft large angle maneuvers. J. Guid. Contr. Dynam. **8**(3), 360–365 (1985)
44. Wiśniewski, R., Blanke, M.: Fully magnetic attitude control for spacecraft subject to gravity gradient. Automatica **35**(7), 1201–1214 (1999)
45. Worth, H.E., Warren, M.: Transit to Tomorrow: Fifty Years of Space Research at the Johns Hopkins University Applied Physics Laboratory. The Johns Hopkins University Applied Physics Laboratory, Laurel (2009)

Chapter 2
Matrices, Vectors, Frames, Transforms

This chapter begins with an overview of matrices and vectors, which are used extensively in attitude analysis. We assume that the reader has some familiarity with this material, so the account is not completely self-contained. The principal objective of this section is to define our notation and conventions.

We next discuss a special category of four-component vectors, which we refer to as quaternions although they differ conceptionally from the quaternions introduced by W. R. Hamilton in 1844. They perform the same function as Hamilton's quaternions in applications, however, and have proved to be extremely useful in attitude analysis.

We then move on to a discussion of rotations in three-dimensional space and the most common parameter sets that have been used to specify these rotations: the Euler axis and angle, the rotation vector, the quaternion, the Rodrigues parameters, the modified Rodrigues parameters, and the Euler angles. The last section in this chapter addresses the representation of attitude errors. A more extensive treatment of the material in this section, including historical references, can be found in Shuster's comprehensive review article [17].

2.1 Matrices

An $m \times n$ matrix A is an array with m rows and n columns of *scalars*:

$$A = \begin{bmatrix} a_{11} & a_{12} & \cdots & a_{1n} \\ a_{21} & a_{22} & \cdots & a_{2n} \\ \vdots & \vdots & \ddots & \vdots \\ a_{m1} & a_{m2} & \cdots & a_{mn} \end{bmatrix} \tag{2.1}$$

F.L. Markley and J.L. Crassidis, *Fundamentals of Spacecraft Attitude Determination and Control*, Space Technology Library 33, DOI 10.1007/978-1-4939-0802-8_2,
© Springer Science+Business Media New York 2014

We will assume that the scalar matrix elements are real numbers. The results in this chapter can be generalized to matrices with complex elements, but we will rarely need to deal with complex matrices. If $m = n$, then the matrix A is *square*. We denote an $m \times n$ matrix with all components equal to zero by $0_{m \times n}$, or sometimes simply by 0 if confusion is unlikely to result.

A *column vector*, or sometimes simply a vector, is an $n \times 1$ matrix

$$\mathbf{x} = \begin{bmatrix} x_1 \\ x_2 \\ \vdots \\ x_n \end{bmatrix} \tag{2.2}$$

We denote an n-component vector with all components equal to zero by $\mathbf{0}_n$, or sometimes simply by $\mathbf{0}$ if confusion is unlikely.

Matrices can be added, subtracted, or multiplied. For addition and subtraction, all matrices must have the same number of rows and columns. The elements of

$$C = A \pm B \tag{2.3}$$

are given by $c_{ij} = a_{ij} \pm b_{ij}$. Matrix addition and subtraction are both commutative, $A \pm B = B \pm A$, and associative, $(A \pm B) \pm C = A \pm (B \pm C)$. The multiplication of two matrices A and B:

$$C = A\,B \tag{2.4}$$

is valid only when the number of columns of A is equal to the number of rows of B (i.e. A and B must be *conformable*). The resulting matrix C will have rows equal to the number of rows of A and columns equal to the number of columns of B. Thus, if A has dimension $m \times n$ and B has dimension $n \times p$, then C will have dimension $m \times p$. The elements of C are given by

$$c_{ij} = \sum_{k=1}^{n} a_{ik} b_{kj} \tag{2.5}$$

for all $i = 1, 2, \ldots, m$ and $j = 1, 2, \ldots, p$. Matrix multiplication is associative, $A(B\,C) = (A\,B)C$, and distributive, $A(B + C) = A\,B + A\,C$, but not commutative in general, $A\,B \neq B\,A$. In those cases for which $A\,B = B\,A$, the matrices A and B are said to *commute*.

The transpose of a matrix, denoted A^T, has rows that are the columns of A and columns that are the rows of A. The transpose of the matrix defined by Eq. (2.1), for example, is

$$A^T = \begin{bmatrix} a_{11} & a_{21} & \cdots & a_{m1} \\ a_{12} & a_{22} & \cdots & a_{m2} \\ \vdots & \vdots & \ddots & \vdots \\ a_{1n} & a_{2n} & \cdots & a_{mn} \end{bmatrix} \tag{2.6}$$

The transpose of a column vector is a *row vector*. The transpose operator has the following properties:

$$(\alpha A)^T = \alpha A^T, \text{ where } \alpha \text{ is a scalar} \tag{2.7a}$$

$$(A + B)^T = A^T + B^T \tag{2.7b}$$

$$(A\,B)^T = B^T A^T \tag{2.7c}$$

If $A = A^T$, then A is a *symmetric* matrix, if $A = -A^T$, then A is a *skew symmetric* matrix.

A *diagonal* matrix is a square matrix with nonzero elements only on the main diagonal and all other elements equal to zero. An $n \times n$ diagonal matrix can be formed from an n-component row or column vector by

$$\text{diag}(\mathbf{x}) = \text{diag}(\mathbf{x}^T) = \text{diag}([x_1\, x_2\, \cdots\, x_n]) \equiv \begin{bmatrix} x_1 & 0 & \cdots & 0 \\ 0 & x_2 & \cdots & 0 \\ \vdots & \vdots & \ddots & \vdots \\ 0 & 0 & \cdots & x_n \end{bmatrix} \tag{2.8}$$

An important special case of a diagonal matrix is the *identity* matrix:

$$I \equiv \text{diag}([1\,1\,\cdots\,1]) \tag{2.9}$$

It has the property that $I\,A = A$ and $B\,I = B$ if the matrices are conformable. We sometimes denote the $n \times n$ identity matrix by I_n if supplying the subscript helps to remove ambiguity.

An *upper triangular* matrix is a matrix in which all the entries below the main diagonal are zero, i.e. $a_{ij} = 0$ for $i < j$. A *lower triangular* matrix has all zeros above the main diagonal, i.e. $a_{ij} = 0$ for $i > j$.

Two useful scalar quantities can be defined for square matrices, the *trace* and the *determinant*. The trace of an $n \times n$ matrix is simply the sum of the diagonal elements:

$$\text{tr}A = \sum_{i=1}^{n} a_{ii} \tag{2.10}$$

Some useful identities involving the matrix trace are given by

$$\mathrm{tr}(\alpha A) = \alpha\, \mathrm{tr} A \tag{2.11a}$$

$$\mathrm{tr}(A + B) = \mathrm{tr} A + \mathrm{tr} B \tag{2.11b}$$

$$\mathrm{tr}(A\, B\, C) = \mathrm{tr}(B\, C\, A) = \mathrm{tr}(C\, A\, B) \tag{2.11c}$$

$$\mathrm{tr}(A\, B) = \mathrm{tr}(B\, A) \tag{2.11d}$$

$$\mathrm{tr}(\mathbf{x}\, \mathbf{y}^T) = \mathbf{x}^T \mathbf{y} \tag{2.11e}$$

$$\mathrm{tr}(A\, \mathbf{y}\, \mathbf{x}^T) = \mathbf{x}^T A\, \mathbf{y} \tag{2.11f}$$

Equation (2.11d)–(2.11f) are special cases of Eq. (2.11c), which shows the cyclic invariance of the trace. The operation $\mathbf{y}\, \mathbf{x}^T$ is known as the *outer product* (note that $\mathbf{y}\, \mathbf{x}^T \neq \mathbf{x}\, \mathbf{y}^T$ in general).

The determinant of an $n \times n$ matrix can be computed using an expansion about any row i or any column j:

$$\det A = \sum_{k=1}^{n} (-1)^{i+k} a_{ik} m_{ik} = \sum_{k=1}^{n} (-1)^{k+j} a_{kj} m_{kj} \tag{2.12}$$

where m_{ij} is the *minor*, which is the determinant of the $(n - 1) \times (n - 1)$ matrix resulting from deleting row i and column j of A. Some useful determinant identities are

$$\det I = 1 \tag{2.13a}$$

$$\det A^T = \det A \tag{2.13b}$$

$$\det(A\, B) = \det A\, \det B \tag{2.13c}$$

$$\det(\alpha\, A) = \alpha^n \det A \tag{2.13d}$$

The elements of the *adjoint* matrix adj A are defined in terms of the minors:

$$[\mathrm{adj}\, A]_{ij} = (-1)^{i+j} m_{ji} \tag{2.14}$$

Note the reversed order of the subscripts on the left and right side of this equation. It can be shown that [21]

$$A(\mathrm{adj}\, A) = (\mathrm{adj}\, A)A = (\det A)I \tag{2.15}$$

Thus we see that a *nonsingular* matrix, which is a square matrix with a nonzero determinant, has an inverse

$$A^{-1} = \frac{\mathrm{adj}\, A}{\det A} \tag{2.16}$$

that satisfies the identities

$$A^{-1}A = A\,A^{-1} = I \tag{2.17a}$$

$$(A^T)^{-1} = (A^{-1})^T \equiv A^{-T} \tag{2.17b}$$

If A and B are $n \times n$ matrices, then the matrix product $A\,B$ is nonsingular if and only if A and B are nonsingular. If these conditions are met, then

$$(A\,B)^{-1} = B^{-1}A^{-1} \tag{2.18}$$

2.2 Vectors

The *dot product, inner product*, or *scalar product* of two vectors of equal dimension, $n \times 1$, is given by

$$\mathbf{x} \cdot \mathbf{y} \equiv \mathbf{x}^T\mathbf{y} = \mathbf{y}^T\mathbf{x} = \sum_{i=1}^{n} x_i\,y_i \tag{2.19}$$

If the dot product is zero, then the vectors are said to be *orthogonal*. A measure of the length of a vector is given by its *Euclidean norm*, which is the square root of the inner product of the vector with itself:

$$\|\mathbf{x}\| \equiv \sqrt{\mathbf{x} \cdot \mathbf{x}} = \left[\sum_{i=1}^{n} x_i^2\right]^{1/2} \tag{2.20}$$

It is easily seen that $\|\alpha\,\mathbf{x}\| = |\alpha|\,\|\mathbf{x}\|$. The norm obeys the inequality

$$\|\mathbf{x}\| \geq 0 \tag{2.21}$$

with equality only for $\mathbf{x} = \mathbf{0}$.

A vector with norm equal to unity is said to be a *unit vector*. Any nonzero vector can be made into a unit vector by dividing it by its norm:

$$\mathbf{x}_{\text{unit}} \equiv \frac{\mathbf{x}}{\|\mathbf{x}\|} \tag{2.22}$$

This is referred to as *normalizing* the vector $\mathbf{x}$.

Figure 2.1a shows two vectors, $\mathbf{x}$ and $\mathbf{y}$, and the *orthogonal projection* of a vector $\mathbf{y}$ onto a vector $\mathbf{x}$. The orthogonal projection of $\mathbf{y}$ onto $\mathbf{x} \neq \mathbf{0}$ is given by

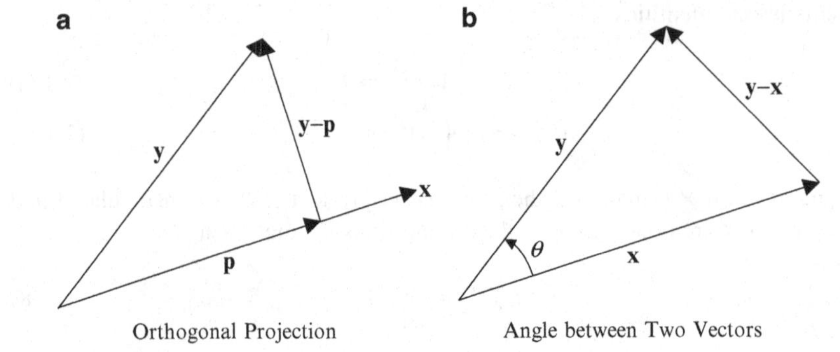

Fig. 2.1 Depiction of an orthogonal projection (**a**) and of the angle between two vectors (**b**)

$$\mathbf{p} = \frac{\mathbf{x} \cdot \mathbf{y}}{\|\mathbf{x}\|^2}\mathbf{x} \tag{2.23}$$

This projection yields $(\mathbf{y} - \mathbf{p}) \cdot \mathbf{x} = 0$. From Eqs. (2.21) and (2.23) we see that

$$0 \leq \|(\mathbf{y} - \mathbf{p})\|^2 = \|\mathbf{y}\|^2 - \frac{(\mathbf{x} \cdot \mathbf{y})^2}{\|\mathbf{x}\|^2} \tag{2.24}$$

This yields the *Cauchy-Schwarz inequality*:

$$|\mathbf{x} \cdot \mathbf{y}| \leq \|\mathbf{x}\| \, \|\mathbf{y}\| \tag{2.25}$$

The Cauchy-Schwarz inequality implies the *triangle inequality*

$$\|\mathbf{x} + \mathbf{y}\| \leq \|\mathbf{x}\| + \|\mathbf{y}\| \tag{2.26}$$

and it allows us to define the angle θ between the vectors $\mathbf{x}$ and $\mathbf{y}$ by

$$\cos\theta = \frac{\mathbf{x} \cdot \mathbf{y}}{\|\mathbf{x}\| \, \|\mathbf{y}\|} \tag{2.27}$$

This angle is illustrated in Fig. 2.1b, as is the vector difference $\mathbf{y} - \mathbf{x}$.

2.3 Jacobian, Gradient, and Hessian

In this section we will introduce notation for partial derivatives with respect to components of a vector $\mathbf{x}$. If $\mathbf{y}(\mathbf{x})$ is an m-component vector function of an n-component vector $\mathbf{x}$, we define the *Jacobian matrix* as the $m \times n$ matrix

$$\frac{\partial \mathbf{y}(\mathbf{x})}{\partial \mathbf{x}} \equiv \begin{bmatrix} \frac{\partial y_1}{\partial x_1} & \frac{\partial y_1}{\partial x_2} & \cdots & \frac{\partial y_1}{\partial x_n} \\ \frac{\partial y_2}{\partial x_1} & \frac{\partial y_2}{\partial x_2} & \cdots & \frac{\partial y_2}{\partial x_n} \\ \vdots & \vdots & \ddots & \vdots \\ \frac{\partial y_m}{\partial x_1} & \frac{\partial y_m}{\partial x_2} & \cdots & \frac{\partial y_m}{\partial x_n} \end{bmatrix} \tag{2.28}$$

In particular, if we have a scalar $f(\mathbf{x})$ in place of the vector function $\mathbf{y}(\mathbf{x})$, this reduces to the $m \times 1$ row vector

$$\frac{\partial f(\mathbf{x})}{\partial \mathbf{x}} \equiv \begin{bmatrix} \frac{\partial f}{\partial x_1} & \frac{\partial f}{\partial x_2} & \cdots & \frac{\partial f}{\partial x_n} \end{bmatrix} \tag{2.29}$$

The transpose of this, a $1 \times m$ column vector, is known as the *gradient* of $f(\mathbf{x})$:

$$\nabla_{\mathbf{x}} f(\mathbf{x}) \equiv \left[\frac{\partial f(\mathbf{x})}{\partial \mathbf{x}} \right]^T \equiv \frac{\partial f(\mathbf{x})}{\partial \mathbf{x}^T} \tag{2.30}$$

The *Hessian* is the $n \times n$ symmetric matrix of second-order partial derivatives of $f(\mathbf{x})$:

$$\frac{\partial^2 f(\mathbf{x})}{\partial \mathbf{x}\, \partial \mathbf{x}^T} \equiv \begin{bmatrix} \frac{\partial^2 f}{\partial x_1 \partial x_1} & \frac{\partial^2 f}{\partial x_1 \partial x_2} & \cdots & \frac{\partial^2 f}{\partial x_1 \partial x_n} \\ \frac{\partial^2 f}{\partial x_2 \partial x_1} & \frac{\partial^2 f}{\partial x_2 \partial x_2} & \cdots & \frac{\partial^2 f}{\partial x_2 \partial x_n} \\ \vdots & \vdots & \ddots & \vdots \\ \frac{\partial^2 f}{\partial x_n \partial x_1} & \frac{\partial^2 f}{\partial x_n \partial x_2} & \cdots & \frac{\partial^2 f}{\partial x_n \partial x_n} \end{bmatrix} \tag{2.31}$$

Some use the term *Hessian* to refer to the determinant of the Hessian matrix. The *Laplacian* is the trace of the Hessian matrix:

$$\nabla^2_{\mathbf{x}} f(\mathbf{x}) \equiv \sum_{i=1}^{n} \frac{\partial^2 f}{\partial x_i^2} = \operatorname{tr}\left(\frac{\partial^2 f(\mathbf{x})}{\partial \mathbf{x}\, \partial \mathbf{x}^T} \right) \tag{2.32}$$

2.4 Orthonormal Bases, Change of Basis

An *orthonormal* set of vectors is a set with inner products obeying

$$\mathbf{x}_i \cdot \mathbf{x}_j = \delta_{ij} \tag{2.33}$$

where the Kronecker delta δ_{ij} is defined as

$$\delta_{ij} = \begin{cases} 0 & \text{if } i \neq j \\ 1 & \text{if } i = j \end{cases} \tag{2.34}$$

If the number of vectors in the set is equal to the dimension of the vectors, then the set is said to constitute an *orthonormal basis*. We will generally denote basis vectors (and some other unit vectors) by the letter $\mathbf{e}$. The orthonormal basis

$$\mathbf{e}_1 = \begin{bmatrix} 1 \\ 0 \\ \vdots \\ 0 \end{bmatrix}, \quad \mathbf{e}_2 = \begin{bmatrix} 0 \\ 1 \\ \vdots \\ 0 \end{bmatrix}, \quad \ldots, \quad \mathbf{e}_n = \begin{bmatrix} 0 \\ 0 \\ \vdots \\ 1 \end{bmatrix} \tag{2.35}$$

is called the *natural basis*. Any vector $\mathbf{x}$ can be written as a linear superposition of basis vectors; using the natural basis gives

$$\mathbf{x} = \begin{bmatrix} x_1 \\ x_2 \\ \vdots \\ x_n \end{bmatrix} = \sum_{j=1}^{n} x_j \mathbf{e}_j \tag{2.36}$$

An orthonormal set of basis vectors defines a *reference frame*. Let us refer to the frame defined by $\mathbf{e}_1, \mathbf{e}_2, \ldots, \mathbf{e}_n$ as frame F; and consider another frame F' defined by the orthonormal basis: $\mathbf{e}'_1, \mathbf{e}'_2, \ldots, \mathbf{e}'_n$. We can express the vector $\mathbf{x}$ in either basis

$$\mathbf{x} = \sum_{j=1}^{n} x_j \mathbf{e}_j = \sum_{k=1}^{n} x'_k \mathbf{e}'_k \tag{2.37}$$

Taking the dot product of this equation with $\mathbf{e}_j$ or $\mathbf{e}'_k$ gives

$$x_j = \mathbf{e}_j \cdot \mathbf{x} = \sum_{k=1}^{n} (\mathbf{e}_j \cdot \mathbf{e}'_k) x'_k \tag{2.38a}$$

$$x'_k = \mathbf{e}'_k \cdot \mathbf{x} = \sum_{j=1}^{n} (\mathbf{e}'_k \cdot \mathbf{e}_j) x_j \tag{2.38b}$$

Substituting the first equality of Eq. (2.38a) into the first equality of Eq. (2.37) gives

$$\mathbf{x} = \sum_{j=1}^{n} \mathbf{e}_j (\mathbf{e}_j^T \mathbf{x}) = \left(\sum_{j=1}^{n} \mathbf{e}_j \mathbf{e}_j^T \right) \mathbf{x} \tag{2.39}$$

It follows that the identity matrix can be expressed in terms of any orthonormal basis by

$$I_n = \sum_{i=1}^{n} \mathbf{e}_i \mathbf{e}_i^T \tag{2.40}$$

When considering transformations among different reference frames, it is very useful (almost indispensable, in fact) to regard $\mathbf{x}$ as an *abstract* vector having an existence in n-dimensional space independent of any particular reference frame, and having *representations* $\mathbf{x}_F$ in reference frame F and $\mathbf{x}_{F'}$ in frame F'. Subscripts will be provided when it is important to indicate the reference frame explicitly or to carefully distinguish between abstract vectors and their representations, but subscripts will often be omitted when confusion is unlikely to arise. Arraying the components of $\mathbf{x}$ in the reference frames F and F' as column vectors gives the representations

$$\mathbf{x}_F = \begin{bmatrix} x_1 \\ x_2 \\ \vdots \\ x_n \end{bmatrix} = \begin{bmatrix} \mathbf{e}_1 \cdot \mathbf{x} \\ \mathbf{e}_2 \cdot \mathbf{x} \\ \vdots \\ \mathbf{e}_n \cdot \mathbf{x} \end{bmatrix} \quad \text{and} \quad \mathbf{x}_{F'} = \begin{bmatrix} x'_1 \\ x'_2 \\ \vdots \\ x'_n \end{bmatrix} = \begin{bmatrix} \mathbf{e}'_1 \cdot \mathbf{x} \\ \mathbf{e}'_2 \cdot \mathbf{x} \\ \vdots \\ \mathbf{e}'_n \cdot \mathbf{x} \end{bmatrix} \tag{2.41}$$

Equation (2.38) can be used to express the relations between $\mathbf{x}_F$ and $\mathbf{x}_{F'}$ as matrix products:

$$\mathbf{x}_F = D_{FF'} \mathbf{x}_{F'} \tag{2.42a}$$

$$\mathbf{x}_{F'} = D_{F'F} \mathbf{x}_F \tag{2.42b}$$

where

$$D_{FF'} = \begin{bmatrix} \mathbf{e}_1 \cdot \mathbf{e}'_1 & \mathbf{e}_1 \cdot \mathbf{e}'_2 & \cdots & \mathbf{e}_1 \cdot \mathbf{e}'_n \\ \mathbf{e}_2 \cdot \mathbf{e}'_1 & \mathbf{e}_2 \cdot \mathbf{e}'_2 & \cdots & \mathbf{e}_2 \cdot \mathbf{e}'_n \\ \vdots & \vdots & \ddots & \vdots \\ \mathbf{e}_n \cdot \mathbf{e}'_1 & \mathbf{e}_n \cdot \mathbf{e}'_2 & \cdots & \mathbf{e}_n \cdot \mathbf{e}'_n \end{bmatrix} \tag{2.43a}$$

$$D_{F'F} = \begin{bmatrix} \mathbf{e}'_1 \cdot \mathbf{e}_1 & \mathbf{e}'_1 \cdot \mathbf{e}_2 & \cdots & \mathbf{e}'_1 \cdot \mathbf{e}_n \\ \mathbf{e}'_2 \cdot \mathbf{e}_1 & \mathbf{e}'_2 \cdot \mathbf{e}_2 & \cdots & \mathbf{e}'_2 \cdot \mathbf{e}_n \\ \vdots & \vdots & \ddots & \vdots \\ \mathbf{e}'_n \cdot \mathbf{e}_1 & \mathbf{e}'_n \cdot \mathbf{e}_2 & \cdots & \mathbf{e}'_n \cdot \mathbf{e}_n \end{bmatrix} \tag{2.43b}$$

The matrices $D_{FF'}$ and $D_{F'F}$ are known as *direction cosine matrices* (DCMs) because their elements are the cosines of the angles between the basis vectors in the two reference frames. These transformation equations hold for any two orthonormal bases. We see immediately that

$$D_{F'F} = D_{FF'}^T \tag{2.44}$$

which means that the matrix transforming vector representations from frame F to frame F' is the transpose of the matrix transforming from F' to F. Another way of stating the content of Eq. (2.43) is that the columns of $D_{FF'}$ are the representations in frame F of the basis vectors of frame F' and *vice versa*:

$$D_{FF'} = \left[\mathbf{e}'_{1F}\ \mathbf{e}'_{2F}\ \cdots\ \mathbf{e}'_{nF}\right] \tag{2.45a}$$

$$D_{F'F} = \left[\mathbf{e}_{1F'}\ \mathbf{e}_{2F'}\ \cdots\ \mathbf{e}_{nF'}\right] \tag{2.45b}$$

We emphasize that the transformations considered here are transformations of the representations of a fixed abstract vector resulting from a change in the reference frame. This is the *passive* interpretation of a transformation, also known as the *alias* sense (from the Latin word for "otherwise," in the sense of "otherwise known as") [17]. The alternative *active* interpretation (also known as the *alibi* sense from the Latin word for "elsewhere") considers the representation in a fixed reference frame of an abstract vector that is rotated from $\mathbf{x}$ to $\mathbf{x}'$. The difference between these two interpretations is illustrated in Fig. 2.2a–c. Figure 2.2a shows the components x_1 and x_2 of the vector $\mathbf{x}$ in frame F. Figure 2.2b shows the components x'_1 and x'_2 of $\mathbf{x}$ in the rotated frame F', illustrating the alias sense. Figure 2.2c shows the components x'_1 and x'_2 of a rotated vector $\mathbf{x}'$ in the unrotated frame F, illustrating the alibi sense. In this example, the values of x'_1 and x'_2 are the same in the alias and alibi interpretations, but the significance of these quantities is completely different in the two cases. They are the components of an unrotated vector in a rotated reference frame in the alias interpretation, while they are the components of a rotated vector in an unrotated reference frame in the alibi interpretation. When necessary, we can use a more precise notation to distinguish these two different interpretations, as in Eq. (2.45); but we can usually avoid this complication because this book will rarely, if ever, employ the active interpretation of transformations. In the example illustrated in Fig. 2.2b,c, it can be seen that the magnitude of the rotation is the same in the two interpretations but the sense of the rotation is opposite. It is important to remember this important difference in the sense of rotation in the two interpretations; it results in some unexpected minus signs, and overlooking them has led to actual errors in flight software.[1]

---

[1]One example is an incorrect sign for the velocity aberration correction for star tracker measurements on the WMAP spacecraft, which fortunately was easily corrected.

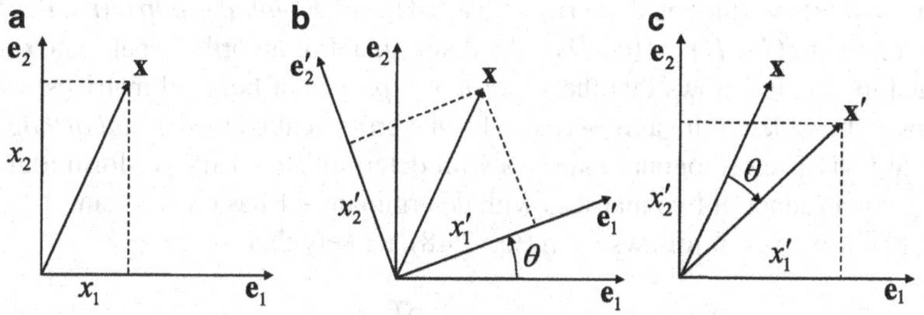

Fig. 2.2 Alias and Alibi interpretations of a transformation. (**a**) Components. (**b**) Alias. (**c**) Alibi

Now consider that we have three reference frames, denoted F, G, and H, and that we transform a vector representation from frame F to frame G and then from frame G to frame H. This is effected by the successive transformations

$$\mathbf{x}_H = D_{HG}\mathbf{x}_G = D_{HG}\left(D_{GF}\mathbf{x}_F\right) = \left(D_{HG}D_{GF}\right)\mathbf{x}_F \tag{2.46}$$

We could have transformed directly from F to H by $\mathbf{x}_H = D_{HF}\mathbf{x}_F$. These transformations must be equivalent for any vector $\mathbf{x}_F$, so it must be true that

$$D_{HF} = D_{HG}D_{GF} \tag{2.47}$$

This says that successive transformations are accomplished by simple matrix multiplication of DCMs, which may appear to be an obvious result. It is not inconceivable, though, that the method for implementing successive transformations could have been more complex.

Transforming from frame F to frame G and back to frame F is effected by the matrix D_{FF}, which must be the identity matrix. But from Eqs. (2.44) and (2.47) this means that

$$I = D_{FG}D_{GF} = D_{FG}D_{FG}^T = D_{GF}^T D_{GF} \tag{2.48}$$

Matrices like DCMs for which $I = D\,D^T = D^T D$ are called *orthogonal* matrices, or sometimes *orthonormal* matrices. The transpose of an orthogonal matrix is equal to its inverse; its columns constitute a set of orthonormal vectors, as do its rows. In fact, the columns and rows of a DCM are just the representations in one reference frame of the basis vectors in the other reference frame. Equation (2.7c) can be used to show that the product of two orthogonal matrices is orthogonal; i.e. if A and B are orthogonal, then

$$(A\,B)(A\,B)^T = A\,B\,B^T A^T = A\,A^T = I \tag{2.49}$$

This means that the set of $n \times n$ orthogonal matrices form a *group*, which requires, among other things, that the product of two elements of the group is also an element.

The group of $n \times n$ orthogonal matrices is called the *orthogonal group* $O(n)$. Because $1 = \det I = \det(D^T D) = (\det D)^2$, the determinant of an orthogonal matrix must be equal to ± 1. It follows that the set of $n \times n$ *proper* orthogonal matrices, which are those whose determinant is $+1$, also form a group, called the *special orthogonal group* $SO(n)$. The orthogonal matrices with determinant -1 do not form a group, because the product of two matrices with determinant -1 has determinant $+1$.

An important result follows from Eq. (2.48), namely that

$$\mathbf{x}_G \cdot \mathbf{y}_G = (D_{GF}\mathbf{x}_F)^T D_{GF}\mathbf{y}_F = \mathbf{x}_F^T D_{GF}^T D_{GF}\mathbf{y}_F = \mathbf{x}_F \cdot \mathbf{y}_F \tag{2.50}$$

This says that the value of the inner product of two vectors is independent of the reference frame in which they are represented, or equivalently that reference frame transformations preserve both lengths of vectors and angles between them. Inserting a matrix between two vectors leads to the relation

$$\mathbf{x}_F^T M_F \mathbf{y}_F = \mathbf{x}_F^T D_{GF}^T D_{GF} M_F D_{GF}^T D_{GF}\mathbf{y}_F = \mathbf{x}_G^T M_G \mathbf{y}_G \tag{2.51}$$

where

$$M_G \equiv D_{GF} M_F D_{GF}^T \tag{2.52}$$

This defines how matrices must transform under reference frame transformations for Eq. (2.51) to hold.

2.5 Vectors in Three Dimensions

The case of three dimensions is especially interesting because we and our vehicles live in three-dimensional space.[2] In three dimensions, the abstract vectors generally represent physical quantities that have both a magnitude and a direction, like displacements or velocities.

We can define a *vector product* or *cross product* for three-component vectors in terms of their components by

$$\mathbf{x} \times \mathbf{y} = \begin{bmatrix} x_2 y_3 - x_3 y_2 \\ x_3 y_1 - x_1 y_3 \\ x_1 y_2 - x_2 y_1 \end{bmatrix} = -\mathbf{y} \times \mathbf{x} \tag{2.53}$$

It is easily seen that the cross product $\mathbf{x} \times \mathbf{y}$ is perpendicular to both $\mathbf{x}$ and $\mathbf{y}$. The cross product can also be obtained using matrix multiplication:

---

[2]This is true in classical physics. Various contemporary physical theories indicate that we live in a space having anywhere from two to eleven dimensions.

$$\mathbf{x} \times \mathbf{y} = [\mathbf{x}\times]\,\mathbf{y} \tag{2.54}$$

where $[\mathbf{x}\times]$ is the *cross product matrix*, defined by

$$[\mathbf{x}\times] \equiv \begin{bmatrix} 0 & -x_3 & x_2 \\ x_3 & 0 & -x_1 \\ -x_2 & x_1 & 0 \end{bmatrix} \tag{2.55}$$

Note that $[\mathbf{x}\times]$ is a skew symmetric matrix.[3]

The cross product and the cross product matrix obey the following relations:

$$\mathbf{x} \cdot (\mathbf{y} \times \mathbf{z}) = (\mathbf{x} \times \mathbf{y}) \cdot \mathbf{z} \tag{2.56a}$$

$$[\mathbf{x}\times]\,[\mathbf{y}\times] = -\,(\mathbf{x} \cdot \mathbf{y})\, I + \mathbf{y}\,\mathbf{x}^T \tag{2.56b}$$

$$[\mathbf{x}\times]\,[\mathbf{y}\times] - [\mathbf{y}\times]\,[\mathbf{x}\times] = \mathbf{y}\,\mathbf{x}^T - \mathbf{x}\,\mathbf{y}^T = [(\mathbf{x} \times \mathbf{y})\times] \tag{2.56c}$$

$$\operatorname{adj}\,[\mathbf{x}\times] = \mathbf{x}\,\mathbf{x}^T \tag{2.56d}$$

It follows from Eq. (2.56b) that

$$\|\mathbf{x} \times \mathbf{y}\|^2 = ([\mathbf{x}\times]\,\mathbf{y})^T\,([\mathbf{x}\times]\,\mathbf{y}) = -\mathbf{y}^T[\mathbf{x}\times]^2\mathbf{y} = \|\mathbf{x}\|^2\|\mathbf{y}\|^2 - (\mathbf{x} \cdot \mathbf{y})^2 \tag{2.57}$$

With Eq. (2.27), this means that

$$\|\mathbf{x} \times \mathbf{y}\| = \|\mathbf{x}\|\;\|\mathbf{y}\|\,\sin\theta \tag{2.58}$$

It is often convenient to express a 3×3 matrix in terms of its columns:

$$M \equiv \begin{bmatrix}\mathbf{a}\;\mathbf{b}\;\mathbf{c}\end{bmatrix} \tag{2.59}$$

With this notation, the determinant is

$$\det M = \mathbf{a} \cdot (\mathbf{b} \times \mathbf{c}) = \mathbf{b} \cdot (\mathbf{c} \times \mathbf{a}) = \mathbf{c} \cdot (\mathbf{a} \times \mathbf{b}) \tag{2.60}$$

and the adjoint is

$$\operatorname{adj} M \equiv \operatorname{adj}\left(\begin{bmatrix}\mathbf{a}\;\mathbf{b}\;\mathbf{c}\end{bmatrix}\right) = \begin{bmatrix} (\mathbf{b} \times \mathbf{c})^T \\ (\mathbf{c} \times \mathbf{a})^T \\ (\mathbf{a} \times \mathbf{b})^T \end{bmatrix} \tag{2.61}$$

We can also derive the useful identity

[3] A vector product of two vectors can be defined only in three dimensions because an $n \times n$ skew-symmetric matrix has exactly n independent parameters only for $n = 3$.

$$M^T[\mathbf{x}\times]M = \begin{bmatrix} \mathbf{a}\cdot(\mathbf{x}\times\mathbf{a}) & \mathbf{a}\cdot(\mathbf{x}\times\mathbf{b}) & \mathbf{a}\cdot(\mathbf{x}\times\mathbf{c}) \\ \mathbf{b}\cdot(\mathbf{x}\times\mathbf{a}) & \mathbf{b}\cdot(\mathbf{x}\times\mathbf{b}) & \mathbf{b}\cdot(\mathbf{x}\times\mathbf{c}) \\ \mathbf{c}\cdot(\mathbf{x}\times\mathbf{a}) & \mathbf{c}\cdot(\mathbf{x}\times\mathbf{b}) & \mathbf{c}\cdot(\mathbf{x}\times\mathbf{c}) \end{bmatrix}$$

$$= \begin{bmatrix} 0 & -(\mathbf{a}\times\mathbf{b})\cdot\mathbf{x} & (\mathbf{c}\times\mathbf{a})\cdot\mathbf{x} \\ (\mathbf{a}\times\mathbf{b})\cdot\mathbf{x} & 0 & -(\mathbf{b}\times\mathbf{c})\cdot\mathbf{x} \\ -(\mathbf{c}\times\mathbf{a})\cdot\mathbf{x} & (\mathbf{b}\times\mathbf{c})\cdot\mathbf{x} & 0 \end{bmatrix} = [\{(\mathrm{adj}\, M)\mathbf{x}\}\times] \tag{2.62}$$

Setting $M = A^T$, where A is a proper orthogonal 3×3 matrix, gives $\mathrm{adj}\, M = A$ and

$$A\,[\mathbf{x}\times]\,A^T = [(A\mathbf{x})\times], \quad \text{for } A \in \mathrm{SO}(3) \tag{2.63}$$

This special case is much more useful than the general case. In the specific case of a reference frame transformation, we have

$$A_{GF}\,[\mathbf{x}_F\times]\,A_{GF}^T = [(A_{GF}\mathbf{x}_F)\times] = [\mathbf{x}_G\times] \tag{2.64}$$

which can be viewed as a special case of Eq. (2.52). This equation can be used to show that

$$\mathbf{x}_G \times \mathbf{y}_G = [(A_{GF}\mathbf{x}_F)\times]A_{GF}\mathbf{y}_F = A_{GF}(\mathbf{x}_F \times \mathbf{y}_F) \tag{2.65}$$

The significance of this is that the cross product of two vectors transforms exactly like any other vector under a reference frame rotation, which is what we want.

The discussion so far has been purely algebraic; it has said nothing about the *right hand rule*. Discussing this rule requires an intuitive picture of vectors in three-dimensional space. First note that the definition of the cross product means that the natural basis vectors defined by Eq. (2.35) satisfy the relation $\mathbf{e}_3 = \mathbf{e}_1 \times \mathbf{e}_2$. Now consider the possible orientation of these three basis vectors in physical space. The orientation of $\mathbf{e}_1$ and $\mathbf{e}_2$ is arbitrary, except that they must be orthogonal; but this leaves us only two choices for $\mathbf{e}_3$, which must be perpendicular to both $\mathbf{e}_1$ and $\mathbf{e}_2$. We choose the orientation of $\mathbf{e}_3$ to satisfy the right hand rule, i.e: we place $\mathbf{e}_1$ and $\mathbf{e}_2$ tail-to-tail, flatten the right hand, extending it in the direction of $\mathbf{e}_1$, curl the fingers toward $\mathbf{e}_2$ through the shortest angle, and choose $\mathbf{e}_3$ to point along the direction indicated by the thumb. This defines a right-handed reference frame, and all cross products will obey the right hand rule if we restrict ourselves to right handed reference frames. To see this explicitly, consider two arbitrary vectors $\mathbf{x}$ and $\mathbf{y}$. There is a reference frame in which

$$\mathbf{x}_F = \|\mathbf{x}\| \begin{bmatrix} 1 \\ 0 \\ 0 \end{bmatrix}, \quad \mathbf{y}_F = \|\mathbf{y}\| \begin{bmatrix} \cos\theta \\ \sin\theta \\ 0 \end{bmatrix} \implies \mathbf{x}_F \times \mathbf{y}_F = \|\mathbf{x}\|\,\|\mathbf{y}\| \begin{bmatrix} 0 \\ 0 \\ \sin\theta \end{bmatrix} \tag{2.66}$$

where θ is the angle between $\mathbf{x}$ and $\mathbf{y}$. This cross product obeys the right-hand rule, and Eq. (2.65) guarantees that the right hand rule will then hold in any reference frame. Note that if A had been an improper orthogonal matrix with determinant -1, an undesirable minus sign would have appeared between the two sides of Eq. (2.63). Seen from this point of view, the problem with improper orthogonal matrices is that they would change a right-handed reference frame into a left-handed one; they would turn the reference frame inside out.

2.6 Some Useful Reference Frames

Several reference frames in three dimensions are of special interest for attitude analysis. We will discuss the most important of these in this section. In general, a reference frame is specified by the location of its origin and the orientation of its coordinate axes, with the orientation being much more important for attitude analysis.

2.6.1 *Spacecraft Body Frame*

A spacecraft body frame is defined by an origin at a specified point in the spacecraft body and three Cartesian axes. A body frame is used to align the various components during spacecraft assembly. Components will generally shift due to the large forces experienced during launch, though, and can also move while on orbit due to thermal deformations. Every effort is made to limit these motions, but they cannot always be neglected. Whether they are negligible or not depends on the pointing accuracy required of the spacecraft. As an additional complication, some components of the spacecraft, such as solar arrays or gimbaled instruments, are moved quite deliberately. Therefore, it is quite common to define the body coordinate system operationally as the orientation of some sufficiently rigid *navigation base*, which is a subsystem of the spacecraft including the most critical attitude sensors and payload instruments. The navigation base often takes the form of a specially constructed *optical bench*, with its attached sensors and payload components. The purpose of attitude estimation and attitude control is to ascertain and to control the orientation of the navigation base relative to some external reference frame.

2.6.2 *Inertial Reference Frames*

An inertial reference frame is a frame in which Newton's laws of motion are valid. It is a well known fact of classical mechanics that any frame moving at constant velocity and without rotation with respect to an inertial frame is also inertial [5].

The existence of these dynamically preferred frames raises the interesting question of whether there is something with respect to which all inertial frames are non-rotating and unaccelerated. Weinberg concludes in [25, p. 474] that " ... inertial frames are any reference frames that move at constant velocity, and without rotation, relative to frames in which the universe appears spherically symmetric." This characterization is consistent with *Mach's Principle*, the hypothesis that inertial frames are somehow determined by the mass of everything in the universe.

Celestial reference frames with their axes fixed relative to distant "fixed" stars are the best realizations of inertial frames. The standard as of this writing is the *International Celestial Reference System* (ICRF) with its axes fixed with respect to the positions of several hundred distant extragalactic sources of radio waves, determined by very long baseline interferometry [8, 11]. The z axis of this frame is aligned with the Earth's North pole, and the x axis with the *vernal equinox*, the intersection of the Earth's equatorial plane with the plane of the Earth's orbit around the Sun, in the direction of the Sun's position relative to the Earth on the first day of spring. Unfortunately, neither the polar axis nor the ecliptic plane is inertially fixed, so the ICRF axes are defined to be the *mean* orientations of the pole and the vernal equinox (the positions with short-period motions removed by dynamic models) at some fixed epoch time. The origin of the ICRF is at the center-of-mass of the solar system.

An approximate inertial frame, known as the *Geocentric Inertial Frame* (GCI) has its origin at the center of mass of the Earth. This frame has a linear acceleration because of the Earth's circular orbit about the Sun, but this is unimportant for attitude analysis. The axes of a "mean of epoch" GCI frame are aligned with the mean North pole and mean vernal equinox at some epoch. The GCI frame is denoted by the triad $\{\mathbf{i}_1, \mathbf{i}_2, \mathbf{i}_3\}$, as shown in Fig. 2.3.

2.6.3 Earth-Centered/Earth-Fixed Frame

The Earth-Centered/Earth-Fixed (ECEF) Frame is denoted by $\{\boldsymbol{\epsilon}_1, \boldsymbol{\epsilon}_2, \boldsymbol{\epsilon}_3\}$ as shown in Fig. 2.3. This frame is similar to the GCI frame with $\boldsymbol{\epsilon}_3 = \mathbf{i}_3$; however, the $\boldsymbol{\epsilon}_1$ axis points in the direction of the Earth's prime meridian, and the $\boldsymbol{\epsilon}_2$ axis completes the right-handed system. Unlike the GCI frame, the ECEF frame rotates with the Earth. The rotation angle is known as the Greenwich Mean Sidereal Time (GMST) angle and is denoted by θ_{GMST} in Fig. 2.3.

The transformation of a position vector $\mathbf{r}$ from its GCI representation $\mathbf{r}_I$ to its ECEF representation $\mathbf{r}_E$ follows

$$\mathbf{r}_E = \begin{bmatrix} x \\ y \\ z \end{bmatrix} = A_{EI}\mathbf{r}_I = \begin{bmatrix} \cos\theta_{\text{GMST}} & \sin\theta_{\text{GMST}} & 0 \\ -\sin\theta_{\text{GMST}} & \cos\theta_{\text{GMST}} & 0 \\ 0 & 0 & 1 \end{bmatrix} \mathbf{r}_I \tag{2.67}$$

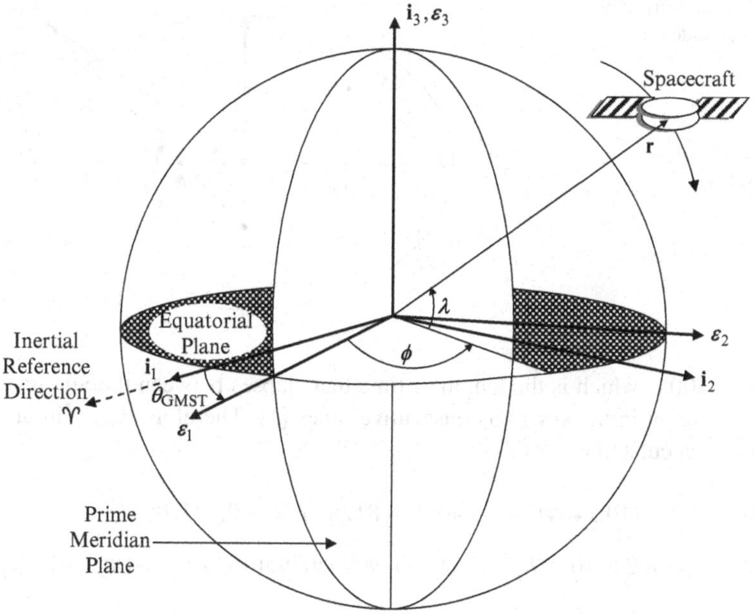

Fig. 2.3 Definitions of various reference frames

Determining the GMST angle requires the Julian date, JD. For a given year Y (between 1901 and 2099), month M, day D, hour h, minute m, and second s, the Julian date is calculated by [24]

$$JD(Y,M,D,h,m,s) = 1{,}721{,}013.5 + 367\,Y - \text{INT}\left\{\frac{7}{4}\left[Y + \text{INT}\left(\frac{M+9}{12}\right)\right]\right\} + \text{INT}\left(\frac{275\,M}{9}\right) + D + \frac{60\,h + m + s/60^*}{1440} \tag{2.68}$$

where INT denotes the integer part and 60* denotes using 61 s for days with a leap second. We need to compute T_0, the number of Julian centuries elapsed from the epoch J2000.0 to zero hours of the date in question:

$$T_0 = \frac{JD(Y,M,D,0,0,0) - 2{,}451{,}545}{36{,}525} \tag{2.69}$$

The GCI coordinate system is fixed relative to the stars, not the Sun, so the GMST angle is the mean *sidereal* time at zero longitude. A *sidereal day* is the length of time that passes between successive crossings of a given projected meridian by a given fixed star in the sky. It is approximately 3 min and 56 s shorter than a *solar*

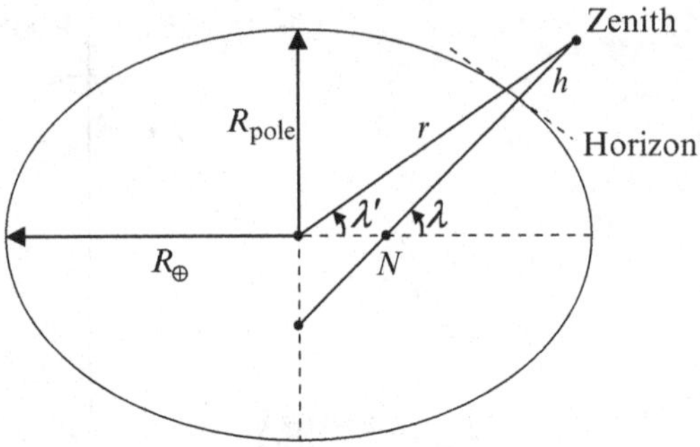

Fig. 2.4 Geocentric and geodetic latitude

day of 86,400 s, which is the length of time that elapses between the Sun reaching its highest point in the sky two consecutive times [1]. Therefore θ_{GMST} in units of seconds is calculated by

$$\begin{aligned}\theta_{GMST} &= 24,110.54841 + 8,640,184.812866\, T_0 + 0.093104\, T_0^2 \\ &\quad - 6.2 \times 10^{-6}\, T_0^3 + 1.002737909350795(3600\, h + 60\, m + s) \qquad (2.70)\end{aligned}$$

This quantity is next reduced to a range from 0 to 86,400 s by adding/subtracting multiples of 86,400. Then θ_{GMST} in degrees is obtained by dividing by 240, because 1 s = 1/240°.

The ECEF position vector can be specified by its magnitude $r \equiv \|\mathbf{r}_E\| = \|\mathbf{r}_I\|$, longitude ϕ, and geocentric latitude λ'. An alternative description in terms of geodetic latitude, λ, is often employed. The Earth's geoid can be approximated by an ellipsoid of revolution about its minor axis, the Earth's rotation axis, as shown in Fig. 2.4 [11]. The geocentric latitude $\lambda' = \sin^{-1}(z/r)$ is the angle between the equatorial plane and the radius vector from the center of the Earth. The geodetic latitude λ is the angle between the equatorial plane and the normal to the reference ellipsoid. The *flattening* of the ellipsoid is given by

$$f = \frac{R_\oplus - R_{pole}}{R_\oplus} \qquad (2.71)$$

where $R_\oplus$ is the equatorial radius of the Earth and R_{pole} is the distance from the center of the Earth to a pole. The eccentricity of the reference ellipsoid is given by

$$e = \sqrt{1 - (1-f)^2} = \sqrt{f(2-f)} \qquad (2.72)$$

Many reference ellipsoid models exist, but for all of them the difference between the equatorial and polar radii is less than 22 km, so that $f \approx 1/298.257$ is a valid approximation. A common ellipsoid model is given by the World Geodetic System 1984 model (WGS-84), with semimajor axis $R_\oplus \equiv a = 6,378,137.0$ m and semiminor axis $R_{pole} \equiv b = 6,356,752.3142$ m. The eccentricity of this ellipsoid is given by $e = 0.0818$.

To determine the ECEF position vector from the geodetic coordinates, latitude λ, longitude ϕ, and height h, we first compute the distance between the z axis and the normal to the ellipsoid, finding [3]

$$N = \frac{R_\oplus}{\sqrt{1 - e^2 \sin^2 \lambda}} \tag{2.73}$$

Then the ECEF position coordinates are given by

$$x = (N + h) \cos \lambda \cos \phi \tag{2.74a}$$

$$y = (N + h) \cos \lambda \sin \phi \tag{2.74b}$$

$$z = [N(1 - e^2) + h] \sin \lambda \tag{2.74c}$$

This gives the following relationship between geocentric and geodetic latitudes

$$\tan \lambda = \frac{N + h}{N(1 - e^2) + h} \tan \lambda' \tag{2.75}$$

which has the first order approximation in the flattening f

$$\lambda = \lambda' + \frac{f R_\oplus}{R_\oplus + h} \sin(2\lambda') \tag{2.76}$$

The difference between geodetic and geocentric latitudes amounts to 12 arcmin at most [11].

The conversion from ECEF to geodetic coordinates is not straightforward, but a closed-form solution is given in [20]. Given, x, y and z in ECEF coordinates, the solution is given by

$$e^2 = 1 - b^2/a^2, \quad \epsilon^2 = a^2/b^2 - 1, \quad \rho = \sqrt{x^2 + y^2} \tag{2.77a}$$

$$p = |z|/\epsilon^2, \quad s = \rho^2/(e^2 \epsilon^2), \quad q = p^2 - b^2 + s \tag{2.77b}$$

$$u = p/\sqrt{q}, \quad v = b^2 u^2/q, \quad P = 27vs/q, \quad Q = (\sqrt{P + 1} + \sqrt{P})^{2/3} \tag{2.77c}$$

$$t = (1 + Q + 1/Q)/6, \quad c = \sqrt{u^2 - 1 + 2t}, \quad w = (c - u)/2 \tag{2.77d}$$

$$d = \text{sign}(z)\sqrt{q}\left[w + (\sqrt{t^2 + v} - uw - t/2 - 1/4)^{1/2}\right] \tag{2.77e}$$

$$N = a\sqrt{1 + \epsilon^2 d^2/b^2}, \quad \lambda = \sin^{-1}[(\epsilon^2 + 1)(d/N)] \tag{2.77f}$$

$$h = \rho \cos \lambda + z \sin \lambda - a^2/N, \quad \phi = \text{atan2}(y, x) \tag{2.77g}$$

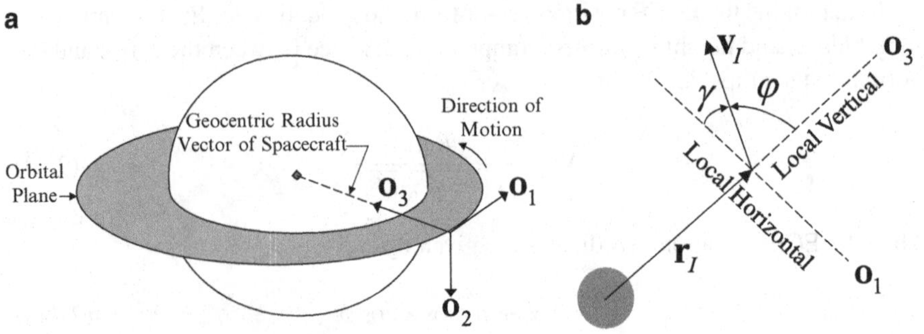

Fig. 2.5 Local-vertical/local-horizontal frame. (**a**) Frame definition. (**b**) Flight path angle

where atan2(y, x) is the standard function giving the argument of the complex number $x + iy$. It should be noted that longitude here is assumed to range between $-180°$ (West) to $+180°$ (East).

2.6.4 *Local-Vertical/Local-Horizontal Frame*

It is often convenient, especially for Earth-pointing spacecraft, to define a reference frame referenced to the spacecraft's orbit, which we will identify by the subscript O. The most common case is the Local-Vertical/Local-Horizontal (LVLH) orbit frame shown in Fig. 2.5a. It has its z axis $\mathbf{o}_3$ pointing along the nadir vector, directly toward the center of the Earth from the spacecraft,[4] and its y axis $\mathbf{o}_2$ pointing along the negative orbit normal, in the direction opposite to the spacecraft's orbital angular velocity. The x axis $\mathbf{o}_1$ completes the right-handed triad. The representations of these vectors in an inertial frame I are

$$\mathbf{o}_{3I} = -\mathbf{r}_I/\|\mathbf{r}_I\| \equiv -g_3\mathbf{r}_I \tag{2.78a}$$

$$\mathbf{o}_{2I} = -(\mathbf{r}_I \times \mathbf{v}_I)/\|\mathbf{r}_I \times \mathbf{v}_I\| \equiv -g_2(\mathbf{r}_I \times \mathbf{v}_I) \tag{2.78b}$$

$$\mathbf{o}_{1I} = \mathbf{o}_{2I} \times \mathbf{o}_{3I} = g_2 g_3(\mathbf{r}_I \times \mathbf{v}_I) \times \mathbf{r}_I = g_2 g_3[\,\|\mathbf{r}_I\|^2\mathbf{v}_I - (\mathbf{r}_I \cdot \mathbf{v}_I)\mathbf{r}_I] \tag{2.78c}$$

where $\mathbf{r}_I$ and $\mathbf{v}_I = \dot{\mathbf{r}}_I$ are the spacecraft position and velocity in the I frame. Note that the x axis is in the direction of the velocity for a circular orbit. The rotation matrix from the O frame to the I frame can be expressed by Eq. (2.45) as

$$A_{IO} = \left[\mathbf{o}_{1I}\ \mathbf{o}_{2I}\ \mathbf{o}_{3I}\right] \tag{2.79}$$

[4]This is the *geocentric* nadir vector. Some spacecraft use the *geodetic* nadir vector, which is normal to the surface of the reference ellipsoid, but we will not consider this complication.

The angle between the velocity vector and the local horizontal ($\mathbf{o}_{1I} - \mathbf{o}_{2I}$) plane is called the *flight-path angle*, denoted by γ in Fig. 2.5b. The specific angular momentum is given by $\mathbf{h}_I = \mathbf{r}_I \times \mathbf{v}_I$, as discussed in Chap. 10. From Eq. (2.58) we have $h \equiv \|\mathbf{h}_|\| = rv \sin\varphi$, where $r \equiv \|\mathbf{r}_I\|$ and $v \equiv \|\mathbf{v}_I\|$. Since $\gamma + \varphi = 90°$, then the flight path angle can be computed using

$$\cos\gamma = \frac{h}{rv} \tag{2.80}$$

The sign of γ is the same as the sign of $\mathbf{r}_I \cdot \mathbf{v}_I$. Notice that $\mathbf{o}_{1I}$ is in the direction of the spacecraft velocity and the flight path angle is always zero for a circular orbit, because $\mathbf{r}_I \cdot \mathbf{v}_I = 0$ in this case.

2.7 Quaternions

We consider a quaternion to be a four-component vector with some additional operations defined on it. A quaternion $\mathbf{q}$ has a three-vector part $\mathbf{q}_{1:3}$ and a scalar part q_4

$$\mathbf{q} = \begin{bmatrix} \mathbf{q}_{1:3} \\ q_4 \end{bmatrix} \quad \text{where} \quad \mathbf{q}_{1:3} = \begin{bmatrix} q_1 \\ q_2 \\ q_3 \end{bmatrix} \tag{2.81}$$

The most important added quaternion operations are two different products of a pair of quaternions $\bar{\mathbf{q}}$ and $\mathbf{q}$

$$\bar{\mathbf{q}} \otimes \mathbf{q} = \begin{bmatrix} q_4 \bar{\mathbf{q}}_{1:3} + \bar{q}_4 \mathbf{q}_{1:3} - \bar{\mathbf{q}}_{1:3} \times \mathbf{q}_{1:3} \\ \bar{q}_4 q_4 - \bar{\mathbf{q}}_{1:3} \cdot \mathbf{q}_{1:3} \end{bmatrix} \tag{2.82a}$$

$$\bar{\mathbf{q}} \odot \mathbf{q} = \begin{bmatrix} q_4 \bar{\mathbf{q}}_{1:3} + \bar{q}_4 \mathbf{q}_{1:3} + \bar{\mathbf{q}}_{1:3} \times \mathbf{q}_{1:3} \\ \bar{q}_4 q_4 - \bar{\mathbf{q}}_{1:3} \cdot \mathbf{q}_{1:3} \end{bmatrix} \tag{2.82b}$$

Notice that these definitions differ only in the sign of the cross product in the vector part, from which it follows that[5]

$$\bar{\mathbf{q}} \otimes \mathbf{q} = \mathbf{q} \odot \bar{\mathbf{q}} \tag{2.83}$$

Our quaternions are conceptually different from those introduced by Hamilton in 1844, before the introduction of vector notation. Hamilton defined a quaternion as $q = q_0 + iq_1 + jq_2 + kq_3$, a hypercomplex extension of a complex number $z = x + iy$, with i, j, and k obeying the relations $i^2 = j^2 = k^2 = -1$, $ij = -ji = k$,

[5]The notation $\bar{\mathbf{q}} \otimes \mathbf{q}$ was introduced in [7], and the notation $\bar{\mathbf{q}} \odot \mathbf{q}$ is a modification of notation introduced in [13].

$jk = -kj = i$, and $ki = -ik = j$. Hamilton's product $\bar{q}q$ corresponds to our product $\bar{\mathbf{q}} \odot \mathbf{q}$, but the product $\bar{\mathbf{q}} \otimes \mathbf{q}$ has proven to be more useful in attitude analysis. Some authors' notations differ from ours in labeling the scalar part of a quaternion q_0 and putting it at the top of the column vector. Care must be taken to thoroughly understand the conventions embodied in any quaternion equation that one chooses to reference.

Quaternion multiplication is associative, $\mathbf{q} \otimes (\bar{\mathbf{q}} \otimes \bar{\bar{\mathbf{q}}}) = (\mathbf{q} \otimes \bar{\mathbf{q}}) \otimes \bar{\bar{\mathbf{q}}}$ and distributive, $\mathbf{q} \otimes (\bar{\mathbf{q}} + \bar{\bar{\mathbf{q}}}) = \mathbf{q} \otimes \bar{\mathbf{q}} + \mathbf{q} \otimes \bar{\bar{\mathbf{q}}}$, but not commutative in general, $\mathbf{q} \otimes \bar{\mathbf{q}} \neq \bar{\mathbf{q}} \otimes \mathbf{q}$. This parallels the situation for matrix multiplication. In those cases for which $\mathbf{q} \otimes \bar{\mathbf{q}} = \bar{\mathbf{q}} \otimes \mathbf{q}$, the quaternions $\mathbf{q}$ and $\bar{\mathbf{q}}$ are said to commute. Analogous equations hold for the product $\bar{\mathbf{q}} \odot \mathbf{q}$.

Quaternion products can be represented by matrix multiplication, very much like the cross product:

$$\mathbf{q} \otimes \bar{\mathbf{q}} = [\mathbf{q} \otimes]\, \bar{\mathbf{q}} = \bar{\mathbf{q}} \odot \mathbf{q} \tag{2.84a}$$

$$\mathbf{q} \odot \bar{\mathbf{q}} = [\mathbf{q} \odot]\, \bar{\mathbf{q}} = \bar{\mathbf{q}} \otimes \mathbf{q} \tag{2.84b}$$

where

$$[\mathbf{q} \otimes] \equiv \begin{bmatrix} q_4 I_3 - [\mathbf{q}_{1:3}\times] & \mathbf{q}_{1:3} \\ -\mathbf{q}_{1:3}^T & q_4 \end{bmatrix} = \begin{bmatrix} \Psi(\mathbf{q}) & \mathbf{q} \end{bmatrix} \tag{2.85}$$

and

$$[\mathbf{q} \odot] \equiv \begin{bmatrix} q_4 I_3 + [\mathbf{q}_{1:3}\times] & \mathbf{q}_{1:3} \\ -\mathbf{q}_{1:3}^T & q_4 \end{bmatrix} = \begin{bmatrix} \Xi(\mathbf{q}) & \mathbf{q} \end{bmatrix} \tag{2.86}$$

with $\Psi(\mathbf{q})$ and $\Xi(\mathbf{q})$ being the 4×3 matrices

$$\Psi(\mathbf{q}) \equiv \begin{bmatrix} q_4 I_3 - [\mathbf{q}_{1:3}\times] \\ -\mathbf{q}_{1:3}^T \end{bmatrix} = \begin{bmatrix} q_4 & q_3 & -q_2 \\ -q_3 & q_4 & q_1 \\ q_2 & -q_1 & q_4 \\ -q_1 & -q_2 & -q_3 \end{bmatrix} \tag{2.87}$$

$$\Xi(\mathbf{q}) \equiv \begin{bmatrix} q_4 I_3 + [\mathbf{q}_{1:3}\times] \\ -\mathbf{q}_{1:3}^T \end{bmatrix} = \begin{bmatrix} q_4 & -q_3 & q_2 \\ q_3 & q_4 & -q_1 \\ -q_2 & q_1 & q_4 \\ -q_1 & -q_2 & -q_3 \end{bmatrix} \tag{2.88}$$

It is easy to show that

$$\Psi^T(\mathbf{q})\Psi(\mathbf{q}) = \Xi^T(\mathbf{q})\Xi(\mathbf{q}) = \|\mathbf{q}\|^2 I_3 \tag{2.89a}$$

$$\Psi(\mathbf{q})\Psi^T(\mathbf{q}) = \Xi(\mathbf{q})\Xi^T(\mathbf{q}) = \|\mathbf{q}\|^2 I_4 - \mathbf{q}\mathbf{q}^T \tag{2.89b}$$

$$\Psi^T(\mathbf{q})\,\mathbf{q} = \Xi^T(\mathbf{q})\,\mathbf{q} = \mathbf{0}_3 \tag{2.89c}$$

from which it follows that $\|\mathbf{q}\|^{-1}[\mathbf{q} \otimes]$ and $\|\mathbf{q}\|^{-1}[\mathbf{q} \odot]$ are orthogonal matrices.

We define the identity quaternion

$$\mathbf{I}_q \equiv \begin{bmatrix} \mathbf{0}_3 \\ 1 \end{bmatrix} \tag{2.90}$$

which obeys $\mathbf{I}_q \otimes \mathbf{q} = \mathbf{q} \otimes \mathbf{I}_q = \mathbf{I}_q \odot \mathbf{q} = \mathbf{q} \odot \mathbf{I}_q = \mathbf{q}$, as required of the identity.

We also define the conjugate $\mathbf{q}^*$ of a quaternion, obtained by changing the sign of the three-vector part:

$$\mathbf{q}^* = \begin{bmatrix} \mathbf{q}_{1:3} \\ q_4 \end{bmatrix}^* \equiv \begin{bmatrix} -\mathbf{q}_{1:3} \\ q_4 \end{bmatrix} \tag{2.91}$$

The product of a quaternion with its conjugate is equal to the square of its norm times the identity quaternion

$$\mathbf{q} \otimes \mathbf{q}^* = \mathbf{q}^* \otimes \mathbf{q} = \mathbf{q} \odot \mathbf{q}^* = \mathbf{q}^* \odot \mathbf{q} = \|\mathbf{q}\|^2 \, \mathbf{I}_q \tag{2.92}$$

The conjugate of the product of two quaternions $\bar{\mathbf{q}}$ and $\mathbf{q}$ is the product of the conjugates in the opposite order $(\mathbf{p} \otimes \mathbf{q})^* = \mathbf{q}^* \otimes \mathbf{p}^*$. This relation, Eq. (2.92), and the associativity of quaternion multiplication can be used to show that

$$\|\mathbf{p} \otimes \mathbf{q}\| = \|\mathbf{p} \odot \mathbf{q}\| = \|\mathbf{p}\| \, \|\mathbf{q}\| \tag{2.93}$$

It is not difficult to see that

$$[\mathbf{q}^* \otimes] = [\mathbf{q} \otimes]^T \quad \text{and} \quad [\mathbf{q}^* \odot] = [\mathbf{q} \odot]^T \tag{2.94}$$

The inverse of any quaternion having nonzero norm is defined by

$$\mathbf{q}^{-1} \equiv \mathbf{q}^* / \|\mathbf{q}\|^2 \tag{2.95}$$

so that $\mathbf{q} \otimes \mathbf{q}^{-1} = \mathbf{q}^{-1} \otimes \mathbf{q} = \mathbf{q} \odot \mathbf{q}^{-1} = \mathbf{q}^{-1} \odot \mathbf{q} = \mathbf{I}_q$, as required by the definition of an inverse. The inverse of the product of two quaternions is the product of the inverses in the opposite order $(\mathbf{p} \otimes \mathbf{q})^{-1} = \mathbf{q}^{-1} \otimes \mathbf{p}^{-1}$.

We will overload the quaternion product notation to allow us to multiply a three-component vector $\mathbf{x}$ and a quaternion, using the definitions

$$\mathbf{x} \otimes \mathbf{q} \equiv \begin{bmatrix} \mathbf{x} \\ 0 \end{bmatrix} \otimes \mathbf{q} = [\mathbf{x} \otimes] \, \mathbf{q} \quad \text{and} \quad \mathbf{q} \otimes \mathbf{x} \equiv \mathbf{q} \otimes \begin{bmatrix} \mathbf{x} \\ 0 \end{bmatrix} \tag{2.96}$$

with analogous definitions for $\mathbf{x} \odot \mathbf{q}$, $[\mathbf{x} \odot]$, and $\mathbf{q} \odot \mathbf{x}$. Note that the matrices

$$[\mathbf{x} \otimes] = \begin{bmatrix} -[\mathbf{x} \times] & \mathbf{x} \\ -\mathbf{x}^T & 0 \end{bmatrix} \equiv \Omega(\mathbf{x}) \quad \text{and} \tag{2.97a}$$

$$[\mathbf{x} \odot] = \begin{bmatrix} [\mathbf{x} \times] & \mathbf{x} \\ -\mathbf{x}^T & 0 \end{bmatrix} \equiv \Gamma(\mathbf{x}) \tag{2.97b}$$

are skew-symmetric. The relation

$$\mathbf{x} \otimes \mathbf{q} = \mathbf{q} \odot \mathbf{x} = \begin{bmatrix} \Xi(\mathbf{q}) & \mathbf{q} \end{bmatrix} \begin{bmatrix} \mathbf{x} \\ 0 \end{bmatrix} = \Xi(\mathbf{q})\,\mathbf{x} \tag{2.98}$$

is often very useful.

2.8 Rotations and Euler's Theorem

The discipline of spacecraft attitude determination is basically the study of methods for estimating the proper orthogonal matrix that transforms vectors from a reference frame fixed in space to a frame fixed in the spacecraft body. Thus it is the study of proper orthogonal 3×3 matrices, or matrices in the group SO(3). We will refer to these as rotation matrices or *attitude* matrices and denote them by the letter A.

Euler's Theorem[6] states one of the most important properties of attitude matrices, namely that any rotation is a rotation about a *fixed* axis. Recall that the transformation of a vector representation $\mathbf{x}_F$ from reference frame F to reference frame G by the attitude matrix A_{GF} is given by

$$A_{GF}\mathbf{x}_F = \mathbf{x}_G \tag{2.99}$$

Euler's Theorem asserts the existence of a vector $\mathbf{e}$ along the direction of the rotation axis that has the same representation in frame G as in frame F. This means that we can substitute $\mathbf{x}_F = \mathbf{x}_G = \mathbf{e}$ in Eq. (2.99) and state Euler's theorem algebraically as

$$A\,\mathbf{e} = \mathbf{e} \tag{2.100}$$

This is a special case of an eigenvalue/eigenvector relationship. An *eigenvector* of a general square matrix M is a nonzero vector $\mathbf{x}$ for which multiplication by M has the same effect as multiplication by a scalar, i.e.

$$M\,\mathbf{x} = \lambda\mathbf{x} \tag{2.101}$$

where the scalar λ is the *eigenvalue* corresponding to the eigenvector $\mathbf{x}$. The solution for $\mathbf{x}$ is only determined up to a scale factor in general, so eigenvectors are almost invariably given as unit vectors. In order for Eq. (2.101) to have a nonzero solution for $\mathbf{x}$, the matrix $(\lambda I - M)$ must be singular.[7] Therefore, from Eq. (2.16) we have

$$\det(\lambda I - M) = \lambda^n + \alpha_1\lambda^{n-1} + \cdots + \alpha_{n-1}\lambda + \alpha_n = 0 \tag{2.102}$$

---

[6] Leonhard Euler (1707–1783) laid the foundations for the analysis of rotations, and his fingerprints are all over the subject. Thus, attaching his name to anything serves poorly for distinguishing it from other results also bearing his name.

[7] Otherwise, we would have $\mathbf{x} = (\lambda I - M)^{-1}\mathbf{0} = \mathbf{0}$.

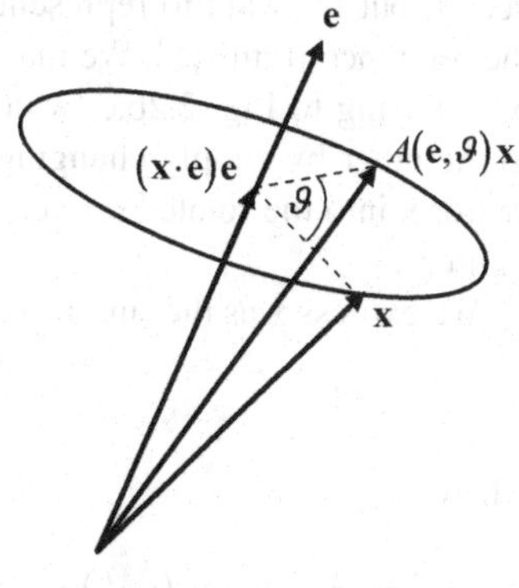

Fig. 2.6 Euler axis/angle rotation

This polynomial equation of degree n is known as the *characteristic equation* of M. The eigenvalues are the n roots of the characteristic equation, counting multiple roots by their multiplicity.

In the language of eigenvalues and eigenvectors, Euler's theorem says that one of the eigenvalues of an attitude matrix has the value $\lambda = 1$. The characteristic equation of a 3×3 matrix M is easily found by explicit computation of the determinant to take the form

$$\lambda^3 - \lambda^2 \text{tr}\, M + \lambda\, \text{tr}(\text{adj}\, M) - \det M = 0 \tag{2.103}$$

For the special case of a proper orthogonal 3×3 matrix, we find the characteristic equation to be

$$0 = \lambda^3 - \lambda^2 \text{tr}\, A + \lambda\, \text{tr}\, A - 1 = (\lambda - 1)[\lambda^2 + \lambda(1 - \text{tr}\, A) + 1] \tag{2.104}$$

This clearly has a root $\lambda = 1$, which proves Euler's theorem with the rotation axis $\mathbf{e}$ being the eigenvector corresponding to this eigenvalue.

2.9 Attitude Representations

2.9.1 Euler Axis/Angle Representation

We have seen that every proper orthogonal 3×3 matrix has a rotation axis specified by a unit vector $\mathbf{e}$. The only other parameter needed to completely specify the matrix is the angle of rotation ϑ about this axis. This axis and angle are known as the *Euler axis* and *Euler angle* of the rotation. We will now show how to parameterize the attitude matrix in terms of these parameters.

Figure 2.6 depicts the rotation of an arbitrary vector $\mathbf{x}$ through an angle ϑ about an axis $\mathbf{e}$. This figure is actually more illustrative of an active (alibi) rotation of the vector, but we want to represent the transformation as a passive (alias) rotation of the reference frame [5]. We make the connection between these two interpretations by referring to Fig. 2.2b,c, which show that we can move from one interpretation to the other by simply changing the direction of the rotation. The mapping of the vector $\mathbf{x}$ into the rotated reference frame is denoted as $A(\mathbf{e}, \vartheta)\mathbf{x}$, as indicated in the figure.

We express $\mathbf{x}$ as the sum of vectors parallel and perpendicular to the rotation axis

$$\mathbf{x} = \mathbf{x}_{\parallel} + \mathbf{x}_{\perp} \tag{2.105}$$

where

$$\mathbf{x}_{\parallel} \equiv (\mathbf{x} \cdot \mathbf{e})\, \mathbf{e} = \left(\mathbf{e}\,\mathbf{e}^T\right) \mathbf{x} \quad \text{and} \quad \mathbf{x}_{\perp} \equiv \mathbf{x} - (\mathbf{x} \cdot \mathbf{e})\mathbf{e} = (I_3 - \mathbf{e}\,\mathbf{e}^T)\mathbf{x} \tag{2.106}$$

The rotation leaves $\mathbf{x}_{\parallel}$ alone, but rotates $\mathbf{x}_{\perp}$ out of the plane defined by $\mathbf{e}$ and $\mathbf{x}$. The cross product $\mathbf{e}\times\mathbf{x}$ is perpendicular to that plane and has magnitude $\|\mathbf{e}\times\mathbf{x}\| = \|\mathbf{x}_{\perp}\|$ by Eq. (2.57), so the result of the norm-preserving rotation by angle ϑ is

$$A(\mathbf{e}, \vartheta)\, \mathbf{x} = \mathbf{x}_{\parallel} + (\cos\vartheta)\, \mathbf{x}_{\perp} - (\sin\vartheta)\, \mathbf{e} \times \mathbf{x} \tag{2.107}$$

The sign of the last term is chosen to agree with the sense of the rotation shown in Fig. 2.2b, which is regarded as a rotation in the positive sense about $\mathbf{e}_3 = \mathbf{e}_1 \times \mathbf{e}_2$, the outward normal from the plane of the figure.

Because $\mathbf{x}$ is an arbitrary vector, Eq. (2.107) means, inserting the definitions of $\mathbf{x}_{\parallel}$ and $\mathbf{x}_{\perp}$, that

$$A(\mathbf{e}, \vartheta) = (\cos\vartheta)\, I_3 - \sin\vartheta\, [\mathbf{e}\times] + (1 - \cos\vartheta)\mathbf{e}\,\mathbf{e}^T \tag{2.108}$$

This is the *Euler axis/angle* parameterization of an attitude matrix. We can also express this, using Eq. (2.56b) as

$$A(\mathbf{e}, \vartheta) = I_3 - \sin\vartheta\, [\mathbf{e}\times] + (1 - \cos\vartheta)[\mathbf{e}\times]^2 \tag{2.109}$$

The attitude matrix is expressed in explicit component form as

$$A(\mathbf{e}, \vartheta) = \begin{bmatrix} c + (1-c)e_1^2 & (1-c)e_1e_2 + s\,e_3 & (1-c)e_1e_3 - s\,e_2 \\ (1-c)e_2e_1 - s\,e_3 & c + (1-c)e_2^2 & (1-c)e_2e_3 + s\,e_1 \\ (1-c)e_3e_1 + s\,e_2 & (1-c)e_3e_2 - s\,e_1 & c + (1-c)e_3^2 \end{bmatrix} \tag{2.110}$$

where we have written $c \equiv \cos\vartheta$ and $s \equiv \sin\vartheta$ for conciseness. The attitude matrix appears to depend on four parameters, but there are only three independent parameters owing to the constraint $\|\mathbf{e}\| = 1$. The nine-component attitude matrix has only three independent parameters because of the orthogonality constraint $AA^T = I$. This matrix constraint is equivalent to six scalar constraints rather than nine because the product AA^T is symmetric.

Equations (2.108)–(2.110) show that the attitude matrix is a periodic function of the rotation angle over an unlimited range with period 2π. Some useful identities satisfied by the Euler axis/angle representation are

$$A(\mathbf{e}, \vartheta) = A(-\mathbf{e}, -\vartheta) \tag{2.111a}$$

$$A^{-1}(\mathbf{e}, \vartheta) = A^T(\mathbf{e}, \vartheta) = A(-\mathbf{e}, \vartheta) = A(\mathbf{e}, -\vartheta) \tag{2.111b}$$

$$A(\mathbf{e}, \pi) = A(-\mathbf{e}, \pi) = 2\,\mathbf{e}\,\mathbf{e}^T - I_3 \tag{2.111c}$$

We now turn to the question of finding the rotation axis and angle corresponding to a given attitude matrix. Noting from Eq. (2.108) that

$$\operatorname{tr} A(\mathbf{e}, \vartheta) = 1 + 2\cos\vartheta \tag{2.112}$$

we see that the rotation angle is given by

$$\vartheta = \cos^{-1}\left(\frac{\operatorname{tr} A(\mathbf{e}, \vartheta) - 1}{2}\right) \tag{2.113}$$

If $\cos\vartheta = 1$ the attitude matrix is $A(\mathbf{e}, \vartheta) = I_3$, and the rotation axis is clearly undefined. If $-1 < \cos\vartheta < 1$, the axis of rotation is given by

$$\mathbf{e} = \frac{1}{2\sin\vartheta}\begin{bmatrix} A_{23}(\mathbf{e}, \vartheta) - A_{32}(\mathbf{e}, \vartheta) \\ A_{31}(\mathbf{e}, \vartheta) - A_{13}(\mathbf{e}, \vartheta) \\ A_{12}(\mathbf{e}, \vartheta) - A_{21}(\mathbf{e}, \vartheta) \end{bmatrix} \tag{2.114}$$

If $\cos\vartheta = -1$ the axis of rotation can be found by normalizing any nonzero column of

$$A(\mathbf{e}, \vartheta) + I_3 = 2\,\mathbf{e}\,\mathbf{e}^T \tag{2.115}$$

because all the columns of this matrix are parallel to **e**. The overall sign of the rotation axis vector is undetermined in this case, but Eq. (2.111c) shows that this sign makes no difference.

The other two eigenvalues of the attitude matrix are the other two roots of Eq. (2.104). Inserting the value of $\operatorname{tr} A(\mathbf{e}, \vartheta)$ into this equation gives

$$0 = \lambda^2 - 2\lambda\cos\vartheta + 1 = \lambda^2 - \lambda\left(e^{i\vartheta} + e^{-i\vartheta}\right) + 1 = \left(\lambda - e^{i\vartheta}\right)\left(\lambda - e^{-i\vartheta}\right) \tag{2.116}$$

These two eigenvalues form a complex conjugate pair on the unit circle in the complex plane, and the corresponding eigenvectors are complex as well. This result can be generalized to proper orthogonal matrices of higher dimensionality. A matrix in SO($2n + 1$) has one eigenvalue equal to $+1$ and n conjugate pairs on the unit circle in the complex plane. A matrix in SO($2n$) has only the n conjugate pairs on the unit circle. Thus Euler's theorem holds in any space of odd dimensionality, but not in a space with an even number of dimensions. If we regard Fig. 2.2b as a rotation in two-dimensional space rather than a projection onto the plane of a rotation in a higher-dimensional space, for example, it is easy to see that there is no invariant vector in the plane. This does not preclude the possibility of a pair of complex conjugate eigenvectors accidentally having the common value +1.

Cross-product and trigonometric identities can be used to find the unsurprising result of successive rotations about the same axis,

$$A(\mathbf{e},\vartheta)A(\mathbf{e},\varphi) = A(\mathbf{e},\varphi)A(\mathbf{e},\vartheta) = A(\mathbf{e},\vartheta+\varphi) \tag{2.117}$$

but this is more easily derived using the quaternion representation of rotations. The composition of rotations about non-parallel axes does not have a simple form in the angle/axis representation. Another useful result holds for two attitude matrices A_0 and $A(\mathbf{e},\vartheta)$. From Eq. (2.108) we have

$$A_0 A(\mathbf{e},\vartheta)A_0^T = (\cos\vartheta)\,I_3 - (\sin\vartheta)A_0[\mathbf{e}\times]A_0^T + (1-\cos\vartheta)A_0\,\mathbf{e}\,\mathbf{e}^T A_0^T \tag{2.118}$$

This can be written, using Eqs. (2.7c) and (2.63), as

$$A_0 A(\mathbf{e},\vartheta)A_0^T = A(A_0\,\mathbf{e},\vartheta) \tag{2.119}$$

which shows that $A_0 A A_0^T$ is a rotation by the same angle as A, but about a rotated axis.

2.9.2 *Rotation Vector Representation*

It is convenient for analysis, but not for computations, to combine the Euler axis and angle into the three-component *rotation vector*

$$\boldsymbol{\vartheta} \equiv \vartheta\,\mathbf{e} \tag{2.120}$$

To express the attitude matrix in terms of the rotation vector, we insert the Taylor series expansions of the sine and cosine into Eq. (2.109), giving[8]

$$A(\mathbf{e},\vartheta) = I_3 - [\mathbf{e}\times]\sum_{i=0}^{\infty}\frac{(-1)^i\vartheta^{2i+1}}{(2i+1)!} - [\mathbf{e}\times]^2\sum_{i=1}^{\infty}\frac{(-1)^i\vartheta^{2i}}{(2n)!} \tag{2.121}$$

Equation (2.56b) can be used to show that $[\mathbf{e}\times]^3 = -[\mathbf{e}\times]$, so we have

$$A(\mathbf{e},\vartheta) = I_3 + \sum_{i=0}^{\infty}\frac{[(-\vartheta\,\mathbf{e})\times]^{2i+1}}{(2i+1)!} + \sum_{i=1}^{\infty}\frac{[(-\vartheta\,\mathbf{e})\times]^{2i}}{(2n)!} = \sum_{n=0}^{\infty}\frac{[(-\vartheta\,\mathbf{e})\times]^{n}}{n!} \tag{2.122}$$

[8]Note that these power series expansions assume that we measure angles in radians.

A function of a square matrix is defined by its Taylor series, so this expression for the attitude matrix in terms of the rotation vector can finally be written as

$$A(\mathbf{e}, \vartheta) = \exp\left(-[\boldsymbol{\vartheta}\times]\right) \tag{2.123}$$

where exp is the matrix exponential.

This is the first three-parameter representation of rotations that we have encountered. It is a very useful representation for the analysis of small rotations; but it is not useful for large rotations, mainly because it obscures the periodicity of the attitude matrix as a function of ϑ. In particular, it is not at all obvious in this representation that a rotation by an angle $\vartheta = 2\pi$ is equivalent to the identity transformation. Equation (2.113) shows that we can restrict the rotation angle to the range $0 \leq \vartheta \leq \pi$, which avoids this problem and gives a 1:1 mapping of rotations with $\vartheta < \pi$ to rotation vectors. The rotation vectors fill a ball of radius π, with the two vectors at the ends of a diameter of the ball representing the same attitude according to Eq. (2.111c). This causes the difficulty that the rotation vector parameterization of a smoothly-varying attitude can jump discontinuously from one side of the ball to the other. This discontinuity can be avoided by giving up the 1:1 mapping by allowing rotation angles greater than π, but this causes other difficulties. It is an unavoidable fact that the rotation group has no global three-component parameterization without singular points [22]. The expense of computing the matrix exponential also renders the rotation vector parameterization impractical for numerical computations.

2.9.3 Quaternion Representation

Substituting $\sin\vartheta = 2\sin(\vartheta/2)\cos(\vartheta/2)$ and $\cos\vartheta = \cos^2(\vartheta/2) - \sin^2(\vartheta/2)$ into Eq. (2.108) and defining the quaternion

$$\mathbf{q}(\mathbf{e}, \vartheta) = \begin{bmatrix} \mathbf{e}\sin(\vartheta/2) \\ \cos(\vartheta/2) \end{bmatrix} \tag{2.124}$$

gives the quaternion representation of the attitude matrix

$$\begin{aligned} A(\mathbf{q}) &= \left(q_4^2 - \|\mathbf{q}_{1:3}\|^2\right) I_3 - 2q_4[\mathbf{q}_{1:3}\times] + 2\,\mathbf{q}_{1:3}\,\mathbf{q}_{1:3}^T \\ &= \begin{bmatrix} q_1^2 - q_2^2 - q_3^2 + q_4^2 & 2(q_1q_2 + q_3q_4) & 2(q_1q_3 - q_2q_4) \\ 2(q_2q_1 - q_3q_4) & -q_1^2 + q_2^2 - q_3^2 + q_4^2 & 2(q_2q_3 + q_1q_4) \\ 2(q_3q_1 + q_2q_4) & 2(q_3q_2 - q_1q_4) & -q_1^2 - q_2^2 + q_3^2 + q_4^2 \end{bmatrix} \end{aligned} \tag{2.125}$$

We are abusing the notation by using both $A(\mathbf{q})$ and $A(\mathbf{e}, \vartheta)$ to denote the attitude matrix, and we will abuse it further when we define other representations. The meaning of the argument of $A(\cdot)$ will always be clear in context, however.

The four parameters of the quaternion representation were first considered by Euler, but their full significance was revealed by Rodrigues.[9] For this reason, they are often referred to as the *Euler symmetric parameters* or the *Euler-Rodrigues* parameters. The beauty of the quaternion representation is that it expresses the attitude matrix as a homogenous quadratic function of the elements of the quaternion, requiring no trigonometric or other transcendental function evaluations. Quaternions are more efficient for specifying rotations than the attitude matrix itself, having only four components instead of nine, and obeying only one constraint, the norm constraint, instead of the six constraints imposed on the attitude matrix by orthogonality.

Quaternions used to parameterize rotations are *unit quaternions*, i.e. quaternions with unit norm, as defined by Eq. (2.124). A unit quaternion always has an inverse, which is identical with its conjugate. Also, the discussion in Sect. 2.7 says that the matrices $[\mathbf{q}\otimes]$ and $[\mathbf{q}\odot]$ for a unit quaternion are orthogonal. We will now derive a useful expression for $[\mathbf{q}\otimes]$ in terms of the rotation vector. Because it is a linear function of $\mathbf{q}$, we have

$$[\mathbf{q}(\mathbf{e},\vartheta)\otimes] = \cos(\vartheta/2)[\mathbf{I}_q\otimes] + \sin(\vartheta/2)[\mathbf{e}\otimes] = \cos(\vartheta/2)I_4 + \sin(\vartheta/2)[\mathbf{e}\otimes] \tag{2.126}$$

Now expand the sine and cosine in Taylor series and use $[\mathbf{e}\otimes]^2 = -I_4$, to get

$$[\mathbf{q}(\mathbf{e},\vartheta)\otimes] = \sum_{i=0}^{\infty}\frac{[(\vartheta\,\mathbf{e}/2)\otimes]^{2i}}{(2n)!} + \sum_{i=0}^{\infty}\frac{[(\vartheta\,\mathbf{e}/2)\otimes]^{2i+1}}{(2i+1)!} = \exp[(\boldsymbol{\vartheta}/2)\otimes] \tag{2.127}$$

This result and its derivation are very similar to Eq. (2.123).

We now want to show how a rotation of a three-component vector $\mathbf{x}$ can be implemented by quaternion multiplication. This is accomplished by the quaternion product

$$\mathbf{q}\otimes\mathbf{x}\otimes\mathbf{q}^* = [\mathbf{q}^*\odot](\mathbf{q}\otimes\mathbf{x}) = [\mathbf{q}\odot]^T[\mathbf{q}\otimes]\begin{bmatrix}\mathbf{x}\\0\end{bmatrix} = \begin{bmatrix}\Xi^T(\mathbf{q})\Psi(\mathbf{q})\,\mathbf{x}\\0\end{bmatrix} \tag{2.128}$$

where we have used several relations from Sect. 2.7. Explicit multiplication and comparison with Eq. (2.125) gives

$$\Xi^T(\mathbf{q})\Psi(\mathbf{q}) = A(\mathbf{q}) \tag{2.129}$$

[9] Olinde Rodrigues (1795–1851) obtained a doctorate in mathematics in 1815, with a thesis containing his well-known formula for the Legendre polynomials. He published nothing mathematical for the next 21 years, devoting himself to banking, the development of the French railways, utopian socialism, writing several pamphlets on banking, and editing an anthology of workers' poetry. Then he published eight papers between 1838 and 1845, including his 1840 paper [14] greatly advancing the state of the art in attitude analysis.

so that

$$\mathbf{q} \otimes \mathbf{x} \otimes \mathbf{q}^* = \begin{bmatrix} A(\mathbf{q})\,\mathbf{x} \\ 0 \end{bmatrix} \tag{2.130}$$

This can be used to derive the rule for performing successive transformations using quaternions. Applying a transformation by a second quaternion $\bar{\mathbf{q}}$ gives

$$\bar{\mathbf{q}} \otimes \left(\mathbf{q} \otimes \mathbf{x} \otimes \mathbf{q}^*\right) \otimes \bar{\mathbf{q}}^* = \bar{\mathbf{q}} \otimes [A(\mathbf{q})\,\mathbf{x}] \otimes \bar{\mathbf{q}}^* = \begin{bmatrix} A(\bar{\mathbf{q}})A(\mathbf{q})\,\mathbf{x} \\ 0 \end{bmatrix} \tag{2.131}$$

This transformation can also be written as

$$(\bar{\mathbf{q}} \otimes \mathbf{q}) \otimes \mathbf{x} \otimes (\bar{\mathbf{q}} \otimes \mathbf{q})^* = \begin{bmatrix} A(\bar{\mathbf{q}} \otimes \mathbf{q})\,\mathbf{x} \\ 0 \end{bmatrix} \tag{2.132}$$

Because this relation must hold for any $\mathbf{x}$, we have proved that

$$A(\bar{\mathbf{q}} \otimes \mathbf{q}) = A(\bar{\mathbf{q}})A(\mathbf{q}) \tag{2.133}$$

Thus the quaternion representation of successive transformations is just the product of the quaternions of the constituent transformations, in the same way that the attitude matrix of the combined transformation is the product of the individual attitude matrices. A simple bilinear composition rule of this type holds only for the attitude matrix and quaternion representations, which is one of the reasons for the popularity of quaternions. With our $\otimes$ quaternion product definition, the order of quaternion multiplication is identical to the order of matrix multiplication. The order would have been reversed if we had used the classical $\odot$ definition of quaternion multiplication. The quaternion equivalent of Eq. (2.117) for successive rotations about the same axis follows from straightforward quaternion multiplication:

$$\mathbf{q}(\mathbf{e}, \vartheta) \otimes \mathbf{q}(\mathbf{e}, \varphi) = \mathbf{q}(\mathbf{e}, \varphi) \otimes \mathbf{q}(\mathbf{e}, \vartheta) = \mathbf{q}(\mathbf{e}, \vartheta + \varphi) \tag{2.134}$$

Unit quaternions reside on the three-dimensional unit sphere S^3 embedded in four-dimensional quaternion space. Equation (2.124) shows that a rotation by 720°, but not a rotation by 360°, is equivalent to the identity transformation in quaternion space, because $\mathbf{q}(\mathbf{e}, \vartheta + 4\pi) = \mathbf{q}(\mathbf{e}, \vartheta)$ but $\mathbf{q}(\mathbf{e}, \vartheta + 2\pi) = -\mathbf{q}(\mathbf{e}, \vartheta)$. The attitude matrix $A(\mathbf{q})$ is a homogenous quadratic function of the elements of the quaternion, though, so $\mathbf{q}$ and $-\mathbf{q}$ give the same attitude matrix. This 2:1 mapping of quaternions to rotations is a minor annoyance that cannot be removed without introducing discontinuities like those that plague all three-parameter attitude representations. Because the quaternions $\mathbf{q}$ and $-\mathbf{q}$ are on opposite hemispheres of S^3, we could get a 1:1 mapping of quaternions to rotations by restricting the quaternions to one hemisphere, which is usually taken to be the hemisphere with $q_4 \geq 0$. This gives rise to the same problem as restricting the rotation vector to $\vartheta \leq \pi$, namely that a smoothly varying quaternion can jump discontinuously from one side to the other of the equator bounding the hemisphere. Restricting the representation to positive q_4 effectively gives a three-parameter representation with $q_4 \equiv \sqrt{1 - \|\mathbf{q}_{1:3}\|^2}$, so it is not surprising that it leads to the same problems as other three-parameter representations.

We finally turn to the problem of extracting a quaternion from an attitude matrix [10]. We construct four four-component vectors from the components of A:

$$\begin{bmatrix} 1 + 2A_{11} - \mathrm{tr}A \\ A_{12} + A_{21} \\ A_{13} + A_{31} \\ A_{23} - A_{32} \end{bmatrix} = 4q_1\mathbf{q}\,, \quad \begin{bmatrix} A_{21} + A_{12} \\ 1 + 2A_{22} - \mathrm{tr}A \\ A_{23} + A_{32} \\ A_{31} - A_{13} \end{bmatrix} = 4q_2\mathbf{q}$$

$$\begin{bmatrix} A_{31} + A_{13} \\ A_{32} + A_{23} \\ 1 + 2A_{33} - \mathrm{tr}A \\ A_{12} - A_{21} \end{bmatrix} = 4q_3\mathbf{q}\,, \quad \begin{bmatrix} A_{23} - A_{32} \\ A_{31} - A_{13} \\ A_{12} - A_{21} \\ 1 + \mathrm{tr}A \end{bmatrix} = 4q_4\mathbf{q} \tag{2.135}$$

The quaternion can be found by normalizing any one of these four vectors. Numerical errors are minimized by choosing the vector with the greatest norm, which is the vector with the largest value of $|q_i|$ on the right side. This can be found by the following procedure. Find the largest of $\mathrm{tr}A$ and A_{ii} for $i = 1, 2, 3$. If $\mathrm{tr}A$ is the largest of these, then $|q_4|$ is the largest of the $|q_i|$, otherwise the largest value of $|q_i|$ is the one with the same index as the largest A_{ii}. The overall sign of the normalized vector is not determined, reflecting the twofold ambiguity of the quaternion representation.

2.9.4 Rodrigues Parameter Representation

The three *Rodrigues parameters* made their appearance in Rodrigues' classic 1840 paper [14]. They were later represented as the "vector semitangent of version" by J. Willard Gibbs, who invented modern vector notation. For this reason, the vector of Rodrigues parameters is often called the *Gibbs vector* and denoted by **g**. They are related to the quaternion by

$$\mathbf{g} = \frac{\mathbf{q}_{1:3}}{q_4} \tag{2.136}$$

which has the inverse

$$\mathbf{q} = \frac{\pm 1}{\sqrt{1 + \|\mathbf{g}\|^2}} \begin{bmatrix} \mathbf{g} \\ 1 \end{bmatrix} \tag{2.137}$$

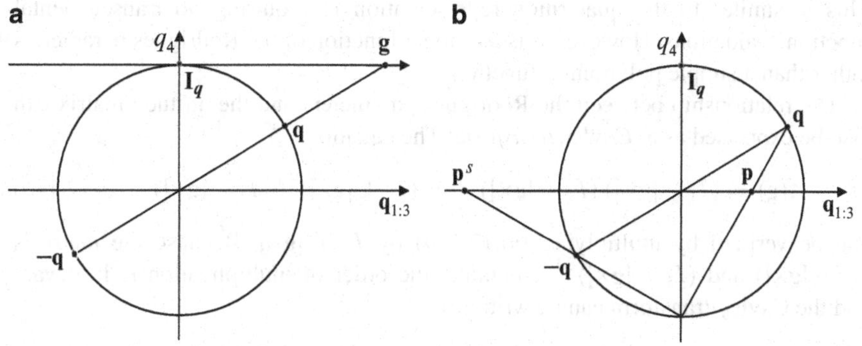

Fig. 2.7 Relationship of the Rodrigues parameters (**a**) and the modified Rodrigues parameters (**b**) to the quaternion

Using Eq. (2.124) to express the quaternion in terms of the Euler axis and angle gives

$$\mathbf{g}(\mathbf{e},\vartheta) = \mathbf{e}\tan(\vartheta/2) \tag{2.138}$$

which explains Gibbs' peculiar terminology.

The mapping from quaternions to Rodrigues parameters is illustrated in Fig. 2.7a. The plane of the figure is the plane containing the origin, $\mathbf{q}$, and the identity quaternion $\mathbf{I}_q$. The circle is the cross-section of the quaternion sphere S^3, so it has unit radius. The vertical axis is the q_4 axis, and the horizontal axis represents the three-dimensional $\mathbf{q}_{1:3}$ hyperplane. The horizontal line passing through $\mathbf{I}_q$ represents the three-dimensional Gibbs vector hyperplane, which is tangent to S^3 at the point $\mathbf{q} = \mathbf{I}_q$. The Gibbs vector $\mathbf{g}$ is the projection of the quaternion from the origin onto the Gibbs vector hyperplane. It is clear from the figure or from Eq. (2.136) that $\mathbf{q}$ and $-\mathbf{q}$ map to the same Gibbs vector, so the Rodrigues parameters provide a 1:1 mapping of rotations. The price paid for this is that the Gibbs vector is infinite for a 180° rotation. Thus this parameterization is not recommended as a global attitude representation, but it provides an excellent representation of small rotations.

Substituting Eq. (2.137) into Eq. (2.125) gives the Rodrigues parameter representation of the attitude matrix

$$A(\mathbf{g}) = \frac{(1-\|\mathbf{g}\|^2)I_3 - 2[\mathbf{g}\times] + 2\,\mathbf{g}\,\mathbf{g}^T}{1+\|\mathbf{g}\|^2} = I_3 + 2\frac{[\mathbf{g}\times]^2 - [\mathbf{g}\times]}{1+\|\mathbf{g}\|^2}$$
$$= \frac{1}{1+\|\mathbf{g}\|^2}\begin{bmatrix} 1+g_1^2-g_2^2-g_3^2 & 2(g_1g_2+g_3) & 2(g_1g_3-g_2) \\ 2(g_2g_1-g_3) & 1-g_1^2+g_2^2-g_3^2 & 2(g_2g_3+g_1) \\ 2(g_3g_1+g_2) & 2(g_3g_2-g_1) & 1-g_1^2-g_2^2+g_3^2 \end{bmatrix} \tag{2.139}$$

This is similar to the quaternion representation in requiring no transcendental function evaluations. However, it is a rational function of the Rodrigues parameters rather than a simple polynomial function.

The relationship between the Rodrigues parameters and the attitude matrix can also be expressed as a *Cayley transform*. The equations

$$A(\mathbf{g}) = (I_3 - [\mathbf{g}\times])\,(I_3 + [\mathbf{g}\times])^{-1} = (I_3 + [\mathbf{g}\times])^{-1}\,(I_3 - [\mathbf{g}\times]) \tag{2.140}$$

can be verified by multiplying Eq. (2.139) by $I_3 + [\mathbf{g}\times]$. Because the matrices $(I_3 - [\mathbf{g}\times])$ and $(I_3 + [\mathbf{g}\times])^{-1}$ commute, the order of multiplication is irrelevant and the Cayley transform can be written as

$$A(\mathbf{g}) = \frac{I_3 - [\mathbf{g}\times]}{I_3 + [\mathbf{g}\times]} \tag{2.141}$$

This is not a useful form for computation, owing to the required matrix inversion; but it can be generalized to higher dimensions. Any $n \times n$ proper orthogonal matrix M can be expressed in terms of an $n \times n$ skew-symmetric matrix S by

$$D = \frac{I_n - S}{I_n + S} \tag{2.142}$$

A skew-symmetric $n \times n$ matrix has $n(n-1)/2$ free parameters, the correct number to parameterize an orthogonal $n \times n$ matrix that must obey $n(n+1)/2$ constraints on its n^2 elements.

The rule for computing the Gibbs vector representing a composite of two rotations is easily derived from Eqs. (2.82), (2.136), and (2.137). The Gibbs vector corresponding to the quaternion product $\bar{\bar{\mathbf{q}}} = \bar{\mathbf{q}} \otimes \mathbf{q}$ is

$$\bar{\bar{\mathbf{g}}} = \frac{\bar{\mathbf{g}} + \mathbf{g} - \bar{\mathbf{g}} \times \mathbf{g}}{1 - \bar{\mathbf{g}} \cdot \mathbf{g}} \tag{2.143}$$

This is not a bilinear function of the constituent Gibbs vectors, so it cannot be represented as a matrix product like quaternion composition.

Extracting the Rodrigues parameters from the attitude matrix is similar to extracting the quaternion using one of the four-component vectors of Eq. (2.135). Instead of normalizing one of those vectors, though, the Rodrigues parameters are found by dividing the first three components of the vector by the fourth component.

2.9.5 Modified Rodrigues Parameters

The modified Rodrigues parameters (MRPs) are the newest of the commonly-employed attitude representations. They were invented by T. F. Wiener in 1962 [26], rediscovered by Marandi and Modi in 1987 [9], and have been championed by Junkins and Schaub [16]. They are related to the quaternion by

$$\mathbf{p} = \frac{\mathbf{q}_{1:3}}{1 + q_4} \tag{2.144}$$

which has the inverse

$$\mathbf{q} = \frac{1}{1+\|\mathbf{p}\|^2}\begin{bmatrix} 2\mathbf{p} \\ 1-\|\mathbf{p}\|^2 \end{bmatrix} \tag{2.145}$$

Using Eq. (2.124) to express the quaternion in terms of the Euler axis and angle gives

$$\mathbf{p}(\mathbf{e},\vartheta) = \mathbf{e}\tan(\vartheta/4) \tag{2.146}$$

It it easy to see that

$$\mathbf{p}(\mathbf{e},\vartheta+2\pi) = -\mathbf{e}\cot(\vartheta/4) = -\frac{\mathbf{p}(\mathbf{e},\vartheta)}{\|\mathbf{p}(\mathbf{e},\vartheta)\|^2} \tag{2.147}$$

but that $\mathbf{p}(\mathbf{e},\vartheta+4\pi) = \mathbf{p}(\mathbf{e},\vartheta)$. Thus the MRP representation is 2:1 just like the quaternion representation.

The *shadow* set of MRPs

$$\mathbf{p}^S \equiv -\frac{\mathbf{p}}{\|\mathbf{p}\|^2} \tag{2.148}$$

represents the same attitude as $\mathbf{p}$, in the same way that $\mathbf{q}$ and $-\mathbf{q}$ represent the same attitude. These two MRP vectors are illustrated in Fig. 2.7b, which shows them as stereographic projections from the point $\mathbf{q} = -\mathbf{I}_q$ onto the MRP hyperplane, which is coincident with the $\mathbf{q}_{1:3}$ hyperplane. It is clear from the figure or from Eq. (2.148) that $\|\mathbf{p}^S\| \geq 1$ if $\|\mathbf{p}\| \leq 1$. Thus, in following the variation of an attitude represented by MRPs, one can always keep the magnitude of the MRP vector from exceeding unity by switching to the shadow MRP when needed. The logic required for this is regarded by many practitioners to be less burdensome than carrying the fourth component of a quaternion and enforcing the quaternion norm constraint, leading them to prefer MRPs for numerical simulation of attitude motion. It is good practice to allow the MRP norm to exceed unity by some amount to avoid "chattering" between the MRP and its shadow in case the norm remains close to unity for an extended period.

The MRP representation of the attitude matrix can be found by substituting Eq. (2.145) into Eq. (2.125). It is easier to note that $A(\mathbf{e},\vartheta) = A^2(\mathbf{e},\vartheta/2)$ from Eq. (2.117), so the MRP representation can be obtained by squaring the Gibbs vector representation for a half-angle rotation:

$$\begin{aligned} A(\mathbf{p}) &= \left(\frac{I_3-[\mathbf{p}\times]}{I_3+[\mathbf{p}\times]}\right)^2 = \left(I_3 + 2\frac{[\mathbf{p}\times]^2-[\mathbf{p}\times]}{1+\|\mathbf{p}\|^2}\right)^2 \\ &= I_3 + \frac{8\,[\mathbf{p}\times]^2 - 4\left(1-\|\mathbf{p}\|^2\right)[\mathbf{p}\times]}{(1+\|\mathbf{p}\|^2)^2} \end{aligned} \tag{2.149}$$

The rule for computing the MRPs representing a composite of two rotations can be derived using Eqs. (2.82), (2.144). and (2.145). The MRP vector corresponding to the quaternion product $\bar{\bar{\mathbf{q}}} = \bar{\mathbf{q}} \otimes \mathbf{q}$ is

$$\bar{\bar{\mathbf{p}}} = \frac{\left(1 - \|\mathbf{p}\|^2\right)\bar{\mathbf{p}} + \left(1 - \|\bar{\mathbf{p}}\|^2\right)\mathbf{p} - 2\,\bar{\mathbf{p}} \times \mathbf{p}}{1 + \|\mathbf{p}\|^2\|\bar{\mathbf{p}}\|^2 - 2\,\bar{\mathbf{p}} \cdot \mathbf{p}} \tag{2.150}$$

This involves more computation than the composition of Gibbs vectors, but not an unreasonable burden.

The most convenient way to extract the MRPs from the attitude matrix is to first extract the quaternion and then compute the MRPs by Eq. (2.144).

Junkins and Schaub have developed a family of attitude representations intermediate between the Rodrigues parameters and the MRPs by choosing a projection point in quaternion space intermediate between the point $\mathbf{q} = \mathbf{0}$ in Fig. 2.7a and the point $\mathbf{q} = -\mathbf{I}_q$ in Fig. 2.7b [15]. Tsiotras, Junkins, and Schaub have investigated higher-order Cayley transforms

$$A(\mathbf{v}) = \left(\frac{I_3 - [\mathbf{v}\times]}{I_3 + [\mathbf{v}\times]}\right)^n \tag{2.151}$$

for $n > 2$ [23]. Neither of these generalizations has found wide application, however.

2.9.6 *Euler Angles*

An *Euler angle* representation expresses a rotation from an initial frame I to a final frame F as the product of three rotations: a rotation first from I to an intermediate frame H, then to a second intermediate frame G, and finally to frame F. The frame indices are generally omitted to simplify the notation, but including them clarifies the form of the overall transformation

$$A_{FI}(\mathbf{e}_\phi, \mathbf{e}_\theta, \mathbf{e}_\psi; \phi, \theta, \psi) \equiv A_{FG}(\mathbf{e}_\psi, \psi) A_{GH}(\mathbf{e}_\theta, \theta) A_{HI}(\mathbf{e}_\phi, \phi) \tag{2.152}$$

The rotation axis vectors of the constituent rotations are constant column vectors, and their subscripts do not label frames explicitly. The column vector $\mathbf{e}_\psi$ is a representation of a rotation axis in frames F and G, $\mathbf{e}_\theta$ is a representation in frames G and H, and $\mathbf{e}_\phi$ is a representation in frames H and I. The rotation angles ϕ, θ, and ψ are the variables used to specify the rotation. The possibility of employing Euler axis sequences with a wide choice of rotation axes was discovered by Davenport [2, 18], generalizing the classical applications of Euler angles that use a more restricted set of rotation axes. We will establish general results for Euler axis sequences using Davenport's general formulation, and then the classical axis sequences will follow as special cases.

We want to be able to represent any attitude matrix by this Euler angle sequence. In particular, we must be able to represent the attitude matrix that transforms $\mathbf{e}_\phi$ into $\mathbf{e}_\psi$, which is to say

$$\mathbf{e}_\psi = A(\mathbf{e}_\phi, \mathbf{e}_\theta, \mathbf{e}_\psi; \phi, \theta, \psi)\mathbf{e}_\phi = A(\mathbf{e}_\psi, \psi)A(\mathbf{e}_\theta, \theta)A(\mathbf{e}_\phi, \phi)\mathbf{e}_\phi \tag{2.153}$$

Multiplying on the left by $A^T(\mathbf{e}_\psi, \psi)$ and noting that $A^T(\mathbf{e}_\psi, \psi)\mathbf{e}_\psi = \mathbf{e}_\psi$ and $A(\mathbf{e}_\phi, \phi)\mathbf{e}_\phi = \mathbf{e}_\phi$, we see that there must be an angle θ_0 for which

$$\mathbf{e}_\psi = A(\mathbf{e}_\theta, \theta_0)\mathbf{e}_\phi \tag{2.154}$$

A little thought (or algebra) shows that this requires $\mathbf{e}_\psi \cdot \mathbf{e}_\theta = \mathbf{e}_\phi \cdot \mathbf{e}_\theta$. We must also be able to represent the attitude matrix that transforms $\mathbf{e}_\phi$ into $-\mathbf{e}_\psi$, which means that there must be an angle θ_1 for which $-\mathbf{e}_\psi = A(\mathbf{e}_\theta, \theta_1)\mathbf{e}_\phi$. It follows from this that $-\mathbf{e}_\psi \cdot \mathbf{e}_\theta = \mathbf{e}_\phi \cdot \mathbf{e}_\theta$. These conditions can be satisfied simultaneously only if the rotation axis $\mathbf{e}_\theta$ is perpendicular to both $\mathbf{e}_\phi$ and $\mathbf{e}_\psi$, so that

$$\mathbf{e}_\psi \cdot \mathbf{e}_\theta = \mathbf{e}_\phi \cdot \mathbf{e}_\theta = 0 \tag{2.155}$$

This relation leads to a more complete description of the orientation of the physical rotation axis vectors. Axis $\mathbf{e}_\phi$ is fixed in the initial reference frame, $\mathbf{e}_\psi$ is fixed in the final reference frame, and $\mathbf{e}_\theta$ is perpendicular to both $\mathbf{e}_\phi$ and $\mathbf{e}_\psi$.

We have shown that Eq. (2.155) is a necessary condition for Eq. (2.152) to represent a general attitude. We will now show that it is a sufficient condition. Note that the angle θ_0 is not a variable, but is defined by the choice of rotation axes. Then using Eqs. (2.111b), (2.117) and (2.119) gives

$$\begin{aligned} A(\mathbf{e}_\phi, \mathbf{e}_\theta, \mathbf{e}_\psi; \phi, \theta, \psi) &= A\left(A(\mathbf{e}_\theta, \theta_0)\mathbf{e}_\phi, \psi\right) A(\mathbf{e}_\theta, \theta)A(\mathbf{e}_\phi, \phi) \\ &= A(\mathbf{e}_\theta, \theta_0)A(\mathbf{e}_\phi, \psi)A(\mathbf{e}_\theta, \theta')A(\mathbf{e}_\phi, \phi) \\ &= A(\mathbf{e}_\theta, \theta_0)A(\mathbf{e}_\phi, \mathbf{e}_\theta, \mathbf{e}_\phi; \phi, \theta', \psi) \end{aligned} \tag{2.156}$$

where $\theta' \equiv \theta - \theta_0$. As $A(\mathbf{e}_\phi, \mathbf{e}_\theta, \mathbf{e}_\psi; \phi, \theta, \psi)$ covers all of SO(3), the product $A^T(\mathbf{e}_\theta, \theta_0)A(\mathbf{e}_\phi, \mathbf{e}_\theta, \mathbf{e}_\psi; \phi, \theta, \psi)$ also covers SO(3), so we only need to show that $A(\mathbf{e}_\phi, \mathbf{e}_\theta, \mathbf{e}_\phi; \phi, \theta', \psi)$ can represent any attitude, or equivalently to show that this rotation can transform the orthonormal basis $\mathbf{e}_\phi$, $\mathbf{e}_\theta$, $\mathbf{e}_\phi \times \mathbf{e}_\theta$ into any other orthonormal basis. It is sufficient to show that $\mathbf{e}'_\phi \equiv A(\mathbf{e}_\phi, \mathbf{e}_\theta, \mathbf{e}_\phi; \phi, \theta', \psi)\mathbf{e}_\phi$ can be any unit vector and $\mathbf{e}'_\theta \equiv A(\mathbf{e}_\phi, \mathbf{e}_\theta, \mathbf{e}_\phi; \phi, \theta', \psi)\mathbf{e}_\theta$ can be any unit vector perpendicular to $\mathbf{e}'_\phi$, because Eq. (2.65) then implies that $\mathbf{e}_\phi \times \mathbf{e}_\theta$ will transform into $\mathbf{e}'_\phi \times \mathbf{e}'_\theta$, completing the orthonormal triad. Now

$$\begin{aligned} \mathbf{e}'_\phi &= A(\mathbf{e}_\phi, \psi)A(\mathbf{e}_\theta, \theta')A(\mathbf{e}_\phi, \phi)\mathbf{e}_\phi = A(\mathbf{e}_\phi, \psi)A(\mathbf{e}_\theta, \theta')\mathbf{e}_\phi \\ &= \cos\theta'\, \mathbf{e}_\phi + \sin\theta' \sin\psi\, \mathbf{e}_\theta + \sin\theta' \cos\psi\, (\mathbf{e}_\phi \times \mathbf{e}_\theta) \end{aligned} \tag{2.157}$$

and we can find values of θ' and ψ to make this equal any desired unit vector. Furthermore, because $A(\mathbf{e}_\phi, \phi)\mathbf{e}_\theta = \cos\phi\,\mathbf{e}_\theta - \sin\phi\,(\mathbf{e}_\phi \times \mathbf{e}_\theta)$, we have

$$\mathbf{e}'_\theta = \cos\phi\, A(\mathbf{e}_\phi, \psi)A(\mathbf{e}_\theta, \theta')\mathbf{e}_\theta - \sin\phi\, A(\mathbf{e}_\phi, \psi)A(\mathbf{e}_\theta, \theta')(\mathbf{e}_\phi \times \mathbf{e}_\theta) \tag{2.158}$$

Rotations preserve orthogonality, so this superposition can represent any unit vector in the plane perpendicular to $\mathbf{e}'_\phi$, completing the proof that any Euler angle sequence obeying Eq. (2.155) is sufficient to represent any attitude matrix.

An Euler angle parameterization has a twofold ambiguity in addition to the usual 2π ambiguity in specifying any angle. To see this ambiguity, insert the product $A(\mathbf{e}_\phi, -\pi)A(\mathbf{e}_\phi, \pi)$ before and after $A(\mathbf{e}_\theta, \theta')$ in the second line of Eq. (2.156) and use some axis/angle representation identities to get

$$\begin{aligned}
A(\mathbf{e}_\phi, \mathbf{e}_\theta, \mathbf{e}_\psi; \phi, \theta, \psi) &= A(\mathbf{e}_\theta, \theta_0)A(\mathbf{e}_\phi, \psi - \pi)A\big(A(\mathbf{e}_\phi, \pi)\mathbf{e}_\theta, \theta - \theta_0\big)A(\mathbf{e}_\phi, \phi + \pi)\\
&= A(\mathbf{e}_\theta, \theta_0)A(\mathbf{e}_\phi, \psi - \pi)A(-\mathbf{e}_\theta, \theta - \theta_0)A(\mathbf{e}_\phi, \phi + \pi)\\
&= A(\mathbf{e}_\theta, \theta_0)A(\mathbf{e}_\phi, \psi - \pi)A(\mathbf{e}_\theta, -\theta_0)A(\mathbf{e}_\theta, 2\,\theta_0 - \theta)A(\mathbf{e}_\phi, \phi + \pi)\\
&= A(\mathbf{e}_\phi, \mathbf{e}_\theta, \mathbf{e}_\psi; \phi + \pi, 2\,\theta_0 - \theta, \psi - \pi)
\end{aligned} \tag{2.159}$$

The rotation axes of the classical Euler angle representation are selected from the set

$$\mathbf{e}_1 = \begin{bmatrix} 1 \\ 0 \\ 0 \end{bmatrix}, \quad \mathbf{e}_2 = \begin{bmatrix} 0 \\ 1 \\ 0 \end{bmatrix}, \quad \mathbf{e}_3 = \begin{bmatrix} 0 \\ 0 \\ 1 \end{bmatrix} \tag{2.160}$$

and a more compact notation is used for these representations

$$A_{ijk}(\phi, \theta, \psi) = A(\mathbf{e}_k, \psi)A(\mathbf{e}_j, \theta)A(\mathbf{e}_i, \phi) \tag{2.161}$$

The possible choice of axes is constrained by the requirements $i \neq j$ and $j \neq k$, as required by Eq. (2.155). This leaves us with six symmetric sets of Euler parameters, with $i-j-k$ equal to:

1−2−1, 1−3−1, 2−3−2, 2−1−2, 3−1−3, and 3−2−3

and six asymmetric sets:

1−2−3, 1−3−2, 2−3−1, 2−1−3, 3−1−2, and 3−2−1

The explicit forms of the attitude matrices for all 12 sets are collected in Chap. 9.

As a specific example of the symmetric sequences, we consider the $3-1-3$ sequence, which is often used for analytical treatments of rigid body motion and for representing the attitude of spinning spacecraft:

$$A_{313}(\phi, \theta, \psi) = A(\mathbf{e}_3, \psi)A(\mathbf{e}_1, \theta)A(\mathbf{e}_3, \phi)$$

$$= \begin{bmatrix} c\psi & s\psi & 0 \\ -s\psi & c\psi & 0 \\ 0 & 0 & 1 \end{bmatrix} \begin{bmatrix} 1 & 0 & 0 \\ 0 & c\theta & s\theta \\ 0 & -s\theta & c\theta \end{bmatrix} \begin{bmatrix} c\phi & s\phi & 0 \\ -s\phi & c\phi & 0 \\ 0 & 0 & 1 \end{bmatrix}$$

$$= \begin{bmatrix} c\psi\, c\phi - s\psi\, c\theta\, s\phi & c\psi\, s\phi + s\psi\, c\theta\, c\phi & s\psi\, s\theta \\ -s\psi\, c\phi - c\psi\, c\theta\, s\phi & -s\psi\, s\phi + c\psi\, c\theta\, c\phi & c\psi\, s\theta \\ s\theta\, s\phi & -s\theta\, c\phi & c\theta \end{bmatrix} \tag{2.162}$$

We have written $c\psi \equiv \cos\psi$, $s\psi \equiv \sin\psi$, and analogous equations for θ and ϕ. The symmetric Euler angle sets all have $\theta_0 = 0$, so the angle ambiguity relation is

$$A_{313}(\phi, \theta, \psi) = A_{313}(\phi + \pi, -\theta, \psi - \pi) \tag{2.163}$$

The asymmetric sets are often called the *Tait-Bryan* angles, although this terminology has been called into question [4]. The three angles in an asymmetric Euler angle sequence are often referred to as *roll*, *pitch*, and *yaw*. This terminology originally described the motions of ships and then was carried over into aircraft and spacecraft. Roll is a rotation about the vehicle body axis that is closest to the vehicle's usual direction of motion, and hence would be perceived as a screwing motion. The roll axis is conventionally assigned index 1. Yaw is a rotation about the vehicle body axis that is usually closest to the direction of local gravity, and hence would be often be perceived as a motion that points the vehicle left or right. The yaw axis is conventionally assigned index 3. Pitch is a rotation about the remaining vehicle body axis, and hence would often be perceived as a motion that points the vehicle up or down. The pitch axis is conventionally assigned index 2. Note that this associates the terms roll, pitch, and yaw with the vehicle axes, while Eq. (2.161) assigns the variables ϕ, θ, and ψ based on the order of rotations in the sequence rather than on the axis indices. Thus there is no definite association between the variables ϕ, θ, and ψ and the axis labels 1, 2, and 3 or the names roll, pitch and yaw. A different convention is followed by many authors who denote roll by ϕ, pitch by θ, and yaw by ψ. As always, the reader consulting any source should be careful to understand the conventions that it follows.

The $3-2-1$ sequence, which is often used to describe the attitude of an Earth-pointing spacecraft, is a specific example of an asymmetric sequence.

$$A_{321}(\phi, \theta, \psi) = A(\mathbf{e}_1, \psi)A(\mathbf{e}_2, \theta)A(\mathbf{e}_3, \phi)$$

$$= \begin{bmatrix} 1 & 0 & 0 \\ 0 & c\psi & s\psi \\ 0 & -s\psi & c\psi \end{bmatrix} \begin{bmatrix} c\theta & 0 & -s\theta \\ 0 & 1 & 0 \\ s\theta & 0 & c\theta \end{bmatrix} \begin{bmatrix} c\phi & s\phi & 0 \\ -s\phi & c\phi & 0 \\ 0 & 0 & 1 \end{bmatrix}$$

$$= \begin{bmatrix} c\theta\, c\phi & c\theta\, s\phi & -s\theta \\ -c\psi\, s\phi + s\psi\, s\theta\, c\phi & c\psi\, c\phi + s\psi\, s\theta\, s\phi & s\psi\, c\theta \\ s\psi\, s\phi + c\psi\, s\theta\, c\phi & -s\psi\, c\phi + c\psi\, s\theta\, s\phi & c\psi\, c\theta \end{bmatrix} \tag{2.164}$$

Equation (2.154) shows that the angle θ_0 for this sequence is equal to $-\pi/2$, so the angle ambiguity relation is

$$A_{321}(\phi, \theta, \psi) = A_{321}(\phi + \pi, \pi - \theta, \psi - \pi) \tag{2.165}$$

In fact, all the Tait-Bryan angle axis sets have $\theta_0 = \pm\pi/2$ so they all obey Eq. (2.165), taking into account the 2π ambiguity in the definition of θ. The Tait-Bryan representations have useful small-angle approximations, as will be shown in Sect. 2.10, but the small angle limits of the symmetric Euler angles are not as useful.

Separate procedures for the different axis sequences are generally used to find the Euler angles representing a given attitude matrix. We will consider the $3-1-3$ and $3-2-1$ sequences as specific examples, and will follow these examples with a general algorithm.

For the symmetric 3–1–3 sequence, the angle θ is computed from the 33 element of A_{313}:

$$\theta = \cos^{-1}([A_{313}]_{33}) \tag{2.166}$$

Unless $[A_{313}]_{33}$ has magnitude unity, two distinct values of θ have the same cosine, corresponding to the two possible signs for $\sin\theta$ and to the twofold ambiguity shown in Eq. (2.163). We are free to select either of these values, but we can avoid the twofold ambiguity by computing θ as the principal value of the inverse cosine, which restricts its range to $0 \le \theta \le \pi$ and gives $\sin\theta \ge 0$. If $\sin\theta \ne 0$, the other two Euler angles can be determined, modulo 2π, by

$$\phi = \text{atan2}(\sigma[A_{313}]_{31}, -\sigma[A_{313}]_{32}) \tag{2.167a}$$

$$\psi = \text{atan2}(\sigma[A_{313}]_{13}, \ \sigma[A_{313}]_{23}) \tag{2.167b}$$

where $\sigma = \pm 1$ is the sign of $\sin\theta$ and atan2(y, x) is the standard function giving the argument of the complex number $x + iy$.

It is obvious that ϕ and ψ cannot be determined from Eqs. (2.167) if θ is equal to 0 or π, since these values give $\sin\theta = 0$. The usual twofold ambiguity is absent in these cases, but the attitude matrix takes the form

$$A_{313}(\phi, (\pi \mp \pi)/2, \psi) = \begin{bmatrix} \cos(\phi \pm \psi) & \sin(\phi \pm \psi) & 0 \\ \mp\sin(\phi \pm \psi) & \pm\cos(\phi \pm \psi) & 0 \\ 0 & 0 & \pm 1 \end{bmatrix} \tag{2.168}$$

It can be seen that only the sum or difference $\phi \pm \psi$ is determined, and not the angles individually. This is known as *gimbal lock*, for reasons that will become apparent when we discuss the kinematics of rotations. Gimbal lock is caused by collinearity of the *physical* rotation axis vectors of the first and third rotations in the sequence. Note that the column vector representations of the rotation axes are always parallel for the symmetric Euler angle sequences, but that does not cause gimbal lock. Gimbal lock occurs for all the symmetric Euler angle sequences when $\sin\theta = 0$.

We could employ a special algorithm for the $\sin\theta = 0$ case, but it is preferable to develop a general algorithm for all values of θ, because we could encounter a loss of precision for small but nonzero values of $\sin\theta$. To this end, we note that

$$[A_{313}]_{11} \pm [A_{313}]_{22} = (1 \pm c\theta)\cos(\phi \pm \psi) \tag{2.169a}$$

$$[A_{313}]_{12} \mp [A_{313}]_{21} = (1 \pm c\theta)\sin(\phi \pm \psi) \tag{2.169b}$$

and that $(1 \pm c\theta)$ is positive if $\cos\theta \neq \mp 1$. Therefore, we find either ϕ or ψ, but not both, from Eq. (2.167). If $\sin\theta = 0$ we can set one of these angles to any convenient value. Then we find the other angle from their sum or difference by

$$\phi + \psi = \text{atan2}\left([A_{313}]_{12} - [A_{313}]_{21}, [A_{313}]_{11} + [A_{313}]_{22}\right) \quad \text{if } [A_{313}]_{33} \geq 0 \tag{2.170a}$$

$$\phi - \psi = \text{atan2}\left([A_{313}]_{12} + [A_{313}]_{21}, [A_{313}]_{11} - [A_{313}]_{22}\right) \quad \text{if } [A_{313}]_{33} < 0 \tag{2.170b}$$

Extraction of the asymmetric Euler angles proceeds in a similar manner. For the $3-2-1$ sequence, the angle θ is computed from the 13 element of A_{321}:

$$\theta = \sin^{-1}\left(-[A_{321}]_{13}\right) \tag{2.171}$$

Unless $[A_{321}]_{13}$ has magnitude unity, two distinct values of θ have the same sine, corresponding to the two possible signs for $\cos\theta$ and to the twofold ambiguity shown in Eq. (2.165). We are free to select either value, but can avoid the ambiguity by computing θ as the principal value of the inverse sine, restricting its range to $|\theta| \leq \pi/2$ and giving $\cos\theta \geq 0$. If $\cos\theta \neq 0$, the other two angles can be determined by

$$\phi = \text{atan2}\left(\sigma'[A_{321}]_{12}, \sigma'[A_{321}]_{11}\right) \tag{2.172a}$$

$$\psi = \text{atan2}\left(\sigma'[A_{321}]_{23}, \sigma'[A_{321}]_{33}\right) \tag{2.172b}$$

where $\sigma' = \pm 1$ is the sign of $\cos\theta$.

Gimbal lock for the asymmetric Euler, or Tait-Bryan, sequences occurs when $\cos\theta = 0$, i.e. when $\theta = \mp\pi/2$. Then the usual twofold ambiguity is absent, and the attitude matrix for the $3-2-1$ example has the form

$$A_{321}(\phi, \mp\pi/2, \psi) = \begin{bmatrix} 0 & 0 & \pm 1 \\ -\sin(\phi \pm \psi) & \cos(\phi \pm \psi) & 0 \\ \mp\cos(\phi \pm \psi) & \mp\sin(\phi \pm \psi) & 0 \end{bmatrix} \tag{2.173}$$

Only the sum or difference $\phi \pm \psi$ is determined, just as for the symmetric Euler axis sequence, again due to collinearity of the physical rotation axis vectors of the first and third rotations. The column vector representations of the rotation axes are always perpendicular for the asymmetric sequences, but that does not prevent gimbal lock from occurring.

In parallel with the 3–1–3 case, we deal with gimbal lock by finding either ϕ or ψ from Eq. (2.172) and then the other from their sum or difference by

$$\phi + \psi = \text{atan2}\,(-[A_{321}]_{32} - [A_{321}]_{21}, [A_{321}]_{22} - [A_{321}]_{31}) \quad \text{if } [A_{321}]_{13} \geq 0 \tag{2.174a}$$

$$\phi - \psi = \text{atan2}\,(\ [A_{321}]_{32} - [A_{321}]_{21}, [A_{321}]_{22} + [A_{321}]_{31}) \quad \text{if } [A_{321}]_{13} < 0 \tag{2.174b}$$

We finally discuss the extraction of the angles for Davenport's general axis sequences. We will accomplish this in a way that uses the results obtained above for extracting the 3–1–3 Euler angles. This technique can be applied to any sequence of conventional Euler or Tait-Bryan angles, as well as to the general Davenport angles, by selecting the rotation axes from the set of coordinate axes $\{\mathbf{e}_1, \mathbf{e}_2, \mathbf{e}_3\}$ [19].

We note that the proper orthogonal matrix

$$C = \begin{bmatrix} \mathbf{e}_\theta^T \\ (\mathbf{e}_\phi \times \mathbf{e}_\theta)^T \\ \mathbf{e}_\phi^T \end{bmatrix} \tag{2.175}$$

has the property that $C\mathbf{e}_\theta = \mathbf{e}_1$ and $C\mathbf{e}_\phi = \mathbf{e}_3$. Then we have from Eq. (2.156)

$$\begin{aligned} CA(\mathbf{e}_\phi, \mathbf{e}_\theta, \mathbf{e}_\psi; \phi, \theta, \psi)C^T &= CA(\mathbf{e}_\theta, \theta_0)C^T CA(\mathbf{e}_\phi, \mathbf{e}_\theta, \mathbf{e}_\phi; \phi, \theta', \psi)C^T \\ &= A(\mathbf{e}_1, \theta_0)CA(\mathbf{e}_\phi, \psi)C^T C(\mathbf{e}_\theta, \theta')C^T CA(\mathbf{e}_\phi, \phi)C^T \\ &= A(\mathbf{e}_1, \theta_0)A_{313}(\phi, \theta', \psi) \end{aligned} \tag{2.176}$$

Thus we can extract ϕ, θ', and ψ from $A^T(\mathbf{e}_1, \theta_0)CA(\mathbf{e}_\phi, \mathbf{e}_\theta, \mathbf{e}_\psi; \phi, \theta, \psi)C^T$ by the standard technique for the $3-1-3$ sequence, and then compute $\theta = \theta' + \theta_0$. An easily-verified special case of Eq. (2.176) is

$$\begin{bmatrix} 0 & 1 & 0 \\ 0 & 0 & 1 \\ 1 & 0 & 0 \end{bmatrix} A_{321}(\phi, \theta, \psi) \begin{bmatrix} 0 & -1 & 0 \\ 1 & 0 & 0 \\ 0 & 0 & 1 \end{bmatrix} = A_{313}(\phi, \theta + \pi/2, \psi) \tag{2.177}$$

Explicit expressions for the Euler angles resulting from successive transformations have been found in some special cases [17], but they have not been widely applied in practice.

2.10 Attitude Error Representations

The attitude matrix represents the rotation A_{BR} from some reference frame R to the spacecraft body frame B. Attitude estimation errors can be represented either as a small rotation $A_{\hat{R}R}$ between R and an estimated reference frame $\hat{R}$

$$A_{BR} = A_{B\hat{R}} A_{\hat{R}R} \tag{2.178}$$

or more commonly as a small rotation $A_{B\hat{B}}$ between B and an estimated body frame $\hat{B}$

$$A_{BR} = A_{B\hat{B}} A_{\hat{B}R} \tag{2.179}$$

The estimated attitude is represented by $A_{B\hat{R}}$ in the former case and by $A_{\hat{B}R}$ in the latter. In either case the matrix representing the errors, $A_{\hat{R}R}$ or $A_{B\hat{B}}$, is expected to be close to the identity matrix.

The most natural representation of attitude errors is in terms of the rotation vector, Eq. (2.123), and its small-angle approximation

$$A(\boldsymbol{\delta\vartheta}) = \exp(-[\boldsymbol{\delta\vartheta}\times]) \approx I_3 - [\boldsymbol{\delta\vartheta}\times] + \frac{1}{2}[\boldsymbol{\delta\vartheta}\times]^2 \tag{2.180}$$

Other attitude parameterizations can be used to represent attitude errors, such as the quaternion, Eq. (2.125),

$$A(\boldsymbol{\delta}\mathbf{q}) \approx I_3 - 2[\boldsymbol{\delta}\mathbf{q}_{1:3}\times] + 2[\boldsymbol{\delta}\mathbf{q}_{1:3}\times]^2 \tag{2.181}$$

the Gibbs vector, Eq. (2.139),

$$A(\boldsymbol{\delta}\mathbf{g}) \approx I_3 - 2[\boldsymbol{\delta}\mathbf{g}\times] + 2[\boldsymbol{\delta}\mathbf{g}\times]^2 \tag{2.182}$$

or the MRPs, Eq. (2.149),

$$A(\boldsymbol{\delta}\mathbf{p}) \approx I_3 - 4[\boldsymbol{\delta}\mathbf{p}\times] + 8[\boldsymbol{\delta}\mathbf{p}\times]^2 \tag{2.183}$$

It is notable that these representations are all equivalent through second order in the errors with the identification

$$\boldsymbol{\delta\vartheta} = 2\boldsymbol{\delta}\mathbf{q}_{1:3} = 2\boldsymbol{\delta}\mathbf{g} = 4\boldsymbol{\delta}\mathbf{p} \tag{2.184}$$

In fact, only the first-order approximation is required for most applications. It must be emphasized that Eqs. (2.180)–(2.184) are only true in the (very useful) approximation of small error angles. Attitude error representations are often used for errors that are not especially small. In that case, the quaternion, Gibbs vector, or MRP representation is often preferred to the rotation vector for computational convenience.

The small-angle approximation of the Tait-Bryan or asymmetric Euler angle representation, Eq. (2.161), is to second order

$$
\begin{aligned}
A_{ijk}(\delta\phi, \delta\theta, \delta\psi) &\approx \left(I_3 - \delta\psi[\mathbf{e}_k\times] + \frac{1}{2}\delta\psi^2[\mathbf{e}_k\times]^2\right) \\
&\quad \times \left(I_3 - \delta\theta[\mathbf{e}_j\times] + \frac{1}{2}\delta\theta^2[\mathbf{e}_j\times]^2\right)\left(I_3 - \delta\phi[\mathbf{e}_i\times] + \frac{1}{2}\delta\phi^2[\mathbf{e}_i\times]^2\right) \\
&\approx I_3 - [\boldsymbol{\delta\vartheta}_1\times] + \frac{1}{2}[\boldsymbol{\delta\vartheta}_1\times]^2 \mp \frac{1}{2}[(\delta\theta\,\delta\psi\,\mathbf{e}_i - \delta\psi\,\delta\phi\,\mathbf{e}_j + \delta\phi\,\delta\,\theta\,\mathbf{e}_k)\times]
\end{aligned} \tag{2.185}
$$

where

$$\boldsymbol{\delta\vartheta}_1 \equiv \delta\phi\,\mathbf{e}_i + \delta\theta\,\mathbf{e}_j + \delta\psi\,\mathbf{e}_k \tag{2.186}$$

The upper sign in the last term in Eq. (2.185), which was derived with the use of Eq. (2.56c), holds if $\{i, j, k\}$ is an even permutation of $\{1, 2, 3\}$, and the lower sign applies if it is an odd permutation. With the identification $\boldsymbol{\delta\vartheta} = \boldsymbol{\delta\vartheta}_1$ Eq. (2.185) agrees with Eq. (2.180) to first order, but not to second order. Agreement to second order would require

$$\boldsymbol{\delta\vartheta} = \left(\delta\phi \pm \frac{\delta\theta\,\delta\psi}{2}\right)\mathbf{e}_i + \left(\delta\theta \mp \frac{\delta\psi\,\delta\phi}{2}\right)\mathbf{e}_j + \left(\delta\psi \pm \frac{\delta\phi\delta\,\theta}{2}\right)\mathbf{e}_k \tag{2.187}$$

This identification has never been used to our knowledge, however, because first-order approximations in the errors are generally adequate.

Problems

2.1. Prove that any real $n \times n$ matrix A can be decomposed into the sum of a symmetric and skew symmetric matrix. Hint: the matrix $A - A^T$ is clearly skew symmetric.

2.2. It is known that a general $n \times n$ symmetric matrix A must possess n mutually orthogonal eigenvectors, even if some of the eigenvalues are repeated. Here, you will prove that this is true when the eigenvalues are not repeated. Let $A\,\mathbf{x}_i = \lambda_i\mathbf{x}_i$ for $i = 1, 2$, where λ_i is the ith eigenvalue and $\mathbf{x}_i$ is the ith eigenvector. Assume that $\lambda_1 \neq \lambda_2$. Start by taking the transpose of $A\,\mathbf{x}_1 = \lambda_1\mathbf{x}_1$ and right multiplying both sides by $\mathbf{x}_2$, and prove that when $A = A^T$ the vectors $\mathbf{x}_1$ and $\mathbf{x}_2$ must be orthogonal.

2.3. Suppose that the characteristic equation of an $n \times n$ matrix A is given by

$$\Delta(\lambda) = \det(\lambda I_n - A) = \lambda^n + a_1\lambda^{n-1} + \cdots + a_{n-1}\lambda + a_n$$

The Cayley-Hamilton theorem states that the matrix A obeys its characteristic equation so that

$$A^n + a_1 A^{n-1} + \cdots + A + a_n I_n = 0$$

Multiplying by A gives

$$A^{n+1} + a_1 A^n + \cdots + A^2 + a_n A = 0$$

This implies that A^{n+1} can be written as a linear combination of A, A^2, $\ldots$, A^n, which in turn can be written as a linear combination of I_n, A, $\ldots$, A^{n-1}. In fact, any A^m with $m > n-1$ can be written using the same linear combination. Use this fact to compute a closed-form expression for the following:

$$\begin{bmatrix} 1 & 0 & 4 \\ 0 & 0 & 1 \\ 0 & -1 & 0 \end{bmatrix}^{100}$$

Note that $f(A)$ is the same linear combination of powers of A as $f(J)$ is of powers of J, where J is a diagonal matrix of the eigenvalues of A.

2.4. Consider the following 2×2 matrix for real a, b and d:

$$A = \begin{bmatrix} a & b \\ b & d \end{bmatrix}$$

Determine the eigenvalues λ_1 and λ_2 of A in terms of a, b and d. Since A is symmetric then the eigenvalue/eigenvector decomposition gives

$$V^T A V = \begin{bmatrix} \lambda_1 & 0 \\ 0 & \lambda_2 \end{bmatrix}$$

where V is an orthogonal matrix. Suppose that V is given by the form

$$V = \begin{bmatrix} c & s \\ -s & c \end{bmatrix}$$

where $s \equiv \sin\theta$ and $c \equiv \cos\theta$ for some angle θ. Find c and s in terms of a, b and d.

2.5. A matrix that is used to reflect an object over a line or plane is called a reflection matrix. Consider the following natural basis, given by Eq. (2.35), for $n = 2$:

$$\mathbf{e}_1 = \begin{bmatrix} 1 \\ 0 \end{bmatrix}, \quad \mathbf{e}_2 = \begin{bmatrix} 0 \\ 1 \end{bmatrix}$$

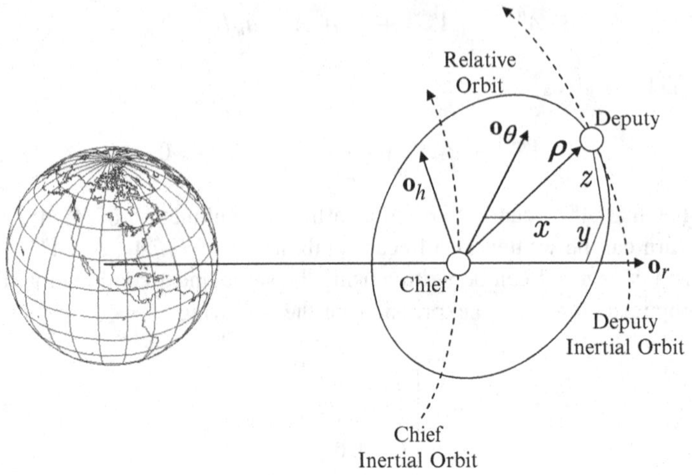

Fig. 2.8 Hill frame

Let ϕ be the counterclockwise angle between some line $\boldsymbol{\ell}$ through the origin and the x-axis. Consider the following matrix that rotates the x-axis onto $\boldsymbol{\ell}$:

$$A = \begin{bmatrix} \cos\phi & -\sin\phi \\ \sin\phi & \cos\phi \end{bmatrix}$$

Noting that $A\,\mathbf{e}_1$ lies on $\boldsymbol{\ell}$ and $A\,\mathbf{e}_1$ is perpendicular to $\boldsymbol{\ell}$, determine the reflection matrix that sends any vector not on $\boldsymbol{\ell}$ to its mirror image about $\boldsymbol{\ell}$. Provide simulation plots for various angles ϕ and any chosen 2×1 vector to reflect.

2.6. Write a computer program that takes some latitude, longitude and height, and converts these quantities to ECEF position using Eq. (2.74). Also, write a computer program that takes ECEF position and converts it to latitude, longitude and height using Eq. (2.77). Pick some latitude, longitude and height. Then compute the ECEF position and convert this position to latitude, longitude and height to ensure that the original quantities are obtained.

2.7. A useful frame for formation flying applications is the "Hill frame" shown in Fig. 2.8 [6]. The frame is given by $\{\mathbf{o}_r, \mathbf{o}_\theta, \mathbf{o}_h\}$, where $\mathbf{o}_r$ points in the chief spacecraft's radial direction, $\mathbf{o}_h$ is along the chief orbit momentum vector and $\mathbf{o}_\theta$ completes the right-handed coordinate system, so that

$$\mathbf{o}_r = \frac{\mathbf{r}_I}{\|\mathbf{r}_I\|}, \quad \mathbf{o}_\theta = \mathbf{o}_h \times \mathbf{o}_r, \quad \mathbf{o}_h = \frac{\mathbf{h}_I}{\|\mathbf{h}_I\|}$$

where $\mathbf{h}_I = \mathbf{r}_I \times \mathbf{v}_I$. Using the dot product approach similar to Eq. (2.43), fully derive the attitude matrix that rotates vectors from the Hill frame to an inertial frame. Also, determine the relationship between the Hill frame and the LVLH frame.

2.8. This is an alternative proof of Euler's Theorem that avoids eigenvalues [12].

a) First assume that the attitude matrix is not symmetric, i.e. that $A \neq A^T$. Define the skew symmetric matrix $S \equiv \frac{1}{2}(A - A^T) = [\mathbf{s}\times]$. Prove that $A\,S\,A^T = S$. Next show that $\mathbf{d} \equiv \mathbf{s}/\|\mathbf{s}\|$ is equivalent to the Euler axis $\mathbf{e}$, to within a sign ambiguity,
b) Now assume that $A = A^T$. Why does the above argument fail in this case? If A is symmetric the orthogonality relation is $A^2 = I_3$. Show that this gives $A(A + I_3) = A + I_3$, so all the columns of $A + I_3$ are unchanged by A. Complete the proof by showing that at least one of these columns is not a column of zeros, and is therefore equivalent to the Euler axis $\mathbf{e}$, to within a sign ambiguity. What Euler angles of rotation, ϑ, correspond to the case of $A = A^T$?

2.9. The generalized Rodrigues parameters [15] (GRPs) can be written as

$$\boldsymbol{\rho} = \frac{f\,\mathbf{q}_{1:3}}{a + q_4}$$

where a is a parameter from 0 to 1 and f is a scale factor.

a) Draw a plot of the rotation angle ϑ verses a that causes the GPRs to become singular.
b) Determine the inverse transformation for $\mathbf{q}_{1:3}$ and q_4 in terms of $\boldsymbol{\rho}$, a and f. Note that your answer for q_4 may seem to provide two solutions, but only one of them is correct.
c) Determine f in terms of a so that the small angle approximation gives $\|\boldsymbol{\rho}\| \approx \vartheta$.

2.10. Show that the square of the quaternion elements can be extracted from the attitude matrix by using the following equations:

$$q_1^2 = \frac{1}{4}(1 + a_{11} - a_{22} - a_{33})$$

$$q_2^2 = \frac{1}{4}(1 + a_{22} - a_{11} - a_{33})$$

$$q_3^2 = \frac{1}{4}(1 + a_{33} - a_{11} - a_{22})$$

$$q_4^2 = \frac{1}{4}(1 + a_{11} + a_{22} + a_{33})$$

2.11. Derive the attitude matrix for a $1-2-3$ sequence. Also, explicitly compute the determinant of this matrix to show that it is $+1$.

2.12. Derive the direct relationship from Euler angles to quaternion for a $1-2-3$ sequence and for a $1-2-1$ sequence. Compare your results to the ones shown in Table 9.5 to make sure they are equivalent.

References

1. Bate, R.R., Mueller, D.D., White, J.E.: Fundamentals of Astrodynamics. Dover Publications, New York (1971)
2. Davenport, P.B.: Rotations about nonorthogonal axes. AIAA J. **11**(6), 853–857 (1973)
3. Farrell, J., Barth, M.: The Global Positioning System & Inertial Navigation. McGraw-Hill, New York (1998)
4. Fraiture, L.: A history of the description of the three-dimensional finite rotation. J. Astronaut. Sci. **57**(1/2), 207–232 (2009)
5. Goldstein, H.: Classical Mechanics, 2nd edn. Addison-Wesley, Reading (1980)
6. Hill, G.W.: Researches in lunar theory. Am. J. Math. **1**(1), 5–26 (1878)
7. Lefferts, E.J., Markley, F.L., Shuster, M.D.: Kalman filtering for spacecraft attitude estimation. J. Guid. Contr. Dynam. **5**(5), 417–429 (1982)
8. Ma, C., Arias, E.F., Eubanks, T.M., Frey, A.L., Gontier, A.M., Jacobs, C.S., Sovers, O.J., Archinal, B.A., Charlot, P.: The International Celestial Reference Frame as realized by very long baseline interferometry. Astron. J. **116**, 516–546 (1998)
9. Marandi, S.R., Modi, V.J.: A preferred coordinate system and the associated orientation representation in attitude dynamics. Acta Astronautica **15**(11), 833–843 (1987)
10. Markley, F.L.: Unit quaternion from rotation matrix. J. Guid. Contr. Dynam. **31**(2), 440–442 (2008)
11. Montenbruck, O., Gill, E.: Satellite Orbits: Models, Methods, and Applications. Springer, Berlin Heidelberg New York (2000)
12. Palais, B., Palais, R.: Euler's fixed point theorem: The axis of a rotation. J. Fixed point Theory Appl. **2**(2), 215–220 (2007)
13. Pittelkau, M.E.: An analysis of the quaternion attitude determination filter. J. Astronaut. Sci. **51**(1) (2003)
14. Rodrigues, O.: Des lois géométriques qui régissent les déplacements d'un système solide dans l'espace, et de la variation des coordonnées provenant de ces déplacements considérés indépendamment des causes qui peuvent les produire. J. de Mathématiques Pures et Appliquées **5**, 380–440 (1840)
15. Schaub, H., Junkins, J.L.: Stereographic orientation parameters for attitude dynamics: A generalization of the Rodrigues parameters. J. Astronaut. Sci. **44**(1), 1–19 (1996)
16. Schaub, H., Junkins, J.L.: Analytical Mechanics of Aerospace Systems, 2nd edn. American Institute of Aeronautics and Astronautics, New York (2009)
17. Shuster, M.D.: A survey of attitude representations. J. Astronaut. Sci. **41**(4), 439–517 (1993)
18. Shuster, M.D., Markley, F.L.: Generalization of the Euler angles. J. Astronaut. Sci. **51**(2), 123–132 (2003)
19. Shuster, M.D., Markley, F.L.: General formula for extracting the Euler angles. J. Guid. Contr. Dynam. **29**(1), 215–217 (2006)
20. Sofair, I.: Improved method for calculating exact geodetic latitude and altitude revisited. J. Guid. Contr. Dynam. **23**(2), 369 (2000)
21. Stewart, G.W.: Introduction to Matrix Computations. Academic Press, New York (1973)
22. Stuelpnagel, J.: On the parametrization of the three-dimensional rotation group. SIAM Rev. **6**(4), 422–430 (1964)
23. Tsiotras, P., Junkins, J.L., Schaub, H.: Higher-order Cayley transforms with applications to attitude representations. J. Guid. Contr. Dynam. **20**(3), 528–534 (1997)

24. Vallado, D.A.: Fundamentals of Astrodynamics and Applications, 3rd edn. Microcosm Press, Hawthorne and Springer, New York (2007)
25. Weinberg, S.: Gravitation and Cosmology: Principles and Applications of the General Theory of Relativity. Wiley, Chichester (1972)
26. Wiener, T.F.: Theoretical analysis of gimballess inertial reference equipment using delta-modulated instruments. Ph.D. dissertation, Massachusetts Institute of Technology, Cambridge (1962)

Chapter 3
Attitude Kinematics and Dynamics

This chapter begins with a discussion of the kinematics of rotations, or attitude kinematics, and then moves on to attitude dynamics. The distinction between kinematics and dynamics is that kinematics covers those aspects of motion that can be analyzed without consideration of forces or torques. When forces and torques are introduced, we are in the realm of dynamics. To make this distinction clear, consider the motion of a point particle in Newtonian physics. If $\mathbf{r}$ denotes position, $\mathbf{v}$ denotes velocity, and time derivatives are indicated by a dot, then the kinematic equation of motion is simply $\dot{\mathbf{r}} = \mathbf{v}$. The dynamic equation of motion is $m\dot{\mathbf{v}} = \mathbf{F}$ or $\dot{\mathbf{p}} = \mathbf{F}$, where $\mathbf{F}$ is the applied force and $\mathbf{p} \equiv m\dot{\mathbf{v}}$ is the translational momentum. Kinematics and dynamics are often subsumed under the single term dynamics by combining the kinematic and dynamic equations in the single relation $m\ddot{\mathbf{r}} = \mathbf{F}$. In fact, it is common in filtering theory to apply the term dynamics to any relation expressing time dependence.

The role of the position vector $\mathbf{r}$ is taken in attitude kinematics by the attitude matrix A or one of its parameterizations, and the role of the velocity $\mathbf{v}$ is taken by the angular velocity $\boldsymbol{\omega}$. The role of translational momentum $\mathbf{p}$ is played by the angular momentum $\mathbf{H}$. The kinematic and dynamic equations of rotational motion are not as simple as those for translational motion. In particular, the angular momentum is not a scalar multiple of the angular velocity in general.

We discuss attitude kinematics first, deriving equations for the time derivatives of the attitude matrix and of its various parameterizations introduced in the previous chapter. We also discuss the time derivatives of representations of a vector in frames that undergo relative rotational motion.

Our treatment of dynamics begins by considering the separation of center-of-mass motion from rotational motion, emphasizing the fundamental role played by angular momentum. We next introduce the concept of a rigid body, an extremely useful approximation for many spacecraft. We summarize the kinematic and dynamic equations of motion and collect them in one place for easy reference, and then analyze the torque-free motion of a rigid body. We follow this with discussions

F.L. Markley and J.L. Crassidis, *Fundamentals of Spacecraft Attitude Determination and Control*, Space Technology Library 33, DOI 10.1007/978-1-4939-0802-8_3,
© Springer Science+Business Media New York 2014

of the modeling of internal and external torques, which include both undesired disturbance torques and deliberately applied control torques. Finally, we examine the special case of Earth-pointing spacecraft.

3.1 Attitude Kinematics

We will discuss the kinematics of the attitude matrix first, since this is the fundamental representation of a rotation. This discussion introduces the concept of the *angular velocity* or *angular rate* vector. We then discuss the kinematics of column vectors in rotating reference frames.

3.1.1 Attitude Matrix

The time dependence of the attitude matrix expressing the rotation from a frame F to a frame G is given by the fundamental definition of a derivative as

$$\begin{aligned}\dot{A}_{GF}(t) &\equiv \lim_{\Delta t \to 0} \frac{A_{GF}(t+\Delta t) - A_{GF}(t)}{\Delta t} \\ &= \lim_{\Delta t \to 0} \frac{A_{GF}(t+\Delta t)A_{FG}(t) - I_3}{\Delta t} A_{GF}(t)\end{aligned} \tag{3.1}$$

because $A_{FG}(t)A_{GF}(t)$ is equal to the identity matrix. As Δt goes to zero, the product $A_{GF}(t+\Delta t)A_{FG}(t)$ differs from the identity matrix by a small rotation that we can represent by a rotation vector,

$$A_{GF}(t+\Delta t)A_{FG}(t) = \exp\left(-[\Delta\boldsymbol{\vartheta}_G^{GF}\times]\right) \approx I_3 - [\Delta\boldsymbol{\vartheta}_G^{GF}\times] \tag{3.2}$$

The higher-order terms omitted in the approximation go to zero faster than Δt as Δt goes to zero. The Δ in front of $\boldsymbol{\vartheta}$ expresses the fact that this is a small rotation, the superscript GF means that it relates to the rotation from frame F to frame G, and the subscript G means that the rotation vector is represented in the frame G. We know that the rotation vector is represented in frame G because $A_{GF}(t+\Delta t)A_{FG}(t)$ is a rotation from frame G at one time to frame G at a different time, and these two frames coincide in the limit that Δt goes to zero. The kinematic relations do not distinguish between the situations where frame F or frame G or both frames are rotating in an absolute sense; they only care about the relative rotation between the two frames. Substituting Eq. (3.2) into Eq. (3.1) gives

$$\dot{A}_{GF}(t) = \lim_{\Delta t \to 0} \frac{-[\Delta\boldsymbol{\vartheta}_G^{GF}\times]}{\Delta t} A_{GF}(t) = -[\boldsymbol{\omega}_G^{GF}(t)\times]A_{GF}(t) \tag{3.3}$$

where the angular rate vector or angular velocity vector is defined by

$$\boldsymbol{\omega}_G^{GF}(t) \equiv \lim_{\Delta t \to 0} \frac{\Delta \boldsymbol{\vartheta}_G^{GF}}{\Delta t} \tag{3.4}$$

The rotation vector is always specified in radians, so the units of the angular velocity are rad/s, assuming that time is measured in seconds.

Equation (3.3) is the fundamental equation of attitude kinematics. We can also write this, omitting the time arguments, as

$$\dot{A}_{GF} = -A_{GF}A_{FG}[\boldsymbol{\omega}_G^{GF}\times]A_{FG}^T = -A_{GF}[A_{FG}\boldsymbol{\omega}_G^{GF}\times] = -A_{GF}[\boldsymbol{\omega}_F^{GF}\times] \tag{3.5}$$

which expresses the kinematics in terms of the components of $\boldsymbol{\omega}^{GF}$ in frame F instead of its components in frame G. Taking the transpose of Eq. (3.3) and remembering that the cross-product matrix is skew-symmetric gives

$$\dot{A}_{FG} = A_{GF}^T[\boldsymbol{\omega}_G^{GF}\times] = A_{FG}[\boldsymbol{\omega}_G^{GF}\times] \tag{3.6}$$

The frame labels are arbitrary, so we can exchange the labels F and G in this expression to get

$$\dot{A}_{GF} = A_{GF}[\boldsymbol{\omega}_F^{FG}\times] \tag{3.7}$$

Comparing with Eq. (3.5) establishes the identity

$$\boldsymbol{\omega}_F^{FG} = -\boldsymbol{\omega}_F^{GF} \tag{3.8}$$

expressing the perfectly reasonable result that the rotational rate of frame F with respect to frame G is the negative of the rate of frame G with respect to frame F. Although we have derived this as a relationship of the components in frame F, the properties of orthogonal transformations show that it must be true for the components in any frame.

The kinematic equations preserve the orthogonality of the attitude matrix because the derivatives of $A_{GF}A_{GF}^T = I_3$ and $A_{GF}^T A_{GF} = I_3$ are

$$\dot{A}_{GF}A_{GF}^T + A_{GF}\dot{A}_{GF}^T = -[\boldsymbol{\omega}_G^{GF}\times]I_3 + I_3[\boldsymbol{\omega}_G^{GF}\times] = 0 \tag{3.9a}$$

$$\dot{A}_{GF}^T A_{GF} + A_{GF}^T\dot{A}_{GF} = -[\boldsymbol{\omega}_F^{FG}\times]I_3 + I_3[\boldsymbol{\omega}_F^{FG}\times] = 0 \tag{3.9b}$$

3.1.2 Vector Addition of Angular Velocity

The specialization of Eq. (2.47) for rotation matrices is

$$A_{HF} = A_{HG}A_{GF} \tag{3.10}$$

Taking the time derivative of this equation gives

$$\begin{aligned}\dot{A}_{HF} &= -[\boldsymbol{\omega}_H^{HF}\times]A_{HF} = \dot{A}_{HG}A_{GF} + A_{HG}\dot{A}_{GF} \\ &= -[\boldsymbol{\omega}_H^{HG}\times]A_{HG}A_{GF} - A_{HG}[\boldsymbol{\omega}_G^{GF}\times]A_{GF} \\ &= -\{[\boldsymbol{\omega}_H^{HG}\times] + A_{HG}[\boldsymbol{\omega}_G^{GF}\times]A_{HG}^T\}A_{HF} \end{aligned} \tag{3.11}$$

which means, by Eq. (2.63), that

$$\boldsymbol{\omega}_H^{HF} = \boldsymbol{\omega}_H^{HG} + A_{HG}\boldsymbol{\omega}_G^{GF} = \boldsymbol{\omega}_H^{HG} + \boldsymbol{\omega}_H^{GF} \tag{3.12}$$

The first form of this equation is more useful in applications, but the second shows that the angular velocity of the motion of frame H relative to frame F is just the vector sum of angular velocity of the motion of H relative to G and the angular velocity of the motion of G relative to F, provided that they are expressed in the same reference frame. Equation (3.8) is a special case of this general result.

3.1.3 Vector Kinematics

Consider the representations of a vector $\mathbf{x}$ in two different frames:

$$\mathbf{x}_G = A_{GF}\mathbf{x}_F \tag{3.13}$$

The time derivative of this equation is

$$\begin{aligned}\dot{\mathbf{x}}_G &= A_{GF}\dot{\mathbf{x}}_F + \dot{A}_{GF}\mathbf{x}_F = A_{GF}\dot{\mathbf{x}}_F - [\boldsymbol{\omega}_G^{GF}\times]A_{GF}\mathbf{x}_F \\ &= A_{GF}\dot{\mathbf{x}}_F - \boldsymbol{\omega}_G^{GF}\times\mathbf{x}_G \end{aligned} \tag{3.14}$$

This is a fundamental equation. It expresses the derivative $\dot{\mathbf{x}}_G$ of a vector in one frame, as seen in that frame, as the sum of two terms: the mapping by orthogonal transformation of the derivative $\dot{\mathbf{x}}_F$ in another frame, as seen in that frame, and an "$\boldsymbol{\omega}\times$" term reflecting the rotational motion between the two frames. Note that no judgement is made as to which frame is moving in an absolute sense; only the relative rotation matters.

Differentiating Eq. (3.14) again gives

$$\begin{aligned}\ddot{\mathbf{x}}_G &= A_{GF}\ddot{\mathbf{x}}_F - [\boldsymbol{\omega}_G^{GF}\times]A_{GF}\dot{\mathbf{x}}_F - \boldsymbol{\omega}_G^{GF}\times\dot{\mathbf{x}}_G - \dot{\boldsymbol{\omega}}_G^{GF}\times\mathbf{x}_G \\ &= A_{GF}\ddot{\mathbf{x}}_F - [\boldsymbol{\omega}_G^{GF}\times](\dot{\mathbf{x}}_G + \boldsymbol{\omega}_G^{GF}\times\mathbf{x}_G) - \boldsymbol{\omega}_G^{GF}\times\dot{\mathbf{x}}_G - \dot{\boldsymbol{\omega}}_G^{GF}\times\mathbf{x}_G \\ &= A_{GF}\ddot{\mathbf{x}}_F - \boldsymbol{\omega}_G^{GF}\times(\boldsymbol{\omega}_G^{GF}\times\mathbf{x}_G) - 2\boldsymbol{\omega}_G^{GF}\times\dot{\mathbf{x}}_G - \dot{\boldsymbol{\omega}}_G^{GF}\times\mathbf{x}_G \end{aligned} \tag{3.15}$$

This equation is more useful for translational motion than for rotational motion. The terms on the right side have special names when $\mathbf{x}$ is a position vector; the second term is the centripetal acceleration, the third term is the Coriolis acceleration, and the last term is sometimes called the Euler acceleration.

3.2 Kinematics of Attitude Parameterizations

As discussed in Chap. 2, attitude simulation and attitude estimation usually employ a representation of the attitude having fewer parameters and fewer constraints than the attitude matrix. We will now derive the kinematic equations for several of these parameterizations. Frame-specifying subscripts are omitted in this section and in most of the book, unless they are necessary to avoid ambiguity. We always understand the attitude rate vector relating to an attitude matrix A_{GF} to be $\boldsymbol{\omega}_G^{GF}$, unless explicitly indicated otherwise.

3.2.1 *Quaternion Kinematics*

The derivation of the kinematic equation for the quaternion is similar to that for the attitude matrix. The derivative is

$$\dot{\mathbf{q}}(t) \equiv \lim_{\Delta t \to 0} \frac{\mathbf{q}(t+\Delta t) - \mathbf{q}(t)}{\Delta t} \tag{3.16}$$

We can use Eq. (2.127) to represent the rotation from $\mathbf{q}(t)$ to $\mathbf{q}(t+\Delta t)$ as the exponential of a rotation vector,

$$\mathbf{q}(t+\Delta t) = \exp[(\Delta\boldsymbol{\vartheta}/2)\otimes]\mathbf{q}(t) \approx \mathbf{q}(t) + [(\Delta\boldsymbol{\vartheta}/2)\otimes]\mathbf{q}(t) \tag{3.17}$$

Inserting this into Eq. (3.16) and taking the limit as Δt goes to zero gives

$$\dot{\mathbf{q}}(t) = \frac{1}{2}[\boldsymbol{\omega}(t)\otimes]\mathbf{q}(t) = \frac{1}{2}\boldsymbol{\omega}(t)\otimes\mathbf{q}(t) = \frac{1}{2}\Omega(\boldsymbol{\omega}(t))\mathbf{q}(t) \tag{3.18}$$

where the angular velocity vector is defined by Eq. (3.4). This result is very similar to Eq. (3.3).

The fact that $[\boldsymbol{\omega}(t)\otimes]$ is a skew-symmetric matrix ensures that Eq. (3.18) preserves the quaternion norm, because the derivative of $\|\mathbf{q}\|^2 = 1$ is

$$\begin{aligned}\dot{\mathbf{q}}^T\mathbf{q} + \mathbf{q}^T\dot{\mathbf{q}} &= \left(\frac{1}{2}[\boldsymbol{\omega}\otimes]\mathbf{q}\right)^T\mathbf{q} + \mathbf{q}^T\left(\frac{1}{2}[\boldsymbol{\omega}\otimes]\mathbf{q}\right)\\ &= \frac{1}{2}\left(-\mathbf{q}^T[\boldsymbol{\omega}\otimes]\mathbf{q} + \mathbf{q}^T[\boldsymbol{\omega}\otimes]\mathbf{q}\right) = 0\end{aligned} \tag{3.19}$$

It is often more convenient to use Eqs. (2.84a) and (2.86) to write the kinematic equation for the quaternion in the form

$$\dot{\mathbf{q}} = \frac{1}{2}\mathbf{q}\odot\boldsymbol{\omega} = \frac{1}{2}\Xi(\mathbf{q})\,\boldsymbol{\omega} \tag{3.20}$$

where $\Xi(\mathbf{q})$ is defined by Eq. (2.88). Then Eq. (2.89a) gives $\boldsymbol{\omega}$ as a function of the quaternion rate as

$$\boldsymbol{\omega} = 2\,\Xi^T(\mathbf{q})\,\dot{\mathbf{q}} \tag{3.21}$$

3.2.2 Rodrigues Parameter Kinematics

The kinematic equation satisfied by the Gibbs vector is most easily obtained from the kinematic equation for the quaternion. Taking the time derivative of Eq. (2.136), inserting Eq. (3.18) for the quaternion derivative, and then substituting Eq. (2.137) gives

$$\dot{\mathbf{g}} = (1/2)\left[\boldsymbol{\omega} + \mathbf{g}\times\boldsymbol{\omega} + (\mathbf{g}\cdot\boldsymbol{\omega})\mathbf{g}\right] = (1/2)\left(I_3 + [\mathbf{g}\times] + \mathbf{g}\mathbf{g}^T\right)\boldsymbol{\omega} \tag{3.22}$$

The inverse of this equation is

$$\boldsymbol{\omega} = 2\left(I_3 + [\mathbf{g}\times] + \mathbf{g}\mathbf{g}^T\right)^{-1}\dot{\mathbf{g}} = 2\left(1 + \|\mathbf{g}\|^2\right)^{-1}(\dot{\mathbf{g}} - \mathbf{g}\times\dot{\mathbf{g}}) \tag{3.23}$$

3.2.3 Modified Rodrigues Parameter Kinematics

The kinematic equation for the modified Rodrigues parameters can also be obtained from the kinematic equation for the quaternion. Taking the time derivative of Eq. (2.144), inserting Eq. (3.18) for the quaternion derivative, and then substituting Eq. (2.145) gives

$$\begin{aligned}\dot{\mathbf{p}} &= (1/4)\left[(1 - \|\mathbf{p}\|^2)\boldsymbol{\omega} + 2\mathbf{p}\times\boldsymbol{\omega} + 2(\mathbf{p}\cdot\boldsymbol{\omega})\mathbf{p}\right] \\ &= (1/4)\left\{(1 - \|\mathbf{p}\|^2)I_3 + 2[\mathbf{p}\times] + 2\mathbf{p}\mathbf{p}^T\right\}\boldsymbol{\omega} \\ &= \frac{1 + \|\mathbf{p}\|^2}{4}\left(I_3 + 2\frac{[\mathbf{p}\times]^2 + [\mathbf{p}\times]}{1 + \|\mathbf{p}\|^2}\right)\boldsymbol{\omega}\end{aligned} \tag{3.24}$$

The matrix in parentheses in the last line of this equation is the transpose of the matrix appearing in Eq. (2.149), which is orthogonal, so the inverse of the kinematic equation for the MRPs is

$$\begin{aligned}\boldsymbol{\omega} &= \frac{4}{1 + \|\mathbf{p}\|^2}\left(I_3 + 2\frac{[\mathbf{p}\times]^2 - [\mathbf{p}\times]}{1 + \|\mathbf{p}\|^2}\right)\dot{\mathbf{p}} \\ &= 4\left(1 + \|\mathbf{p}\|^2\right)^{-2}\left[(1 - \|\mathbf{p}\|^2)\dot{\mathbf{p}} - 2\mathbf{p}\times\dot{\mathbf{p}} + 2(\mathbf{p}\cdot\dot{\mathbf{p}})\mathbf{p}\right]\end{aligned} \tag{3.25}$$

3.2.4 Rotation Vector Kinematics

The kinematic equation for the rotation vector $\boldsymbol{\vartheta} = \mathbf{e}\,\vartheta$ is also most easily obtained from the kinematic equation for the quaternion. Taking the time derivative of

$$\boldsymbol{\vartheta} = 2(\cos^{-1} q_4)\frac{\mathbf{q}_{1:3}}{\|\mathbf{q}_{1:3}\|} \tag{3.26}$$

and substituting Eq. (3.18) gives

$$\dot{\boldsymbol{\vartheta}} = \left\{ \frac{\mathbf{q}_{1:3}\mathbf{q}_{1:3}^T}{\|\mathbf{q}_{1:3}\|^2} + \frac{\cos^{-1} q_4}{\|\mathbf{q}_{1:3}\|^3} \left(\|\mathbf{q}_{1:3}\|^2 I_3 - \mathbf{q}_{1:3}\mathbf{q}_{1:3}^T \right) (q_4 I_3 + [\mathbf{q}_{1:3}\times]) \right\} \boldsymbol{\omega} \tag{3.27}$$

after replacing $\sqrt{1 - q_4^2}$ by $\|\mathbf{q}_{1:3}\|$. We now substitute Eq. (2.124) for the quaternion, use the identities $\mathbf{e}\mathbf{e}^T = [\mathbf{e}\times]^2 + I_3$ and $[\mathbf{e}\times]^3 = -[\mathbf{e}\times]$, and collect terms to yield the desired kinematic equation

$$\dot{\boldsymbol{\vartheta}} = \boldsymbol{\omega} + \frac{1}{2}\boldsymbol{\vartheta} \times \boldsymbol{\omega} + \frac{1}{\vartheta^2}\left(1 - \frac{\vartheta}{2}\cot\frac{\vartheta}{2}\right) \boldsymbol{\vartheta} \times (\boldsymbol{\vartheta} \times \boldsymbol{\omega}) \tag{3.28}$$

The coefficient of the last term is singular for ϑ equal to any nonzero multiple of 2π, making the rotation vector parameterization unsuitable for numerical simulations. This coefficient is difficult to evaluate numerically for $\vartheta \approx 0$, but it is nonsingular in this limit, and expanding the cotangent in a power series leads to the small angle approximation

$$\dot{\boldsymbol{\vartheta}} \approx \boldsymbol{\omega} + (1/2)\, \boldsymbol{\vartheta} \times \boldsymbol{\omega} + (1/12)\, \boldsymbol{\vartheta} \times (\boldsymbol{\vartheta} \times \boldsymbol{\omega}) \tag{3.29}$$

The inverse of Eq. (3.28) is

$$\boldsymbol{\omega} = \dot{\boldsymbol{\vartheta}} - \frac{1 - \cos\vartheta}{\vartheta^2} \boldsymbol{\vartheta} \times \dot{\boldsymbol{\vartheta}} + \frac{\vartheta - \sin\vartheta}{\vartheta^3} \boldsymbol{\vartheta} \times (\boldsymbol{\vartheta} \times \dot{\boldsymbol{\vartheta}}) \tag{3.30}$$

This is easily verified with the use of the relation $[\mathbf{e}\times]^3 = -[\mathbf{e}\times]$ and some trigonometric identities.

3.2.5 Euler Angle Kinematics

The product rule for differentiation gives the time derivative of a generalized Euler angle representation of a rotation, defined by Eq. (2.152), as

$$\begin{aligned} \dot{A}(\mathbf{e}_\phi, \mathbf{e}_\theta, \mathbf{e}_\psi; \phi, \theta, \psi) &= \dot{A}(\mathbf{e}_\psi, \psi) A(\mathbf{e}_\theta, \theta) A(\mathbf{e}_\phi, \phi) \\ &\quad + A(\mathbf{e}_\psi, \psi) \dot{A}(\mathbf{e}_\theta, \theta) A(\mathbf{e}_\phi, \phi) + A(\mathbf{e}_\psi, \psi) A(\mathbf{e}_\theta, \theta) \dot{A}(\mathbf{e}_\phi, \phi) \end{aligned} \tag{3.31}$$

which can be seen to be a specific application of the methods of Sect. 3.1.2. Differentiating Eq. (2.109) and remembering that $\mathbf{e}_\theta$ is constant gives

$$\dot{A}(\mathbf{e}_\theta, \theta) = \dot{\theta}(-\cos\theta\, [\mathbf{e}_\theta\times] + \sin\theta [\mathbf{e}_\theta\times]^2) = -\dot{\theta}[\mathbf{e}_\theta\times] A(\mathbf{e}_\theta, \theta) \tag{3.32}$$

Analogous relations hold for $\dot{A}(\mathbf{e}_\psi, \psi)$ and $\dot{A}(\mathbf{e}_\phi, \phi)$, so

$$\begin{aligned}\dot{A}(\mathbf{e}_\phi, \mathbf{e}_\theta, \mathbf{e}_\psi; \phi, \theta, \psi) &= -\dot{\psi}[\mathbf{e}_\psi \times] A(\mathbf{e}_\psi, \psi) A(\mathbf{e}_\theta, \theta) A(\mathbf{e}_\phi, \phi) \\ &\quad -\dot{\theta} A(\mathbf{e}_\psi, \psi)[\mathbf{e}_\theta \times] A(\mathbf{e}_\theta, \theta) A(\mathbf{e}_\phi, \phi) \\ &\quad -\dot{\phi} A(\mathbf{e}_\psi, \psi) A(\mathbf{e}_\theta, \theta)[\mathbf{e}_\phi \times] A(\mathbf{e}_\phi, \phi) \\ &= -\{\dot{\psi}[\mathbf{e}_\psi \times] + \dot{\theta} A(\mathbf{e}_\psi, \psi)[\mathbf{e}_\theta \times] A^T(\mathbf{e}_\psi, \psi) \\ &\quad +\dot{\phi} A(\mathbf{e}_\psi, \psi) A(\mathbf{e}_\theta, \theta)[\mathbf{e}_\phi \times] A^T(\mathbf{e}_\theta, \theta) \\ &\quad \times A^T(\mathbf{e}_\psi, \psi)\} A(\mathbf{e}_\phi, \mathbf{e}_\theta, \mathbf{e}_\psi; \phi, \theta, \psi)\end{aligned} \tag{3.33}$$

But we know that $\dot{A} = -[\boldsymbol{\omega}\times]A$, so using Eq. (2.63) gives

$$\boldsymbol{\omega} = \dot{\psi}\,\mathbf{e}_\psi + \dot{\theta} A(\mathbf{e}_\psi, \psi)\mathbf{e}_\theta + \dot{\phi} A(\mathbf{e}_\psi, \psi) A(\mathbf{e}_\theta, \theta)\mathbf{e}_\phi \tag{3.34}$$

where $\boldsymbol{\omega}$, $\dot{\psi}$, $\dot{\theta}$, and $\dot{\phi}$ are expressed all in rad/s. This expression is intuitively plausible. It expresses the total angular velocity as the sum of the components $\dot{\psi}\,\mathbf{e}_\psi$, $\dot{\theta}\,\mathbf{e}_\theta$, and $\dot{\phi}\,\mathbf{e}_\phi$, all mapped into the frame into which $A(\mathbf{e}_\phi, \mathbf{e}_\theta, \mathbf{e}_\psi; \phi, \theta, \psi)$ transforms. The axis $\mathbf{e}_\psi$ is already in this frame, but $\mathbf{e}_\theta$ and $\mathbf{e}_\phi$ must be rotated by the matrices $A(\mathbf{e}_\psi, \psi)$ and $A(\mathbf{e}_\psi, \psi)A(\mathbf{e}_\theta, \theta)$, respectively.

We have obtained an expression for the angular velocity as a function of the Euler angle rates, but the inverse expression for the Euler angle rates in terms of the angular velocity is more useful. Because $\mathbf{e}_\psi$ is not affected by the rotation $A(\mathbf{e}_\psi, \psi)$, we can write Eq. (3.34) as

$$\boldsymbol{\omega} = A(\mathbf{e}_\psi, \psi)[\dot{\psi}\,\mathbf{e}_\psi + \dot{\theta}\,\mathbf{e}_\theta + \dot{\phi} A(\mathbf{e}_\theta, \theta)\mathbf{e}_\phi] = A(\mathbf{e}_\psi, \psi) M \begin{bmatrix} \dot{\phi} \\ \dot{\theta} \\ \dot{\psi} \end{bmatrix} \tag{3.35}$$

where M is the 3×3 matrix

$$\begin{aligned}M &\equiv \left[A(\mathbf{e}_\theta, \theta)\mathbf{e}_\phi \quad \mathbf{e}_\theta \quad \mathbf{e}_\psi\right] = \left[A(\mathbf{e}_\theta, \theta) A^T(\mathbf{e}_\theta, \theta_0)\mathbf{e}_\psi \quad \mathbf{e}_\theta \quad \mathbf{e}_\psi\right] \\ &= \left[\cos(\theta - \theta_0)\mathbf{e}_\psi + \sin(\theta - \theta_0)(\mathbf{e}_\psi \times \mathbf{e}_\theta) \quad \mathbf{e}_\theta \quad \mathbf{e}_\psi\right]\end{aligned} \tag{3.36}$$

We have used Eqs. (2.154), (2.155), and (2.117) in the above. The determinant of M is

$$\det M = [\cos(\theta - \theta_0)\mathbf{e}_\psi + \sin(\theta - \theta_0)(\mathbf{e}_\psi \times \mathbf{e}_\theta)] \cdot (\mathbf{e}_\theta \times \mathbf{e}_\psi) = -\sin(\theta - \theta_0) \tag{3.37}$$

because $\mathbf{e}_\psi$ and $\mathbf{e}_\theta$ are orthogonal. Equations (2.16) and (2.61) give the inverse of Eq. (3.35) as

$$\begin{bmatrix} \dot{\phi} \\ \dot{\theta} \\ \dot{\psi} \end{bmatrix} = M^{-1} A^T(\mathbf{e}_\psi, \psi)\boldsymbol{\omega} = B(\theta, \psi)\boldsymbol{\omega} \tag{3.38}$$

where

$$B(\theta,\psi) \equiv \frac{1}{\sin(\theta-\theta_0)} \begin{bmatrix} (\mathbf{e}_\psi \times \mathbf{e}_\theta)^T \\ \sin(\theta-\theta_0)\mathbf{e}_\theta^T \\ \sin(\theta-\theta_0)\mathbf{e}_\psi^T - \cos(\theta-\theta_0)(\mathbf{e}_\psi \times \mathbf{e}_\theta)^T \end{bmatrix} A^T(\mathbf{e}_\psi,\psi) \tag{3.39}$$

Although hidden by the notation, $B(\theta,\psi)$ also depends on $\mathbf{e}_\psi$, $\mathbf{e}_\theta$, and $\mathbf{e}_\phi$, the last implicitly through θ_0; but it is independent of ϕ.

Equations (3.38) and (3.39) exhibit the gimbal-lock phenomenon; both $\dot{\psi}$ and $\dot{\phi}$ become infinitely large as $\sin(\theta-\theta_0) \to 0$ unless the angular velocity vector $\boldsymbol{\omega}$ is orthogonal to $A(\mathbf{e}_\psi,\psi)(\mathbf{e}_\psi \times \mathbf{e}_\theta)$. Gimbal lock is a real physical phenomenon for gimbaled platforms, where the Euler or Bryan-Tait angles are the actual physical angles of the gimbal mechanisms, which clearly cannot attain infinite rates.

Using $A(\mathbf{e}_\psi,\psi)\mathbf{e}_\psi = \mathbf{e}_\psi$ and $A(\mathbf{e}_\psi,\psi)\mathbf{e}_\theta = \cos\psi\, \mathbf{e}_\theta - \sin\psi(\mathbf{e}_\psi \times \mathbf{e}_\theta)$ allows us to express Eq. (3.39) as

$$B(\theta,\psi) = \begin{bmatrix} \csc(\theta-\theta_0)[\cos\psi(\mathbf{e}_\psi \times \mathbf{e}_\theta) + \sin\psi\, \mathbf{e}_\theta]^T \\ \cos\psi\, \mathbf{e}_\theta^T - \sin\psi(\mathbf{e}_\psi \times \mathbf{e}_\theta)^T \\ \mathbf{e}_\psi^T - \cot(\theta-\theta_0)[\cos\psi(\mathbf{e}_\psi \times \mathbf{e}_\theta) + \sin\psi\, \mathbf{e}_\theta]^T \end{bmatrix} \tag{3.40}$$

Now let us look at some specific examples of conventional Euler angle sequences. The axes for the 3–1–3 sequence are $\mathbf{e}_\psi = \mathbf{e}_3$, $\mathbf{e}_\theta = \mathbf{e}_1$, and $\mathbf{e}_\psi \times \mathbf{e}_\theta = \mathbf{e}_2$, and $\theta_0 = 0$, so Eq. (3.35) gives

$$\boldsymbol{\omega} \equiv \begin{bmatrix} \omega_1 \\ \omega_2 \\ \omega_3 \end{bmatrix} = A(\mathbf{e}_3,\psi) \begin{bmatrix} 0 & 1 & 0 \\ 0 & 0 & \sin\theta \\ 1 & 0 & \cos\theta \end{bmatrix} \begin{bmatrix} \dot{\psi} \\ \dot{\theta} \\ \dot{\phi} \end{bmatrix} = \begin{bmatrix} \dot{\phi}\sin\theta\sin\psi + \dot{\theta}\cos\psi \\ \dot{\phi}\sin\theta\cos\psi - \dot{\theta}\sin\psi \\ \dot{\psi} + \dot{\phi}\cos\theta \end{bmatrix} \tag{3.41}$$

and Eq. (3.40) gives

$$B(\theta,\psi) = \begin{bmatrix} \csc\theta\sin\psi & \csc\theta\cos\psi & 0 \\ \cos\psi & -\sin\psi & 0 \\ -\cot\theta\sin\psi & -\cot\theta\cos\psi & 1 \end{bmatrix} \tag{3.42}$$

The axes for the $3-2-1$ sequence are $\mathbf{e}_\psi = \mathbf{e}_1$, $\mathbf{e}_\theta = \mathbf{e}_2$, and $\mathbf{e}_\psi \times \mathbf{e}_\theta = \mathbf{e}_3$, and $\theta_0 = -\pi/2$, so Eq. (3.35) gives

$$\boldsymbol{\omega} = A(\mathbf{e}_1,\psi) \begin{bmatrix} 1 & 0 & -\sin\theta \\ 0 & 1 & 0 \\ 0 & 0 & \cos\theta \end{bmatrix} \begin{bmatrix} \dot{\psi} \\ \dot{\theta} \\ \dot{\phi} \end{bmatrix} = \begin{bmatrix} \dot{\psi} - \dot{\phi}\sin\theta \\ \dot{\phi}\cos\theta\sin\psi + \dot{\theta}\cos\psi \\ \dot{\phi}\cos\theta\cos\psi - \dot{\theta}\sin\psi \end{bmatrix} \tag{3.43}$$

and Eq. (3.40) gives

$$B(\theta,\psi) = \begin{bmatrix} 0 & \sec\theta\sin\psi & \sec\theta\cos\psi \\ 0 & \cos\psi & -\sin\psi \\ 1 & \tan\theta\sin\psi & \tan\theta\cos\psi \end{bmatrix} \tag{3.44}$$

Expressions for $B(\theta, \psi)$ and its inverse for all the conventional Euler and Tait-Bryan representations can be derived in a parallel fashion, and explicit formulas for them can be found in Chap. 9.

3.2.6 Attitude Error Kinematics

The analysis of attitude error kinematics uses the results of Sects. 2.10 and 3.1.2. To first order in the attitude error, the angular velocity of the attitude error is given by Eqs. (3.5) and (2.180) as

$$[\boldsymbol{\omega}(\boldsymbol{\delta\vartheta})\times] = -\dot{A}(\boldsymbol{\delta\vartheta})A^T(\boldsymbol{\delta\vartheta}) = [\boldsymbol{\delta\dot{\vartheta}}\times]\,(I_3 + [\boldsymbol{\delta\vartheta}\times]) = [\boldsymbol{\delta\dot{\vartheta}}\times] \tag{3.45}$$

Thus we have simply $\boldsymbol{\omega}(\boldsymbol{\delta\vartheta}) = \boldsymbol{\delta\dot{\vartheta}}$.

Consider first the case where the attitude errors are defined in the spacecraft body frame. Equation (2.179) gives $A_{BR} = A_{B\hat{B}}A_{\hat{B}R}$, so Eq. (3.12) gives

$$\boldsymbol{\omega}_B^{BR} = \boldsymbol{\omega}_B^{B\hat{B}} + A_{B\hat{B}}\,\boldsymbol{\omega}_{\hat{B}}^{\hat{B}R} \tag{3.46}$$

In this case the attitude error rate is $\boldsymbol{\omega}(\boldsymbol{\delta\vartheta}_B) = \boldsymbol{\omega}_B^{B\hat{B}} = \boldsymbol{\delta\dot{\vartheta}}_B$, so we have

$$\boldsymbol{\delta\dot{\vartheta}}_B = \boldsymbol{\omega}_B^{BR} - A_{B\hat{B}}\,\boldsymbol{\omega}_{\hat{B}}^{\hat{B}R} = \boldsymbol{\omega} - A_{B\hat{B}}\,\hat{\boldsymbol{\omega}} \tag{3.47}$$

where the second form uses the compact notation of $\boldsymbol{\omega}$ for the true body rate, $\boldsymbol{\omega}_B^{BR}$, and $\hat{\boldsymbol{\omega}}$ for the estimated body rate, $\boldsymbol{\omega}_{\hat{B}}^{\hat{B}R}$. Inserting Eq. (2.180) for $A_{B\hat{B}}$ gives

$$\boldsymbol{\delta\dot{\vartheta}}_B = \boldsymbol{\omega} - (I_3 - [\boldsymbol{\delta\vartheta}_B\times])\hat{\boldsymbol{\omega}} = (\boldsymbol{\omega} - \hat{\boldsymbol{\omega}}) - \hat{\boldsymbol{\omega}} \times \boldsymbol{\delta\vartheta}_B \tag{3.48}$$

The final form of this equation is the one most commonly applied. It appears to be inconsistent because it involves the difference of $\boldsymbol{\omega}$ and $\hat{\boldsymbol{\omega}}$, which are referenced to two different coordinate frames. Equation (3.47) shows that the attitude error rate is really defined consistently; the apparent contradiction arises from inserting the approximate form for $A_{B\hat{B}}$. In fact, it is just this approximation that gives rise to the $\hat{\boldsymbol{\omega}} \times \boldsymbol{\delta\vartheta}$ term in the kinematic equation.

Now consider the case where the attitude errors are defined in the reference frame. Equation (2.178) gives $A_{BR} = A_{B\hat{R}}A_{\hat{R}R}$, so Eq. (3.12) gives

$$\boldsymbol{\omega}_B^{BR} = \boldsymbol{\omega}_B^{B\hat{R}} + A_{B\hat{R}}\,\boldsymbol{\omega}_{\hat{R}}^{\hat{R}R} \tag{3.49}$$

In this case the attitude error rate is $\boldsymbol{\omega}(\boldsymbol{\delta\vartheta}_R) = \boldsymbol{\omega}_{\hat{R}}^{\hat{R}R} = \boldsymbol{\delta\dot{\vartheta}}_R$, so

$$\boldsymbol{\delta\dot{\vartheta}}_R = A_{B\hat{R}}^T(\boldsymbol{\omega}_B^{BR} - \boldsymbol{\omega}_B^{B\hat{R}}) = \hat{A}^T(\boldsymbol{\omega} - \hat{\boldsymbol{\omega}}) \tag{3.50}$$

where in this case $\hat{A} = A_{B\hat{R}}$ is the estimated attitude and $\hat{\boldsymbol{\omega}} = \boldsymbol{\omega}_B^{B\hat{R}}$ is the estimated body rate.

The error kinematics in these two cases look quite different. Equations (3.47) or (3.48) involves the attitude error, while Eq. (3.50) involves the attitude estimate. The difference arises because either attitude dynamics or strapdown gyro measurements give the components of the body rate in the spacecraft body frame. If the attitude errors are specified in the body frame, there are two slightly different body frames, the true frame and the estimated frame. If the attitude errors are specified in the reference frame, on the other hand, there is only one body frame. The difference between the two formulations is purely kinematic, it has nothing to do with either frame being inertial or non-inertial.[1]

3.3 Attitude Dynamics

We now turn to attitude dynamics, emphasizing the fundamental role of angular momentum. We begin by defining the center of mass of a collection of mass points and showing how the rotational motion of this system can be treated separately from the motion of the center of mass. We then specialize to the case of a rigid body, defining the moment of inertia tensor and expressing the angular momentum and rotational kinetic energy in terms of the moment of inertia and the angular velocity. We collect the rotational equations of motion, kinematic and dynamic, in one place for easy reference, and then discuss the torque-free motion of a rigid body. Then we show how to include internal and external torques in the dynamics.

3.3.1 Angular Momentum and Kinetic Energy

We will consider a spacecraft (or anything else) to be made up of a collection of n point masses. The angular momentum with respect to the origin 0 of an inertial coordinate frame is defined in terms of the masses m_i, positions $\mathbf{r}^{i0}$, and velocities $\mathbf{v}^{i0} = \dot{\mathbf{r}}^{i0}$ of the points relative to 0 by

$$\mathbf{H}^0 \equiv \sum_{i=1}^{n} \mathbf{r}^{i0} \times m_i \mathbf{v}^{i0} \tag{3.51}$$

Newton's second law of motion tells us that $m_i \dot{\mathbf{v}}_I^{i0} = \mathbf{F}_I^i$ in an inertial reference frame, and $\mathbf{v}^{i0} \times \mathbf{v}^{i0} = 0$, so the angular momentum obeys the equation

$$\dot{\mathbf{H}}_I^0 = \sum_{i=1}^{n} \mathbf{r}_I^{i0} \times \mathbf{F}_I^i = \sum_{i=1}^{n} \mathbf{r}_I^{i0} \times \left(\sum_{j=1}^{n} \mathbf{F}_I^{ij} + \mathbf{F}_I^{iext} \right) \tag{3.52}$$

[1]The situation would be different if the gyros were used to stabilize a platform to serve as an inertially fixed reference, but such platforms are now used only infrequently in space.

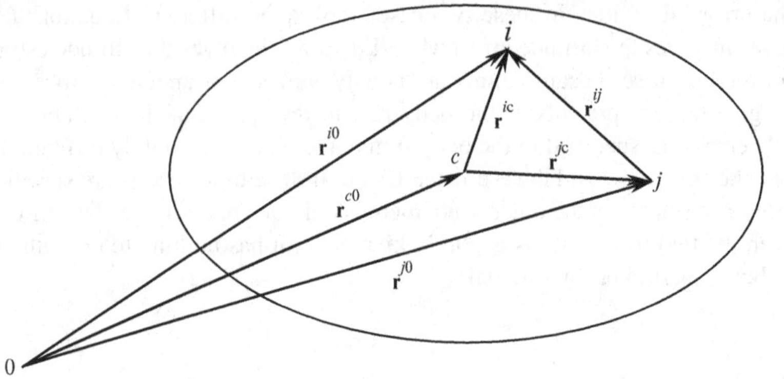

Fig. 3.1 Center of mass, c, and two representative mass points

where $\mathbf{F}^{ij}$ is the force exerted on mass point i by mass point j and $\mathbf{F}^{iext}$ is the force exerted on i by everything external to the system of mass points.[2] Any pair of mass points with indices k and ℓ appears twice in the double sum over i and j, as $\mathbf{r}_I^{k0} \times \mathbf{F}_I^{k\ell}$ and as $\mathbf{r}_I^{\ell 0} \times \mathbf{F}_I^{\ell k}$. Newton's third law of motion says that $\mathbf{F}^{\ell k} = -\mathbf{F}^{k\ell}$, so these two terms sum to $(\mathbf{r}_I^{k0} - \mathbf{r}_I^{\ell 0}) \times \mathbf{F}_I^{k\ell} = \mathbf{r}_I^{k\ell} \times \mathbf{F}_I^{k\ell}$. This geometry is illustrated in Fig. 3.1. We assume that the force between two mass points acts along the line between them, an assumption known as the *strong law of action and reaction* [10], so the cross product vanishes and

$$\dot{\mathbf{H}}_I^0 = \sum_{i=1}^{n} \mathbf{r}_I^{i0} \times \mathbf{F}_I^{iext} \equiv \mathbf{L}_I^0 \tag{3.53}$$

where $\mathbf{L}_I^0$ is the net torque about 0 exerted on the collection of mass points by all the external forces.[3] Note that internal forces give no contribution to the net torque.

We will now define the center of mass of the collection of mass points and prove the important result that the motion of the center of mass and the motion of the mass points about their center of mass are uncoupled. The location of the center of mass c with respect to the origin 0 is defined as

$$\mathbf{r}^{c0} \equiv \left(\sum_{i=1}^{n} m_i \mathbf{r}^{i0} \right) \Big/ M \tag{3.54}$$

where $M \equiv \sum_{i=1}^{n} m_i$ is the total mass of the collection of mass points. It follows from this definition that

$$\sum_{i=1}^{n} m_i \mathbf{r}^{ic} = \sum_{i=1}^{n} m_i (\mathbf{r}^{i0} - \mathbf{r}^{c0}) = M\mathbf{r}^{c0} - M\mathbf{r}^{c0} = \mathbf{0} \tag{3.55}$$

[2] We assume that $\mathbf{F}^{ii}$, the force exerted by a mass point on itself, is zero.

[3] Warning! Many authors use the letter $\mathbf{L}$ to denote angular momentum.

where these vectors are illustrated in Fig. 3.1. Using this result and its derivative, the angular momentum about the arbitrary point 0 can be written as the sum of the angular momentum about the center of mass and the angular momentum of the center-of-mass motion, namely

$$\begin{aligned}\mathbf{H}^0 &= \sum_{i=1}^{n} m_i(\mathbf{r}^{ic} + \mathbf{r}^{c0}) \times (\mathbf{v}^{ic} + \mathbf{v}^{c0}) = \sum_{i=1}^{n} m_i \mathbf{r}^{ic} \times \mathbf{v}^{ic} + \mathbf{r}^{c0} \times M\mathbf{v}^{c0} \\ &= \mathbf{H}^c + \mathbf{r}^{c0} \times M\mathbf{v}^{c0}\end{aligned} \tag{3.56}$$

The net torque can similarly be expressed as

$$\begin{aligned}\mathbf{L}^0 &= \sum_{i=1}^{n} (\mathbf{r}^{ic} + \mathbf{r}^{c0}) \times \mathbf{F}^{iext} = \sum_{i=1}^{n} \mathbf{r}^{ic} \times \mathbf{F}^{iext} + \mathbf{r}^{c0} \times \sum_{i=1}^{n} \mathbf{F}^{iext} \\ &= \mathbf{L}^c + \mathbf{r}^{c0} \times \mathbf{F}\end{aligned} \tag{3.57}$$

where $\mathbf{F}$ is the net external force on the collection of point masses. Now in an inertial frame, we have $\mathbf{F}_I = M\dot{\mathbf{v}}_I^{c0}$, so substituting Eqs. (3.56) and (3.57) into Eq. (3.53) and canceling the term due to motion of the center of mass gives *Euler's equation*

$$\dot{\mathbf{H}}_I^c = \mathbf{L}_I^c \tag{3.58}$$

This is the fundamental equation of attitude dynamics. It is important to note that this result holds even if the center of mass undergoes acceleration, as it usually does. Equation (3.58) only holds in a non-rotating frame, however.[4]

The separation into contributions from center-of-mass motion and motion about the center of mass also holds for the kinetic energy. We have

$$\begin{aligned}E_k^0 &\equiv \frac{1}{2}\sum_{i=1}^{n} m_i \|\mathbf{v}^{i0}\|^2 = \frac{1}{2}\sum_{i=1}^{n} m_i (\mathbf{v}^{ic} + \mathbf{v}^{c0}) \cdot (\mathbf{v}^{ic} + \mathbf{v}^{c0}) \\ &= \frac{1}{2}\sum_{i=1}^{n} m_i \|\mathbf{v}^{ic}\|^2 + \frac{1}{2}M\|\mathbf{v}^{c0}\|^2 = E_k^c + \frac{1}{2}M\|\mathbf{v}^{c0}\|^2\end{aligned} \tag{3.59}$$

The derivative of the kinetic energy is

$$\dot{E}_k^0 = \sum_{i=1}^{n} m_i \mathbf{v}_I^{i0} \cdot \dot{\mathbf{v}}_I^{i0} = \sum_{i=1}^{n} (\mathbf{v}_I^{ic} + \mathbf{v}_I^{c0}) \cdot \mathbf{F}_I^i = \sum_{i=1}^{n} \mathbf{v}_I^{ic} \cdot \mathbf{F}_I^i + \mathbf{v}_I^{c0} \cdot \mathbf{F}_I \tag{3.60}$$

The internal forces cancel in the $\mathbf{v}_I^{c0} \cdot \mathbf{F}_I^i$ sum due to the *weak* form of Newton's third law of motion, i.e. this cancelation does not require the force between two mass points to be along the line joining them. The internal forces do not cancel in

[4]This is discussed in Sect. 2.6.2.

the other term, though. This means that internal forces cannot affect the energy of center-of-mass motion, but they can modify the energy of rotation about the center of mass. It is apparent that the last term of Eq. (3.60) is equal to the derivative of the last term of Eq. (3.59), from which it follows that

$$\dot{E}_k^c = \sum_{i=1}^{n} \mathbf{v}_I^{ic} \cdot \mathbf{F}_I^i \tag{3.61}$$

3.3.2 *Rigid Body Dynamics*

A rigid body is defined by the existence of a reference frame, called the *body frame* B in which all the vectors $\mathbf{r}_B^{ic}$ are constant. The body frame is not unique; any frame related to it by a constant orthogonal transformation is also a body frame. The constancy of the vectors in the body frame and Eq. (3.14) allow us to obtain the velocity of a mass point in the inertial frame as

$$\mathbf{v}_I^{ic} = \dot{\mathbf{r}}_I^{ic} = A_{IB}\dot{\mathbf{r}}_B^{ic} - \boldsymbol{\omega}_I^{IB} \times \mathbf{r}_I^{ic} = \boldsymbol{\omega}_I^{BI} \times \mathbf{r}_I^{ic} \tag{3.62}$$

where we have used Eq. (3.8) to obtain the last form. Substituting this into Eq. (3.56) gives

$$\mathbf{H}_I^c = \sum_{i=1}^{n} m_i \mathbf{r}_I^{ic} \times \mathbf{v}_I^{ic} = \sum_{i=1}^{n} m_i \mathbf{r}_I^{ic} \times (\boldsymbol{\omega}_I^{BI} \times \mathbf{r}_I^{ic}) = -\sum_{i=1}^{n} m_i [\mathbf{r}_I^{ic}\times]^2 \boldsymbol{\omega}_I^{BI} = J_I^c\, \boldsymbol{\omega}_I^{BI} \tag{3.63}$$

The last equality defines the representation in the frame I of the symmetric 3×3 matrix J_I^c known as the *moment of inertia tensor*, or MOI.[5] The MOI in Eq. (3.63) is specific to frame I, but we can define it in a general frame by

$$J^c \equiv -\sum_{i=1}^{n} m_i [\mathbf{r}^{ic}\times]^2 = \sum_{i=1}^{n} m_i \left[\|\mathbf{r}^{ic}\|^2 I_3 - \mathbf{r}^{ic}(\mathbf{r}^{ic})^T\right] \tag{3.64}$$

which defines the MOI in the frame in which the vectors $\mathbf{r}^{ic}$ are represented, whatever that frame is. In fact the MOI is almost always expressed in the body frame

$$J_B^c = -\sum_{i=1}^{n} m_i [\mathbf{r}_B^{ic}\times]^2 = \sum_{i=1}^{n} m_i \left[\|\mathbf{r}_B^{ic}\|^2 I_3 - \mathbf{r}_B^{ic}(\mathbf{r}_B^{ic})^T\right] \tag{3.65}$$

[5]The MOI is more specifically a *second-rank tensor*, which means that it transforms by Eq. (2.52) under reference frame transformations. This follows directly from the fact that $\mathbf{r}_B^{ic} = A_{BI}\mathbf{r}_I^{ic}$.

because it is constant in that frame. The angular momentum in the body frame is given, using Eqs. (2.52) and (2.42), by

$$\mathbf{H}_B^c = A_{BI}\mathbf{H}_I^c = A_{BI}J_I^c\,\boldsymbol{\omega}_I^{BI} = A_{BI}J_I^c A_{BI}^T A_{BI}\boldsymbol{\omega}_I^{BI} = J_B^c\,\boldsymbol{\omega}_B^{BI}. \tag{3.66}$$

The *parallel axis theorem* can be used to find the MOI of a rigid body about an arbitrary point p in terms of its MOI about its center of mass:

$$\begin{aligned} J^p &= \sum_{i=1}^{n} m_i\left[\|\mathbf{r}^{ip}\|^2 I_3 - \mathbf{r}^{ip}(\mathbf{r}^{ip})^T\right] \\ &= \sum_{i=1}^{n} m_i\left[\|(\mathbf{r}^{ic}+\mathbf{r}^{cp})\|^2 I_3 - (\mathbf{r}^{ic}+\mathbf{r}^{cp})(\mathbf{r}^{ic}+\mathbf{r}^{cp})^T\right] \\ &= M\left[\|\mathbf{r}^{cp}\|^2 I_3 - \mathbf{r}^{cp}(\mathbf{r}^{cp})^T\right] + J^c \end{aligned} \tag{3.67}$$

with Eq. (3.55) causing the other terms in the sum to vanish. This allows us to express the MOI of a large body in terms of the MOIs of m subassemblies as

$$J^c = \sum_{k=1}^{m}\left\{M_k\left[\|\mathbf{r}^{c_k c}\|^2 I_3 - \mathbf{r}^{c_k c}(\mathbf{r}^{c_k c})^T\right] + J^{c_k}\right\} \tag{3.68}$$

where c is the center of mass of the whole system and M_k, c_k, and J^{c_k} denote the mass, location of the center of mass, and MOI about c_k of the kth subassembly. This is the method used to compute MOIs in practice, where the subassemblies can be structural elements, electronics boxes, reaction wheel assemblies, star trackers, etc. The MOIs of the subassemblies can be directly measured or they can be computed by breaking the subassemblies down into sub-subassemblies.

The matrix elements of the MOI tensor are given explicitly by

$$[J^c]_{11} = \sum_{i=1}^{n} m_i\left[(r_2^{ic})^2 + (r_3^{ic})^2\right] \tag{3.69a}$$

$$[J^c]_{22} = \sum_{i=1}^{n} m_i\left[(r_3^{ic})^2 + (r_1^{ic})^2\right] \tag{3.69b}$$

$$[J^c]_{33} = \sum_{i=1}^{n} m_i\left[(r_1^{ic})^2 + (r_2^{ic})^2\right] \tag{3.69c}$$

$$[J^c]_{k\ell} = -\sum_{i=1}^{n} m_i\, r_k^{ic}\, r_\ell^{ic}, \quad \text{for} \quad k \neq \ell \tag{3.69d}$$

The off-diagonal elements, $[J^c]_{k\ell}$, or their negatives, $-[J^c]_{k\ell}$, are often referred to as the *products of inertia*.[6]

Being a real symmetric 3×3 matrix, the MOI tensor has three orthogonal eigenvectors $\mathbf{e}_B^k$ and three real eigenvalues J_k satisfying the relation

$$J_B^c \mathbf{e}_B^k = J_k \mathbf{e}_B^k, \quad \text{for} \quad k = 1, 2, 3 \tag{3.70}$$

The unit vectors $\mathbf{e}_B^k$ are called the *principal axes* and the scalars J_k are known as the *principal moments of inertia*. The transformation of the MOI tensor from an arbitrary body frame to the principal axis frame is given by

$$A_{BP}^T J_B^c A_{BP} = J_P^c = \text{diag}\left([J_1 \; J_2 \; J_3]\right) \tag{3.71}$$

where $A_{BP} = [\mathbf{e}_B^1 \; \mathbf{e}_B^2 \; \mathbf{e}_B^3]$. In the principal axis frame Eq. (3.69) has the form

$$J_1 = \sum_{i=1}^{n} m_i \left[(r_{P2}^{ic})^2 + (r_{P3}^{ic})^2\right] \tag{3.72a}$$

$$J_2 = \sum_{i=1}^{n} m_i \left[(r_{P3}^{ic})^2 + (r_{P1}^{ic})^2\right] \tag{3.72b}$$

$$J_3 = \sum_{i=1}^{n} m_i \left[(r_{P1}^{ic})^2 + (r_{P2}^{ic})^2\right] \tag{3.72c}$$

$$0 = \sum_{i=1}^{n} m_i \, r_{Pk}^{ic} \, r_{P\ell}^{ic}, \quad \text{for} \quad k \neq \ell \tag{3.72d}$$

The sum in the last of these equations must have balancing positive and negative contributions, expressing the intuitive idea that the mass is distributed symmetrically about the principal axes. In particular, any axis of rotational symmetry of a mass distribution is a principal axis. The first three equations show that the principal moments are all positive, unless the mass is all concentrated on a mathematical straight line, which is impossible for a real physical body. It can also be seen from these equations that the principal moments of inertia satisfy the triangle inequalities

$$J_k \leq J_\ell + J_m \tag{3.73}$$

where equality holds only if all the mass is concentrated in the $\ell - m$ plane, and where k, ℓ, and m are any permutation of the indices $1, 2, 3$.

---

[6]Be warned that notation varies. The majority of authors use I for the MOI, but we reserve this notation for the identity matrix. Some denote our J_{kk} by I_{kk} and our $J_{k\ell}$ for $k \neq \ell$ by $-I_{k\ell}$, but this unfortunate notation should be shunned.

It also follows from this analysis that the MOI tensor of any real physical body has an inverse in any body frame, and we can write Eq. (3.66) as

$$\boldsymbol{\omega}_B^{BI} = (J_B^c)^{-1}\mathbf{H}_B^c \tag{3.74}$$

Implicit in Eqs. (3.66) and (3.74) is the key fact that the angular momentum and angular velocity of a rigid body are parallel if and only if the body rotates about a principal axis.

The rotational kinetic energy of a rigid body can also be expressed in terms of the MOI tensor. Substituting Eq. (3.62) into Eq. (3.59) gives

$$\begin{aligned} E_k^c &= \frac{1}{2}\sum_{i=1}^{n} m_i \|\mathbf{v}^{ic}\|^2 = \frac{1}{2}\sum_{i=1}^{n} m_i (\boldsymbol{\omega}_I^{BI} \times \mathbf{r}_I^{ic})^T (\boldsymbol{\omega}_I^{BI} \times \mathbf{r}_I^{ic}) \\ &= \frac{1}{2}\sum_{i=1}^{n} m_i \left([\mathbf{r}_I^{ic}\times]\boldsymbol{\omega}_I^{BI}\right)^T \left([\mathbf{r}_I^{ic}\times]\boldsymbol{\omega}_I^{BI}\right) = \frac{1}{2}(\boldsymbol{\omega}_I^{BI})^T J_I^c\, \boldsymbol{\omega}_I^{BI} \end{aligned} \tag{3.75}$$

The rotational kinetic energy can be computed in any frame F by

$$E_k^c = \frac{1}{2}(\boldsymbol{\omega}_F^{BI})^T J_F^c\, \boldsymbol{\omega}_F^{BI} = \frac{1}{2}\boldsymbol{\omega}_F^{BI} \cdot \mathbf{H}_F^c = \frac{1}{2}(\mathbf{H}_F^c)^T (J_F^c)^{-1}\, \mathbf{H}_F^c \tag{3.76}$$

The time derivative of Eq. (3.75) is, with Eqs. (3.63) and (3.58),

$$\dot{E}_k^c = (\boldsymbol{\omega}_I^{BI})^T J_I^c\, \dot{\boldsymbol{\omega}}_I^{BI} = \boldsymbol{\omega}_I^{BI} \cdot \mathbf{L}_I^c = \boldsymbol{\omega}_B^{BI} \cdot \mathbf{L}_B^c \tag{3.77}$$

showing that only external torques can modify the energy of rotation of a *rigid* body about its center of mass. This can be seen more explicitly by substituting Eq. (3.62) into Eq. (3.61), giving

$$\dot{E}_k^c = \sum_{i=1}^{n} (\boldsymbol{\omega}_I^{BI} \times \mathbf{r}_I^{ic}) \cdot \left(\sum_{j=1}^{n} \mathbf{F}_I^{ij} + \mathbf{F}_I^{iext} \right) = \boldsymbol{\omega}_I^{BI} \cdot \mathbf{L}_I^c \tag{3.78}$$

The sum over the internal forces $\mathbf{F}_I^{ij}$ vanishes for the same reason that the corresponding sum in Eq. (3.52) gave zero contribution, and the final equality follows from applying Eqs. (2.56a) and (3.57).

We can now collect in one place the basic equations needed to model the attitude motion of a rigid body. We will assume that the attitude is parameterized by a quaternion, but any other representation could be used instead:

$$\dot{\mathbf{H}}_I^c = \mathbf{L}_I^c \tag{3.79a}$$

$$\mathbf{H}_B^c = A(\mathbf{q}_{BI})\mathbf{H}_I^c \tag{3.79b}$$

$$\boldsymbol{\omega}_B^{BI} = (J_B^c)^{-1}\mathbf{H}_B^c \tag{3.79c}$$

$$\dot{\mathbf{q}}_{BI} = \frac{1}{2}\boldsymbol{\omega}_B^{BI} \otimes \mathbf{q}_{BI} \tag{3.79d}$$

The external torques are often more easily computed in the body frame, so Eq. (3.14) is often employed to replace the first two of these equations by

$$\dot{\mathbf{H}}_B^c = \mathbf{L}_B^c - \boldsymbol{\omega}_B^{BI} \times \mathbf{H}_B^c \tag{3.80}$$

A further reduction in the number of equations can be achieved by combining the above equation with Eq. (3.66) to obtain *Euler's rotational equation*

$$\dot{\boldsymbol{\omega}}_B^{BI} = (J_B^c)^{-1}\left[\mathbf{L}_B^c - \boldsymbol{\omega}_B^{BI} \times (J_B^c\,\boldsymbol{\omega}_B^{BI})\right] \tag{3.81}$$

This equation and a kinematics equation, such as the quaternion kinematics equation, provide a complete description of the motion of a rigid body.

Use of the principal axis reference frame is generally not especially advantageous for numerical integration of the equations of motion, since a computer can easily deal with a full 3×3 inertia tensor. The principal axis frame is almost invariably employed for analytical studies of attitude motion, however, and we will use it in the next subsection.

3.3.3 Rigid Body Motion

We now begin a discussion of the motion produced by the rigid body dynamic and kinematic equations. This section will consider the qualitative aspects of the motion, without solving the kinematic and dynamic differential equations explicitly. Then Sect. 3.3.4 will discuss in detail the solutions in the absence of torques. We will simplify the notation by omitting the superscript c, with the understanding that we always treat motion with respect to the center of mass unless explicitly indicated otherwise. We will denote the components of $\mathbf{H}_B$ in a principal axis frame by H_1, H_2, H_3 and the components of $\boldsymbol{\omega}_B^{BI}$ by $\omega_1, \omega_2, \omega_3$. There are many discussions of rigid body motion in the literature, including notable ones by Goldstein [10], Kaplan [17], Markley [20], and Hughes [14].

3.3.3.1 Spin Stabilization

Equation (3.58) tells us that $\mathbf{H}_I$ is constant if there are no external torques. This has led to the technique of *spin stabilization* of spacecraft. The basic idea is to stabilize the pointing direction of one axis by spinning the spacecraft about that axis.

If a torque of magnitude L transverse to the direction of the spin angular momentum acts on a spinning spacecraft, the change in angular momentum over a time interval Δt will be

$$\|\Delta \mathbf{H}_I\| = L\Delta t \approx H\Delta\vartheta \tag{3.82}$$

where H is the magnitude of the angular momentum and $\Delta\vartheta$ is the angle (measured in radians) over which it rotates. It is clear that a larger amount of angular momentum will result in a smaller angular motion for a given level of disturbance torque. Spin stabilization was widely employed early in the space program, then was largely displaced by active control methods, but has made a comeback in the era of microsatellites and nanosatellites.

It is important for spin stabilization that the angular momentum also have a constant direction in the body frame. Equation (3.80) says that this will be the case if $\boldsymbol{\omega}_B^{BI} \times \mathbf{H}_B = \mathbf{0}$, which requires the rotation axis to be a principal axis of the inertia tensor. Thus we want the spacecraft to rotate about a principal axis. If the spacecraft spins about a principal axis, but this does not align precisely with the desired pointing axis, *coning* results, with the pointing axis rotating at the spin rate around a cone centered on the inertially fixed angular momentum. The only way to eliminate coning is to carefully balance the spacecraft. This is generally accomplished by adding balance weights before launch similar to balancing an automobile tire, but some spacecraft have carried movable weights into orbit to allow compensation for inertia shifts caused by launch forces or to provide better balance than can be measured on the ground.

If the angular momentum is not perfectly aligned with a principal axis of inertia, the angular momentum will not be constant in the body frame. The resulting motion of the angular momentum, and of the angular velocity, is called *nutation*.[7] For torque-free motion, $\|\mathbf{H}_B\| = \|\mathbf{H}_I\| \equiv H$ is constant during nutation, so the angular momentum vector moves on the surface of a sphere of radius H in the body frame.

Investigating the stability of the motion for angular velocity close to, but not exactly on, a principal axis shows that all principal axes are not created equal for the purpose of spin stabilization. We will see that spin stabilization should always be about the principal axis with the largest or smallest principal moment of inertia, known as the *major* or *minor* principal axis, respectively, with a strong preference for the major principal axis. If there are no torques, the component form of Eq. (3.80) in the principal axis frame is

$$\dot{H}_1 = (J_3^{-1} - J_2^{-1})\, H_2\, H_3 = [(J_2 - J_3)/(J_2 J_3)]\, H_2\, H_3 \tag{3.83a}$$

$$\dot{H}_2 = (J_1^{-1} - J_3^{-1})\, H_3\, H_1 = [(J_3 - J_1)/(J_3 J_1)]\, H_3\, H_1 \tag{3.83b}$$

$$\dot{H}_3 = (J_2^{-1} - J_1^{-1})\, H_1\, H_2 = [(J_1 - J_2)/(J_1 J_2)]\, H_1\, H_2 \tag{3.83c}$$

[7]This is the usual aerospace meaning of the term *nutation*; it is used quite differently in other applications of classical mechanics to refer to a wobbling or nodding motion.

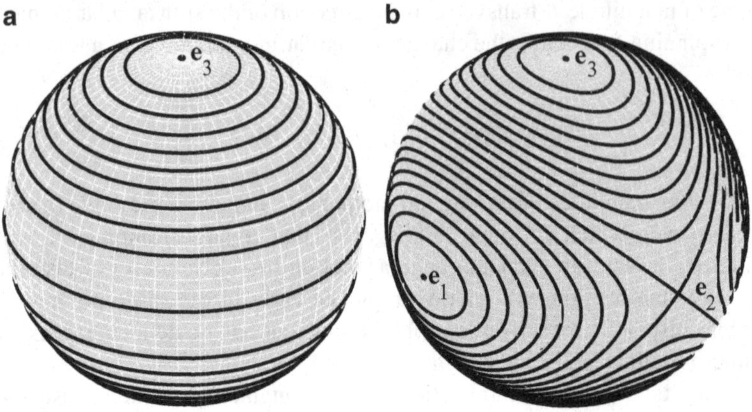

Fig. 3.2 Constant energy paths on the angular momentum sphere. (**a**) Axisymmetric inertia ratios $J_1 : J_2 : J_3 = 4 : 4 : 6$. (**b**) Triaxial inertia ratios $J_1 : J_2 : J_3 = 3 : 4 : 6$

We will choose the spin axis to be close to the $\mathbf{e}_3$ axis for the purpose of the following argument, which means that H_1 and H_2 are both much less than H_3. Thus their product on the right side of Eq. (3.83c) is negligibly small, and we can approximate H_3 as being constant. Then differentiating Eq. (3.83a) and substituting Eq. (3.83b) gives

$$\ddot{H}_1(t) = [(J_2 - J_3)(J_3 - J_1)/(J_1 J_2)](H_3/J_3)^2 H_1(t) \tag{3.84}$$

If J_3 is the largest or smallest principal moment, the product $(J_2 - J_3)(J_3 - J_1)$ is negative, resulting in a periodic motion of H_1. Then Eq. (3.83a) shows that H_2 also undergoes periodic motion with the same period. This is nutation about the $\mathbf{e}_3$ axis. The product $(J_2 - J_3)(J_3 - J_1)$ is positive if J_3 is intermediate between J_1 and J_2, however, and H_1 and H_2 both grow exponentially.[8] Thus rotation about an intermediate principal axis is inherently unstable. It is also possible that J_3 is equal to one of the other principal moments. If $J_3 = J_1$, for instance, then Eq. (3.83b) tells us that H_2 is exactly constant, and Eq. (3.83a) says that H_1 grows linearly with time. This motion is also unstable, although not exponentially unstable.

This behavior is illustrated in Fig. 3.2a,b, which show the torque-free motion of $\mathbf{H}_B$ on the sphere of radius H for two different inertia tensors. The rotational kinetic energy in a principal axis frame is given by

$$2E_k = H_1^2/J_1 + H_2^2/J_2 + H_3^2/J_3 \tag{3.85}$$

[8] Exponential growth continues until the right side of Eq. (3.83c) is no longer negligible.

For a given angular momentum magnitude, the kinetic energy is restricted to the range

$$H^2/J_{\max} \leq 2E_k \leq H^2/J_{\min} \tag{3.86}$$

where $J_{\min}$ and $J_{\max}$ are the minimum and maximum principal moments of inertia. Conservation of energy confines the motion of $\mathbf{H}_B$ to a path on the sphere and Fig. 3.2a,b show these paths for energy levels with even spacing $0.025H^2/J_{\max}$ in the range allowed by Eq. (3.86).

We will first discuss the simpler axially symmetric, or *axisymmetric* case, with two principal moments equal. We will take $\mathbf{e}_3$ to be the axis of symmetry of the inertia tensor in this case, which is shown in Fig. 3.2a. The axes $\mathbf{e}_1$ and $\mathbf{e}_2$ are not shown on the figure because they can be any two axes in the equatorial plane that form a right-handed orthogonal triad with $\mathbf{e}_3$. With $J_1 = J_2 \equiv J_t$, where the subscript t denotes "transverse," Eq. (3.85) becomes

$$2E_k = (H_1^2 + H_2^2)/J_t + H_3^2/J_3 \tag{3.87}$$

It follows that

$$H_3 = \pm \left[\frac{J_3(H^2 - 2J_t E_k)}{J_3 - J_t} \right]^{1/2} \tag{3.88a}$$

$$H_t \equiv \left[H_1^2 + H_2^2\right]^{1/2} = \left[\frac{J_t(2J_3 E_k - H^2)}{J_3 - J_t} \right]^{1/2} \tag{3.88b}$$

This shows that nutation of an axisymmetric body is motion of the angular momentum vector in the body frame along circles of constant radius H_t a constant distance of H_3 above or below the equatorial plane, as shown in Fig. 3.2a. This extends the analysis of Eq. (3.84) to nutation with any allowable magnitude about the major or minor principal axis $\pm\mathbf{e}_3$. The figure also shows that there are no small nutation paths close to the $\mathbf{e}_1 - \mathbf{e}_2$ plane.

Figure 3.2b illustrates the *triaxial* case, where no two principal moments are equal. We have labeled the principal axes for this discussion so that $J_1 < J_2 < J_3$. The paths followed by nutational motion about the major and minor axes are not simple circles in this case; their exact form will be found in Sect. 3.3.4. We see that there is no stable motion about the intermediate axis $\pm\mathbf{e}_2$ in this case. Instead, we find motion along two great circles passing through $\mathbf{e}_2$. They are called *separatrices*, because they separate nutational motion about $\pm\mathbf{e}_1$ from nutational motion about $\pm\mathbf{e}_3$. The separatrices are the curves for $2E_k = H^2/J_2$, so they satisfy the equation

$$(H_3^2 + H_2^2 + H_1^2)/J_2 = H_1^2/J_1 + H_2^2/J_2 + H_3^2/J_3 \tag{3.89}$$

which gives

$$\frac{H_3}{H_1} = \left[\frac{J_3(J_2 - J_1)}{J_1(J_3 - J_2)} \right]^{1/2} \tag{3.90}$$

Fig. 3.3 William Pickering, James Van Allen, and Wernher von Braun holding a Full-Scale Model of Explorer 1. Source: NASA

3.3.3.2 Energy Dissipation

We have seen that stable rotation is possible around either a major or minor principal axis of inertia. However, internal forces in a body that is not completely rigid can lead to energy dissipation. If external torques are absent, $\mathbf{H}_I$ will be constant, but the rotational energy will decrease to its minimum value of $H^2/J_{\max}$, resulting in stable rotation about a major principal axis. If the intended spin axis is a minor axis, energy dissipation will result in *flat spin*, an undesirable rotation about an axis perpendicular to the preferred axis. A famous example of this is the pencil-shaped Explorer 1, the first Earth satellite successfully launched by the United States, which was intended to spin about its longitudinal axis. Its entry into flat spin was attributed to energy dissipation in the flexible turnstile antenna array, comprising the four wires attached to the fuselage just aft of Wernher von Braun's right hand in Fig. 3.3 [5].

Steady spin about the major principal axis is commonly the desired outcome, with spin-stabilized spacecraft designed to be more nearly disc-shaped than pencil-shaped; and passive *nutation dampers* are often placed on these spacecraft to produce this result. For triaxial inertia, energy dissipation can result in motion starting with increasing nutation about $\pm\mathbf{e}_1$, then crossing one of the separatrices, followed by decreasing nutation about either $\mathbf{e}_3$ or $-\mathbf{e}_3$, depending on where a separatrix is crossed. Because $\mathbf{H}_I$ is fixed in the inertial frame, these two outcomes result in the pointing axis being oriented in opposite directions in inertial space. It is usually the case that only one of these pointing directions is satisfactory, which requires careful control of the energy damping to ensure the correct crossing of the separatrix.

3.3.3.3 Poinsot's Construction

The discussion up to this point has not provided a picture of the motion in the inertial frame. *Poinsot's construction* supplies this picture by focusing on angular velocity rather than angular momentum [10, 14, 17]. Conservation of energy in torque-free motion restricts angular velocity to the surface of the *inertia ellipsoid* defined by

$$2E_k = (\boldsymbol{\omega}_B^{BI})^T J_B\, \boldsymbol{\omega}_B^{BI} = J_1\omega_1^2 + J_2\omega_2^2 + J_3\omega_3^2 \tag{3.91}$$

with semimajor axes of length $\sqrt{2E_k/J_1}$, $\sqrt{2E_k/J_2}$, $\sqrt{2E_k/J_3}$. A small change $\Delta\boldsymbol{\omega}$ in the angular velocity would result in an energy change

$$\Delta E_k = (\boldsymbol{\omega}_B^{BI})^T J_B^c\, \Delta\boldsymbol{\omega} = \mathbf{H}_B \cdot \Delta\boldsymbol{\omega} \tag{3.92}$$

which is zero if $\Delta\boldsymbol{\omega}$ is perpendicular to the angular momentum. Changes $\Delta\boldsymbol{\omega}$ that do not change E_k are in the plane tangent to the inertia ellipsoid, leading to the conclusion that the normal to the inertia ellipsoid at any $\boldsymbol{\omega}_B^{BI}$ is in the direction of the angular momentum $\mathbf{H}_B = J_B\boldsymbol{\omega}_B^{BI}$.

Each closed path on the momentum sphere illustrated in Fig. 3.2a or b maps onto a closed path called a *polhode* on the inertia ellipsoid. As the angular velocity moves along the polhode, the inertia ellipsoid (which is fixed in the rigid body) moves in such away that the normal to its surface at the position of the instantaneous angular velocity maintains a fixed direction (that of $\mathbf{H}_I$) in inertial space. But Poinsot tells us more. Writing the energy equation in the form $2E_k = \boldsymbol{\omega}_I^{BI} \cdot \mathbf{H}_I$ reveals that the component of $\boldsymbol{\omega}_I^{BI}$ parallel to $\mathbf{H}_I$ has the constant value $2E_k/H$. Thus the tip of the vector $\boldsymbol{\omega}_I^{BI}$ always lies in a fixed plane normal to $\mathbf{H}_I$ at a distance of $2E_k/H$ from the center of the inertia ellipsoid. The path followed by the angular velocity in this *invariant plane* is called the *herpolhode*. Thus the inertia ellipsoid rolls on the invariant plane with its center at a fixed point a distance $2E_k/H$ above the plane and with the tip of the angular velocity vector as the point of contact. Since the angular velocity is the instantaneous axis of rotation, there is no slippage at the contact point, and Poinsot's construction can be succinctly summarized by the statement that the polhode rolls without slipping on the herpolhode lying in the invariant plane.[9] Goldstein [10], Kaplan [17], and Hughes [14] have pictures illustrating Poinsot's construction.

Poinsot's construction is not especially easy to visualize, and there is a simpler picture for the familiar axial symmetry case. In any case, we now turn to analytic solutions of the equations of motion, which are often more useful than pictures for estimation and control applications.

---

[9] A statement aptly characterized by Goldstein [10] as "jabberwockian."

3.3.4 *Torque-Free Motion of a Rigid Body*

Analytic solutions of the rigid body equations of motion are customarily expressed in terms of angular velocity rather than angular momentum. Thus we will consider the component form of Eq. (3.81) in the principal axis frame in the absence of torques:

$$\dot{\omega}_1(t) = [(J_2 - J_3)/J_1]\,\omega_2(t)\,\omega_3(t) \tag{3.93a}$$

$$\dot{\omega}_2(t) = [(J_3 - J_1)/J_2]\,\omega_3(t)\,\omega_1(t) \tag{3.93b}$$

$$\dot{\omega}_3(t) = [(J_1 - J_2)/J_3]\,\omega_1(t)\,\omega_2(t) \tag{3.93c}$$

As we saw in Sect. 3.3.3, the rate of rotation about one of the principal axes is constant if the moments of inertia about the other two principal axes are equal. We will treat this simpler case of axial symmetry first, and then turn to the case of a triaxial inertia tensor.

3.3.4.1 Axial Symmetry

Taking $\mathbf{e}_3$ to be the axis of symmetry of the inertia tensor, we have

$$\dot{\omega}_1(t) = (1 - J_3/J_t)\,\omega_2(t)\,\omega_3(t) \tag{3.94a}$$

$$\dot{\omega}_2(t) = -(1 - J_3/J_t)\,\omega_3(t)\,\omega_1(t) \tag{3.94b}$$

$$\dot{\omega}_3(t) = 0 \tag{3.94c}$$

This means that ω_3 is constant, so we can omit the time argument. Note that ω_3 is exactly constant for axial symmetry, as opposed to its approximate constancy in the stability analysis of Sect. 3.3.3. It follows from Eq. (3.88a) that

$$\omega_3 = \frac{H_3}{J_3} = \pm\left[\frac{H^2 - 2J_t E_k}{J_3(J_3 - J_t)}\right]^{1/2} \tag{3.95}$$

Differentiating Eq. (3.94a) and substituting Eq. (3.94b) gives

$$\ddot{\omega}_1(t) = -(1 - J_3/J_t)^2 \omega_3^2\, \omega_1(t) \tag{3.96}$$

which has the solution

$$\omega_1(t) = \omega_t \sin(\psi_0 + \omega_p t) \tag{3.97}$$

where ψ_0 is a constant initial phase,

$$\omega_p \equiv (1 - J_3/J_t)\omega_3 \tag{3.98}$$

and, from Eq. (3.88b),

$$\omega_t = \frac{H_t}{J_t} = \left[\frac{2J_3E_k - H^2}{J_t(J_3 - J_t)}\right]^{1/2} \tag{3.99}$$

Then $\omega_2(t)$ is given by Eq. (3.94a) as

$$\omega_2(t) = \omega_t \cos(\psi_0 + \omega_p t) \tag{3.100}$$

The angular rate ω_p is called the *body nutation rate* because it is the rate at which the angular velocity vector rotates about the symmetry axis in the body frame. Note that ω_p has the same sign as ω_3 if $J_t > J_3$ and the opposite sign if $J_t < J_3$. If $J_t = J_3$, there is no nutation, only steady rotation, since all axes are principal axes in that case. The angular momentum rotates about the symmetry axis in the body frame at the same rate, because the three vectors $\mathbf{H}, \boldsymbol{\omega}$, and $\mathbf{e}_3$ are coplanar in the case of triaxial symmetry, as is easily seen from the relation

$$\mathbf{H} = J_t \begin{bmatrix} \omega_1 \\ \omega_2 \\ 0 \end{bmatrix} + J_3 \begin{bmatrix} 0 \\ 0 \\ \omega_3 \end{bmatrix} = J_t\boldsymbol{\omega} + (J_3 - J_t)\omega_3\mathbf{e}_3 \tag{3.101}$$

The speed at which the angular momentum vector moves over the sphere of radius H in the body frame can also be computed by

$$\|\dot{\mathbf{H}}_B\|^2 = \|\boldsymbol{\omega}_B^{BI} \times \mathbf{H}_B\|^2 = H^2\|\boldsymbol{\omega}_B^{BI}\|^2 - (\boldsymbol{\omega}_B^{BI} \cdot \mathbf{H}_B)^2 = H^2\|\boldsymbol{\omega}_B^{BI}\|^2 - (2E_k)^2 \tag{3.102}$$

Substituting Eqs. (3.95), (3.99), (3.97) and (3.100) and performing some straightforward algebra gives

$$\|\dot{\mathbf{H}}_B\| = |(J_3 - J_t)\omega_t\omega_3| = H_t|\omega_p| \tag{3.103}$$

It is easy to see that these equations give $\|\dot{\mathbf{H}}_B\| = 0$ for rotation about a principal axis, for which either ω_t or ω_3 is zero.

By using the addition formulas for the sine and cosine, we can express the transverse components of the angular velocity in terms of their initial values:

$$\omega_1(t) = \omega_{01}\cos\omega_p t + \omega_{02}\sin\omega_p t \tag{3.104a}$$

$$\omega_2(t) = \omega_{02}\cos\omega_p t - \omega_{01}\sin\omega_p t \tag{3.104b}$$

where $\omega_{01} = \omega_t \sin\psi_0$ and $\omega_{02} = \omega_t \cos\psi_0$.

We must now solve the kinematic equations of motion in order to have a complete mathematical description of the torque-free motion. We will follow the almost universal practice for torque-free motion of specifying the attitude by 3–1–3 Euler angles. This gives us three first-order equations to integrate, which would lead to three constants of integration in addition to the three we have already found: H, E_k, and ψ_0. Two of these constants can be thought of as specifying the fixed direction of $\mathbf{H}_I$, and we eliminate these by choosing the inertial reference frame with its third axis in the direction of $\mathbf{H}_I$. Then the angular momentum in the body frame is given by

$$\mathbf{H}_B = \begin{bmatrix} J_t\omega_1 \\ J_t\omega_2 \\ J_3\omega_3 \end{bmatrix} = A_{BI}\mathbf{H}_I = A_{313}(\phi, \theta, \psi) \begin{bmatrix} 0 \\ 0 \\ H \end{bmatrix} = H \begin{bmatrix} \sin\psi \sin\theta \\ \cos\psi \sin\theta \\ \cos\theta \end{bmatrix} \tag{3.105}$$

where Eq. (2.162) provides the last equality. We immediately see that the *nutation angle* θ has constant value between 0 and π given by

$$\theta = \cos^{-1}(J_3\,\omega_3/H) \tag{3.106}$$

and comparison of the first two components of Eq. (3.105) with Eqs. (3.97) and (3.100) shows that

$$\sin\theta = J_t\,\omega_t/H \geq 0 \tag{3.107}$$

and

$$\psi = \psi_0 + \omega_p t \tag{3.108}$$

up to an irrelevant multiple of 2π. This explains our choice of the notation ψ_0 in Eq. (3.97). The third Euler angle is found by integrating Eq. (3.38) using Eq. (3.42):

$$\dot{\phi} = \csc\theta(\omega_1 \sin\psi + \omega_2 \cos\psi) = \omega_t \csc\theta = H/J_t \equiv \omega_\ell \tag{3.109}$$

so that

$$\phi = \phi_0 + \omega_\ell t \tag{3.110}$$

The angular rate ω_ℓ is called the *inertial nutation rate* because it is the rate at which the angular velocity vector and the symmetry axis of the rigid body rotate about the fixed angular momentum vector in the inertial frame. Equation (3.98) can be used to write Eq. (3.101) as

$$\boldsymbol{\omega} = \mathbf{H}/J_t + (1 - J_3/J_t)\omega_3\mathbf{e}_3 = \mathbf{H}/J_t + \omega_p\mathbf{e}_3 \tag{3.111}$$

This shows that the angular velocity in the axisymmetric case can be expressed as the sum of two (nonorthogonal) vectors of magnitude ω_ℓ and ω_p.

We complete the analysis of axisymmetric motion by finding the components of the angular velocity in the inertial frame. These are given by

$$\boldsymbol{\omega}_I^{BI} = A_{IB}\boldsymbol{\omega}_B^{BI} = A_{313}^T(\phi,\theta,\psi)\begin{bmatrix}\omega_1\\ \omega_2\\ \omega_3\end{bmatrix}$$

$$= A(\mathbf{e}_3,-\phi)A(\mathbf{e}_1,-\theta)A(\mathbf{e}_3,-\psi)\begin{bmatrix}\omega_t \sin\psi\\ \omega_t\cos\psi\\ \omega_3\end{bmatrix} \tag{3.112}$$

Performing the matrix multiplications and applying Eqs. (2.108), (3.106), and (3.107) gives

$$A(\mathbf{e}_1,-\theta)A(\mathbf{e}_3,-\psi)\begin{bmatrix}\omega_t \sin\psi\\ \omega_t\cos\psi\\ \omega_3\end{bmatrix} = A(\mathbf{e}_1,-\theta)\begin{bmatrix}0\\ \omega_t\\ \omega_3\end{bmatrix}$$

$$= \begin{bmatrix}0\\ \omega_t\cos\theta - \omega_3\sin\theta\\ \omega_3\cos\theta + \omega_t\sin\theta\end{bmatrix} = \frac{1}{H}\begin{bmatrix}0\\ (J_3 - J_t)\omega_t\omega_3\\ 2E_k\end{bmatrix} \tag{3.113}$$

Thus the component of $\boldsymbol{\omega}_I^{BI}$ parallel to $\mathbf{H}_I$ has the constant value $2E_k/H$, as we saw in our discussion of Poinsot's construction; and the component transverse to $\mathbf{H}_I$ has a magnitude agreeing with Eq. (3.103) and rotates around a circular herpolhode at the inertial nutation rate.

We now present a pictorial view of this motion, which is the specialization of Poinsot's construction to the axisymmetric case. The angular velocity vector precesses at rate ω_ℓ and at an angle θ from the angular momentum vector around a *space cone* fixed in the inertial frame. At the same time, the angular velocity vector precesses at rate ω_p and at an angle β from the $\mathbf{e}_3$ axis around a *body cone* fixed in the body frame. These cones are illustrated in Fig. 3.4a for the prolate case and in Fig. 3.4b for the oblate case.[10] The cones are tangent at the angular velocity because $\mathbf{H}$, $\boldsymbol{\omega}$, and $\mathbf{e}_3$ are coplanar, and there is no slippage along the line of tangency because that is the axis around which the rotation takes place. Thus, as viewed from the inertial frame, the body cone (and the spacecraft which is fixed to it) rolls without slipping around the fixed space cone; while, as viewed from the body frame, the space cone (and the universe which is fixed to it) rolls without slipping around the fixed body cone.

The body cone angle β obeys

$$\cos\beta = \omega_3/\|\boldsymbol{\omega}\| \quad \text{and} \quad \sin\beta = \omega_t/\|\boldsymbol{\omega}\| \tag{3.114}$$

[10] The figure illustrates the case of $\omega_3 > 0$, but a corresponding figure for $\omega_3 < 0$ shows that all the discussion of this section holds in that case also.

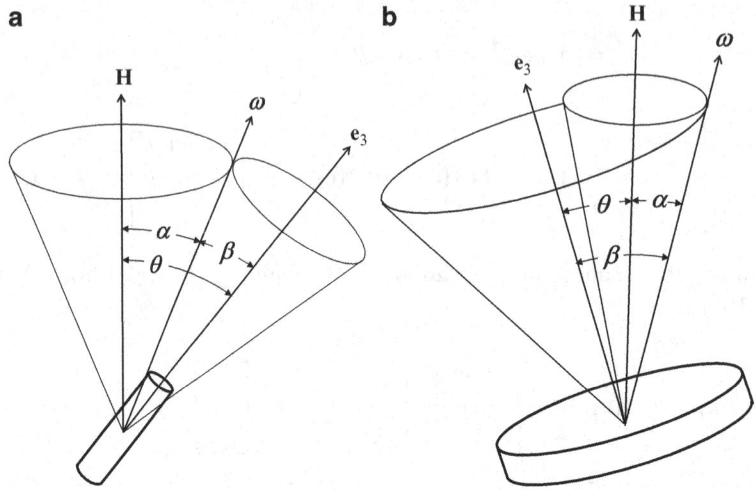

Fig. 3.4 Torque-free motion of an axisymmetric rigid body. (**a**) Prolate rigid body $J_3 < J_t$. (**b**) Oblate rigid body $J_3 > J_t$

and comparison with Eqs. (3.106) and (3.107) shows that the space cone rolls on the outside of the body cone for a prolate rigid body and on the inside for an oblate body, as shown in the figure. This explains why ω_p has the same sign as ω_3 in the prolate case and the opposite sign in the oblate case, in agreement with Eq. (3.98). Finally, we note from the figure that α, the angle between $\mathbf{H}$ and $\boldsymbol{\omega}$, is equal to $|\theta - \beta|$. This relation can be verified algebraically by applying Eqs. (3.106), (3.107) and (3.114) to get

$$H\,\|\boldsymbol{\omega}\|(\cos\beta\cos\theta + \sin\beta\sin\theta) = 2E_k = \mathbf{H}\cdot\boldsymbol{\omega} = H\,\|\boldsymbol{\omega}\|\cos\alpha \qquad (3.115)$$

3.3.4.2 Triaxial Symmetry

We will designate the intermediate principal moment of inertia by J_2 in the triaxial case. We saw in the discussion of Fig. 3.2b that the separatrices divide the motion into two regimes. For $2E_k < H^2/J_2$ the motion has the form of nutation about the principal axis having the largest moment of inertia, and $2E_k > H^2/J_2$ results in nutation about the principal axis with the smallest moment of inertia. Motion along a separatrix is a limiting case of either of these regimes. In order that one analytic formulation will cover both of these cases, we label the principal axes so that nutation is always about $\pm\mathbf{e}_3$. Thus

$$J_1 < J_2 < J_3 \quad \text{if} \quad 2E_k \le H^2/J_2, \qquad (3.116a)$$

$$J_1 > J_2 > J_3 \quad \text{if} \quad 2E_k \ge H^2/J_2 \qquad (3.116b)$$

No component of the angular velocity is constant in the triaxial symmetry case, so the solutions cannot be expressed in terms of sines and cosines. The closed-form solutions are expressed in terms of the Jacobian elliptic functions [1] $\mathrm{sn}(u|m)$, $\mathrm{cn}(u|m)$, and $\mathrm{dn}(u|m)$ with *argument* u and *parameter* m.[11] We define two signs, s_1 and s_3 equal to ± 1, a dimensionless constant

$$\kappa \equiv \left[\frac{J_1(J_3 - J_1)}{J_2(J_3 - J_2)}\right]^{1/2} \tag{3.117}$$

and the angular rates

$$\omega_{1m} \equiv \left[\frac{2J_3 E_k - H^2}{J_1(J_3 - J_1)}\right]^{1/2} \tag{3.118a}$$

$$\omega_{3m} \equiv \left[\frac{H^2 - 2J_1 E_k}{J_3(J_3 - J_1)}\right]^{1/2} \tag{3.118b}$$

Then

$$\omega_1(t) = s_1\, \omega_{1m}\, \mathrm{cn}\left(u_0 + \omega_p t\, |m\right) \tag{3.119a}$$

$$\omega_2(t) = -s_1 \kappa\, \omega_{1m}\, \mathrm{sn}\left(u_0 + \omega_p t\, |m\right) \tag{3.119b}$$

$$\omega_3(t) = s_3\, \omega_{3m}\, \mathrm{dn}\left(u_0 + \omega_p t\, |m\right) \tag{3.119c}$$

where

$$\omega_p = [(J_2 - J_3)/J_1]\kappa s_3\, \omega_{3m} \tag{3.120}$$

and

$$m = \frac{(J_2 - J_1)(2J_3 E_k - H^2)}{(J_3 - J_2)(H^2 - 2J_1 E_k)} \tag{3.121}$$

Note that $m = 0$ for $2E_k = H^2/J_3$, $m = 1$ for $2E_k = H^2/J_2$, and Eq. (3.116) restricts m to always lie between these limits. Figure 3.5a shows the Jacobian elliptic functions for $m = 0.7$, which is the value for $2J_2 E_k = (16/17)H^2$ with the inertia ratios used in Fig. 3.2b.

The Jacobian elliptic functions obey the differential equations

$$\mathrm{sn}'(u|m) = \mathrm{cn}(u|m)\mathrm{dn}(u|m) \tag{3.122a}$$

$$\mathrm{cn}'(u|m) = -\mathrm{sn}(u|m)\mathrm{dn}(u|m) \tag{3.122b}$$

$$\mathrm{dn}'(u|m) = -m\, \mathrm{sn}(u|m)\mathrm{cn}(u|m) \tag{3.122c}$$

---

[11] Some authors, including Hughes [14], use the *modulus* $k \equiv m^{1/2}$ in place of the parameter.

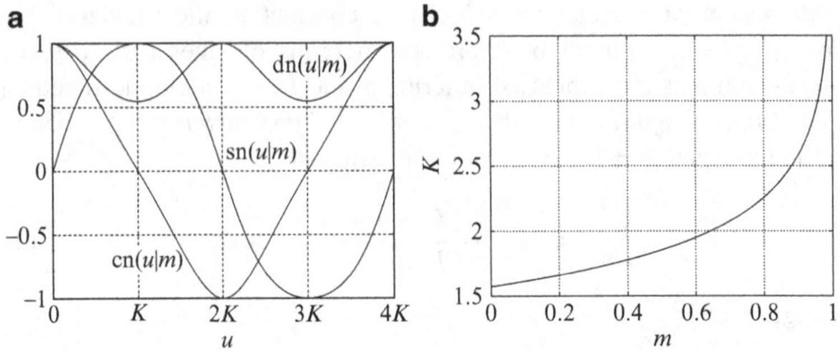

Fig. 3.5 Jacobian elliptic functions. (**a**) Elliptic functions for $m = 0.7$. (**b**) Dependence of the quarter-period K on m

where the prime denotes differentiation with respect to the argument. These equations can be used to show that Eq. (3.119) obeys Eq. (3.93). The Jacobian elliptic functions also satisfy the quadratic relations

$$\mathrm{cn}^2(u|m) + \mathrm{sn}^2(u|m) = \mathrm{dn}^2(u|m) + m\,\mathrm{sn}^2(u|m) = 1, \tag{3.123}$$

which verify that E_k and H are the rotational kinetic energy and angular momentum magnitude, respectively, of the solutions. The Jacobian elliptic functions are related to the usual trigonometric functions by

$$\mathrm{sn}(u|m) = \sin\varphi \tag{3.124a}$$

$$\mathrm{cn}(u|m) = \cos\varphi \tag{3.124b}$$

$$\mathrm{dn}(u|m) = (1 - m\sin^2\varphi)^{1/2} \tag{3.124c}$$

where φ is an implicit function of u and m defined by the integral relation

$$u = \int_0^{\varphi} \frac{d\theta}{(1 - m\sin^2\theta)^{1/2}} \tag{3.125}$$

These equations lead directly to Eqs. (3.122) and (3.123). The quarter-period of the sine or cosine function is $\pi/2$, so the quarter-period of the Jacobian elliptic functions is

$$K(m) = \int_0^{\pi/2} \frac{d\theta}{(1 - m\sin^2\theta)^{1/2}} \tag{3.126}$$

This integral is known as a *complete elliptic integral of the first kind* [1]. Figure 3.5b shows that the quarter-period is an increasing function of m, equal to $\pi/2$ for $m = 0$ and becoming infinite as m goes to 1.

We can use the addition formulas for the Jacobian elliptic functions to write the triaxial solutions in terms of initial values. The addition formulas are, with the parameter m omitted for notational convenience,

$$\mathrm{sn}(u+v) = \frac{\mathrm{sn}u\,\mathrm{cn}v\,\mathrm{dn}v + \mathrm{sn}v\,\mathrm{cn}u\,\mathrm{dn}u}{1 - m\,\mathrm{sn}^2 u\,\mathrm{sn}^2 v} \tag{3.127a}$$

$$\mathrm{cn}(u+v) = \frac{\mathrm{cn}u\,\mathrm{cn}v - \mathrm{sn}u\,\mathrm{dn}u\,\mathrm{sn}v\,\mathrm{dn}v}{1 - m\,\mathrm{sn}^2 u\,\mathrm{sn}^2 v} \tag{3.127b}$$

$$\mathrm{dn}(u+v) = \frac{\mathrm{dn}u\,\mathrm{dn}v - m\,\mathrm{sn}u\,\mathrm{cn}u\,\mathrm{sn}v\,\mathrm{cn}v}{1 - m\,\mathrm{sn}^2 u\,\mathrm{sn}^2 v} \tag{3.127c}$$

The angular rates can then be written in terms of the initial values as

$$\omega_1(t) = \frac{\omega_{01}\,\mathrm{cn}(\omega_p t) + [(J_2 - J_3)/J_1](\omega_{02}\omega_{03}/\omega_p)\mathrm{sn}(\omega_p t)\mathrm{dn}(\omega_p t)}{1 - [(J_2 - J_3)/J_1][(J_1 - J_2)/J_3](\omega_{02}/\omega_p)^2\mathrm{sn}^2(\omega_p t)} \tag{3.128a}$$

$$\omega_2(t) = \frac{\omega_{02}\,\mathrm{cn}(\omega_p t)\mathrm{dn}(\omega_p t) + [(J_3 - J_1)/J_2](\omega_{03}\omega_{01}/\omega_p)\mathrm{sn}(\omega_p t)}{1 - [(J_2 - J_3)/J_1][(J_1 - J_2)/J_3](\omega_{02}/\omega_p)^2\mathrm{sn}^2(\omega_p t)} \tag{3.128b}$$

$$\omega_3(t) = \frac{\omega_{03}\,\mathrm{dn}(\omega_p t) + [(J_1 - J_2)/J_3](\omega_{01}\omega_{02}/\omega_p)\mathrm{sn}(\omega_p t)\mathrm{cn}(\omega_p t)}{1 - [(J_2 - J_3)/J_1][(J_1 - J_2)/J_3](\omega_{02}/\omega_p)^2\mathrm{sn}^2(\omega_p t)} \tag{3.128c}$$

Comparison with Eq. (3.93) suggests the alternative form

$$\omega_1(t) = \frac{\omega_{01}\,\mathrm{cn}(\omega_p t) + (\dot{\omega}_{01}/\omega_p)\mathrm{sn}(\omega_p t)\mathrm{dn}(\omega_p t)}{1 - \omega_p^{-2}(\dot{\omega}_{01}/\omega_{03})(\dot{\omega}_{03}/\omega_{01})\mathrm{sn}^2(\omega_p t)} \tag{3.129a}$$

$$\omega_2(t) = \frac{\omega_{02}\,\mathrm{cn}(\omega_p t)\mathrm{dn}(\omega_p t) + (\dot{\omega}_{02}/\omega_p)\mathrm{sn}(\omega_p t)}{1 - \omega_p^{-2}(\dot{\omega}_{01}/\omega_{03})(\dot{\omega}_{03}/\omega_{01})\mathrm{sn}^2(\omega_p t)} \tag{3.129b}$$

$$\omega_3(t) = \frac{\omega_{03}\,\mathrm{dn}(\omega_p t) + (\dot{\omega}_{03}/\omega_p)\mathrm{sn}(\omega_p t)\mathrm{cn}(\omega_p t)}{1 - \omega_p^{-2}(\dot{\omega}_{01}/\omega_{03})(\dot{\omega}_{03}/\omega_{01})\mathrm{sn}^2(\omega_p t)} \tag{3.129c}$$

but neither form is especially convenient for computation.

Equations (3.124) and (3.125) show that the Jacobian elliptic functions are equal to the usual trigonometric functions for $m = 0$:

$$\mathrm{sn}(u|0) = \sin u \tag{3.130a}$$

$$\mathrm{cn}(u|0) = \cos u \tag{3.130b}$$

$$\mathrm{dn}(u|0) = 1 \tag{3.130c}$$

They are equal to hyperbolic functions for $m = 1$:

$$\mathrm{sn}(u|1) = \tanh u \tag{3.131a}$$

$$\mathrm{cn}(u|1) = \mathrm{sech}\, u \tag{3.131b}$$

$$\mathrm{dn}(u|1) = \mathrm{sech}\, u \tag{3.131c}$$

which give the solution for motion on the separatrices with $2E_k = H^2/J_2$. These solutions are not periodic, because the quarter-period is infinite for $m = 1$; they approach pure rotation about the principal axis $\mathbf{e}_2$ in the infinite past and future.

The speed at which the angular momentum vector moves over the sphere of radius H in the body frame is also given by Eq. (3.102) in the triaxial case, but substituting Eq. (3.119) shows that the speed is not constant in this case. However, nice expressions can be found for the maximum and minimum speeds. The maximum, reached when $\omega_2 = 0$ or equivalently where the path of the angular momentum crosses the plane perpendicular to $\mathbf{e}_2$, is

$$\|\dot{\mathbf{H}}_B\|_{\max} = (J_3 - J_1)\omega_{1m}\omega_{3m} \tag{3.132}$$

The minimum, reached when $|\omega_2|$ obtains its maximum value on the path, is

$$\|\dot{\mathbf{H}}_B\|^2_{\min} = (H^2 - 2J_2E_k)(2J_3E_k - H^2)/(J_2J_3) \tag{3.133}$$

The maximum speed has the same form as for an axisymmetric body, but the minimum speed is specific to the triaxial case. It is not difficult to see that these equations give $\|\dot{\mathbf{H}}_B\| = 0$ for rotation about a principal axis, for which $2E_k = H^2/J_k$.

Solution of the kinematic equations proceeds in parallel with the axisymmetric case. We again specify the attitude by 3–1–3 Euler angles and choose the inertial reference frame with its third axis in the direction of $\mathbf{H}_I$. Then Eq. (3.105) holds, and substituting Eq. (3.119) gives

$$\begin{bmatrix} J_1\,\omega_1 \\ J_2\,\omega_2 \\ J_3\,\omega_3 \end{bmatrix} = \begin{bmatrix} s_1 J_1\,\omega_{1m}\,\mathrm{cn}\left(u_0 + \omega_p t\;|m\right) \\ -s_1 J_2 \kappa\,\omega_{1m}\,\mathrm{sn}\left(u_0 + \omega_p t\;|m\right) \\ s_3 J_3\,\omega_{3m}\,\mathrm{dn}\left(u_0 + \omega_p t\;|m\right) \end{bmatrix} = H \begin{bmatrix} \sin\psi\sin\theta \\ \cos\psi\sin\theta \\ \cos\theta \end{bmatrix} \tag{3.134}$$

Thus

$$\cos\theta = (s_3 J_3\,\omega_{3m}/H)\mathrm{dn}\left(u_0 + \omega_p t\;|m\right) \tag{3.135a}$$

$$\psi = \mathrm{atan2}\left(s_1 J_1 \mathrm{cn}\left(u_0 + \omega_p t\;|m\right), -s_1 J_2 \kappa\,\mathrm{sn}\left(u_0 + \omega_p t\;|m\right)\right) \tag{3.135b}$$

The third Euler angle is found by integrating Eq. (3.38) using Eq. (3.42):

$$\begin{aligned} \dot{\phi} &= \csc\theta(\omega_1\sin\psi + \omega_2\cos\psi) = H\frac{J_1\omega_1^2 + J_2\omega_2^2}{J_1^2\omega_1^2 + J_2^2\omega_2^2} \\ &= H\frac{J_3 - J_2 + (J_2 - J_1)\mathrm{sn}^2(u_0 + \omega_p t\;|m)}{J_1(J_3 - J_2) + J_3(J_2 - J_1)\mathrm{sn}^2(u_0 + \omega_p t\;|m)} \end{aligned} \tag{3.136}$$

The quantities θ, $\dot{\psi}$, and $\dot{\phi}$ are all time-varying for triaxial inertia.

We complete the analysis of motion for triaxial inertia by finding the components of the angular velocity in the inertial frame. These are given by Eq. (3.112). Performing the matrix multiplications gives

$$A(\mathbf{e}_1, -\theta)A(\mathbf{e}_3, -\psi)\begin{bmatrix}\omega_1\\ \omega_2\\ \omega_3\end{bmatrix} = A(\mathbf{e}_1, -\theta)\begin{bmatrix}(J_2 - J_1)\,\omega_1\omega_2/(H\sin\theta)\\ (J_1\omega_1^2 + J_2\omega_2^2)/(H\sin\theta)\\ \omega_3\end{bmatrix}$$

$$= \frac{1}{H}\begin{bmatrix}(J_2 - J_1)\,\omega_1\omega_2\csc\theta\\ (J_1^2\omega_1^2 + J_2^2\omega_2^2)^{-1/2}J_1(J_3 - J_1)\omega_{1m}^2\omega_3\\ 2E_k\end{bmatrix} \quad (3.137)$$

The component of $\boldsymbol{\omega}_I^{BI}$ parallel to $\mathbf{H}_I$ has the constant value $2E_k/H$, as required by Poinsot's construction and as we saw for axial symmetry. The component of $\boldsymbol{\omega}_I^{BI}$ transverse to $\mathbf{H}_I$ has a varying magnitude, so the herpolhode is not a circle. In fact, the herpolhode is not a closed curve in general, because the motion of ϕ is not commensurate with that of θ and ψ.

It is not difficult to see that all the equations for the triaxial case reduce to the axisymmetric results, with $\psi_0 = (u_0 + s_1\pi/2)$, if $J_1 = J_2$.

3.3.5 *Internal Torques*

Although some spacecraft can be modeled as a single rigid body, many are more complex. These complexities can be of several types. The first is that the spacecraft consists of a number of rigid bodies connected by joints having one, two, or three degrees of rotational freedom, and sometimes admitting sliding motion as well. Powerful and general commercial software packages, many employing Kane's method [15, 16] are available to analyze the dynamics of these systems. We will only treat the simple cases of reaction wheels and control moment gyros, the more complex systems being beyond the scope of this text.

The rigid model of a spacecraft will also be inadequate if we cannot ignore flexibility, which is always present at some level. Analysis of flexible body dynamics generally uses finite element methods [30], which are almost always applied to the analysis of large spacecraft but are also beyond the scope of this text.

A third complication often encountered is fluid motion, or *slosh* of liquid fuels or cryogens. We will only give a basic introduction to the treatment of slosh, with references to the literature.

These internal torques are also known as *momentum exchange* torques because they result in, or are a result of, the exchange of angular momentum between components of complex spacecraft without a change in the net system momentum of the entire spacecraft. As the listing above indicates, some internal torques constitute undesirable disturbances, while others are provided by control mechanisms.

3.3.5.1 Reaction Wheels and Control Moment Gyros

Let us consider a spacecraft with n reaction wheels or control moment gyros, labeled by an index ℓ. Each wheel rotates about its spin axis with an angular velocity ω_ℓ^w with respect to the body. The wheel is axially symmetric about its spin axis, so the spin axis is a principal axis with principal moment of inertia $J_\ell^\parallel$, and every axis perpendicular to the spin axis is a principal axis with moment $J_\ell^\perp$.[12] Thus the inertia tensor of the ℓth wheel in the body reference frame is

$$J_\ell^w = J_\ell^\perp (I_3 - \mathbf{w}_\ell \mathbf{w}_\ell^T) + J_\ell^\parallel \mathbf{w}_\ell \mathbf{w}_\ell^T \tag{3.138}$$

where the unit vector $\mathbf{w}_\ell$ defines the spin axis in the body frame. Letting $\tilde{J}_B$ represent the moment of inertia of the spacecraft *without* the wheels, the body frame representation of total angular momentum with respect to inertial space of the spacecraft with its wheels is

$$\begin{aligned}\mathbf{H}_B &= \tilde{J}_B \boldsymbol{\omega}_B^{BI} + \sum_{\ell=1}^{n} J_\ell^w (\boldsymbol{\omega}_B^{BI} + \omega_\ell^w \mathbf{w}_\ell) \\ &= J_B \boldsymbol{\omega}_B^{BI} + \mathbf{H}_B^w \end{aligned} \tag{3.139}$$

where

$$J_B \equiv \tilde{J}_B + \sum_{\ell=1}^{n} J_\ell^\perp (I_3 - \mathbf{w}_\ell \mathbf{w}_\ell^T) \tag{3.140}$$

and

$$\mathbf{H}_B^w = \sum_{\ell=1}^{n} J_\ell^\parallel (\mathbf{w}_\ell \cdot \boldsymbol{\omega}_B^{BI} + \omega_\ell^w) \mathbf{w}_\ell \equiv \sum_{\ell=1}^{n} H_\ell^w \mathbf{w}_\ell \tag{3.141}$$

The body moment of inertia J_B is defined to include the inertia of the wheels transverse to their spin axes, but not their inertia along their spin axes. Note that $\mathbf{H}_B^w$ denotes only the angular momentum of the wheels along their spin axes; the momentum of their transverse rotation is included in $J_B \boldsymbol{\omega}_B^{BI}$. The difference between reaction wheels and control moment gyros is that $\mathbf{w}_\ell$ is fixed and ω_ℓ^w is changing for reaction wheels, while ω_ℓ^w is constant and $\mathbf{w}_\ell$ is moved by a gimbal (or gimbals) for control moment gyros. Some control wheels have been designed to vary both $\mathbf{w}_\ell$ and ω_ℓ^w, but these have not been widely employed.

[12]This describes an ideal wheel. The effects of deviations from this ideal case will be discussed in Sect. 4.8.

The rotational dynamics of the spacecraft with wheels are still described by Eq. (3.80) but with Eq. (3.79c) replaced by

$$\boldsymbol{\omega}_B^{BI} = J_B^{-1}(\mathbf{H}_B - \mathbf{H}_B^w) \tag{3.142}$$

where the superscript c is understood but has been omitted for economy of notation. For many attitude determination problems, $\boldsymbol{\omega}_B^{BI}$ and $\mathbf{H}_B^w$ can be computed from Eqs. (3.141) and (3.142) using tachometer data for reaction wheels or axis orientation data for control moment gyros.[13] To complete the dynamics analysis, however, a dynamic equation for $\mathbf{H}_B^w$ is required.

The rotational kinetic energy of a spacecraft with reaction wheels or control moment gyros is given by

$$\begin{aligned} E_k &= \frac{1}{2}(\boldsymbol{\omega}_B^{BI})^T \tilde{J}_B \,\boldsymbol{\omega}_B^{BI} + \frac{1}{2}\sum_{\ell=1}^{n}(\boldsymbol{\omega}_B^{BI} + \omega_\ell^w \mathbf{w}_\ell)^T J_\ell^w (\boldsymbol{\omega}_B^{BI} + \omega_\ell^w \mathbf{w}_\ell) \\ &= \frac{1}{2}(\boldsymbol{\omega}_B^{BI})^T J_B \,\boldsymbol{\omega}_B^{BI} + \frac{1}{2}\sum_{\ell=1}^{n} J_\ell^{\|}(\mathbf{w}_\ell \cdot \boldsymbol{\omega}_B^{BI} + \omega_\ell^w)^2 \\ &= \frac{1}{2}(\boldsymbol{\omega}_B^{BI})^T J_B \,\boldsymbol{\omega}_B^{BI} + \frac{1}{2}\sum_{\ell=1}^{n}(J_\ell^{\|})^{-1}(H_\ell^w)^2 \end{aligned} \tag{3.143}$$

Now let us specialize to reaction wheels, which have been more commonly employed in small to medium size spacecraft than control moment gyros. The representations in the body frame of the angular momenta of the wheels are the terms in the sum in Eq. (3.139). It follows that the equation of motion of the ℓth wheel in the body frame, using Eq. (3.138), is

$$\begin{aligned} J_\ell^w(\dot{\boldsymbol{\omega}}_B^{BI} + \dot{\omega}_\ell^w \mathbf{w}_\ell) &= \tilde{\mathbf{L}}_\ell^w - \boldsymbol{\omega}_B^{BI} \times [J_\ell^w(\boldsymbol{\omega}_B^{BI} + \omega_\ell^w \mathbf{w}_\ell)] \\ &= \tilde{\mathbf{L}}_\ell^w + [J_\ell^{\perp}(\mathbf{w}_\ell \cdot \boldsymbol{\omega}_B^{BI}) - H_\ell^w](\boldsymbol{\omega}_B^{BI} \times \mathbf{w}_\ell) \end{aligned} \tag{3.144}$$

where $\tilde{\mathbf{L}}_\ell^w$ is the applied torque. The other term on the right side of this equation is perpendicular to the spin axis; it is provided by the wheel bearings and is not under our direct control. Thus we are only interested in the spin axis component of the torque, which is denoted by L_ℓ^w and is given by

$$L_\ell^w = \mathbf{w}_\ell^T J_\ell^w(\dot{\boldsymbol{\omega}}_B^{BI} + \dot{\omega}_\ell^w \mathbf{w}_\ell) = J_\ell^{\|}(\mathbf{w}_\ell \cdot \dot{\boldsymbol{\omega}}_B^{BI} + \dot{\omega}_\ell^w) = \dot{H}_\ell^w \tag{3.145}$$

Then Eq. (3.141) gives[14]

[13] Giving $\boldsymbol{\omega}_B^{BI} = (J_B + \sum_{\ell=1}^n J_\ell^{\|}\mathbf{w}_\ell\mathbf{w}_\ell^T)^{-1}(\mathbf{H}_B - \sum_{\ell=1}^n J_\ell^{\|}\omega_\ell^w\mathbf{w}_\ell)$ and $\mathbf{H}_B^w = \mathbf{H}_B - J_B\boldsymbol{\omega}_B^{BI}$.

[14] Including an $\boldsymbol{\omega}_B^{BI} \times \mathbf{H}_B^w$ term in this equation would be double-counting, because this term was already accounted for in Eq. (3.144). It can be seen explicitly in Eq. (3.147).

$$\dot{\mathbf{H}}_B^w = \sum_{\ell=1}^{n} \dot{H}_\ell^w \mathbf{w}_\ell = \sum_{\ell=1}^{n} L_\ell^w \mathbf{w}_\ell \equiv \mathbf{L}_B^w \tag{3.146}$$

We find the generalization of Eq. (3.81) to include reaction wheels by substituting this and Eq. (3.139) into Eq. (3.80), yielding

$$\dot{\boldsymbol{\omega}}_B^{BI} = J_B^{-1}[\mathbf{L}_B - \mathbf{L}_B^w - \boldsymbol{\omega}_B^{BI} \times (J_B \boldsymbol{\omega}_B^{BI} + \mathbf{H}_B^w)] \tag{3.147}$$

The negative sign before $\mathbf{L}_B^w$ on the right side reflects Newton's third law of motion.

The rate of change of the kinetic energy is given by the derivative of Eq. (3.143),

$$\begin{aligned} \dot{E}_k &= \boldsymbol{\omega}_B^{BI} \cdot [\mathbf{L}_B - \mathbf{L}_B^w - \boldsymbol{\omega}_B^{BI} \times \mathbf{H}_B] + \sum_{\ell=1}^{n} (\mathbf{w}_\ell \cdot \boldsymbol{\omega}_B^{BI} + \omega_\ell^w) L_\ell^w \\ &= \boldsymbol{\omega}_B^{BI} \cdot \mathbf{L}_B + \sum_{\ell=1}^{n} \omega_\ell^w L_\ell^w \end{aligned} \tag{3.148}$$

This shows that the change in rotational kinetic energy is the sum of work done by the external torques and by the internal torques, a concrete illustration of the observation made near the end of Sect. 3.3.1 that internal forces can modify the energy of rotational motion.

3.3.5.2 Slosh

Spacecraft often contain fluids, the most common being propellants or cryogens for cooling scientific instruments. The motion of these fluids, commonly called *slosh*, is often a source of undesirable attitude disturbances. Slosh is very difficult to analyze, and we will only provide a very brief introduction to some of the issues; the monograph by Dodge provides a much more thorough discussion [8].

Fuel slosh can be a major problem for spin-stabilized spacecraft. In particular, the energy dissipated by sloshing liquid can destabilize the motion of a spacecraft spinning about an axis of minimum MOI, as discussed in Sect. 3.3.3.2. Slosh is also a problem for repointing maneuvers of fine-pointing spacecraft. A strong coupling of the fluid motion to the rigid spacecraft can cause large pointing errors. Weak coupling of the fluid motion, on the other hand, leads to smaller pointing errors, but the perturbations take a correspondingly longer time to damp out. The resulting long resettling times after attitude maneuvers decrease the time available for fine-pointing observations.

The main dynamic effects of slosh result from the motion of the center of mass of the moving fluid. This is usually represented by the mechanical model of a pendulum or a mass on a spring to represent the motion of the fluid, along with some mechanism to provide damping. Dodge presents methods for computing the frequencies and damping coefficients for fluid containers of various shapes, including a variety of baffles, diaphragms, and fluid management devices. The *fill factor* of the container, the ratio of the amount of fluid contained to the capacity of the full tank, is a critical factor. Slosh is much less of a factor for a nearly full tank, because the center of mass cannot move much, or for a nearly empty tank, because there is less moving mass.

Fluid motion in microgravity is different from motion under acceleration forces. The relative importance of inertia forces, gravity forces, and capillary forces on fluid motion are characterized by the *Bond number*, the *Weber number*, and the *Froude number*, while the *Reynolds number* determines the importance of viscosity. All these considerations are important for providing a slosh model from first principles, but it is quite common for the parameters in a slosh model to be determined empirically.

3.3.6 External Torques

External torques involve an interaction with entities external to the spacecraft. As opposed to internal torques, external torques change the overall momentum of the spacecraft. In common with internal torques, external torques include both undesirable disturbance torques and torques deliberately applied for control.

3.3.6.1 Gravity-Gradient Torque

Any nonsymmetrical rigid body in a gravity field is subject to a gravity-gradient torque. We compute this torque by summing the contributions of the gravitational forces on the various point masses constituting the rigid body as in Eq. (3.57). The gravitational force on the ith particle is

$$\mathbf{F}^{iext} = m_i \mathbf{g}(\mathbf{r}^{i0}) = m_i \; \nabla_{\mathbf{r}} U(\mathbf{r})|_{\mathbf{r}=\mathbf{r}^{i0}} \tag{3.149}$$

where $U(\mathbf{r})$ is the gravity potential.[15] Assuming that only first-order variations in the gravitational field over the rigid body are significant allows us to expand in a power series, retaining only the first two terms:

$$\mathbf{F}^{iext} = m_i \mathbf{g}(\mathbf{r}^{ic} + \mathbf{r}^{c0}) = m_i [\mathbf{g}(\mathbf{r}^{c0}) + G(\mathbf{r}^{c0})\mathbf{r}^{ic}] \tag{3.150}$$

where

$$G(\mathbf{r}^{c0}) \equiv \left.\frac{\partial \mathbf{g}(\mathbf{r})}{\partial \mathbf{r}}\right|_{\mathbf{r}=\mathbf{r}^{c0}} = \left.\frac{\partial^2 U(\mathbf{r})}{\partial \mathbf{r}\, \partial \mathbf{r}^T}\right|_{\mathbf{r}=\mathbf{r}^{c0}} \tag{3.151}$$

[15] As is customary in astrodynamics, this is the *negative* of the potential energy.

is the gravity-gradient tensor, evaluated at the center of mass. It follows from Eq. (3.151) that the gravity-gradient tensor is a symmetric 3×3 matrix.

Substituting into Eq. (3.57) gives the gravity-gradient torque about the center of mass as

$$\begin{aligned} \mathbf{L}^c_{gg} &= \sum_{i=1}^{n} m_i \mathbf{r}^{ic} \times \left[\mathbf{g}(\mathbf{r}^{c0}) + G(\mathbf{r}^{c0})\mathbf{r}^{ic}\right] \\ &= \left(\sum_{i=1}^{n} m_i \mathbf{r}^{ic}\right) \times \mathbf{g}(\mathbf{r}^{c0}) + \sum_{i=1}^{n} m_i \mathbf{r}^{ic} \times \left[G(\mathbf{r}^{c0})\mathbf{r}^{ic}\right] \\ &= \sum_{i=1}^{n} m_i \mathbf{r}^{ic} \times \left[G(\mathbf{r}^{c0})\mathbf{r}^{ic}\right] \end{aligned} \tag{3.152}$$

using the property of the center of mass expressed in Eq. (3.55).

It is usually adequate to approximate the gravity field as spherically symmetric for computing gravity-gradient torques. In this case, we have

$$\mathbf{g}(\mathbf{r}) = -\frac{\mu \mathbf{r}}{r^3} \tag{3.153}$$

and therefore

$$G(\mathbf{r}) = -\frac{\mu}{r^3}\left(I_3 - 3\frac{\mathbf{r}\mathbf{r}^T}{r^2}\right) \tag{3.154}$$

where μ is the gravitational parameter of the central body, the product of its mass and Newton's universal gravitational constant, $\mathbf{r}$ is the radius vector from the center of the central body, and $r \equiv \|\mathbf{r}\|$. Inserting this into Eq. (3.152), letting $\mathbf{r}^{c0} = -r\mathbf{n}$, where $\mathbf{n}$ is the body frame representation of a nadir-pointing unit vector, and using various identities for dot and cross products gives

$$\begin{aligned} \mathbf{L}^c_{gg} &= -\frac{\mu}{r^3}\sum_{i=1}^{n} m_i \mathbf{r}^{ic} \times \left(\mathbf{r}^{ic} - 3\mathbf{n}\mathbf{n}^T\mathbf{r}^{ic}\right) = -\frac{3\mu}{r^3}\mathbf{n} \times \sum_{i=1}^{n} m_i (\mathbf{r}^{ic})(\mathbf{r}^{ic})^T\mathbf{n} \\ &= \frac{3\mu}{r^3}\mathbf{n} \times \sum_{i=1}^{n} m_i \left[\|\mathbf{r}^{ic}\|^2 I_3 - (\mathbf{r}^{ic})(\mathbf{r}^{ic})^T\right]\mathbf{n} = \frac{3\mu}{r^3}\mathbf{n} \times (J^c\mathbf{n}) \end{aligned} \tag{3.155}$$

where J^c is the moment of inertia tensor about the center of mass. Several properties of the gravity-gradient torque are apparent from this equation: its magnitude is inversely proportional to the cube of the distance from the center of the central body, its direction is perpendicular to the radius vector from the central body, and it vanishes if the radius vector is along any principal axis of inertia.

To investigate the gravity-gradient torque in more detail, consider a spacecraft orbiting the Earth (or any celestial body) and oriented to the nadir. We represent the attitude relative to the LVLH frame[16] by a 3–2–1 Euler angle sequence of yaw $= \phi$, pitch $= \theta$, and roll $= \psi$ angles. The nadir vector is the $\mathbf{e}_3$ axis in the LVLH frame, so the nadir vector in the body frame is given by Eq. (2.164) as

$$\mathbf{n} = A_{321}(\phi, \theta, \psi) \begin{bmatrix} 0 \\ 0 \\ 1 \end{bmatrix} = \begin{bmatrix} -\sin\theta \\ \cos\theta \sin\psi \\ \cos\theta \cos\psi \end{bmatrix} \tag{3.156}$$

Substituting this into Eq. (3.155) and assuming that the body frame is a principal axis frame gives

$$\mathbf{L}^c_{gg} = \frac{3\mu}{r^3} \begin{bmatrix} (J_3 - J_2)\cos^2\theta \cos\psi \sin\psi \\ (J_3 - J_1)\cos\theta \sin\theta \cos\psi \\ (J_1 - J_2)\cos\theta \sin\theta \sin\psi \end{bmatrix} \tag{3.157}$$

We notice that the torque does not depend on the yaw angle, which is one of the reasons for choosing a $3-2-1$ sequence to specify the attitude.[17] The yaw angle can be quite large for some missions, with 180° yaw maneuvers used to keep one side of the spacecraft cool by facing it away from the Sun. The roll and pitch angles are usually small, though, and almost certainly less than 90° in magnitude. It is clear that the gravity-gradient torque vanishes if the roll and pitch are both zero, so this is an equilibrium configuration. For small pitch and roll angles, the third (yaw) component of the gravity-gradient torque is small, so we will not be concerned with it. Comparing Eqs. (3.156) and (3.157) shows that the first (roll) component of the gravity-gradient torque will drive the roll angle toward zero if $J_3 < J_2$ and away from zero if $J_3 > J_2$. Similarly, the second (pitch) component of the gravity-gradient torque will drive the pitch angle toward zero if $J_3 < J_1$ and away from zero if $J_3 > J_1$. Thus the equilibrium at zero roll and pitch is a stable equilibrium if J_3 is the smallest principal moment, otherwise the equilibrium is unstable for rotations about one or both axes. Another way of stating this is that gravity-gradient torque will tend to align a spacecraft with its principal axis of minimum inertia aligned with the nadir vector.

Gravity-gradient torques are often used for passive stabilization of a spacecraft. A *gravity-gradient boom* with a mass at the end can be deployed along the positive or negative yaw axis to increase the J_1 and J_2 moments of inertia relative to the J_3 inertia about the desired nadir-pointing axis. The boom deployment must be carefully timed to avoid an inverted orientation, with the desired nadir-pointing axis pointing in the zenith direction. Pendular motions, known as *libration*, can be damped out by energy-dissipating *libration dampers*, which are very similar in design and function to nutation dampers. Finally, we note that the gravity-gradient torque cannot provide stability against rotations around the nadir vector. These are controlled either by active means or by employing a momentum wheel to provide a

---

[16]The LVLH frame is defined in Sect. 2.6.4.

[17]A $3-1-2$ sequence would serve as well.

momentum bias along the pitch axis. Issues arising from the deployment of gravity-gradient booms are described on pp. 669–677 of [19].

To handle cases more general than a spherically symmetric gravity field, we can use Eq. (2.56c) to find an expression for the gravity-gradient torque for a general gravity-gradient tensor. This identity gives, suppressing the argument of G for notational convenience,

$$\begin{aligned}
[\mathbf{L}^c_{gg}\times] &= \sum_{i=1}^{n} m_i[(\mathbf{r}^{ic}\times G\mathbf{r}^{ic})\times] = \sum_{i=1}^{n} m_i[G\mathbf{r}^{ic}(\mathbf{r}^{ic})^T - \mathbf{r}^{ic}(G\mathbf{r}^{ic})^T] \\
&= \sum_{i=1}^{n} m_i\left[\|\mathbf{r}^{ic}\|^2 I_3 - (\mathbf{r}^{ic})(\mathbf{r}^{ic})^T\right]G^T - G\sum_{i=1}^{n} m_i\left[\|\mathbf{r}^{ic}\|^2 I_3 - (\mathbf{r}^{ic})(\mathbf{r}^{ic})^T\right] \\
&= (GJ^c)^T - GJ^c \qquad (3.158)
\end{aligned}$$

making use of Eq. (2.7c) and the fact that both G and J^c are symmetric.

3.3.6.2 Magnetic Torque

The torque generated by a magnetic dipole $\mathbf{m}$ in a magnetic field $\mathbf{B}$ is

$$\mathbf{L}_{\text{mag}} = \mathbf{m}\times\mathbf{B} \qquad (3.159)$$

The most basic source of a magnetic dipole is a current loop. A current of I amperes flowing in a planar loop of area A produces a dipole moment of magnitude $m = IA$ in the direction normal to the plane of the loop and satisfying a right-hand rule. It follows from this definition that the natural unit for the dipole moment is Am^2. When $\mathbf{m}$ is in Am^2 and the magnetic field is specified in Tesla, Eq. (3.159) gives the torque in Nm. If there are N turns of wire in the loop, the dipole moment has magnitude $m = NIA$. The dipole moment can be significantly increased by wrapping the wire loops around a ferromagnetic core, as will be discussed in Sect. 4.10.

Magnetic control torques are used almost exclusively in near-Earth orbits, where the magnitude of the Earth's magnetic field is roughly in the range of 20–50 μT. Commercially available torquers can provide dipole moments from 1 to 1,000 Am^2, so the resulting magnetic control torques range from 2×10^{-5} to 0.05 Nm. As described in Sect. 11.1, the field strength falls off as the inverse cube of the distance from the center of the Earth, so magnetic control has rarely been employed in higher orbits, but it has sometimes been used even in geosynchronous orbits. Undesirable magnetic dipoles can lead to magnetic disturbance torques, which are generally several orders of magnitude smaller than the above estimates of control torques.

It is customary to express $\mathbf{L}_{\text{mag}}$, $\mathbf{m}$, and $\mathbf{B}$ in body-frame coordinates. There are two ways to determine the body-frame magnetic field. The first employs measurements by an onboard three-axis magnetometer (TAM), as described in Sect. 4.5. The other method is to compute the magnetic field vector, $\mathbf{R}$, in reference-frame coordinates from a model, either a simple dipole model, the International Geomagnetic Reference Field (IGRF), or some truncated version of the IGRF, as described in Sect. 11.1. The attitude matrix then rotates the field to the body frame by $\mathbf{B} = A\mathbf{R}$. In addition to an attitude estimate, this method requires an onboard ephemeris, which in modern spacecraft would be provided by GPS, as well as an onboard magnetic field model.

One advantage of magnetic torques is that they produce no force, so they do not perturb the spacecraft's orbit. A significant disadvantage is that the torques are constrained to lie in the plane orthogonal to the magnetic field, as is clear from Eq. (3.159), so only two out of three axes can be controlled at a given time instant. However, full three-axis control is available over a complete orbit provided that the spacecraft's orbital plane does not coincide with the geomagnetic equatorial plane and does not contain the magnetic poles [3]. The Earth's rotation causes this geometry to change, so any simulation involving magnetic control should be at least 24 h in length to ensure that unfavorable magnetic field geometry does not cause a problem at some point.[18]

Because magnetic torques cannot provide three-axis control at any instant of time, they are generally employed in conjunction with some other form of attitude control. This can be passive control, such as spin stabilization [9, 11, 13, 23] or gravity-gradient stabilization [7,28,29], but it is more common to employ magnetic control in conjunction with reaction wheels. In this application, the wheels provide the actual pointing and maneuvering torques, and magnetic torques are used to unload the secular angular momentum buildup in the wheels. Reference [6] provides an analysis of the orbit-averaged behavior of magnetic control for unloading angular momentum. Some specific examples of magnetic control laws can be found in Chap. 7.

3.3.6.3 Aerodynamic Torque

For objects in low-Earth orbit, atmospheric drag is a significant source of perturbing torque. Aerodynamic torque is generally computed by modeling the spacecraft as a collection of N flat plates of area S_i and outward normal unit vector $\mathbf{n}_B^i$ expressed in the spacecraft body-fixed coordinate system. The torque depends on the velocity of the spacecraft relative to the atmosphere. This is not simply the velocity of the spacecraft in the GCI frame, because the atmosphere is not stationary in that frame. The most common assumption is that the atmosphere co-rotates with the Earth. The relative velocity in the GCI frame is then given by

$$\mathbf{v}_{\text{rel}I} = \mathbf{v}_I + [\boldsymbol{\omega}_{\oplus I}\times]\,\mathbf{r}_I \tag{3.160}$$

[18]This point was often emphasized by Henry Hoffman of Goddard Space Flight Center.

where $\mathbf{r}_I$ and $\mathbf{v}_I$ are the position and velocity of the spacecraft expressed in the GCI coordinate frame. The Earth's angular velocity vector is $\boldsymbol{\omega}_{\oplus I} = \omega_\oplus [0\ \ 0\ \ 1]^T$ with $\omega_\oplus = 0.000\ 072\ 921\ 158\ 553$ rad/s. Inserting this $\boldsymbol{\omega}_{\oplus I}$ gives the relative velocity in the body frame as

$$\mathbf{v}_{\text{rel}\,B} = A \begin{bmatrix} \dot{x} + \omega_\oplus y \\ \dot{y} - \omega_\oplus x \\ \dot{z} \end{bmatrix} \tag{3.161}$$

where A is the attitude matrix. The inclination of the ith plate to the relative velocity is given by

$$\cos\theta^i_{\text{aero}} = \frac{\mathbf{n}^i_B \cdot \mathbf{v}_{\text{rel}\,B}}{\|\mathbf{v}_{\text{rel}}\|} \tag{3.162}$$

The aerodynamic force on the ith plate in the flat plate model is

$$\mathbf{F}^i_{\text{aero}} = -\frac{1}{2}\rho\, C_D \|\mathbf{v}_{\text{rel}}\| \mathbf{v}_{\text{rel}\,B}\, S_i \max\left(\cos\theta^i_{\text{aero}}, 0\right) \tag{3.163}$$

where ρ is the atmospheric density and C_D, is a dimensionless *drag coefficient.* The drag coefficient is determined empirically, and is usually in the range between 1.5 and 2.5. Methods for computing the atmospheric density are presented in Sect. 11.2.

The aerodynamic torque on the spacecraft is then

$$\mathbf{L}^i_{\text{aero}} = \sum_{i=1}^{N} \mathbf{r}^i \times \mathbf{F}^i_{\text{aero}} \tag{3.164}$$

where $\mathbf{r}^i$ is the vector from the spacecraft center of mass to the center of pressure of the ith plate. Note this algorithm does not account for potential self-shielding that would exist on concave spacecraft.

In principle, aerodynamic torques could be used for attitude control, either for passive control like the feathers on an arrow, or even for active control by providing movable surfaces. Applications of this concept have been exceedingly rare, however.

3.3.6.4 Solar Radiation Pressure Torque

Solar radiation pressure (SRP) is another source of disturbance torque. In low-Earth orbit, the effect of SRP is dominated by aerodynamics, but SRP torques will generally dominate aerodynamic torques in higher altitude orbits ($\geq$800 km). The SRP torque is zero when the spacecraft is in the shadow of the Earth or any other body, of course. In contrast to the case of aerodynamic torques, movable surfaces have been used on some spacecraft in geosynchronous orbits to balance the SRP torques. In most applications, the surfaces have been moved by daily commands, and not controlled autonomously or in real time by an onboard computer [12].

As for aerodynamic torque, we model the surface of the spacecraft as a collection of N flat plates of area S_i, outward normal $\mathbf{n}_B^i$ in the body coordinate frame, specular reflection coefficient R_{spec}^i, diffuse reflection coefficient R_{diff}^i, and absorption coefficient R_{abs}^i. Diffuse reflection is assumed to be Lambertian, which means that the intensity of the reflected light in any direction is proportional to the cosine of the angle between the reflection direction and the normal. The coefficients sum to unity; $R_{\text{spec}}^i + R_{\text{diff}}^i + R_{\text{abs}}^i = 1$.

The spacecraft-to-Sun unit vector in the body frame is

$$\mathbf{s} = A\,\mathbf{e}_{\text{sat}\odot} \tag{3.165}$$

where A is the attitude matrix and $\mathbf{e}_{\text{sat}\odot}$ is the spacecraft-to-Sun vector in the GCI frame. The angle between the Sun vector and the normal to the ith plate is given by

$$\cos\theta_{\text{SRP}}^i = \mathbf{n}_B^i \cdot \mathbf{s} \tag{3.166}$$

The SRP force on the ith plate can then be expressed as [26]

$$\mathbf{F}_{\text{SRP}} = -P_{\odot} S_i \left[2\left(\frac{R_{\text{diff}}^i}{3} + R_{\text{spec}}^i \cos\theta_{\text{SRP}}^i \right)\mathbf{n}_B^i + (1 - R_{\text{spec}}^i)\mathbf{s} \right] \max\left(\cos\theta_{\text{SRP}}^i, 0\right) \tag{3.167}$$

where $P_{\odot}$ is the solar radiation pressure. Section 11.3 present methods for computing the Sun position, solar radiation pressure, and conditions for shadowing.

The SRP torque on the spacecraft is then

$$\mathbf{L}_{\text{SRP}}^i = \sum_{i=1}^{N} \mathbf{r}^i \times \mathbf{F}_{\text{SRP}}^i \tag{3.168}$$

where $\mathbf{r}^i$ is the vector from the spacecraft center of mass to the center of pressure of the SRP on the ith plate.

This formulation has several limitations. First, the Sun is not the only source of radiation, although it is by far the largest for Earth-orbiting spacecraft. Reflected light from the Earth or the Moon, called *albedo*, can be significant if very precise dynamical modeling is required; and models incorporating this effect have been developed [4].

Secondly, the torque due to thermal radiation emitted from the spacecraft has been ignored. A spacecraft is usually in a long-term energy balance, so all the absorbed radiation is emitted as thermal radiation, although not necessarily at the same time or from the same surface as its absorption. Accurate modeling of thermal radiation requires knowledge of the absolute temperature T_i and emissivity ϵ^i (a dimensionless constant between 0 and 1) of each surface. Then the thermal radiation flux from the surface is given by the Stefan-Boltzmann law

$$\mathscr{F}_{\text{thermal}}^i = \epsilon^i \sigma T_i^4 \tag{3.169}$$

where $\sigma = 5.67 \times 10^{-8}\,\mathrm{Wm^{-2}K^{-4}}$ is the Stefan-Boltzmann constant. If the thermal radiation from every surface is Lambertian, it gives rise to a net torque

$$\mathbf{L}_{\text{thermal}} = -\frac{2}{3}\sum_{i=1}^{N} \mathscr{F}^{i}_{\text{thermal}} S_i (\mathbf{r}^i \times \mathbf{n}^i_B) \tag{3.170}$$

Thermal radiation torque can usually be neglected because the thermal flux is emitted roughly equally in all directions, so that the net torque is small.

Finally, Eqs. (3.168) and (3.170) ignore potential self-shadowing of concave spacecraft. If the configuration of the spacecraft is known a priori, self-shadowing can be taken into account by replacing S_i with the area of the flat plate that is visible to the Sun after accounting for shadowing. Modeling the effects of reflected radiation or thermal radiation from one surface striking another surface is an additional complication. Another drawback to Eq. (3.167) is that it is only valid for a collection of flat surfaces with uniquely defined outward normals. Most real spacecraft have some curved surfaces, and accurately approximating these surfaces by a collection of flat plates causes the size of the model to grow, increasing the computational burden.

3.3.6.5 Mass-Expulsion Torques

Translational momentum is the product $m\mathbf{v}$ of mass and velocity. We generally think of a force producing a rate of change of momentum $m\dot{\mathbf{v}}$. Mass-expulsion forces, on the other hand, are the result of a change of momentum $\mathbf{F}_{\text{mexp}} = -\dot{m}\mathbf{v}_{\text{rel}}$, where $\dot{m}$ is the rate at which mass is expelled, and $\mathbf{v}_{\text{rel}}$ is the velocity of the expelled mass relative to the spacecraft. Newton's third law of motion gives the negative sign, because this is a reaction force on the spacecraft. Another way to state this is to say that the mass-expulsion force reflects the conservation of momentum of the system consisting of the spacecraft and the expelled mass. Mass-expulsion forces provided by thrusters can be used to adjust the trajectories of spacecraft that require such corrections.

A mass-expulsion force will generally be accompanied by a torque

$$\mathbf{L}_{\text{mexp}} = \mathbf{r} \times \mathbf{F}_{\text{mexp}} = -\dot{m}\,\mathbf{r} \times \mathbf{v}_{\text{rel}} \tag{3.171}$$

where $\mathbf{r}$ is the vector from the spacecraft center of mass to the point where the mass is expelled. Undesirable mass-expulsion torques during orbit maneuvers can be minimized either by requiring the line of action of the thrust to pass through the spacecraft's center of mass or by using multiple thrusters whose torques cancel. It is impossible in practice to ensure exact cancelation, however.

Thrusters can also be used specifically as sources of torque. Attitude control thrusters are generally much smaller than orbit adjustment thrusters, because attitude control requires less force. An advantage of using thrusters for attitude control is that they can be used anywhere, unlike magnetic torquers or gravity-gradient booms that require an ambient magnetic or gravitational field. They have the disadvantage of requiring expendable propellant, which can often be the element limiting the lifetime of a mission. Another disadvantage is that attitude control thrusters are accompanied by orbit-perturbing forces unless their forces are arranged in exact couples, equal and opposite pairs, which are impossible to attain in practice.

Use of thrusters for attitude control will be considered in more detail in Sect. 4.11 and Chap. 7, so this section will concentrate on mass-expulsion torques as a disturbance source. One source already mentioned is residual torques from orbit maneuvers, which can result from thruster misalignments or from impingement of thruster plumes on the spacecraft structure. Other common sources are outgassing of water vapor from the spacecraft structure during early stages of a mission or venting of cryogens. These can be minimized by arranging the vents to provide cancelation of the torques. Disturbances can also result from leaks of fuel, fuel pressurizing agents, or air from pressurized compartments, as on the International Space Station [18].

The Wilkinson Microwave Anisotropy Probe (WMAP) provided an interesting example of a mass-expulsion torque [25]. WMAP had a warm Sun-facing side and a cold side separated by a Sun shield of radius 2.5 m (see Fig. 7.4). Shortly after launch, WMAP executed three highly elliptical orbits with periods of approximately 7 days and with perigees on the sunlit side of the Earth. WMAP's attitude was inertially fixed prior to planned orbit adjustments at the perigees. About 40 min before the first perigee passage, the spacecraft angular momentum began to increase, peaking at about 2 Nms approximately 20 min before perigee and then decreasing. Various mechanisms for this anomalous torque were considered and rejected. The final explanation was that water vapor outgassing from the spacecraft had condensed on the cold side of the Sun shield as ice while WMAP was on the dark side of the Earth and had sublimated, first from one half of the Sun shield and then from the other, when the cold side was subjected to reflected sunlight from the Earth near perigee. The average velocity of the sublimated water molecules was estimated, assuming a temperature of 150 K, to be

$$v_{\text{rel}} = \sqrt{2k_B T / m_{H_2O}} = 370 \text{ m/s} \tag{3.172}$$

where $k_B = 1.38 \times 10^{-23}$ J/K is the Boltzmann constant and $m_{H_2O} = 3 \times 10^{-26}$ kg is the mass of a water molecule. Assuming an average lever arm of $R = 1.1$ m, the quantity of sublimated ice required to explain the anomalous torque is only $\Delta m = \Delta H / R\, v_{\text{rel}} = 5$ g.

3.3.7 Angular Momentum for Health Monitoring

We have emphasized that internal momentum-exchange torques can lead to rapid variation of a spacecraft's angular velocity, but not of the system angular momentum. This insight led to an application of angular momentum conservation for spacecraft failure detection [21]. A computer onboard the Hubble Space Telescope (HST) calculates the total system angular momentum $\mathbf{H}_B$ by means of Eqs. (3.139)–(3.141), with the angular velocity being sensed by the gyros and the wheel angular momentum computed using wheel tachometer data. The high torque of the reaction wheels can cause both the wheel angular momentum and the body angular momentum to change rapidly, but their vector sum only changes slowly. Subtracting easily computable gyroscopic, magnetic, and gravity-gradient torques from the rate of change of the computed system momentum gives an apparent disturbance torque:

$$\mathbf{L}_{\text{disturbance}} = \dot{\mathbf{H}}_B + \boldsymbol{\omega} \times \mathbf{H}_B - \mathbf{m} \times \mathbf{B} - \mathbf{L}_{gg} \tag{3.173}$$

with all the vectors computed in the spacecraft body frame. A large value of this disturbance torque indicates a failure of either a reaction wheel tachometer or a gyro. A tachometer failure could be identified by an independent test, so this angular momentum test was implemented onboard HST to identify gyro failures. The test initiated entry to a gyroless safehold mode three times in late 2002 and early 2003 [22].

3.3.8 Dynamics of Earth-Pointing Spacecraft

A great many spacecraft are pointed at the Earth to study its weather, climate, and resources. Thus it is useful to consider the special case of Earth-pointing spacecraft, whose body axes are closely aligned with the LVLH frame defined in Sect. 2.6.4. The attitude A_{BO} specifying the orientation of the spacecraft body axes to the axes of the LVLH frame, denoted by index O, is conveniently described by a $3-2-1$ Euler sequence of yaw $= \phi$, pitch $= \theta$, and roll $= \psi$ angles.

The dynamic equations give the motion relative to an inertial frame, so we write $A_{BI} = A_{BO}A_{OI}$ and use the Eq. (3.12) with the appropriate assignment of frame indices to find $\boldsymbol{\omega}_B^{BI}$. The matrix A_{BO} is given by Eq. (2.164), so Eq. (3.43) gives the components of $\boldsymbol{\omega}_B^{BO}$. The matrix A_{IO} is given by Eq. (2.79), so Eq. (3.3) can be used to find the angular velocity of the O frame with respect to the I frame:

$$-[\boldsymbol{\omega}_O^{OI}\times] = \dot{A}_{OI}A_{OI}^T = \dot{A}_{IO}^T A_{IO} = \begin{bmatrix} \dot{\mathbf{o}}_{1I}^T \\ \dot{\mathbf{o}}_{2I}^T \\ \dot{\mathbf{o}}_{3I}^T \end{bmatrix} \begin{bmatrix} \mathbf{o}_{1I} & \mathbf{o}_{2I} & \mathbf{o}_{3I} \end{bmatrix}$$

$$= \begin{bmatrix} \dot{\mathbf{o}}_{1I}\cdot\mathbf{o}_{1I} & \dot{\mathbf{o}}_{1I}\cdot\mathbf{o}_{2I} & \dot{\mathbf{o}}_{1I}\cdot\mathbf{o}_{3I} \\ \dot{\mathbf{o}}_{2I}\cdot\mathbf{o}_{1I} & \dot{\mathbf{o}}_{2I}\cdot\mathbf{o}_{2I} & \dot{\mathbf{o}}_{2I}\cdot\mathbf{o}_{3I} \\ \dot{\mathbf{o}}_{3I}\cdot\mathbf{o}_{1I} & \dot{\mathbf{o}}_{3I}\cdot\mathbf{o}_{2I} & \dot{\mathbf{o}}_{3I}\cdot\mathbf{o}_{3I} \end{bmatrix} \tag{3.174}$$

Considering the derivatives of $\mathbf{o}_{iI}\cdot\mathbf{o}_{jI} = \delta_{ij}$ confirms that this 3×3 matrix is skew-symmetric. The inner products are frame-independent, but the subscript I is needed to specify the frame used for differentiation. We substitute Eq. (2.78) and carry out some tedious but straightforward vector algebra to find the angular velocity of the LVLH frame relative to the inertial frame as

$$\boldsymbol{\omega}_O^{OI} = \begin{bmatrix} -\dot{\mathbf{o}}_{3I} \cdot \mathbf{o}_{2I} \\ \dot{\mathbf{o}}_{3I} \cdot \mathbf{o}_{1I} \\ -\dot{\mathbf{o}}_{2I} \cdot \mathbf{o}_{1I} \end{bmatrix} = \begin{bmatrix} (\dot{g}_3 \mathbf{r}_I + g_3 \mathbf{v}_I) \cdot \mathbf{o}_{2I} \\ -(\dot{g}_3 \mathbf{r}_I + g_3 \mathbf{v}_I) \cdot \mathbf{o}_{1I} \\ (\dot{g}_2\, \mathbf{r}_I \times \mathbf{v}_I + g_2\, \mathbf{r}_I \times \dot{\mathbf{v}}_I) \cdot \mathbf{o}_{1I} \end{bmatrix}$$

$$= \begin{bmatrix} 0 \\ -\|\mathbf{r}_I \times \mathbf{v}_I\| / \|\mathbf{r}_I\|^2 \\ \|\mathbf{r}_I\| (\mathbf{o}_{2I} \cdot \dot{\mathbf{v}}_I) / \|\mathbf{r}_I \times \mathbf{v}_I\| \end{bmatrix} \tag{3.175}$$

The roll component of this angular velocity is zero, and the yaw component is also zero if the spacecraft's acceleration $\dot{\mathbf{v}}_I$ is perpendicular to $\mathbf{o}_{2I}$, as it is for a purely central force. As shown in Sect. 10.4.3, a non-central force causes the orbit plane to precess, producing a small but finite yaw rotation rate. The pitch component of $\boldsymbol{\omega}_O^{OI}$ is by far the largest.

We can obtain the attitude motion in the general case by solving the dynamic equations from Sect. 3.3.2 with $\boldsymbol{\omega}_B^{BI}$ given by the procedure described above. However, it is useful to study the special case of uncontrolled attitude motion in a nearly circular Keplerian orbit with only small excursions from perfect alignment with the LVLH coordinate frame. Applying Eqs. (10.32), (10.39), (10.40), (10.15), and (10.43) of Chap. 10 to the pitch component of $\boldsymbol{\omega}_O^{OI}$ shows that

$$\|\mathbf{r}_I \times \mathbf{v}_I\| / \|\mathbf{r}_I\|^2 = n(1 - e^2)^{-3/2} (1 + e \cos \nu)^2 \tag{3.176}$$

where n is the mean motion, e is the eccentricity, ν is the true anomaly. Now we ignore all terms of higher than first order in the Euler angles, their rates, and the eccentricity, and we approximate the cosines of the Euler angles by unity and their sines by the angles themselves. With these approximations, Eqs. (3.12), (3.43), (2.164), and (3.176) give

$$\boldsymbol{\omega}_B^{BI} = \boldsymbol{\omega}_B^{BO} - \|\mathbf{r}_I \times \mathbf{v}_I\| / \|\mathbf{r}_I\|^2 \begin{bmatrix} \cos\theta \sin\phi \\ \cos\psi \cos\phi + \sin\psi \sin\theta \sin\phi \\ -\sin\psi \cos\phi + \cos\psi \sin\theta \sin\phi \end{bmatrix}$$

$$\approx \begin{bmatrix} \dot{\psi} \\ \dot{\theta} \\ \dot{\phi} \end{bmatrix} - n(1 + 2e \cos \nu) \begin{bmatrix} \phi \\ 1 \\ -\psi \end{bmatrix} \approx \begin{bmatrix} \dot{\psi} - n\phi \\ \dot{\theta} - n(1 + 2e\cos\nu) \\ \dot{\phi} + n\psi \end{bmatrix} \tag{3.177}$$

The first and third components of this equation exhibit the phenomenon of roll/yaw coupling. If we assume that the roll and yaw components of $\boldsymbol{\omega}_B^{BI}$ are exactly zero, we have $\dot{\psi} = n\phi$ and $\dot{\phi} = -n\psi$, with the solution

$$\psi(t) = \psi(t_0) \cos n(t - t_0) + \phi(t_0) \sin n(t - t_0) \tag{3.178a}$$

$$\phi(t) = \phi(t_0) \cos n(t - t_0) - \psi(t_0) \sin n(t - t_0) \tag{3.178b}$$

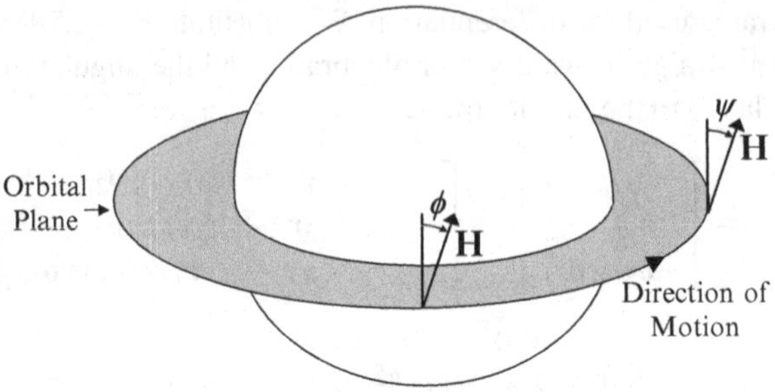

Fig. 3.6 Roll/yaw coupling

Thus a positive yaw becomes a positive roll one quarter orbit later and a negative yaw a quarter orbit after that, etc., which is why roll/yaw coupling is also known as quarter-orbit coupling. Figure 3.6 illustrates this effect, where **H** indicates the orientation of the spacecraft's total rotational angular momentum, which is assumed to be directed along the spacecraft's negative pitch axis but not exactly perpendicular to the orbit plane. Angular momentum conservation keeps the direction of this momentum fixed in inertial space, thereby ensuring that the roll and yaw components of $\boldsymbol{\omega}_B^{BI}$ remain small.

Let us now investigate the dynamic equations of motion, assuming that a reaction wheel, or possibly a combination of several wheels, provides a constant angular momentum bias h along the pitch axis. This is a common method for enhancing the roll/yaw coupling effect. Differentiating Eq. (3.177) and substituting Eq. (3.147) with $\mathbf{L}_B^w = 0$ as required by Eq. (3.146) gives

$$\begin{bmatrix} \ddot{\psi} - n\dot{\phi} \\ \ddot{\theta} + 2en^2 \sin\nu \\ \ddot{\phi} + n\dot{\psi} \end{bmatrix} = J_B^{-1}[\mathbf{L}_B - \boldsymbol{\omega}_B^{BI} \times (J_B \boldsymbol{\omega}_B^{BI} + h\mathbf{e}_2)] \tag{3.179}$$

This makes the further approximation $\dot{\nu} = n$, ignoring terms that would be of order e^2 in Eq. (3.179). Next assume that the body axes are principal axes and that the only significant external torque is the gravity-gradient torque, given by Eq. (3.157). Applying Eqs. (10.20) and (10.43) of Chap. 10 shows that

$$\frac{\mu}{r^3} = n^2 \left(\frac{1 + e\cos\nu}{1 - e^2} \right)^3 \tag{3.180}$$

We make the small-angle approximation for the trigonometric functions and ignore products of the eccentricity and the small angles, giving

$$\mathbf{L}_B = \mathbf{L}_{gg} = 3n^2 \begin{bmatrix} (J_3 - J_2)\psi \\ (J_3 - J_1)\theta \\ 0 \end{bmatrix} \tag{3.181}$$

Substituting Eq. (3.177) for the components of $\boldsymbol{\omega}_B^{BI}$ on the right side of Eq. (3.179), ignoring second-order terms in small quantities, and collecting terms gives

$$J_1\ddot{\psi} = n[4n(J_3 - J_2) + h]\psi + [n(J_1 - J_2 + J_3) + h]\dot{\phi} \tag{3.182a}$$

$$J_2\ddot{\theta} = 3n^2(J_3 - J_1)\theta - 2en^2 J_2 \sin\nu \tag{3.182b}$$

$$J_3\ddot{\phi} = n[n(J_1 - J_2) + h]\phi - [n(J_1 - J_2 + J_3) + h]\dot{\psi} \tag{3.182c}$$

The first property of these equations to notice is that pitch motion is decoupled from the roll and yaw motion. The pitch equation gives unstable motion with linear growth if $J_1 = J_3$ and unstable motion with exponential growth if $J_1 < J_3$, so pitch stability demands that $J_1 > J_3$. With the approximation $\dot{\nu} = n$, the solution for $J_1 > J_3$ is found to be

$$\theta = \theta_{\text{lib}} \cos(\omega_{\text{lib}} t + \alpha) + \frac{2en^2}{n^2 - \omega_{\text{lib}}^2} \sin\nu \tag{3.183}$$

where θ_{lib} and α are constants of integration, and

$$\omega_{\text{lib}} = n\sqrt{3(J_1 - J_3)/J_2} \tag{3.184}$$

is the libration frequency. The first term on the right side of Eq. (3.183) is the libration term, describing a pendular motion at the libration frequency. The second term gives a sinusoidal error at the orbit rate in a non-circular orbit, a result of the conflicting tendencies of rotational inertia to keep the pitch rate constant and of gravity-gradient torque to keep the yaw axis pointing along the nadir. This term grows very large near the pitch resonance case of $\omega_{\text{lib}} = n$, which must be avoided.[19] Its amplitude for the GEOS-3 spacecraft, with $e = 0.0054$, was 0.03° [27].

Now consider the roll/yaw dynamics expressed in Eqs. (3.182a) and (3.182c). The general solution of these two coupled second-order linear differential equation is a superposition of four components of the form

$$\begin{bmatrix} \psi(t) \\ \phi(t) \end{bmatrix} = \begin{bmatrix} \psi(0) \\ \phi(0) \end{bmatrix} e^{st} \tag{3.185}$$

with coefficients satisfying initial conditions. Substituting Eq. (3.185) into the roll/yaw dynamics gives a result expressible in matrix form as

$$\begin{bmatrix} J_1 s^2 + n[4n(J_2 - J_3) - h] & [n(J_2 - J_1 - J_3) - h]s \\ -[n(J_2 - J_1 - J_3) - h]s & J_3 s^2 + n[n(J_2 - J_1) - h] \end{bmatrix} \begin{bmatrix} \psi(0) \\ \phi(0) \end{bmatrix} e^{st} = 0 \tag{3.186}$$

[19]In the exact resonance case, Eq. (3.182b) has the growing solution $\theta = ent \cos\nu$.

This has a nontrivial solution only if the determinant of the 2×2 matrix is zero, which means that

$$J_1 J_3 s^4 + b s^2 + c = 0 \tag{3.187}$$

where

$$b = [n(J_2 - J_1 - J_3) - h]^2 + nJ_3[4n(J_2 - J_3) - h] + nJ_1[n(J_2 - J_1) - h] \tag{3.188a}$$

$$c = n^2[4n(J_2 - J_3) - h][n(J_2 - J_1) - h] \tag{3.188b}$$

Stable motion in roll and yaw requires that none of the roots of Eq. (3.187) has a positive real part. It is clear that if s is a root, then $-s$ is also a root, so stability requires all the roots to be purely imaginary numbers. The well-known solution of Eq. (3.187) is

$$2J_1 J_3 s^2 = -b \pm \sqrt{b^2 - 4J_1 J_3 c} \tag{3.189}$$

Both of the solutions, s^2, of Eq. (3.189) must be real and negative, with purely imaginary square roots, for roll/yaw stability. This will hold if and only if

$$c > 0 \quad \text{and} \quad b \geq 2\sqrt{J_1 J_3 c} \tag{3.190}$$

The first of these conditions is easier to satisfy; we see from Eq. (3.188b) that it requires

$$h > n \max\left(4(J_2 - J_3), (J_2 - J_1)\right) \quad \text{or} \tag{3.191a}$$

$$h < n \min\left(4(J_2 - J_3), (J_2 - J_1)\right) \tag{3.191b}$$

The second condition is harder to analyze, but a large enough positive or negative momentum bias h can provide roll/yaw stabilization for any moments of inertia, because b tends asymptotically to h^2, while c is asymptotic to n^2h^2. In this asymptotic limit Eq. (3.189) becomes

$$\begin{aligned} 2J_1 J_3 s^2 &= -h^2\left(1 \pm \sqrt{1 - 4J_1 J_3 n^2/h^2}\right) \approx -h^2[1 \pm (1 - 2J_1 J_3 n^2/h^2)] \\ &\approx \begin{cases} -2h^2 \\ -2J_1 J_3 n^2 \end{cases} \end{aligned} \tag{3.192}$$

Comparison of the first root, $\omega = \sqrt{-s^2} = h/\sqrt{J_1 J_3}$, with Eq. (3.109) identifies this as a nutation frequency. The second root gives $\omega = n$, the frequency of the quarter-orbit roll/yaw coupling. These two frequencies do not separate as cleanly for general moments of inertia and bias angular momentum.

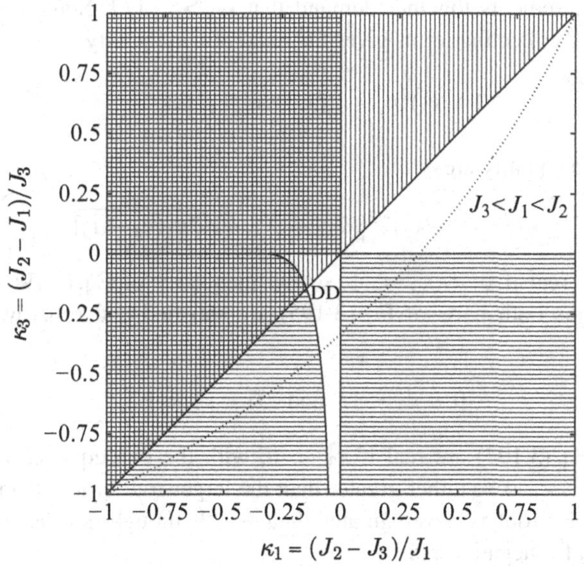

Fig. 3.7 Stability region map for gravity-gradient stabilization. *Unshaded areas* are regions of stable motion. The label DD identifies the DeBra-Delp region. *Areas with horizontal hatching* are unstable in roll and yaw; those with *vertical hatching* are unstable in pitch. The *dotted line* is the locus of the pitch orbital eccentricity resonance

We now investigate stability in the case of $h = 0$, a spacecraft with only gravity-gradient stabilization, which has been analyzed in detail. In this case, Eq. (3.188) reduces to

$$b = J_1 J_3 n^2 (1 + 3\kappa_1 + \kappa_1 \kappa_3) \tag{3.193a}$$

$$c = 4 J_1 J_3 n^4 \kappa_1 \kappa_3 \tag{3.193b}$$

where

$$\kappa_1 \equiv (J_2 - J_3)/J_1 \quad \text{and} \quad \kappa_3 \equiv (J_2 - J_1)/J_3 \tag{3.194}$$

The triangle inequality for principal moments of inertia, Eq. (3.73), implies that $-1 < \kappa_1 < 1$ and $-1 < \kappa_3 < 1$. Note that $\kappa_1 = 0$ for axially-symmetric inertia about the roll axis, $\kappa_3 = 0$ for symmetry about the yaw axis, and $\kappa_1 = \kappa_3$ for symmetry about the pitch axis.

Figure 3.7 shows the stability regions in terms of these variables in the $h = 0$ case. The requirements for roll/yaw stability in this case are simply

$$\kappa_1 \kappa_3 > 0 \quad \text{and} \quad 1 + 3\kappa_1 + \kappa_1 \kappa_3 \geq 4\sqrt{\kappa_1 \kappa_3} \tag{3.195}$$

These two requirements together demand that $\kappa_1 > -1/3$. Squaring the second requirement and collecting terms gives the quadratic inequality

$$(3+\kappa_3)^2\kappa_1^2 + 2(3-7\kappa_3)\kappa_1 + 1 \geq 0 \tag{3.196}$$

The roots of the equality are

$$\kappa_1 = (3+\kappa_3)^{-2}[(7\kappa_3 - 3 \pm 4\sqrt{3\kappa_3(\kappa_3-1)})] \tag{3.197}$$

Neither root is real if $0 < \kappa_3 < 1$, so the inequality of Eq. (3.196) is satisfied for any κ_1. Thus both parts of Eq. (3.195) are satisfied and roll/yaw stability is assured for

$$0 < \kappa_3 < 1 \quad \text{and} \quad 0 < \kappa_1 < 1 \tag{3.198}$$

The roots of Eq. (3.197) are real if $\kappa_3 < 0$, with the consequence that roll/yaw stability requires κ_1 to be either greater than the larger root or less than the smaller root. The smaller root is never greater than $-1/3$, though, so the only roll/yaw stability region for negative κ_3 is

$$-1 < \kappa_3 < 0 \quad \text{and} \quad (3+\kappa_3)^{-2}[(7\kappa_3 - 3 + 4\sqrt{3\kappa_3(\kappa_3-1)})] \leq \kappa_1 < 0 \tag{3.199}$$

This region, called the DeBra-Delp region [2], has $J_2 < J_3 < J_1$. It has rarely been employed in practice; almost all gravity-gradient-stabilized spacecraft are in the range of inertia values specified by Eq. (3.198), i.e. $J_3 < J_1 < J_2$.

It is always possible, and sometimes necessary, to supplement passive stabilization with active control using thrusters, magnetic torquers or reaction wheels. Active control effort can be minimized by starting with a stable configuration, however. A single pitch wheel can be used to control pitch and even to stabilize the pitch dynamics if $J_1 \leq J_3$. Roll/yaw coupling means that only two wheels are needed for three-axis control. Quite often the two wheels are slightly misaligned from the pitch axis so they can provide the momentum bias and also torques on two axes by commanding them in the same or opposite directions. GOES I-M, for instance, had two 51 Nms wheels with their spin axes tilted 1.66° in the positive and negative yaw directions from the negative pitch axis [24]. A much smaller 2.1 Nms wheel with its spin axis along yaw protected against failure of one of the two larger wheels.

Problems

3.1. Consider the Hill frame shown in Fig. 2.8. The deputy position expressed in Hill frame coordinates, denoted by $\mathbf{r}_{d_{\text{hill}}}$, is given by

$$\mathbf{r}_{d_{\text{hill}}} = \mathbf{r}_{c_{\text{hill}}} + \boldsymbol{\rho}_{\text{hill}} = (r_c + x)\mathbf{o}_r + y\,\mathbf{o}_\theta + z\,\mathbf{o}_h$$

where $\mathbf{r}_{c_{\text{hill}}}$ is the chief position vector expressed in Hill frame coordinates, $\boldsymbol{\rho}_{\text{hill}}$ is the relative position vector expressed in Hill frame coordinates and r_c is the chief position magnitude. The vectors expressed in inertial coordinates are given by

$$\mathbf{r}_{d_I} = \mathbf{r}_{c_I} + A_{I\,\text{hill}}\,\boldsymbol{\rho}_{\text{hill}}$$

Using Eq. (3.14) derive an expression for the derivative of $\mathbf{r}_{d_I}$ in terms of $\mathbf{r}_{c_I}$, $A_{I\,\text{hill}}$, $\boldsymbol{\rho}_{\text{hill}}$, and/or their derivatives, and the derivative of the true anomaly of the chief.

3.2. Another way to derive the quaternion kinematics in Eq. (3.20) is to use the attitude matrix kinematics. Using the quaternion representation of the attitude matrix given by Eq. (2.125) and the attitude matrix kinematics equation given by $\dot{A} = -[\boldsymbol{\omega}\times]A$, derive the quaternion kinematics equation.

3.3. Beginning with Eq. (3.21) and the definition of the quaternion in Eq. (2.124) show that $\|\boldsymbol{\omega}\| = \dot{\vartheta}$ if the axis of rotation is fixed.

3.4. Prove that the inverse of Eq. (3.28) is given by Eq. (3.30).

3.5. Another way to derive the Euler angle kinematics in Eq. (3.38) is to use the attitude matrix kinematics. Using the Euler angle representation of the attitude matrix for a $1-2-3$ sequence in Table 9.2 and the attitude matrix kinematics equation given by $\dot{A} = -[\boldsymbol{\omega}\times]A$, derive the matrix $B(\theta, \psi)$ in Eq. (3.38) for a 1–2–3 sequence. Also, derive an analytical expression for $B^{-1}(\theta, \psi)$ for a 1–2–3 sequence.

3.6. Consider the following inertia matrix:

$$J_B^c = \begin{bmatrix} 100 & 0 & 0 \\ 0 & 50 & 0 \\ 0 & 0 & 25 \end{bmatrix} \text{ kg-m}^2$$

Numerically integrate Eq. (3.81) with $\mathbf{L}_B^c = \mathbf{0}_3$ for a time span of 2.5 h with the following initial conditions:

a) $\boldsymbol{\omega}_B^{BI}(0) = [0\ \ 0.01\ \ 0]^T$ rad/s
b) $\boldsymbol{\omega}_B^{BI}(0) = [0.01\ \ 0.0001\ \ 0.0001]^T$ rad/s
c) $\boldsymbol{\omega}_B^{BI}(0) = [0.0001\ \ 0.01\ \ 0.0001]^T$ rad/s

Explain the differences between these cases.

3.7. Consider the angular velocity solution for the axially symmetric case given in Eq. (3.104). Show that the closed-form solution for the quaternion is given by

$$\mathbf{q}(t) = \mathbf{z}(t) \otimes \mathbf{q}_0$$

where $\mathbf{z}(t)$ is given by

$$\mathbf{z}(t) = \begin{bmatrix} h_{01}\cos(\alpha)\sin(\beta) + h_{02}\sin(\alpha)\sin(\beta) \\ h_{02}\cos(\alpha)\sin(\beta) - h_{01}\sin(\alpha)\sin(\beta) \\ h_{03}\cos(\alpha)\sin(\beta) + \sin(\alpha)\cos(\beta) \\ \cos(\alpha)\cos(\beta) - h_{03}\sin(\alpha)\sin(\beta) \end{bmatrix}$$

with the definitions $\alpha = \frac{1}{2}\omega_p t$, $\beta = \frac{1}{2}\omega_\ell t$, $\mathbf{h}_0 = \mathbf{H}_0/\|\mathbf{H}_0\| = [h_{01}\ h_{02}\ h_{03}]^T$, where $\omega_\ell = H/J_t$ is the inertial nutation rate and $\mathbf{H}_0 = J\boldsymbol{\omega}_0$ is the initial angular momentum vector.

3.8. Instead of Eq. (3.139), we could write

$$\mathbf{H}_B = J_B^*\boldsymbol{\omega}_B^{BI} + \mathbf{H}_B^*$$

where

$$J_B^* \equiv \tilde{J}_B + \sum_{\ell=1}^{n} J_\ell^w \quad \text{and} \quad \mathbf{H}_B^* = \sum_{\ell=1}^{n} J_\ell^{\|}\omega_\ell^w \mathbf{w}_\ell$$

This form is sometimes more convenient for attitude estimation problems where the wheel speeds are known better than the applied torques, but it is less well suited to dynamics analysis. Show that the equivalent of Eq. (3.147) in this formulation is

$$\dot{\boldsymbol{\omega}}_B^{BI} = (J_B^*)^{-1}[\mathbf{L}_B - \dot{\mathbf{H}}_B^* - \boldsymbol{\omega}_B^{BI} \times (J_B^*\boldsymbol{\omega}_B^{BI} + \mathbf{H}_B^*)]$$

Then show that the dynamics of a spacecraft with constant $\mathbf{H}_B^*$ are basically the same as the dynamics for constant $\mathbf{H}_B^w$ using J_B^* rather than J_B for the MOI.

3.9. Prove the relationship shown in Eq. (3.175).

3.10. The roll/yaw motion of a gravity-gradient-stabilized spacecraft would exhibit a resonant motion analogous to the pitch motion exhibited by Eq. (3.183) if the solution of Eq. (3.189) gave $s^2 = -n^2$. Show that this roll/yaw resonance condition for $h = 0$ is equivalent to $\kappa_1(1 - \kappa_3) = 0$, and thus that the roll/yaw resonance condition is never satisfied inside a region of stable motion.

3.11. Show that the condition for pitch resonance with orbit eccentricity, $\omega_{\text{lib}} = n$, is equivalent to $\kappa_1 = (1 + 3\kappa_3)/(3 + \kappa_3)$, which is the curve plotted in Fig. 3.7.

3.12. If the principal moments of inertia are all equal, both the gyroscopic and gravity-gradient torques will vanish. Find the solutions of Eq. (3.189) in this case without any assumptions on h except that it is nonzero.

References

1. Abramowitz, M., Stegun, I.A.: Handbook of Mathematical Functions with Formulas, Graphs and Mathematical Tables. Applied Mathematics Series - 55. National Bureau of Standards, Washington, DC (1964)
2. BeDra, D.B., Delp, R.H.: Rigid body attitude stability and natural frequencies in a circular orbit. J. Astronaut. Sci. **8**(1), 1–14 (1961)
3. Bhat, S.P., Dham, A.S.: Controllability of spacecraft attitude under magnetic actuation. In: Proceedings of the 42nd IEEE Conference on Decision and Control, pp. 2383–2388. Maui (2003)
4. Borderies, N., Longaretti, P.: A new treatment of the albedo radiation pressure in the case of a uniform albedo and of a spherical satellite. Celestial Mech. Dyn. Astron. **49**(1), 69–98 (1990)
5. Bracewell, R.N., Garriott, O.K.: Rotation of artificial Earth satellites. Nature **182**(4638), 760–762 (1958)
6. Camillo, P.J., Markley, F.L.: Orbit-averaged behavior of magnetic control laws for momentum unloading. J. Guid. Contr. **3**(6), 563–568 (1980)
7. Damaren, C.J.: Comments on 'fully magnetic attitude control for spacecraft subject to gravity gradient'. Automatica **38**(12), 2189 (2002)
8. Dodge, F.T.: The New "Dynamic Behavior of Liquids in Moving Containers". Southwest Research Inst., San Antonio (2000). URL http://books.google.com/books?id=RltitwAACAAJ
9. Gambhir, B.L., Sood, D.R.: Spin axis magnetic coil maneuvers. In: Wertz, J.R. (ed.) Spacecraft Attitude Determination and Control, chap. 16. Kluwer Academic, Dordrecht (1978)
10. Goldstein, H.: Classical Mechanics, 2nd edn. Addison-Wesley, Reading (1980)
11. Grell, M.G., Shuster, M.D.: Spin plane magnetic coil maneuvers. In: Wertz, J.R. (ed.) Spacecraft Attitude Determination and Control, chap. 16. Kluwer Academic, Dordrecht (1978)
12. Harvie III, E., Rowe, J., Tsui, Y.K.J.: Performance analysis of the GOES trim tab solar pressure torque angular momentum control. In: Proceedings of SPIE, GOES-8 and Beyond, vol. 2812. Denver (1996)
13. Hecht, E., Manger, W.P.: Magnetic attitude control of the Tiros satellites. Appl. Math. Mech. **7**, 127–135 (1964)
14. Hughes, P.C.: Spacecraft Attitude Dynamics. Wiley, New York (1986)
15. Kane, T.R., Levinson, D.A.: Dynamics: Theory and Application. McGraw-Hill, New York (1985)
16. Kane, T.R., Likens, P.W., Levinson, D.A.: Spacecraft Dynamics. McGraw-Hill, New York (1983)
17. Kaplan, M.H.: Modern Spacecraft Dynamics and Control. Wiley, New York (1976)
18. Kim, J.W., Crassidis, J.L., Vadali, S.R., Dershowitz, A.L.: International Space Station leak localization using attitude respone data. J. Guid. Contr. Dynam. **29**(5), 1041–1050 (2006)
19. Lerner, G.M.: Attitude acquisition. In: Wertz, J.R. (ed.) Spacecraft Attitude Determination and Control, chap. 19. Kluwer Academic, Dordrecht (1978)
20. Markley, F.L.: Attitude dynamics. In: Wertz, J.R. (ed.) Spacecraft Attitude Determination and Control, chap. 16. Kluwer Academic, Dordrecht (1978)
21. Markley, F.L., Kennedy, K.R., Nelson, J.D., Moy, E.W.: Autonomous spacecraft gyro failure detection based on conservation of angular momentum. J. Guid. Contr. Dynam. **17**(6), 1385–1387 (1994)
22. Markley, F.L., Nelson, J.D.: Zero-gyro safemode controller for the Hubble Space Telescope. J. Guid. Contr. Dynam. **17**(4), 815–822 (1994)
23. Parkinson, B.W., Kasdin, N.J.: A magnetic attitude control system for precision pointing of the rolling GP-B spacecraft. Acta Astronautica **21**(6–7), 477–486 (1990)
24. Space Systems Loral: GOES I–M data book (1996). http://goes.gsfc.nasa.gov/text/goes.databook.html (accessed Jan. 27, 2014)

25. Starin, S.R., O'Donnell Jr., J.R., Ward, D.K., Wollack, E.J., Bay, P.M., Fink, D.R.: Anomalous force on the Wilkinson Microwave Anisotropy Probe. J. Spacecraft Rockets **41**(6), 1056–1062 (2004)
26. Vallado, D.A.: Fundamentals of Astrodynamics and Applications, 3rd edn. Microcosm Press, Hawthorne and Springer, New York (2007)
27. Wertz, J.R. (ed.): Spacecraft Attitude Determination and Control. Kluwer Academic, Dordrecht (1978)
28. Wiśniewski, R.: Linear time-varying approach to satellite attitude control using only electromagnetic actuation. J. Guid. Contr. Dynam. **23**(4), 640–647 (2000)
29. Wiśniewski, R., Blanke, M.: Fully magnetic attitude control for spacecraft subject to gravity gradient. Automatica **35**(7), 1201–1214 (1999)
30. Zienkiewicz, O.C., Taylor, R.L., Zhu, J.Z.: Finite Element Method - Its Basis and Fundamentals, 6th edn. Elsevier Butterworth-Heineman, Oxford (2005)

Chapter 4
Sensors and Actuators

This chapter will discuss several kinds of sensors and actuators used to determine and control spacecraft attitude [26, 44, 54, 66]. The history of attitude sensor development has emphasized increased resolution and accuracy as well as decreased size, weight, and power (often abbreviated as SWaP). Actuator technologies have also been scaled down to be appropriate for microsatellites and cubesats. We begin with a brief introduction to redundancy considerations, and then consider some specific sensors and actuators.

4.1 Redundancy

The space environment is stressful, and failures of ACS components have sometimes led to the degradation or premature termination of space missions. A requirement for many missions is the ability to survive the failure of any one component, a *single-point failure*, without any loss of capability. This is often accomplished by providing redundant components. Some designs leave the redundant equipment unpowered until a failure of the operating unit occurs, a configuration usually referred to as a *cold backup* configuration. If a primary unit fails in a cold backup configuration, there is some delay from the time that the backup component is powered on until it is available for use. A *warm backup* configuration, in which the redundant device is powered on but not used, allows a quicker recovery in the event of failure of the primary component. A *hot backup* configuration, with the redundant component being used all the time, is often preferred for actuators such as reaction wheels or CMGs, where the increased control authority is useful.

Redundant components can be connected in a *block-redundant* or a *cross-strapped* fashion. This is illustrated schematically in Fig. 4.1, where Sensors A and B might be a star tracker and a gyro, Actuators A and B could be two reaction wheels, and the numerical subscripts represent redundant components. An actual control system would likely incorporate more sensors and actuators,

F.L. Markley and J.L. Crassidis, *Fundamentals of Spacecraft Attitude Determination and Control*, Space Technology Library 33, DOI 10.1007/978-1-4939-0802-8_4,

© Springer Science+Business Media New York 2014

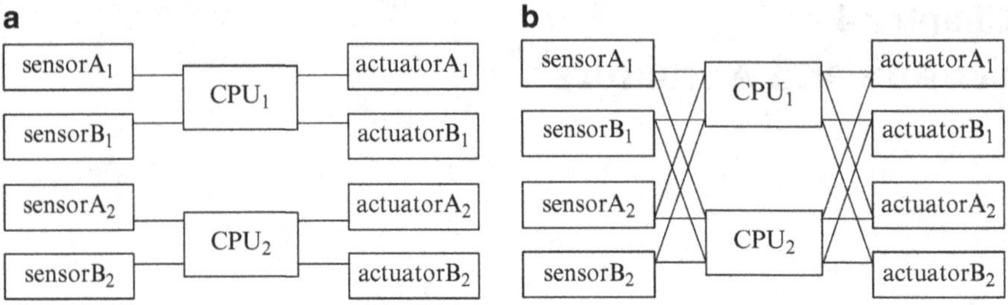

Fig. 4.1 Alternative redundancy configurations. (**a**) Block redundant. (**b**) Cross strapped

of course. The block-redundant system of Fig. 4.1a has two entirely separate *strings*; a single-string system would just be the upper half of the figure. A failure in the active string of a block-redundant system causes a switch to all the components of another string. A failure in the cross-strapped system of Fig. 4.1b would generally result in replacing only the failed unit with its backup. It can be seen that the block-redundant and cross-strapped configurations are both single-fault-tolerant. The block-redundant system would only tolerate a second fault if it were in the same string as the first fault. We can see that cross-strapping usually leads to a more robust system, because it can accommodate a greater range of multiple failures. There are instances, however, where cross-strapping can reduce reliability by allowing a fault in one component to propagate through the system.

Cross-strapping increases system complexity and cost. The increased expense is especially prominent in the testing phase, where it is desirable to test all the paths through the control system. It can be seen from the figure that the number of paths in a block-redundant system increases linearly with the degree of redundancy, but the number of paths increases exponentially in a fully cross-strapped system. It is often cost-effective to build a system with limited redundancy and/or partial cross-strapping. A careful reliability analysis must be performed as part of the design process, to assess the degree of redundancy and cross-strapping needed to provide the desired probability of completing the mission successfully.

Hardware redundancy is not the only option for protecting against single-point failures. Instead, it is often possible to provide the function of a failed component by using an entirely different component or set of components. This often involves extra computation, and is referred to as *analytic redundancy*. One example is the provision for attitude determination of the Tropical Rainfall Measuring Mission (TRMM) spacecraft using gyros, Sun sensors, and a three-axis magnetometer in place of the horizon sensor, which is described in Sect. 5.8. This contingency mode was designed to cope with potential degradation of the horizon sensor, which never occurred, but it enabled raising the TRMM altitude above the operational altitude of the horizon sensor in order to conserve propellant and extend the mission.

4.2 Star Trackers

4.2.1 Overview

Reference [66] describes several types of star trackers, many of which are no longer used. Beginning around 1990, they were superseded by solid state star trackers that track many stars simultaneously [43]. We will discuss only these state-of-the-art trackers, many of which autonomously match the tracked stars with stars in an internal catalog and use one of the methods described in Chap. 5 to compute the star tracker's attitude with respect to a celestial reference frame. A typical tracker has an update rate between 0.5 and 10 Hz, a mass of about 3 kg and a power requirement on the order of 10 W. It provides accuracy of a few arcseconds in the boresight pointing direction, with larger errors for rotation about the boresight. References [43] and [34] review the operation and performance of star trackers, and [64] provides a detailed description of the star tracker used by the WMAP spacecraft.

A star tracker is basically a digital camera with a focal plane populated by either CCD (charge-coupled device) or CMOS (complementary metal-oxide semiconductor) pixels (picture elements). CCDs have lower noise, but CMOS has several advantages. It is the same technology used for microprocessors, so the pixels can include some data processing capabilities on the focal plane itself. Sensors taking advantage of this capability are known as active pixel sensors (APS). CMOS is more resistant to radiation damage than CCDs, and also provides the capability of reading out different pixels at different rates, which is not feasible with CCDs.

Figure 4.2 shows the geometry of a star tracker, which is basically the geometry of a pinhole camera. The x, y, and z axes constitute a right-handed coordinate system with its origin at the vertex of the optical system and its z axis along the optical axis, the tracker's boresight. The focal plane is a distance f, the focal length of the optics, behind the vertex. The optics are slightly defocused so a star image covers several pixels. This enables the location of the centroid of a star image, computed as the "center of mass" of the photoelectrons in an $n \times n$ block of pixels, to

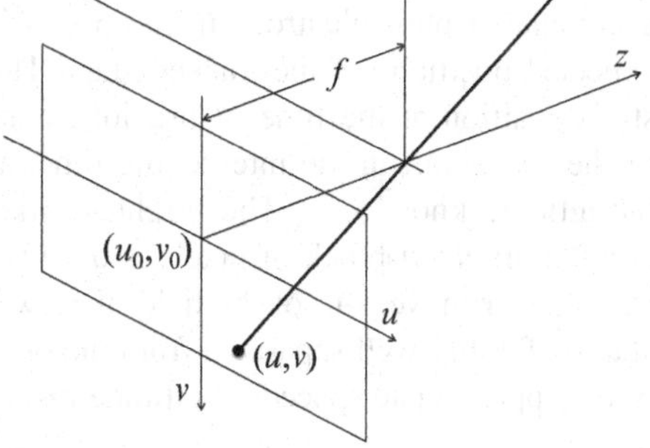

Fig. 4.2 Star tracker geometry

be determined to an accuracy of a fraction of a pixel [34]. Optimal defocusing gives a point spread function (PSF) with a width between 1.5 and 2 pixels, with n typically equal to 2 or 3 [56, 57]. The resulting accuracy, κ_{cent}, of the centroid depends on the star brightness, the exposure time, and various noise sources. Centroiding to an accuracy of $\kappa_{\text{cent}} = 0.1$ pixel widths is generally achieved. Computing a weighted center of mass may reduce the effect of noise and improve this accuracy [46].

The focal plane has a (u, v) coordinate system whose origin can be at its center or at one of its corners. The center of the focal plane, at the point where the z axis pierces it, is designated by (u_0, v_0). The unit vector $\mathbf{s}$ from the spacecraft to a star can be computed from the focal plane coordinates of the centroid of its image as

$$\mathbf{s} = \frac{1}{\sqrt{f^2 + (u - u_0)^2 + (v - v_0)^2}} \begin{bmatrix} u - u_0 \\ v - v_0 \\ f \end{bmatrix} \tag{4.1}$$

It is conventional to define $\alpha \equiv \tan^{-1}(s_2/s_3)$ and $\beta \equiv \tan^{-1}(s_1/s_3)$ so that

$$\mathbf{s} = \frac{1}{\sqrt{1 + \tan^2\beta + \tan^2\alpha}} \begin{bmatrix} \tan\beta \\ \tan\alpha \\ 1 \end{bmatrix} \tag{4.2}$$

The focal plane coordinates of the centroid in terms of the star vector are

$$u = u_0 + f s_1/s_3 = u_0 + f \tan\beta \tag{4.3a}$$

$$v = v_0 + f s_2/s_3 = u_0 + f \tan\alpha \tag{4.3b}$$

4.2.2 *Modes of Operation*

A star tracker has two modes of operation: tracking mode and initial attitude acquisition. We will first discuss the more straightforward tracking mode, in which the tracker is following several stars that have already been matched with catalogued stars. After a fixed integration time, the star tracker reads out the number of accumulated photoelectrons in the pixels in regions of interest (ROI) around the expected positions of the tracked stars. The location of each ROI is based on the star's position at the time of previous readout and the estimated attitude motion of the spacecraft in the intervening time, and its size depends on the accuracy of the attitude knowledge. The brightest pixels in each ROI are identified, and the appropriate $n \times n$ block of pixels is used to compute the centroid of each star. If a tracked star moves out of the field of view (FOV), the tracker searches for another star, preferably well separated from the other tracked stars. The a priori knowledge of the approximate spacecraft attitude makes this search relatively easy.

Initial attitude acquisition mode is more picturesquely known as lost-in-space mode [62]. In this case the tracker searches the entire FOV for the brightest clusters of pixels, and computes at least three centroids. The arc length separation between these stars, their brightness, and some other computed properties are used to match them with entries in the star catalog. This can be accomplished in a few seconds using sophisticated algorithms for pattern matching and for rapidly searching the catalog. Reference [62] provides a very useful survey of these methods and a complete guide to the literature.

4.2.3 Field of View, Resolution, Update Rate

We will consider some trades in the design of star trackers, following the discussions by McQuerry, et al. [43] and Liebe [34]. The resolution of the star tracker depends on the number of pixels, the size of the FOV, and the accuracy of the centroiding. We will consider a square focal plane of size $N_{\rm pixels} \times N_{\rm pixels}$, typical values being 512×512 or $1{,}024 \times 1{,}024$. Assuming that the focal plane assembly is centered on the optical axis, it images a spherical quadrilateral on the celestial sphere bounded by the four great circles given by Eq. (4.2) with $\alpha = \pm\alpha_{\max}$ and $\beta = \pm\beta_{\max}$. The area of this FOV on the sphere is given by spherical geometry[1] as

$$\Omega_{\rm rectangle} = 4\sin^{-1}(\sin\alpha_{\max}\sin\beta_{\max}) \approx (2\alpha_{\max})(2\beta_{\max}) \text{ steradians} \tag{4.4}$$

The approximation is for small angles, and it is quite good, having an error of only 1 % for $\alpha_{\max} = \beta_{\max} = 10°$. Optical distortions often make it desirable to ignore stars in the corners of the FOV, reducing the useful area for $\alpha_{\max} = \beta_{\max}$ to that of a small circle of radius $\alpha_{\max}$,

$$\Omega_{\rm circle} = 2\pi(1 - \cos\alpha_{\max}) \approx \pi\alpha_{\max}^2 \text{ steradians} \tag{4.5}$$

Each pixel of a square focal plane subtends an angle $2\beta_{\max}/N_{\rm pixels}$ in the small angle approximation, so the resolution of the tracker is $2\,\kappa_{\rm cent}\beta_{\max}/N_{\rm pixels}$. Higher resolution can be obtained by decreasing the size of the FOV, increasing the number of pixels in the focal plane, or improving the centroiding. If the physical size of a pixel and the field of view are held constant, adding pixels requires a larger focal plane and thus a proportionally larger focal length, increasing the weight of the optics. Pixel sizes have historically decreased, however, allowing more pixels in a smaller focal plane.

Star trackers have several sources of errors. Optical distortions can be reduced by calibration, and temperature-dependent errors can be minimized by controlling the temperature of both the focal plane and the optics. *Shot noise* results from the

[1] This can be derived, for example, from Eqs. (A-12) and (A-22) in [66].

random nature of photons, which obey Poisson statistics. Thus if a pixel accumulates n_e photoelectrons on average, this number will have a fluctuation with standard deviation $\sqrt{n_e}$. *Dark current*, the accumulation of electrons in the absence of light, can also be minimized by cooling the focal plane, but this increases the power demands of the tracker. Sometimes the tracker software must be modified to ignore *hot pixels*, i.e. pixels with anomalously large dark currents. The effects of shot noise and dark currents can be minimized by gathering more light to increase the number of photoelectrons, which requires either increasing the aperture of the optics, with concomitant weight gain, or increasing the integration time, i.e. the time allowed for photoelectrons to accumulate before counting them. Increasing integration time obviously slows the attainable update rate of star tracker data and also makes it more difficult for the tracker to deal with spacecraft attitude motion.

If N_{stars} are tracked simultaneously, averaging of random errors reduces the errors in the attitude estimate about the two axes perpendicular to the boresight by a factor of $N_{\text{stars}}^{-1/2}$, with resulting accuracy

$$\Delta\theta_{\text{cross-boresight}} = \frac{2\,\kappa_{\text{cent}}\beta_{\max}}{N_{\text{pixels}}\sqrt{N_{\text{stars}}}} \tag{4.6}$$

For $\kappa_{\text{cent}} = 0.1$, $N_{\text{pixels}} = 1{,}024$, and $N_{\text{stars}} = 5$, this gives 3.1 arcsec for a $20°\times20°$ FOV, and 1.3 arcsec for an $8° \times 8°$ FOV. The rotation angle around the tracker's boresight, often called roll, cannot be determined with equal accuracy. Its estimation requires some separation between the tracked stars to provide a lever arm. The methods of Sect. 5.5.2 can be used to show that roll accuracy is reduced relative to the cross-axis accuracy by the root-mean-square (RMS) distance of stars from the boresight (in radians).[2] This distance for a large collection of stars uniformly distributed over a square FOV is

$$\beta_{\text{RMS}} = \sqrt{\frac{\int_{-\beta_{\max}}^{\beta_{\max}}\int_{-\beta_{\max}}^{\beta_{\max}}(x^2+y^2)dxdy}{\int_{-\beta_{\max}}^{\beta_{\max}}\int_{-\beta_{\max}}^{\beta_{\max}}dxdy}} = \frac{2\beta_{\max}}{\sqrt{6}} \tag{4.7}$$

Using this RMS value gives the error in the rotation about the boresight as

$$\Delta\theta_{\text{roll}} = \frac{\sqrt{6}\,\kappa_{\text{cent}}}{N_{\text{pixels}}\sqrt{N_{\text{stars}}}} \tag{4.8}$$

This error is independent of the size of the FOV; for $\kappa_{\text{cent}} = 0.1$, $N_P = 1{,}024$, and $N_{\text{stars}} = 5$, it is equal to 22 arcsec. We see that a large-FOV star tracker and a small-FOV tracker with the same focal plane will produce equally accurate roll attitude estimates, but Eq. (4.6) shows that the tracker with the smaller FOV will provide better measurements of the cross-axis attitude.

[2] An example calculation is set as an exercise in Chap. 5.

4.2.4 Star Catalogs

Both for lost-in-space star identification and for averaging of random errors, it is desirable to track at least four stars, and preferably more. This drives the size of the star catalog. The star availability requirement is usually stated as the probability that at least N stars will be available in the tracker's FOV. The number of stars in the FOV can be assumed to follow a Poisson distribution, which says that the probability of finding N stars in the FOV is given by

$$P(N) = e^{-\bar{N}} \frac{\bar{N}^N}{N!} \tag{4.9}$$

where $\bar{N}$ is the average number of stars in the FOV. Representative values following from this assumption are that $\bar{N} = 6.75$ gives a 90 % probability of finding 4 stars in the FOV, $\bar{N} = 8$ gives a 90 % probability of finding 5 stars in the FOV, $\bar{N} = 10$ gives a 99 % probability of finding 4 stars in the FOV, and $\bar{N} = 11.7$ gives a 99 % probability of finding 5 stars in the FOV.

These values can immediately be used to estimate the required size of the star catalog. The size of the celestial sphere is $4\pi(180/\pi)^2 = 41{,}253$ deg^2, so the star catalog should contain $11.7 \times 41{,}253/64 > 7{,}500$ stars to offer a 99 % probability of finding 5 stars in a $8^\circ \times 8^\circ$ FOV. On the other hand, only $6.75 \times 41{,}253/400 \approx 700$ stars are needed to offer a 90 % probability of finding 4 stars in a $20^\circ \times 20^\circ$ FOV.

The required size of the catalog can then be used to estimate the magnitude range of stars that must be tracked. Haworth [24] has counted the number of stars in the Tycho star catalog [60] in visual magnitude ranges from –0.5 to 11.5. His values for the number of stars of magnitude less than M_V for $3.5 \leq M_V \leq 10.5$ can be fitted to within 3 % by the simple relation

$$N(M_V) = 3.9 \exp(1.258 M_V - 0.011 M_V^2) \tag{4.10}$$

The M_V^2 term in this expression gives a curvature to the plot of differential star counts as shown, for example, in Figure 2 of [4], but neither Eq. (4.10) nor Haworth's estimates are reliable for magnitudes greater than $M_V = 10.5$. According to Eq. (4.10), a catalog containing 7,542 stars must include stars as dim as $M_V = 6.4$, while a catalog with 696 stars need only extend to $M_V = 4.3$, a magnitude signifying $10^{0.4(6.4-4.3)} \approx 7$ times as much energy flux as $M_V = 6.4$.

These magnitude estimates are only approximate for two reasons. Firstly, the magnitude of a star depends on the spectral response of the detector, and the response of a CCD or APS pixel is not the same as that of the human eye. For this reason, star catalogs are created using *instrument magnitudes*, which are specific to the detector technology employed by the tracker. Secondly, stars are not uniformly distributed on the celestial sphere, but tend to be concentrated in our galaxy, the Milky Way. Because star densities are lower at the galactic poles, it is often necessary to fill in a star catalog with dimmer stars in these regions.

4.2.5 Proper Motion, Parallax, and Aberration

Star catalogs give the apparent positions of stars in the inertial ICRF frame with its origin at the center-of-mass of the solar system, as discussed in Sect. 2.6.2. Stars are not all absolutely stationary in this frame, and a few exhibit *proper motion* of as much as several arseconds per year. Star catalogs include the proper motion of any star for which it is significant, so the star location can be corrected for it when the catalog is accessed. Stars with near neighbors, including double stars, are generally omitted from star catalogs.

Parallax, the change in the apparent location of a star due to the change in the position of the observer, is negligible for all but the nearest stars. For a satellite in near-Earth orbit, the maximum parallax is equal to the radius of the Earth's orbit divided by the distance to the star. For the nearest star, Proxima Centauri at a distance of 4.24 light years, the maximum parallax is 3.7 μrad, or 0.77 arcsec. However, Proxima Centauri would likely be excluded from a star catalog because it appears to form a triple star system with α Centauri A and B [65]. In fact, some star catalogs only include stars at distances greater than 100 light years from the solar system, so that their parallax is less than 0.03 arcsec near the Earth, and less than 1 arcsec anywhere inside the orbit of Neptune [59].

Stellar aberration is a change in the apparent position of a star due to the velocity of the observer. It is also known as astronomical aberration or Bradley aberration, after James Bradley who showed in 1729 that it was a result of the finite speed of light. Stellar aberration is not to be confused with distortions in optical systems such as spherical aberration or chromatic aberration, which are completely unrelated. We will explain aberration using Bradley's classical argument, which gives a result correct to the first order in v/c, the ratio of the observer's velocity to the speed of light. A complete explanation requires special relativity, but our analysis will avoid this refinement [59].

Figure 4.3a illustrates Bradley's argument in the ICRF frame, in which the star tracker is moving with velocity $\mathbf{v}$ and the star is assumed to be stationary.[3] The vertex V and focal plane FP of the tracker are shown at times t_1, when the starlight passes through the effective pinhole at V, and t_2 when it strikes the focal plane; the tracker is displaced a distance $\mathbf{v}\Delta t \equiv \mathbf{v}(t_2 - t_1)$ between these times. Light travels in a straight line with speed c in an inertial reference frame, so the light from the star at position S^{true} passes through $V(t_1)$ to $FP(t_2)$. The vector from $FP(t_2)$ to $V(t_1)$ is $c\,\mathbf{s}^{\text{true}}\Delta t$, as indicated on the figure. The star appears to the star tracker to have come through $V(t_2)$ to $FP(t_2)$, so the apparent direction to the star is the direction of the vector sum $(c\,\mathbf{s}^{\text{true}} + \mathbf{v})\Delta t$. Thus the apparent unit vector to the star is

$$\mathbf{s}^{\text{apparent}} = \frac{\mathbf{s}^{\text{true}} + \mathbf{v}/c}{\|\mathbf{s}^{\text{true}} + \mathbf{v}/c\|} \tag{4.11}$$

[3] Proper motion is accounted for separately.

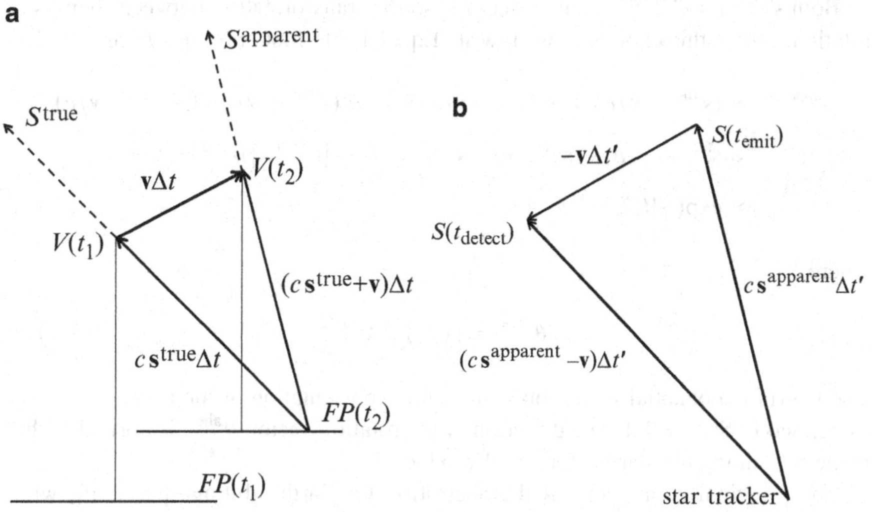

Fig. 4.3 Stellar aberration. (**a**) Celestial reference frame. (**b**) Star tracker frame

It can be seen that the apparent star vector is tilted forward from the true star vector in the direction of the velocity. A moving telescope with a very small field of view must be tilted forward to catch the photons, much as someone walking in the rain must tilt an umbrella forward to catch the raindrops.

Figure 4.3b illustrates an alternative derivation in an inertial reference frame moving at velocity $\mathbf{v}$ with respect to the ICRF, where $\mathbf{v}$ is the instantaneous velocity of the star tracker at the time the star is detected. The star tracker is stationary in this frame, and the star moves with velocity $-\mathbf{v}$. The apparent position of the star is the direction from the tracker to the position of the star $S(t_{emit})$ when the light was emitted. The star moves a distance $-\mathbf{v}\Delta t' \equiv -\mathbf{v}(t_{detect} - t_{emit})$ during the time the light travels the distance $c\Delta t'$ from the star to the tracker. Thus the catalogued position of the star at time t_{detect} is in the direction of $(c\,\mathbf{s}^{apparent} - \mathbf{v})\Delta t'$, namely

$$\mathbf{s}^{true} = \frac{\mathbf{s}^{apparent} - \mathbf{v}/c}{\|\mathbf{s}^{apparent} - \mathbf{v}/c\|} \tag{4.12}$$

Equations (4.11) and (4.12) agree to order v/c, but disagree in order $(v/c)^2$, which is not surprising because they were derived using completely classical arguments. Equation (4.12) has the appearance of a correction for the actual motion of the star during the light transmission time, but that is not at all what it represents. The star tracker certainly did not have a constant velocity while the light was traveling from the star, and it is unlikely that the tracker even existed for most of that time.

Both $\mathbf{s}^{\text{true}}$ and $\mathbf{s}^{\text{apparent}}$ are unit vectors, so the transformation between them is a rotation. To see this explicitly, we rewrite Eq. (4.11) to first order in v/c as

$$\begin{aligned}\mathbf{s}^{\text{apparent}} &\approx (\mathbf{s}^{\text{true}} + \mathbf{v}/c)(1 + 2\mathbf{s}^{\text{true}} \cdot \mathbf{v}/c)^{-1/2} \approx (\mathbf{s}^{\text{true}} + \mathbf{v}/c)(1 - \mathbf{s}^{\text{true}} \cdot \mathbf{v}/c)\\ &\approx \mathbf{s}^{\text{true}} + \mathbf{v}/c - (\mathbf{s}^{\text{true}} \cdot \mathbf{v}/c)\mathbf{s}^{\text{true}} = (I_3 - [\boldsymbol{\theta}^{\text{aber}}\times])\mathbf{s}^{\text{true}}\\ &\approx \exp(-[\boldsymbol{\theta}^{\text{aber}}\times])\mathbf{s}^{\text{true}}\end{aligned} \tag{4.13}$$

where

$$\boldsymbol{\theta}^{\text{aber}} \equiv (\mathbf{v}/c) \times \mathbf{s}^{\text{true}} \tag{4.14}$$

The matrix exponential is the rotation vector representation of the rotation matrix, as defined in Sect. 2.9.2. The direction of the rotation vector, $\boldsymbol{\theta}^{\text{aber}}$, is normal to the plane containing the star vector and the velocity.

The principal component of the aberration for Earth-orbiting spacecraft, with an amplitude of approximately 20 arcsec, is due to the motion of the Earth about the Sun.[4] Aberration arising from the motion of the spacecraft about the Earth is less than 5 arcsec. Aberration can be neglected when matching stars for attitude initialization, but it must be taken into account in determining the attitude of any spacecraft with fine pointing requirements. A straightforward procedure is to use Eq. (4.1) to compute an aberrated vector to a star from its centroid, then use Eq. (4.12) to compute the true vector, which can then be used for precise attitude estimation.

The difference in the aberration of two stars is

$$\boldsymbol{\theta}_1^{\text{aber}} - \boldsymbol{\theta}_2^{\text{aber}} = (\mathbf{v}/c) \times (\mathbf{s}_1^{\text{true}} - \mathbf{s}_2^{\text{true}}) \tag{4.15}$$

The maximum magnitude of this *differential aberration* is $2(v/c)\sin(\theta_{12}/2)$, where θ_{12} is the angular separation of the star vectors. The differential aberration between the boresight direction and any star is a maximum of 25 % of the full aberration for a tracker with a $20° \times 20°$ FOV, and 10 % of the full aberration for a tracker with an $8° \times 8°$ FOV. Tracked stars are usually not at the extreme edge of the focal plane, though, and they tend to be more or less uniformly distributed around the focal plane, causing the average of their differential aberrations with respect to the boresight to be small. Thus aberration is often computed assuming that all stars are on the boresight of the tracker, i.e. at the point (u_0, v_0) in the focal plane.

With the approximation that the aberration angle is the same for all stars, an autonomous star tracker can compute its attitude with respect to the celestial reference frame using the apparent star vectors without any aberration correction. The resulting attitude matrix, A^{aber}, maps the star vectors from the celestial frame

[4]This is $v/c \approx 100$ μrad, so $(v/c)^2 \approx 0.002$ arcsec, which indicates that a fully relativistic analysis is not required.

to their aberrated representations in the body frame, and thus it maps all reference frame vectors to tracker frame values rotated by $\boldsymbol{\theta}^{\text{aber}}$. The true attitude matrix that maps the reference vectors to their true values is found by rotating by the *negative* of the aberration vector

$$A^{\text{true}} = \exp([\boldsymbol{\theta}_T^{\text{aber}}\times])A^{\text{aber}} = A^{\text{aber}}\exp([\boldsymbol{\theta}_I^{\text{aber}}\times]) \tag{4.16}$$

where the aberration angle is computed assuming the stars are all along the tracker's boresight. Note that the aberration correction is applied on the left or the right side of the attitude matrix depending on whether $\boldsymbol{\theta}^{\text{aber}}$ is represented in the tracker frame T or the celestial frame I. Equation (2.63) establishes the equivalence of these two procedures. If the star tracker uses the aberrated measurements to compute an attitude quaternion $\mathbf{q}^{\text{aber}}$, the true quaternion is given to order v/c by

$$\mathbf{q}^{\text{true}} = \frac{1}{\sqrt{1+\|\boldsymbol{\theta}_T^{\text{aber}}\|^2}}\begin{bmatrix} -\boldsymbol{\theta}_T^{\text{aber}} \\ 1 \end{bmatrix} \otimes \mathbf{q}^{\text{aber}} = \frac{1}{\sqrt{1+\|\boldsymbol{\theta}_I^{\text{aber}}\|^2}}\mathbf{q}^{\text{aber}} \otimes \begin{bmatrix} -\boldsymbol{\theta}_I^{\text{aber}} \\ 1 \end{bmatrix} \tag{4.17}$$

If the star tracker is used to point a telescope or other instrument at an object that is closely aligned with the tracker's boresight, i.e. if the tracker is co-boresighted with the instrument, we can often omit the aberration correction of both the tracker's attitude and the science target's location, since these will cancel. Other instruments, such as magnetometers or Sun sensors, will appear to be misaligned in this case, but a 20 arcsec error may be insignificant compared with other error sources in their measurements.

4.3 Sun Sensors

Sun sensors fall into two classes, *coarse* Sun sensors (CSSs) and *fine* or *digital* Sun sensors (DSSs). The most common form of a CSS is a photocell (an *eye*) or an assembly of photocells. To a good approximation, the output of a photocell is an electric current directly proportional to the intensity of the light falling on it. This may include light from Earth albedo or glint off nearby components of the spacecraft, which can pull CSS outputs off the true Sun direction by as much as 20° in extreme cases [5,6]. The output of a CSS eye basically gives the average direction to sources of light energy falling on it. It is usually an adequate approximation to ignore these effects and treat the Sun as a point source, giving

$$I_j = \begin{cases} I_{\max}(\mathbf{n}_j \cdot \mathbf{s}) & \text{for } \mathbf{n}_j \cdot \mathbf{s} > 0 \\ 0 & \text{for } \mathbf{n}_j \cdot \mathbf{s} \le 0 \end{cases} \tag{4.18}$$

where $\mathbf{n}_j$ is the unit vector in the direction of the eye's outward normal and $\mathbf{s}$ is the unit vector in the direction from the spacecraft to the Sun. The response as a function of angle can be calibrated, and compensation for albedo and glint can be applied if necessary.

A second eye, with normal vector $\mathbf{n}_{-j} = -\mathbf{n}_j$, would have the output

$$I_{-j} = \begin{cases} I_{\max}(\mathbf{n}_{-j} \cdot \mathbf{s}) = -I_{\max}(\mathbf{n}_j \cdot \mathbf{s}) & \text{for } \mathbf{n}_{-j} \cdot \mathbf{s} = -\mathbf{n}_j \cdot \mathbf{s} > 0 \\ 0 & \text{for } \mathbf{n}_{-j} \cdot \mathbf{s} = -\mathbf{n}_j \cdot \mathbf{s} \leq 0 \end{cases} \tag{4.19}$$

so differencing the two outputs gives

$$I_j - I_{-j} = I_{\max}(\mathbf{n}_j \cdot \mathbf{s}) \quad \text{for all } \mathbf{n}_j \cdot \mathbf{s} \tag{4.20}$$

Six CSS eyes, with normal vectors $\pm\mathbf{n}_j$, $\pm\mathbf{n}_k$, and $\pm\mathbf{n}_\ell$, give

$$\begin{bmatrix} I_j - I_{-j} \\ I_k - I_{-k} \\ I_\ell - I_{-\ell} \end{bmatrix} = I_{\max} \begin{bmatrix} \mathbf{n}_j \cdot \mathbf{s} \\ \mathbf{n}_k \cdot \mathbf{s} \\ \mathbf{n}_\ell \cdot \mathbf{s} \end{bmatrix} = I_{\max} \begin{bmatrix} \mathbf{n}_j^T \\ \mathbf{n}_k^T \\ \mathbf{n}_\ell^T \end{bmatrix} \mathbf{s} \tag{4.21}$$

If $\mathbf{n}_j$, $\mathbf{n}_k$, and $\mathbf{n}_\ell$ are not coplanar, the Sun unit vector can be computed as

$$\begin{aligned} \mathbf{s} &= \frac{1}{I_{\max}} \begin{bmatrix} \mathbf{n}_j^T \\ \mathbf{n}_k^T \\ \mathbf{n}_\ell^T \end{bmatrix}^{-1} \begin{bmatrix} I_j - I_{-j} \\ I_k - I_{-k} \\ I_\ell - I_{-\ell} \end{bmatrix} \\ &= \frac{(I_j - I_{-j})(\mathbf{n}_k \times \mathbf{n}_\ell) + (I_k - I_{-k})(\mathbf{n}_\ell \times \mathbf{n}_j) + (I_\ell - I_{-\ell})(\mathbf{n}_j \times \mathbf{n}_k)}{I_{\max}[\mathbf{n}_j \cdot (\mathbf{n}_k \times \mathbf{n}_\ell)]} \end{aligned} \tag{4.22}$$

The eye currents are slowly varying functions of the Sun direction near normal incidence, so the Sun vector is least well determined when the Sun vector is aligned with one of the eye normal vectors. A six-eye CSS configuration (actually two redundant sets of six eyes) was chosen for WMAP, for which the expected position of the Sun was 22.5° from the $-z$ axis (see Fig. 7.4). The eyes were located at the outer edges of the six solar panels pointing alternately 35.26° up and 35.26° down from the x–y plane. Thus the unit normal vectors formed an orthogonal triad, perpendicular to the faces of a cube with one of its body diagonals along the $-z$ axis.

Fine Sun sensors have historically relied on arrays of slits (or reticles) with linear photosensitive surfaces behind them. Sun sensors of this design are described in Wertz [66] and in Sect. 7.7 of this text. A newer DSS design is basically a star tracker with a pinhole in place of the light-gathering optics, which are not needed because of the Sun's brightness. The resolutions of DSSs range from 1 deg to better than 1 arc min. They have medium-size fields of view, a typical value being 128° × 128°, so CSSs are usually employed to maneuver the spacecraft to move the Sun into the DSS FOV.

Two-slit Sun sensors for spinning spacecraft are also described by Wertz [66]. A *command* slit parallel to the spin axis notes the time at which the slit azimuth is in the Sun direction. A second *measurement* slit perpendicular to the spin axis directs the sunlight to a linear array of detectors to determine the elevation of the Sun unit vector with respect to the spin axis.

4.4 Horizon Sensors

Horizon sensors have been used on many Earth-orbiting spacecraft, especially on Earth-pointing spacecraft. The Earth has a finite size and cannot be treated as a point to any degree of accuracy, so a horizon sensor detects points on the Earth's horizon, as its name implies. The appearance of the Earth in visible wavelengths is quite complicated; aside from having oceans, vegetation, and deserts, it has phases like the Moon. The appearance is more uniform at infrared wavelengths, especially in the narrow 14–16 μm emission band of the CO_2 molecule, so almost all horizon sensors are designed to detect infrared radiation in this range [66].

Horizon sensors are fundamentally of two types: *static* sensors that look in fixed directions, and *scanning* sensors that move a small FOV of a detector across the Earth. Static sensors, by their nature, are limited to Earth-pointing spacecraft at small pitch and roll angles and in a limited altitude range. Static horizon sensors looking at four points roughly equally spaced around the Earth's horizon were used by TRMM and by the Television Infrared Observation Satellite (TIROS) and Defense Meteorological Satellite Program (DMSP) spacecraft. Scanning sensors can use an oscillatory or rotational scan, with rotation provided by incorporating the horizon scanner into a reaction wheel, by providing a separate mechanism, or by using the spin of a spin-stabilized spacecraft.

Unlike star trackers and Sun sensors that detect objects with definite positions in a reference frame, horizon sensors detect a point on the Earth's horizon with an a priori unknown location. Multiple horizon crossings can be used to find a nadir vector by means discussed in [66], which contains a wealth of information on horizon sensors. The nadir computed from horizon sensor measurements on an oblate Earth is a better approximation to the geodetic nadir normal to the surface of the ellipsoidal figure of the Earth than to the geocentric nadir pointing to the center of the Earth. As shown in Sect. 2.6.3, these directions can differ by as much as 12 arc min. The accuracy of horizon sensors is limited to about 0.1°, mainly by the ability to accurately model the height of the CO_2 layer in the Earth's atmosphere.

4.5 Magnetometers

Most spacecraft magnetometers are fluxgate magnetometers [66], which are relatively small, lightweight, rugged, and inexpensive. They have no moving parts and do not require a clear field of view. They do require a well-modeled magnetic field if they are to be used as attitude sensors, which practically limits their use in this capacity to low-Earth orbit. A magnetic field model is not needed if magnetometer measurements are used only to compute magnetic torque commands.

Magnetometers measure the sum of the ambient field that is of interest and any local fields produced by the spacecraft. Local fields can be produced by ferromagnetic materials or by current loops in solar arrays, electric motors, payload instruments, or most especially attitude control torquers. If the local fields are known, they can be compensated for. If they are not known, the magnetometers can be located far from the sources of magnetic contamination, on a deployable boom if necessary, to take advantage of the $1/r^3$ falloff of a magnetic dipole field (see Sect. 11.1).

4.6 Global Positioning System

Using the GPS for satellite navigation is now widespread. The GPS constellation was originally developed to permit a wide variety of user vehicles an accurate means of determining position for autonomous navigation [51]. The original constellation included 24 space vehicles (SVs) in known semi-synchronous (12-h) orbits, providing a minimum of six SVs in view for ground-based navigation. The underlying principle involves geometric triangulation with the GPS SVs as known reference points to determine the user's position to a high degree of accuracy. A minimum of four SVs is required so that, in addition to the three-dimensional position of the user, the time of the solution can be determined and in turn employed to correct the user's clock. As of December 2012, 8 more satellites have been added to the constellation, which was also changed to a nonuniform arrangement. This allows nearly 9 SVs to be available at any time and location on the Earth, which provides considerable redundancy and improvements in user location. The number of available GPS satellites for space-based applications depends on the user spacecraft's altitude.

The fundamental signal in GPS is the pseudo-random code (PRC), which is a complicated binary sequence of pulses. Each SV has its own complex PRC, which guarantees that the receiver will not be confused with another SV's signal. The GPS satellites transmit signals on two carrier frequencies: L1 at 1575.42 MHz and L2 at 1227.60 MHz. The modulated PRC at the L1 carrier is called the Coarse Acquisition (C/A) code, which repeats every 1,023 bits and modulates at a 1 MHz rate. The C/A code is the basis for civilian GPS use. Another PRC is called the Precise (P) code, which repeats on a 7-day cycle and modulates both the L1 and L2 carriers at a 10 MHz rate. This code is intended for military users and can be encrypted. Position location is made possible by comparing how late in time the SV's PRC appears relative to the receiver's code. Multiplying the travel time by the speed of light, one obtains the distance to the SV. This requires very accurate timing in the receiver, which is provided by using a fourth SV to correct a "clock bias" in the receiver's internal clock.

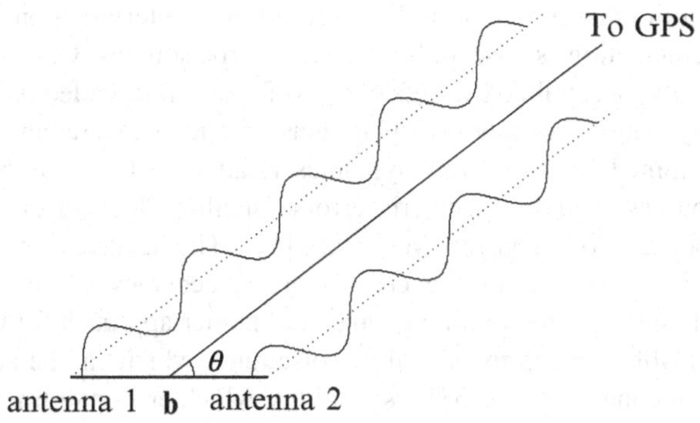

Fig. 4.4 GPS waves and wavefront angle

Attitude determination using phase differences of GPS signals received by antennas located at different locations is a later development, even though the original concept was proposed in 1976 by Spinney [61]. The first practical application for GPS attitude determination was shown by Cohen in 1993 [15]. Cohen and Trimble Navigation, Ltd. designed the TANS Vector sensor system, which was then primarily used for airborne applications and tracked up to 6 satellites on 4 separate antennas. The Radar Calibration (RADCAL) satellite was the first satellite to provide spaced-based GPS measurements that determined spacecraft attitude using post-processed data. The first realtime space-based attitude determination application was the Cryogenic Infrared Spectrometers and Telescopes for the Atmosphere/Shuttle Pallet Satellite (CRISTA-SPAS) based on algorithms developed by Lightsey [35].

Attitude determination using GPS signals is based on measuring the magnitude of the carrier wavelength directly. At least two antennas are required to form a "baseline" vector **b** expressed in body-frame coordinates, as depicted in Fig. 4.4. It is assumed that the baseline distance is significantly smaller than the distance to the GPS satellites so that planar waves are given to the receiver. The different path lengths from each GPS satellite to the antennas at the two ends of the baseline create a phase difference of the received signals. Note that an integer number of cycles may be present in this phase difference if the baseline distance, given by $\|\mathbf{b}\|$, is larger than the GPS carrier wavelength, which is 19 cm for the L1 carrier. This leads to the classic "integer ambiguity" problem [16]. There are several steps required to perform attitude determination. A short list of the general ones is as follows: (1) performing a "self survey" to determine several system parameters, such as integer ambiguities, line biases, and baselines [50], (2) determining the sightline vectors to all available GPS satellites, (3) maintaining a count of the integer cycles as the attitude changes, and (4) using the known multiple baseline vectors, sightline vectors, and phase differences (with integers removed) to determine the attitude, which is discussed in Sect. 5.9.

Even though the theory behind GPS-based attitude determination is sound, its practical implementation is susceptible to many error sources. One of the largest error sources is its susceptibility to reflections off spacecraft, called *multipath* [10]. This essentially causes a slowly-varying bias in the measurements with time constants determined by spacecraft dynamics relative to the GPS constellation. Other error sources include ephemeris errors, satellite clock errors, ionosphere errors, troposphere errors, and receiver errors [52]. The accuracy of GPS attitude determination is a function of the error sources, accuracy of the self survey, lengths of the baselines, and satellite geometry. Another approach for GPS attitude determination involves using the signal-to-noise ratio, which can be accomplished using a single antenna [3, 36, 39, 58]. Pseudolites, which are small transceivers that are used to create local GPS-like signals, can also be used for attitude determination. However, this leads to more complicated solutions because planar wavefronts cannot be assumed in general [49].

A GPS receiver built for terrestrial applications will not work properly for space applications in general. Lightsey [35] discusses the issues involved with modifying a GPS receiver for space. The first issue is that higher vehicle velocities exist in space than for ground and air applications. For example, for low-Earth orbit applications this results in Doppler shifts that are more than 10 times greater than those observed from the ground and Doppler shift rates that can be more than 100 times higher. The conclusion from these facts is that the carrier and code tracking loops must be redesigned for space applications. The next issue is the "full sky pointing" problem. For ground applications one side of the vehicle generally points in the same direction at all times. This is vastly different for rotating spacecraft, which may cause significant outages if all the hemispherical antennas are pointed in the same direction. This can be mitigated by using a carrier phase measurement correction, which Lightsey defines as a design of a "non-aligned" antenna array. Another issue is that the GPS electronics, especially the processor units, must be redesigned to handle the harsh space environment. Other issues include vibrational effects on the antennas, which can change their baseline length, and larger multipath errors than ground applications due to more reflections from the metallic spacecraft components. All of these issues can decrease the attitude performance if they are not properly accounted for in the GPS receiver design.

4.6.1 GPS Satellites

The onboard GPS satellite information is usually given by a GPS ephemeris. For simulation purposes a less-precise almanac is used,[5] which provides orbital element information, including the time of applicability, t_a, eccentricity, e, inclination, i,

[5]The U.S. Coast Guard Navigation Center maintains a website that contains GPS almanacs, and as of this writing this website is given by http://www.navcen.uscg.gov/.

Table 4.1 Equations to compute GPS ECEF positions over time

$a = \sqrt{a^2}$	Semimajor Axis
$n = \sqrt{\dfrac{\mu}{a^3}}$	Computed Mean Motion
$t_k = t - t_a$	Time Since Applicability
$M_k = M_0 + t_k\, n$	Mean Anomaly
$E_k = M_k + e \sin E_k$	Solve Kepler's Equation for E_k
$\tan \dfrac{v_k}{2} = \sqrt{\dfrac{1+e}{1-e}} \tan \dfrac{E_k}{2}$	True Anomaly
$\Omega_k = \Omega_0 + \dot{\Omega}\, t_k - \omega_e\, t$	Corrected Ascending Node
$\lambda_k = v_k + \omega$	Argument of Latitude
$r_k = a\,(1 - e \cos E_k)$	Orbital Radius
$\mathbf{r}_k^0 = \begin{bmatrix} r_k \cos \lambda_k \\ r_k \sin \lambda_k \end{bmatrix}$	Orbit Plane Position
$\mathbf{r}_k^E = \begin{bmatrix} \cos \Omega_k & -\cos i \sin \Omega_k \\ \sin \Omega_k & \cos i \cos \Omega_k \\ 0 & \sin i \end{bmatrix} \mathbf{r}_k^0$	ECEF Position

semimajor axis, a, right ascension, Ω_0, rate of right ascension, $\dot{\Omega}$, argument of perigee, ω, and mean anomaly, M_0. See Chap. 10 for a discussion of the orbital elements. We should note that the right ascension is given with respect to the prime meridian, which allows us to compute the ECEF position directly. It should be noted that GPS time is based on the atomic standard time and is continuous without the leap seconds that Universal Time (UT) uses to account for the non-smooth rotation of the Earth. GPS epoch is midnight of January 6, 1980, and GPS time is conventionally represented in weeks and seconds from this epoch. The GPS week is represented by an integer from 0 to 1,023. A rollover occurred on August 22, 1999, so that 1,024 needs to be added for references past this date.

To simulate the GPS sightline vector a simple algorithm using GPS almanac data is sufficient. The broadcast ephemeris, which contains more parameters such as amplitude of second-order harmonic perturbations [32], should be used if more accuracy is needed. For simulation purposes, counting the days past GPS epoch to determine UT is adequate (ignoring leap seconds, but not leap days). The position vector of the GPS satellite is denoted by $\mathbf{r}^E$ and the position of the GPS receiver is denoted by $\mathbf{p}^E$. Table 4.1 gives the equations necessary to determine the GPS ECEF positions. The variable $\omega_e = 7.292115 \times 10^{-5}$ rad/s is the Earth's rotation rate given from WGS-84, $\mu = 3.98600441 \times 10^{14}$ m^3/s^2 is the Earth's gravitational parameter, and t_k is the time past the time of applicability (the subscript k denotes the kth time-step). The sightline vector is computed using

$$\mathbf{s}_j = \frac{\mathbf{r}_j^E - \mathbf{p}^E}{\|\mathbf{r}_j^E - \mathbf{p}^E\|} \tag{4.23}$$

where the subscript j now denotes the jth available GPS satellite.

4.7 Gyroscopes

Gyroscopes on spacecraft are almost exclusively used in *strapdown* mode, which means that the gyro is solidly attached to the spacecraft rather than being used to control a separate gimbaled platform. Gyros generally fall into two broad classes: rate gyros that read angular rates, and rate-integrating gyros (RIGs) that measure integrated rates or angular displacements [19,44,54]. Gyros can also be classified by the physical mechanisms they use: spinning-mass gyros, optical gyros, or Coriolis vibratory gyros (CVGs).

Before the 1980s, all operational gyros were spinning-mass gyros, which depend on the tendency of the angular momentum of a rotating mass to remain fixed in inertial space. These are used in torque rebalance mode, meaning that the angular momentum is held constant in the spacecraft frame. Equation (3.80) tells us that the torque required to keep the gyro's angular momentum $\mathbf{H}_B$ constant is

$$\mathbf{L}_B = \boldsymbol{\omega}_B^{BI} \times \mathbf{H}_B \tag{4.24}$$

The two components of $\boldsymbol{\omega}_B^{BI}$ along the input axes perpendicular to $\mathbf{H}_B$ determine the required two components of $\mathbf{L}_B$ along the output axes, given by the cross products of the input axes with the direction of $\mathbf{H}_B$. Single-axis gyros supply and sense one of these torques electromagnetically, depending on mechanical pivots to provide the other torque, while two-axis gyros provide and sense both of these torques electromagnetically.

The most accurate spacecraft gyros have been single-axis floated gyros, as depicted in Fig. 4.5. These have the gyro rotor and its drive motor contained in a cylindrical "can," with the rotor's spin direction along one of its diameters. The perpendicular diameter is the input axis, and the output axis is directed from one end of the can to the other. The can is supported by pivots at its ends and by the buoyancy of a fluid having a density equal to the total mass of the can and its contents divided by its volume. Neutral buoyancy allows the pivots to control two

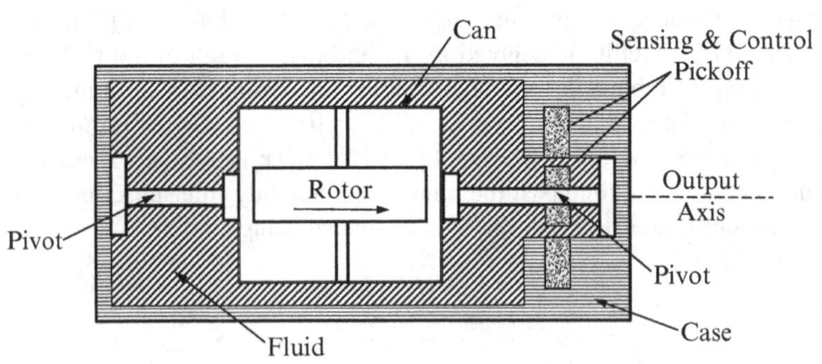

Fig. 4.5 Single-axis floated gyro

degrees of freedom of the can while leaving it essentially free to rotate about its output axis. This rotation is sensed and controlled electromagnetically to measure the angular rate about the input axis. In order to minimize perturbations on the can, the power and signals to drive and sense the gyro rotation are carried through the flotation fluid by thin flexible leads.

Two-axis gyros are called dry tuned-rotor gyros (DTGs), because they are not floated [44]. The motor is rigidly fixed to the spacecraft, and the rotor is attached to the shaft by gimbals and torsion springs carefully designed and tuned to the spin speed of the rotor to minimize their effective restoring torque, allowing the rotor to flex in two directions. The rotor deflection is sensed on two axes and controlled on the two cross-product axes electromagnetically. DTGs have come very close to the performance of floated gyros, but never achieved performance equivalent to the best floated gyros. However, their relative simplicity and ruggedness led to their use on a great number of spacecraft in the three decades beginning around 1978.

The most common failure modes of spinning-mass gyros are due to problems with the rotor bearings. The floated gyros on HST, which use bromotrifluoroethylene (BTFE) as the flotation fluid to provide the required high density, experience a different failure mode. Corrosion of the silver-copper alloy flex leads by the BTFE has caused multiple gyro failures requiring on-orbit replacement during Space Shuttle servicing missions [29].

The problems with spinning-mass gyros, including the difficulty and expense of their fabrication, has led to the development of alternative rate-sensing technologies, specifically optical gyros and CVGs. Optical gyros depend on the Sagnac effect, which applies to light traveling in opposite directions around a closed path that is rotating with respect to inertial space [14, 44]. Consider a circular path of radius R in a plane that is rotating with clockwise angular velocity ω about an axis perpendicular to the plane. Light making a complete circuit clockwise must travel a distance

$$L^+ = c\,t^+ = 2\pi R + R\,\omega\,t^+ \tag{4.25}$$

where t^+ is the time to complete the circuit and c is the speed of light. For light traveling counterclockwise, we have

$$L^- = c\,t^- = 2\pi R - R\,\omega\,t^- \tag{4.26}$$

The difference in the path lengths is

$$\begin{aligned} \Delta L = L^+ - L^- = c(t^+ - t^-) &= c\left(\frac{2\pi R}{c - R\,\omega} - \frac{2\pi R}{c + R\,\omega}\right) \\ &= \frac{4\pi c R^2\,\omega}{c^2 - R^2\,\omega^2} \approx \frac{4\pi R^2\,\omega}{c} = \frac{4A\,\omega}{c} \end{aligned} \tag{4.27}$$

This equation generalizes to noncircular and nonplanar paths, with A being the area enclosed by the path projected onto the plane perpendicular to the angular velocity [55].

The Sagnac path difference can in principle be detected as a phase difference $\Delta\phi = 2\pi \Delta L/\lambda$ rad, where λ is the wavelength of the light. However, detecting a rate of 0.001 deg/h = 4.85 nrad/s using helium-neon laser light with a wavelength of 632.8 nm in a circular loop with $R = 0.1$ m requires measuring a phase difference of only 2×10^{-11} rad, which is impossible. There are two ways to circumvent this problem. The first is to make the light travel around the closed path many times, by using many turns of an optical fiber. An interferometric fiber optic gyro (IFOG) using 5,000 turns for a total fiber length of 3,142 m only has to measure a phase difference of 10^{-7} rad for the example above, which is feasible.[6] We will discuss IFOGs later.

The second way to circumvent the problem of measuring small phase differences is to use an active laser medium in the path to replace the phase measurement by a frequency measurement. This is the basic idea of a ring laser gyro (RLG). The rotation splits the laser wavelength and frequency into clockwise and counterclockwise modes

$$\nu^{\pm} = \frac{c}{\lambda^{\pm}} = \frac{kc}{L^{\pm}} \tag{4.28}$$

where k is the integer number of wavelengths in the cavity. Then[7]

$$\Delta\nu = \nu^{-} - \nu^{+} = \frac{kc}{L^{-}} - \frac{kc}{L^{+}} = \frac{kc\Delta L}{L^{+}L^{-}} \approx \frac{kc\Delta L}{L^{2}} = \frac{4A\,\omega}{L\lambda} \tag{4.29}$$

where we have substituted Eq. (4.27) and $k = L/\lambda$ to obtain the last form. This form holds for a path of any shape, where L is the perimeter and A is the enclosed area projected onto the plane perpendicular to the rotation. Most RLGs have paths in the shape of an equilateral triangle with reflecting mirrors at the vertices. If ℓ is the length of one side then $L = 3\ell$, $A = (\sqrt{3}/4)\ell^2$, and

$$\Delta\nu = \frac{\ell\,\omega}{\sqrt{3}\lambda} \tag{4.30}$$

With $\lambda = 632.8$ nm and $\ell = 0.1$ m, this gives $\Delta\nu = (9.1 \times 10^4)\omega$, which is equal to 442 μHz for $\omega = 0.001$ deg/h. The relative ease of detecting small frequency differences has led to the widespread use of RLGs on aircraft and, to a lesser extent, on spacecraft.

An RLG has difficulty in detecting very slow rotations, because nonlinear interactions between the clockwise and counterclockwise oscillations cause their frequencies to lock in to each other. This is usually solved by introducing a sinusoidal dither of the gyro about its input axis, with typical frequency of 400 Hz

[6] Equation (4.27) assumes that the light travels through a vacuum. The fiber's index of refraction complicates the analysis, but does not change the order-of-magnitude estimates.

[7] Note that ν is measured in Hz, and ω in rad/s.

and amplitude of 0.1° [44]. Using laser oscillations with four frequencies traveling in a non-planar path can avoid lock-in without dithering. Although RLGs have established themselves as the best rate sensing technology in terms of performance and reliability in a wide range of applications, they have not been perfect. A slow decrease of laser intensity, thought to be a combined effect of degradation of the mirror surfaces and of the lasing medium, has been observed on many RLGs. This is less of a problem on aircraft, which are serviced periodically, than on spacecraft.

Although the basic concept of the IFOG is simple, the implementation is not. Reference [45] provides many details. By the mid 1980s, the IFOG had become competitive with the RLG, despite its lower sensitivity, typically about 0.1 deg/h as contrasted with 0.001 deg/h achieved by RLGs. The reason for the IFOG's success includes some desirable features like light weight, small size, limited power consumption, projected long lifetime and, not least, lower price. Continuing development has led to increased sensitivity, although a market for navigation-grade IFOGs has not yet materialized; economic forces have concentrated IFOG development to low price, medium sensitivity applications.

CVGs detect the motion of a vibrational mode in a structure caused by Coriolis forces. The first CVG to be developed was the hemispheric resonator gyroscope (HRG), which uses a flexural mode in a thin wine-glass-shaped resonator anchored by a thick stem. Coriolis forces arising from rotation about the stem axis cause a slow precession of a standing wave around the resonator with an angular rate that differs from the input rate. An HRG has two modes of operation: a rebalance mode that applies forces to keep the standing wave pattern fixed and a less accurate "whole-angle" mode that measures the free motion of the pattern. HRGs have achieved accuracy equal to that of the best DTGs and have found many spacecraft applications.

CVGs manufactured with micro-electro mechanical systems (MEMS) technology are the newest development. These are packaged similarly to other integrated circuits with a single part often including gyroscopic sensors for multiple axes. The resonant structures of MEMS gyroscopes are lithographically constructed tuning forks, vibrating wheels, or resonant solids of various designs. MEMS gyros have low cost, low mass, and low power requirements, but also low performance and a short lifetime compared to other gyros. MEMS gyros using components qualified for the radiation environment of space have only recently become available.

4.7.1 Gyro Measurement Model

A widely used three-axis continuous-time mathematical model for a rate-integrating gyro is given by [20]

$$\boldsymbol{\omega}(t) = \boldsymbol{\omega}^{\text{true}}(t) + \boldsymbol{\beta}^{\text{true}}(t) + \boldsymbol{\eta}_v(t) \tag{4.31a}$$

$$\dot{\boldsymbol{\beta}}^{\text{true}}(t) = \boldsymbol{\eta}_u(t) \tag{4.31b}$$

where $\boldsymbol{\omega}^{\text{true}}$ is the true rate, $\boldsymbol{\omega}$ is the measured rate, $\boldsymbol{\beta}^{\text{true}}$ is the true bias or drift, and $\boldsymbol{\eta}_v$ and $\boldsymbol{\eta}_u$ are independent zero-mean Gaussian white-noise processes with

$$E\left\{\boldsymbol{\eta}_v(t)\boldsymbol{\eta}_v^T(\tau)\right\} = \sigma_v^2\delta(t-\tau)I_3 \tag{4.32a}$$

$$E\left\{\boldsymbol{\eta}_u(t)\boldsymbol{\eta}_u^T(\tau)\right\} = \sigma_u^2\delta(t-\tau)I_3 \tag{4.32b}$$

where $\delta(t-\tau)$ is the Dirac delta function defined as

$$\begin{aligned} \delta(t-\tau) &= 0 \quad \text{if } t \neq \tau \\ \int_{-\infty}^{\infty} \delta(t-\tau)d\tau &= 1 \end{aligned} \tag{4.33}$$

A more general gyro model includes scale factors and misalignments, which can also be estimated in realtime [48, 53]. The general gyro model including scaling factors and misalignments is given by

$$\boldsymbol{\omega} = \left(I_3 + S^{\text{true}}\right)\boldsymbol{\omega}^{\text{true}} + \boldsymbol{\beta}^{\text{true}} + \boldsymbol{\eta}_v \tag{4.34}$$

with S^{true} denoting the matrix of scale factors and misalignments.

Both Eq. (4.31) and Eq. (4.34) represent continuous-time gyro models but in practice discrete-time gyro measurements are employed and therefore discrete-time models are required. Because the white-noise processes are assumed to be uncorrelated we can treat each axis separately. Dividing the single-axis version of Eq. (4.31a) by the gyro sampling interval, Δt, and integrating gives

$$\frac{1}{\Delta t}\int_{t_0}^{t_0+\Delta t} \omega(t)\,dt = \frac{1}{\Delta t}\int_{t_0}^{t_0+\Delta t} \left[\omega^{\text{true}}(t) + \beta^{\text{true}}(t) + \eta_v(t)\right]\,dt \tag{4.35}$$

Approximating the measured and true values as constant over the interval yields[8]

$$\omega(t_0+\Delta t) = \omega^{\text{true}}(t_0+\Delta t) + \frac{1}{\Delta t}\int_{t_0}^{t_0+\Delta t} \left[\beta^{\text{true}}(t) + \eta_v(t)\right]\,dt \tag{4.36}$$

Integrating Eq. (4.31b) gives

$$\beta^{\text{true}}(t_0+\Delta t) = \beta^{\text{true}}(t_0) + \int_{t_0}^{t_0+\Delta t} \eta_u(t)\,dt \tag{4.37}$$

The variance of the gyro drift bias is given by

$$E\left\{[\beta^{\text{true}}(t_0+\Delta t)]^2\right\} = E\left\{\left[\beta^{\text{true}}(t_0) + \int_{t_0}^{t_0+\Delta t} \eta_u(t)\,dt\right]^2\right\} \tag{4.38}$$

[8]Note that we cannot make the same assumption for the stochastic variables.

Using $E\left\{\eta_u(t)\eta_u(\tau)\right\} = \sigma_u^2\,\delta(t-\tau)$ gives

$$E\left\{[\beta^{\text{true}}(t_0+\Delta t)]^2\right\} = E\left\{[\beta^{\text{true}}(t_0)]^2\right\} + \sigma_u^2\,\Delta t \tag{4.39}$$

Therefore, the bias can be simulated using

$$\beta_m^{\text{true}}(t_0+\Delta t) = \beta_m^{\text{true}}(t_0) + \sigma_u\,\Delta t^{1/2} N_u \tag{4.40}$$

where the subscript m denotes a modeled quantity and N_u is a zero-mean random variable with unit variance.

The bias at time t is given by

$$\beta^{\text{true}}(t) = \beta^{\text{true}}(t_0) + \int_{t_0}^{t} \eta_u(\tau)\,d\tau \tag{4.41}$$

Substituting Eq. (4.41) into Eq. (4.36) gives

$$\omega(t_0+\Delta t) = z + \frac{1}{\Delta t}\int_{t_0}^{t_0+\Delta t}\int_{t_0}^{t} \eta_u(\tau)\,d\tau\,dt + \frac{1}{\Delta t}\int_{t_0}^{t_0+\Delta t} \eta_v(t)\,dt \tag{4.42}$$

where $z \equiv \omega^{\text{true}}(t_0+\Delta t) + \beta^{\text{true}}(t_0)$. The correlation between the drift and rate measurement is given by

$$\begin{aligned}
E\left\{\beta^{\text{true}}\,(t_0+\Delta t)\,\omega(t_0+\Delta t)\right\} = E\Bigg\{&\left[\beta^{\text{true}}(t_0) + \int_{t_0}^{t_0+\Delta t} \eta_u(\tau)\,d\tau\right] \\
&\times\Bigg[\omega^{\text{true}}(t_0+\Delta t) + \beta^{\text{true}}(t_0) \\
&+ \frac{1}{\Delta t}\int_{t_0}^{t_0+\Delta t}\int_{t_0}^{t} \eta_u(\zeta)\,d\zeta\,dt + \frac{1}{\Delta t}\int_{t_0}^{t_0+\Delta t} \eta_v(t)\,dt\Bigg]\Bigg\}
\end{aligned} \tag{4.43}$$

Since $\eta_u(t)$ and $\eta_v(t)$ are uncorrelated we have

$$\begin{aligned}
E\left\{\beta^{\text{true}}(t_0+\Delta t)\,\omega(t_0+\Delta t)\right\} &= E\left\{z\,\beta^{\text{true}}(t_0)\right\} \\
&\quad + \frac{\sigma_u^2}{\Delta t}\int_{t_0}^{t_0+\Delta t}\int_{t_0}^{t_0+\Delta t}\int_{t_0}^{t} \delta(\tau-\zeta)\,d\zeta\,d\tau\,dt \\
&= E\left\{z\,\beta^{\text{true}}(t_0)\right\} + \frac{\sigma_u^2}{\Delta t}\int_{t_0}^{t_0+\Delta t} (t-t_0)\,dt \\
&= E\left\{z\,\beta^{\text{true}}(t_0)\right\} + \frac{1}{2}\sigma_u^2\,\Delta t
\end{aligned} \tag{4.44}$$

Equation (4.44) can be satisfied by modeling the gyro measurement using

$$\omega_m(t_0+\Delta t) = \omega_m^{\text{true}}(t_0+\Delta t) + \beta_m^{\text{true}}(t_0) + \frac{1}{2}\sigma_u\,\Delta t^{1/2} N_u + c\,N_v \tag{4.45}$$

where the quantity c is yet to be determined and N_v is a zero-mean random variable with unit variance. Note that Eq. (4.45) can be proven by evaluating $E\{\beta_m^{\text{true}}(t_0+\Delta t)\,\omega_m(t_0+\Delta t)\}$. Solving Eq. (4.40) for N_u and substituting the resultant into Eq. (4.45) yields

$$\omega_m(t_0+\Delta t) = \omega_m^{\text{true}}(t_0+\Delta t) + \frac{1}{2}[\beta_m^{\text{true}}(t_0+\Delta t) + \beta_m^{\text{true}}(t_0)] + c\,N_v \tag{4.46}$$

Note that $\frac{1}{2}[\beta_m^{\text{true}}(t_0+\Delta t) + \beta_m^{\text{true}}(t_0)]$ is the "average" of the bias at the two times. This term is present due to the fact that the trapezoid rule for integration is exact for linear systems. To evaluate c we compute the variance of the rate measurement:

$$\begin{aligned} &E\left\{\omega^2\,(t_0+\Delta t)\right\} \\ &\quad= E\left\{\left[z + \frac{1}{\Delta t}\int_{t_0}^{t_0+\Delta t}\int_{t_0}^{\tau}\eta_u(\upsilon)\,d\upsilon\,d\tau + \frac{1}{\Delta t}\int_{t_0}^{t_0+\Delta t}\eta_v(\tau)\,d\tau\right]\right. \\ &\quad\quad \times\left.\left[z + \frac{1}{\Delta t}\int_{t_0}^{t_0+\Delta t}\int_{t_0}^{t}\eta_u(\zeta)\,d\zeta\,dt + \frac{1}{\Delta t}\int_{t_0}^{t_0+\Delta t}\eta_v(t)\,dt\right]\right\} \end{aligned} \tag{4.47}$$

Since $\eta_u(t)$ and $\eta_v(t)$ are uncorrelated and using $E\{\eta_v(t)\eta_v(\tau)\} = \sigma_v^2\,\delta(t-\tau)$, then Eq. (4.47) simplifies to

$$\begin{aligned} E\left\{\omega^2(t_0+\Delta t)\right\} = &\frac{\sigma_u^2}{\Delta t^2}\int_{t_0}^{t_0+\Delta t}\int_{t_0}^{t_0+\Delta t}\int_{t_0}^{t}\int_{t_0}^{\tau}\delta(\upsilon-\zeta)\,d\upsilon\,d\zeta\,d\tau\,dt \\ &+ E\left\{z^2\right\} + \frac{\sigma_v^2}{\Delta t^2}\int_{t_0}^{t_0+\Delta t}\int_{t_0}^{t_0+\Delta t}\delta(t-\tau)\,d\tau\,dt \end{aligned} \tag{4.48}$$

The quadruple integral in this equation is equal to

$$\begin{aligned} &\int_{t_0}^{t_0+\Delta t}\int_{t_0}^{t_0+\Delta t}\min(\tau-t_0,\,t-t_0)\,d\tau\,dt = \int_{t_0}^{t_0+\Delta t}\int_{t_0}^{t_0+\Delta t}\min(x,\,y)\,dx\,dy \\ &= \int_0^{\Delta t}\left(\int_0^{y} x\,dx + \int_y^{\Delta t} y\,dx\right)dy = \int_0^{\Delta t}\left[\frac{1}{2}y^2 + y(\Delta t - y)\right]dy = \frac{1}{3}\Delta t^3 \end{aligned} \tag{4.49}$$

Therefore, Eq. (4.48) reduces down to

$$E\left\{\omega^2(t_0+\Delta t)\right\} = E\left\{z^2\right\} + \frac{1}{3}\sigma_u^2\,\Delta t + \frac{\sigma_v^2}{\Delta t} \tag{4.50}$$

The variance of the modeled rate measurement in Eq. (4.45) is given by

$$E\left\{\omega_m^2(t_0+\Delta t)\right\} = E\left\{z_m^2\right\} + \frac{1}{4}\sigma_u^2\,\Delta t + c^2 \tag{4.51}$$

Comparing Eq. (4.51) to Eq. (4.50) gives

$$c^2 = \frac{\sigma_v^2}{\Delta t} + \frac{1}{12}\sigma_u^2\,\Delta t \tag{4.52}$$

Hence, the modeled rate measurement is given by

$$\begin{aligned}\omega_m(t_0+\Delta t) &= \omega_m^{\text{true}}(t_0+\Delta t) + \frac{1}{2}[\beta_m^{\text{true}}(t_0+\Delta t) + \beta_m^{\text{true}}(t_0)] \\ &+ \left[\frac{\sigma_v^2}{\Delta t} + \frac{1}{12}\sigma_u^2\,\Delta t\right]^{1/2} N_v\end{aligned} \tag{4.53}$$

Generalizing Eqs. (4.40) and (4.53) for all times, dropping the subscript m, and considering all three axes gives the following formulas for the discrete-time rate and bias equations:

$$\boldsymbol{\omega}_{k+1} = \boldsymbol{\omega}_{k+1}^{\text{true}} + \frac{1}{2}(\boldsymbol{\beta}_{k+1}^{\text{true}} + \boldsymbol{\beta}_k^{\text{true}}) + \left(\frac{\sigma_v^2}{\Delta t} + \frac{1}{12}\sigma_u^2\,\Delta t\right)^{1/2} \mathbf{N}_{v_k} \tag{4.54a}$$

$$\boldsymbol{\beta}_{k+1}^{\text{true}} = \boldsymbol{\beta}_k^{\text{true}} + \sigma_u\,\Delta t^{1/2}\mathbf{N}_{u_k} \tag{4.54b}$$

where the subscript k denotes the kth time-step, and $\mathbf{N}_{v_k}$ and $\mathbf{N}_{u_k}$ are zero-mean Gaussian white-noise processes with covariance each given by the identity matrix. Replacing $\boldsymbol{\omega}_{k+1}^{\text{true}}$ with $(I_3 + S^{\text{true}})\,\boldsymbol{\omega}_{k+1}^{\text{true}}$ in Eq. (4.54a) provides the discrete-time model for Eq. (4.34).

4.8 Reaction Wheels

Reaction wheels are used as the primary attitude control actuators on most spacecraft. Momentum-bias spacecraft may use one or two reaction wheels, but full three-axis attitude control requires three or more wheels. Although numerous reaction wheels have operated flawlessly for decades, reaction wheel failures have been a problem on many space missions. Providing extra reaction wheels for redundancy gives some protection, but failure of one reaction wheel is often followed by failure of other wheels of the same design on the same spacecraft. For example, four reaction wheels failed on the Far Ultraviolet Spectroscopy Explorer (FUSE) spacecraft in 2001, 2002, 2004, and 2007; and two each on Hayabusa in July and October 2005, Dawn in 2010 and 2012, and Kepler in 2012 and 2013 [17].

4.8.1 Reaction Wheel Characteristics

Reference [8] provides an overview of reaction wheel design. A reaction wheel assembly contains a rotating flywheel, typically supported by ball bearings, an internal brushless DC electric motor, and associated electronics. Reaction wheels are produced with a wide range of capabilities: maximum torque from 0.01 to 1.0 Nm, maximum angular momentum from 2 to 250 Nms, and maximum rotational speeds from 1,000 to 6,000 rpm. A vendor will frequently use the same flywheel and bearing assembly to construct a wheel with high angular momentum capacity and a different wheel with high torque capability. The back electromotive force developed at the high speeds required to provide high angular momentum makes it difficult to provide high torque in the same wheel.

The motor drives of some reaction wheels accept a torque command. Other reaction wheels have an internal closed-loop controller that holds the reaction wheel speed at a commanded value. The control electronics of some wheels are integral with the wheel unit, while others locate the controller in a separate electronics box. A reaction wheel is generally provided with a digital or analog tachometer, or both, to provide a reading of the wheel's rotational speed.

Wheel friction, or drag, is usually modeled as a sum of viscous and Coulomb components

$$L^w_{drag} = -\tau_v\, \omega^w - \tau_c\, \text{sign}(\omega^w) \tag{4.55}$$

where the coefficients τ_v and τ_c are empirically determined and can exhibit temperature dependence [8]. This model is adequate in many applications, but does not give a good description of friction at low speeds, especially when crossing zero speed. A dynamic model of friction incorporating hysteresis is required to represent these phenomena adequately. Reference [1] provides an excellent review of the issues involved in characterizing and modeling friction, and of the historical development of friction modeling. The first dynamic friction model to find widespread applications was the Dahl model, introduced in 1968 [18]. Many models have been proposed since, including the LuGre model developed by control groups in Lund and Grenoble [2, 12]. This modification of the Dahl model allows frictional forces to be larger at low speeds, an effect known as the Stribeck effect or as stiction. Increasing drag torque is often an indication of potential wheel failure, so it is a general practice to carefully monitor the drag torque on reaction wheels on orbiting spacecraft.

4.8.2 Reaction Wheel Disturbances

Although reaction wheels are invaluable for providing fine pointing control of many spacecraft, they are also one of the major sources of attitude disturbances. References [9] and [31] provide excellent overviews of the various categories of reaction wheel disturbances, which can be classified as radial forces, axial forces, radial moments, and axial moments. Disturbance forces perturb the spacecraft's attitude by creating an $\mathbf{r}\times\mathbf{F}$ torque, where $\mathbf{r}$ is the vector from the spacecraft's center of mass to the point of application of the force. Thus the effects of reaction wheel force disturbances can be minimized by locating the wheel close to the spacecraft's center of mass. The effects of disturbance moments are independent of the location of the wheel.

Deviations of the wheels from the assumption of perfect balance made in Sect. 3.3.5.1 are a principal contributor to disturbances. These deviations are classified as *static imbalance* or *dynamic imbalance*. Static imbalance is the condition that the wheel's center of mass is not on the axis of rotation. In this case, the spacecraft must provide, through the bearing assembly, the centripetal force needed to continuously accelerate the center of mass in a circular motion about the axis of rotation. If M^w is the mass of the rotor, ω^w is the magnitude of its angular velocity, $\mathbf{w}$ is a unit vector along its axis of rotation, and $\mathbf{x}$ is a vector from the axis to the wheel's center of mass, the static imbalance force is

$$\mathbf{F}_s = M^w(\omega^w)^2\mathbf{w}\times(\mathbf{w}\times\mathbf{x}) = -M^w(\omega^w)^2\begin{bmatrix} x_1 \\ x_2 \\ 0 \end{bmatrix} \tag{4.56}$$

where the last form assumes that $\mathbf{w} = [0\ \ 0\ \ 1]^T$. This is a radial force, constant in the wheel frame and rotating with angular velocity ω^w in the spacecraft frame. It has magnitude $F_s = U_s(\omega^w)^2$, where the static imbalance $U_s \equiv M^w\sqrt{x_1^2 + x_2^2}$ is proportional to the perpendicular distance from the axis to the center of mass. The SI unit for static imbalance would be kg-m, but it is usually given in g-cm (or in ounce-inches) because it is so small.

Dynamic imbalance is the condition that the axis of rotation of the wheel is not a principal axis. In this case, keeping the wheel rotating at a constant rate ω^w about the spin axis, which we assume to be $\mathbf{w} = [0\ 0\ 1]^T$, requires a torque

$$\mathbf{L}_d = \boldsymbol{\omega}\times\mathbf{H} = (\omega^w)^2\mathbf{w}\times(J\mathbf{w}) = (\omega^w)^2\begin{bmatrix} -J_{23} \\ J_{13} \\ 0 \end{bmatrix} \tag{4.57}$$

This is a radial moment, constant in the wheel frame and rotating with angular velocity ω^w in the spacecraft frame. It has magnitude $L_d = U_d(\omega^w)^2$, where $U_d \equiv \sqrt{J_{23}^2 + J_{13}^2}$ is called the dynamic imbalance. It is generally specified in g-cm^2, although the proper SI unit would be kg-m^2.

In addition to static and dynamic imbalance, a reaction wheel also exhibits structural dynamic modes resulting from compliance in the bearings. These modes fall into the three classes illustrated in Fig. 4.6: an axial translation mode, modes corresponding to radial translation in two orthogonal directions, and rocking modes about two orthogonal rocking axes. If the bearing compliance is isotropic, the two radial translation modes will have equal frequencies.

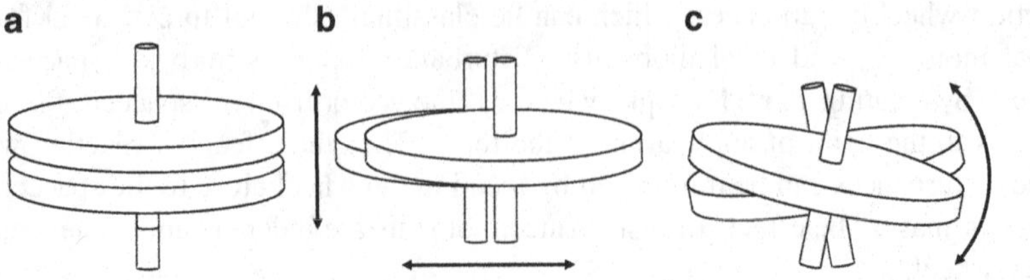

Fig. 4.6 Structural dynamic modes of a reaction wheel. (**a**) Axial translation. (**b**) Radial translation. (**c**) Rocking

The rocking modes interact with the rotation of the wheel to give *whirl* modes. Assume that the rotor's actual spin axis is misaligned by a small rotation vector $\boldsymbol{\delta\vartheta}$. Then the wheel's angular momentum in a reaction wheel frame in which the nominal spin axis of the wheel is $[0\ 0\ 1]^T$ is

$$\mathbf{H} = \text{diag}\left(\left[J^{\perp}\ J^{\perp}\ J^{\parallel}\right]\right)\boldsymbol{\delta\dot{\vartheta}} + (I_3 - [\boldsymbol{\delta\vartheta}\times])\left[0\ 0\ J^{\parallel}\omega^w\right]^T$$
$$= \begin{bmatrix} J^{\perp}\delta\dot{\vartheta}_1 - J^{\parallel}\omega^w\delta\vartheta_2 \\ J^{\perp}\delta\dot{\vartheta}_2 + J^{\parallel}\omega^w\delta\vartheta_1 \\ J^{\parallel}(\delta\dot{\vartheta}_3 + \omega^w) \end{bmatrix} \quad (4.58)$$

to first order in $\boldsymbol{\delta\vartheta}$, where the $\boldsymbol{\delta\dot{\vartheta}}$ term comes from Eq. (3.45). Note that $\delta\vartheta_3$ only appears in this equation in the form of its derivative. We can absorb $\delta\dot{\vartheta}_3$ into ω^w to eliminate $\delta\vartheta_3$ completely from the analysis. We can treat the wheel frame as inertial, because its motion is negligible compared to the motion of the rotor. Thus the equations of motion are

$$\dot{\mathbf{H}} = \begin{bmatrix} J^{\perp}\delta\ddot{\vartheta}_1 - J^{\parallel}\omega^w\delta\dot{\vartheta}_2 \\ J^{\perp}\delta\ddot{\vartheta}_2 + J^{\parallel}\omega^w\delta\dot{\vartheta}_1 \\ J^{\parallel}\dot{\omega}^w \end{bmatrix} = \mathbf{L} = -J^{\perp}\omega_0^2\begin{bmatrix} \delta\vartheta_1 \\ \delta\vartheta_2 \\ 0 \end{bmatrix} \quad (4.59)$$

The last equality assumes isotropic elastic restoring forces with coefficients on both axes which we choose to write as $J^{\perp}\omega_0^2$. The third component of the equation says that ω^w is constant, which is not surprising. We are looking for periodic solutions, so we write

$$\begin{bmatrix} \delta\vartheta_1(t) \\ \delta\vartheta_2(t) \end{bmatrix} = \begin{bmatrix} \delta\vartheta_1(0) \\ \delta\vartheta_2(0) \end{bmatrix} e^{i\omega t} \quad (4.60)$$

Substituting this into Eq. (4.59) gives

$$\mathbf{0} = \begin{bmatrix} J^{\perp}(-\omega^2 + \omega_0^2) & -iJ^{\parallel}\omega^w\omega \\ iJ^{\parallel}\omega^w\omega & J^{\perp}(-\omega^2 + \omega_0^2) \end{bmatrix}\begin{bmatrix} \delta\vartheta_1(0) \\ \delta\vartheta_2(0) \end{bmatrix} \quad (4.61)$$

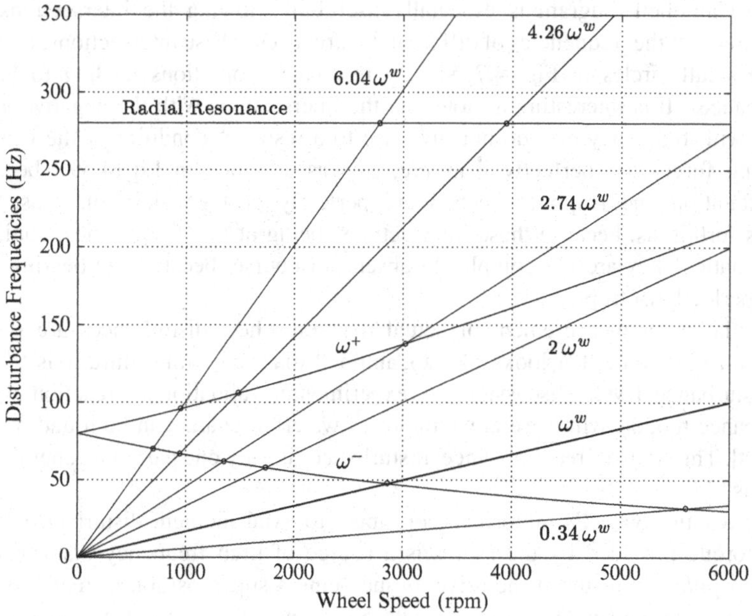

Fig. 4.7 Frequencies of radial moment disturbances

The determinant of the 2 × 2 matrix must vanish for a solution to exist, which gives

$$0 = J^{\perp}(-\omega^2 + \omega_0^2) \pm J^{\parallel}\omega^w\omega = -J^{\perp}(\omega^2 - \omega_0^2 \mp 2\gamma\omega^w\omega) \tag{4.62}$$

where we have written $J^{\parallel} = 2\gamma J^{\perp}$. Note that $\gamma < 1$ by the triangle inequality for principal moments, and that $\gamma = 1$ for an ideal infinitely thin disk. The frequencies of the positive and negative whirl modes are the roots of this quadratic equation;

$$\omega^{\pm} = \sqrt{(\gamma\omega^w)^2 + \omega_0^2} \pm \gamma\omega^w \tag{4.63}$$

We choose the positive sign of the square root because frequencies are conventionally defined to be positive. The higher frequency mode is a forward whirl, where the whirl is in the same direction as the wheel rotation, while the lower frequency mode is a negative whirl, with motion opposed to the wheel rotation [9].

It is useful to plot the frequencies of the disturbances as a function of the reaction wheel speed in a *Campbell diagram.* Figure 4.7 is an example showing the radial moment disturbance frequencies of a notional reaction wheel. The radial moment diagram is the most interesting case because it exhibits the whirl modes. The fundamental mode $\omega = \omega^w$ is the disturbance arising from dynamic imbalance. The first harmonic $\omega = 2\omega^w$ is also apparent, as are frequencies at non-integral multiples of ω^w arising from the bearing disturbances. The specific multiples shown in the figure are those shown in [9], which also has a table relating the frequencies of bearing disturbances to the physical properties of the bearings.

The Campbell diagram is especially useful in showing the intersections, or "collisions" of the frequencies of different disturbances. These intersections, marked by the small circles in Fig. 4.7, signify resonance conditions leading to larger disturbances. It is interesting to note that the intersections with the negative whirl mode, with frequency ω^-, do not give rise to a resonant condition if the bearing restoring forces are perfectly isotropic, as was assumed in Eq. (4.59), because the excitation and response vectors are perfectly orthogonal in that case [9]. This is fortuitous, because these intersections, being at low frequencies, could be problematic. They are not completely absent, of course, because the bearings are never perfectly isotropic.

Simulation or measurement of actual reaction wheel disturbances are shown on a *waterfall chart* that looks like a Campbell diagram with a third axis to plot the disturbance level. Discussions of experimental validation of reaction wheel disturbance models with several examples of waterfall charts can be found in [42] and [38]. The study of reaction wheel disturbances is only one part of a general jitter analysis [37].

The reaction wheel's motor can contribute to axial moment disturbances. The drive torque provided by the motor is a source of high frequency disturbances. *Torque ripple* is a result of the drive torque being a superposition of rectified sine waves. The torque ripple of a motor with a greater number of poles is at a higher frequency, where it is less problematic. *Cogging* torques, a result of the magnets in the rotor moving past a ferromagnetic stator, are present regardless of whether a torque is applied, but are absent from reaction wheels that have no ferromagnetic materials in the stator [9]. Migration of lubricant in the bearings can give rise to low frequency torque noise, which can have a significant effect on spacecraft pointing. Light oils are less likely to produce torque noise than grease, but grease is less prone than oil to migration during storage at one g and during launch.

The effects of reaction wheel disturbances on spacecraft jitter are often mitigated by mounting the reaction wheels on flexible isolators. These must be carefully analyzed, however, because the interaction of the wheels' rocking modes with the isolator modes can lead to potentially troublesome mode collisions [11].

4.8.3 Redundant Wheel Configurations

Spacecraft are often equipped with more than the three reaction wheels, both for redundancy and to provide greater torque and momentum storage capability. Redundant reaction wheels are quite commonly placed in a hot backup configuration for this purpose. The torque provided by six reaction wheels allows the NASA Swift spacecraft to complete 50° slew maneuvers in 75 s in order to collect data on transient gamma ray bursts. Similar configurations also allow for longer time intervals between momentum unloads for spacecraft, such as the James Webb Space Telescope (JWST), that cannot unload angular momentum continuously.

The momentum transfer maneuver has been studied in detail using multiple wheels. This problem is related to the initial acquisition after launch since the wheel or wheels are at rest while the spacecraft has some momentum imposed by the launch vehicle in a given axis. Reference [28] studied the strategy to transfer this initial momentum to the wheel(s) as well as to avoid occurrence of a singular state, and [63] introduced a near minimum time eigenaxis rotation maneuver using a set of three reaction wheels without momentum bias for a nadir pointing spacecraft in a circular low-Earth orbit. The combined eigenaxis maneuvers (using reaction wheels and thrusters) are presented in [13] to reduce the time spent on large rotations. The reaction wheel optimization problem has been extensively analyzed in the literature (e.g. see [22,23], and [41]). In the first two of these references the assembly configuration has been selected to minimize the power consumption of the wheels in their task of absorbing the secular portion of the environmental torques.

Our analysis will implicitly assume that all the reaction wheels on a spacecraft are identical, which is usually, but not universally, the case. We arrange the applied torque and the wheel angular momenta of the individual wheels, L_i^w and H_i^w, in column vectors $\mathbf{L}_W^w$ and $\mathbf{H}_W^w$, where the subscript W denotes the n-dimensional wheel frame. The transformation from the wheel frame to the body frame is given by the $3 \times n$ distribution matrix $\mathscr{W}_n$, whose columns are unit vectors in the body frame, $\mathbf{w}_i$, along the spin axes of the wheels:

$$\mathscr{W}_n = \begin{bmatrix} \mathbf{w}_1 \; \mathbf{w}_2 \; \cdots \; \mathbf{w}_n \end{bmatrix} \tag{4.64}$$

The total wheel torque and angular momentum in the body frame are given by

$$\mathbf{L}_B^w = \mathscr{W}_n \left[L_1^w \; L_2^w \; \cdots \; L_n^w \right]^T \equiv \mathscr{W}_n \mathbf{L}_W^w \tag{4.65a}$$

$$\mathbf{H}_B^w = \mathscr{W}_n \left[H_1^w \; H_2^w \; \cdots \; H_n^w \right]^T \equiv \mathscr{W}_n \mathbf{H}_W^w \tag{4.65b}$$

Because the equations for torque and angular momentum are parallel, we will concentrate on angular momentum in the following. Every result we find for angular momentum has a corresponding relation for torque. It is important to note that $\mathbf{H}_W^w$ and $\mathbf{H}_B^w$ will not have the same magnitude in general, because $\mathscr{W}_n$ is generally not an orthogonal matrix.

We really want to invert Eqs. (4.65), that is to transform a desired body frame angular momentum or torque vector into the wheel frame. This is trivial for $n = 3$, we simply set $\mathbf{H}_W^w = \mathscr{W}_3^{-1}\mathbf{H}_B^w$, assuming that the distribution matrix is invertible. Invertibility requires the distribution matrix to have full rank, rank three in this case, and is equivalent to the requirement that the spin axis unit vectors $\mathbf{w}_i$ do not all lie in a plane. This will be true for any reaction wheel configuration designed to provide three-axis control.

There is no unique way to distribute the torque or angular momentum for $n > 3$, and we will consider two approaches: the pseudoinverse and minimax methods. Equation (4.65b) demands that the difference, $\Delta\mathbf{H}_W^w$, between two distinct values of $\mathbf{H}_W^w$ that produce the same $\mathbf{H}_B^w$ must satisfy the equation $\mathscr{W}_n \Delta\mathbf{H}_W^w = \mathbf{0}$. This says that $\Delta\mathbf{H}_W^w$ must lie in the null space of the matrix $\mathscr{W}_n$, which is $(n-3)$-dimensional if $\mathscr{W}_n$ has full rank, as we assume.

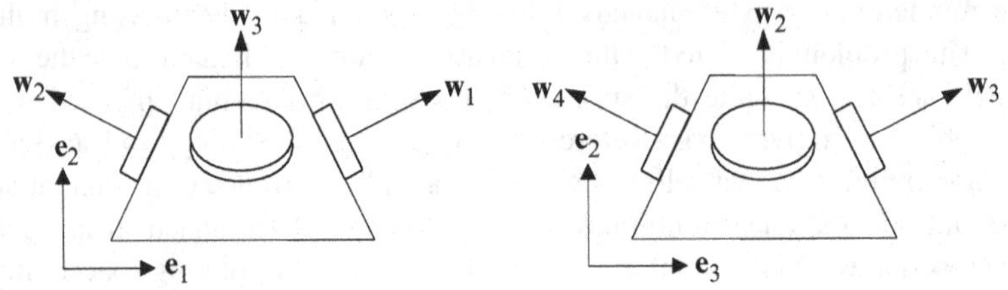

Fig. 4.8 Pyramid configuration

Let us now look at some specific configurations of redundant reaction wheels. We will consider two popular four-wheel configurations: a pyramid configuration and a configuration known as the NASA standard configuration that was used on the Multimission Modular Spacecraft (MMS). Figure 4.8 shows the pyramid configuration. This configuration assumes symmetry on each coordinate plane. Its transformation matrix from the reaction wheel coordinate system to the body coordinate system is given by

$$\mathscr{W}_4 = \begin{bmatrix} a & -a & 0 & 0 \\ b & b & c & c \\ 0 & 0 & d & -d \end{bmatrix} \tag{4.66}$$

where $a^2+b^2 = c^2+d^2 = 1$. This matrix assumes a preferential direction along the body $\mathbf{e}_2$ axis, with each wheel contributing to this direction. For $a = d$ and $b = c$, the spin axes make the same angle with the $(\mathbf{e}_1, \mathbf{e}_3)$ plane. This layout is the most common configuration for Earth-pointing spacecraft in which a constant rate along the pitch axis is required to point the instruments toward the Earth's surface. The preferred axis of the transformation in Eq. (4.66) or of any other distribution matrix can be changed by multiplying $\mathscr{W}_n$ on the left by the appropriate rotation matrix. The null space of $\mathscr{W}_4$ is one-dimensional, consisting of scalar multiples of the unit vector

$$\mathbf{n}_4 = \frac{1}{\sqrt{2(b^2+c^2)}} \begin{bmatrix} c \\ c \\ -b \\ -b \end{bmatrix} \tag{4.67}$$

The NASA Standard four-wheel configuration places three orthogonal wheel spin axes along the body axes and the fourth, skew, wheel axis with projected components along each body axis, as shown in Fig. 4.9. The transformation matrix for this configuration is given by

$$\mathscr{W}_N = \begin{bmatrix} 1 & 0 & 0 & \alpha \\ 0 & 1 & 0 & \beta \\ 0 & 0 & 1 & \gamma \end{bmatrix} \tag{4.68}$$

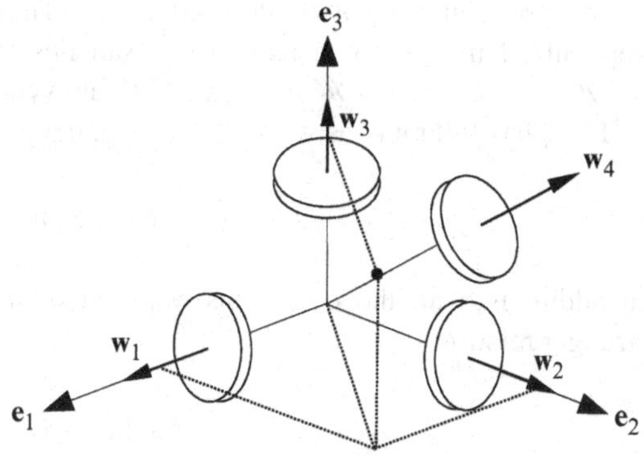

Fig. 4.9 NASA standard configuration

where $\alpha^2 + \beta^2 + \gamma^2 = 1$. The null space of $\mathscr{W}_N$ is along the unit vector

$$\mathbf{n}_N = \frac{1}{\sqrt{1 + \alpha^2 + \beta^2 + \gamma^2}} \begin{bmatrix} \alpha \\ \beta \\ \gamma \\ -1 \end{bmatrix} \tag{4.69}$$

For the special, but very common, case given by $\alpha = \beta = \gamma = 1/\sqrt{3}$, the fourth wheel's spin axis is at an angle of 54.7° with respect to the coordinate axes.

We will also consider two six-wheel configurations, both with distribution matrix

$$\mathscr{W}_6 = \begin{bmatrix} b & c & b & c & b & c \\ 0 & \sqrt{3}d/2 & \sqrt{3}a/2 & 0 & -\sqrt{3}a/2 & -\sqrt{3}d/2 \\ a & d/2 & -a/2 & -d & -a/2 & d/2 \end{bmatrix} \tag{4.70}$$

where $a^2 + b^2 = c^2 + d^2 = 1$. The null space of $\mathscr{W}_6$ is three-dimensional, requiring three linearly independent vectors to span it. The usual case, which we will refer to as the pyramid configuration, has $b = c$ and $a = d$, as on Swift and JWST, so the spin axis directions are evenly distributed around a cone centered on the $\mathbf{e}_1$ axis. The other six-wheel configuration attains the maximum symmetry possible with six wheels by choosing the parameters $b = \sin\alpha$, $a = \cos\alpha$, $c = \sin\beta$, and $d = \cos\beta$, with $\tan\alpha = (3 + \sqrt{5})/4$ and $\tan\beta = (3 - \sqrt{5})/4$, so that the wheel spin axis vectors are normal to the faces of a regular dodecahedron.[9]

4.8.3.1 Pseudoinverse Distribution Law

The method chosen to distribute torque or angular momentum among redundant wheels is specific to the mission's requirements, but the most common method is to

[9]The authors thank John P. Downing for suggesting this configuration.

use the pseudoinverse of $\mathscr{W}$, denoted by $\mathscr{W}^{\dagger}$. The pseudoinverse, or Moore-Penrose generalized inverse, of a real matrix satisfies the conditions: $\mathscr{W}\mathscr{W}^{\dagger}\mathscr{W} = \mathscr{W}$, $\mathscr{W}^{\dagger}\mathscr{W}\mathscr{W}^{\dagger} = \mathscr{W}^{\dagger}$, and $\mathscr{W}\mathscr{W}^{\dagger}$ and $\mathscr{W}^{\dagger}\mathscr{W}$ are symmetric [21,25].

If $\mathscr{W}_n$ has full rank, as we are assuming, the pseudoinverse is given by

$$\mathscr{W}_n^{\dagger} = \mathscr{W}_n^T \left(\mathscr{W}_n \mathscr{W}_n^T\right)^{-1} \tag{4.71}$$

In addition to all the conditions required of a pseudoinverse, this satisfies the stronger relation

$$\mathscr{W}_n \mathscr{W}_n^{\dagger} = I_3 \tag{4.72}$$

Note, however, that $\mathscr{W}_n^{\dagger}\mathscr{W}_n \neq I_n$ for $n > 3$. Equation (4.72) guarantees that Eq. (4.65b) is satisfied if we set

$$\mathbf{H}_W^w = \mathscr{W}_n^{\dagger} \mathbf{H}_B^w \tag{4.73}$$

The most general wheel angular momentum vector $\mathbf{H}_W^w$ satisfying Eq. (4.65b) is given by

$$\mathbf{H}_W^w = \mathscr{W}_n^{\dagger} \mathbf{H}_B^w + \mathbf{n}_n = \mathscr{W}_n^T \left(\mathscr{W}_n \mathscr{W}_n^T\right)^{-1} \mathbf{H}_B^w + \mathbf{n}_n \tag{4.74}$$

where $\mathbf{n}_n$ is an arbitrary vector in the null space of $\mathscr{W}_n$. This gives

$$\begin{aligned} \|\mathbf{H}_W^w\|^2 &= \|\mathscr{W}_n^T \left(\mathscr{W}_n \mathscr{W}_n^T\right)^{-1} \mathbf{H}_B^w\|^2 + 2(\mathbf{H}_B^w)^T \left(\mathscr{W}_n \mathscr{W}_n^T\right)^{-1} \mathscr{W}_n \mathbf{n}_n + \|\mathbf{n}_n\|^2 \\ &= (\mathbf{H}_B^w)^T \left(\mathscr{W}_n \mathscr{W}_n^T\right)^{-1} \mathbf{H}_B^w + \|\mathbf{n}_n\|^2 \end{aligned} \tag{4.75}$$

by the definition of the null space. We see that the Euclidean norm, or L_2 norm, of the wheel momentum vector is minimized by setting $\mathbf{n}_n = \mathbf{0}$, i.e. by using the pseudoinverse to distribute the angular momenta.

Now let us consider some specific cases. The pseudoinverse of $\mathscr{W}_4$ for the four-wheel pyramid is given by Eq. (4.71) as

$$\mathscr{W}_4^{\dagger} = \frac{1}{2} \begin{bmatrix} 1/a & b/(b^2+c^2) & 0 \\ -1/a & b/(b^2+c^2) & 0 \\ 0 & c/(b^2+c^2) & 1/d \\ 0 & c/(b^2+c^2) & -1/d \end{bmatrix} \tag{4.76}$$

It is easy to see that $\mathscr{W}_4^{\dagger}$ satisfies Eq. (4.72) and that

$$\mathscr{W}_4^{\dagger} \mathscr{W}_4 = I_4 - \mathbf{n}_4 \mathbf{n}_4^T \tag{4.77}$$

The pseudoinverse of $\mathscr{W}_{\rm N}$ for the NASA standard configuration is given by

$$\mathscr{W}_{\rm N}^{\dagger} = \frac{1}{1+\alpha^2+\beta^2+\gamma^2}\begin{bmatrix} 1+\beta^2+\gamma^2 & -\alpha\beta & -\alpha\gamma \\ -\alpha\beta & 1+\alpha^2+\gamma^2 & -\beta\gamma \\ -\alpha\gamma & -\beta\gamma & 1+\alpha^2+\beta^2 \\ \alpha & \beta & \gamma \end{bmatrix} \tag{4.78}$$

which satisfies Eq. (4.72) and

$$\mathscr{W}_{\rm N}^{\dagger}\mathscr{W}_{\rm N} = I_4 - \mathbf{n}_{\rm N}\mathbf{n}_{\rm N}^T \tag{4.79}$$

The pseudoinverse of $\mathscr{W}_6$ for the six-wheel configurations is given by

$$\mathscr{W}_6^{\dagger} = \frac{1}{3}\begin{bmatrix} b/(b^2+c^2) & 0 & 2a/(a^2+d^2) \\ c/(b^2+c^2) & \sqrt{3}d/(a^2+d^2) & d/(a^2+d^2) \\ b/(b^2+c^2) & \sqrt{3}a/(a^2+d^2) & -a/(a^2+d^2) \\ c/(b^2+c^2) & 0 & -2d/(a^2+d^2) \\ b/(b^2+c^2) & -\sqrt{3}a/(a^2+d^2) & -a/(a^2+d^2) \\ c/(b^2+c^2) & -\sqrt{3}d/(a^2+d^2) & d/(a^2+d^2) \end{bmatrix} \tag{4.80}$$

We have seen that the pseudoinverse method for distributing torque or momentum among redundant reaction wheels minimizes the Euclidean norm of the torque or momentum vector in the wheel frame. That is, it minimizes the sum of the squares of the individual wheel torques or angular momenta. This may be optimal from an energy standpoint, but it does not necessarily result in using the full capability of the reaction wheel array. The pseudoinverse has the advantage of being relatively simple to implement, and it also has a subtler advantage. If we distribute the angular momentum according to Eq. (4.73), then the time derivative of this equation gives

$$\mathbf{L}_W^w = \dot{\mathbf{H}}_W^w = \mathscr{W}_n^{\dagger}\dot{\mathbf{H}}_B^w = \mathscr{W}_n^{\dagger}\mathbf{L}_B^w \tag{4.81}$$

showing that the pseudoinverse momentum and torque distribution laws are mutually consistent.

4.8.3.2 Minimax Distribution Law

The ultimate capability of an assembly of reaction wheels is attained by minimizing the maximum effort expended by any wheel, rather than by minimizing the sum of squares of the efforts expended by the reaction wheels. Since the L_∞ norm of a vector is defined to be the magnitude of its largest component, this minimax optimization is also referred to as the L_∞ method [41]. We will continue to emphasize optimization of angular momentum, since the results obtained can easily be adapted to optimize torque distribution.

We will only consider reaction wheel configurations where no three of the wheel spin axis vectors, $\mathbf{w}_i$, lie in a plane. This is not a restrictive assumption, as almost all reaction wheel assemblies satisfy it. The following argument shows that minimizing the largest magnitude of individual wheel angular momentum in an array of n reaction wheels will result in $n-2$ of the wheels all having momentum of the same magnitude. If this were not the case, we could reduce the magnitude of the momentum of the wheels having the largest momentum by distributing it among the other wheels. This process could be continued until there were only two wheels remaining with less than the maximum magnitude. The process cannot be carried further because a general three-component excess momentum vector cannot be distributed between only two wheels.

Let H_0 denote the maximum wheel momentum magnitude. We will consider the problem of maximizing the body momentum $\mathbf{H}_B^w$ for a given value of H_0 because this is equivalent to minimizing H_0 for a given $\mathbf{H}_B^w$. Let us first maximize the angular momentum in the direction $\mathbf{w}_i \times \mathbf{w}_j$. This angular momentum does not depend on H_i^w and H_j^w, so the maximum in the direction $\mathbf{w}_i \times \mathbf{w}_j$ is obtained by assigning the maximum magnitude H_0 to all the other wheels with the signs given by

$$H_k^w = H_0 \operatorname{sign}(s_{ijk}) \quad \text{for} \quad k \neq i, j \tag{4.82}$$

where

$$s_{ijk} \equiv (\mathbf{w}_i \times \mathbf{w}_j) \cdot \mathbf{w}_k \tag{4.83}$$

With this distribution, the angular momentum in the body frame is

$$\mathbf{H}_B^w = H_i^w \mathbf{w}_i + H_j^w \mathbf{w}_j + H_0 \mathbf{v}_{ij} \tag{4.84}$$

with

$$\mathbf{v}_{ij} \equiv \sum_{\substack{k=1 \\ k \neq i,j}}^{n} \mathbf{w}_k \operatorname{sign}(s_{ijk}) \tag{4.85}$$

Because the angular momenta of wheels i and j can have any magnitude less than or equal to H_0, the range of angular momenta described by Eq. (4.84) covers a rhombus with sides of length $2H_0$ parallel to $\mathbf{w}_i$ and $\mathbf{w}_j$. The distance in the normal direction from the origin to this rhombus is $H_0\, d_{ij}$, where

$$d_{ij} \equiv \|\mathbf{w}_i \times \mathbf{w}_j\|^{-1}[(\mathbf{w}_i \times \mathbf{w}_j) \cdot \mathbf{v}_{ij}] = \|\mathbf{w}_i \times \mathbf{w}_j\|^{-1} \sum_{\substack{k=1 \\ k \neq i,j}}^{n} |s_{ijk}| \tag{4.86}$$

The rhombi for all $n(n-1)$ pairs of spin axis indices form an $n(n-1)$-faced polyhedron. It is symmetrical about the origin, with its ji face opposite to its ij face, because $\mathbf{v}_{ij} = -\mathbf{v}_{ji}$. At the vertices of the polyhedron all the wheel momenta have

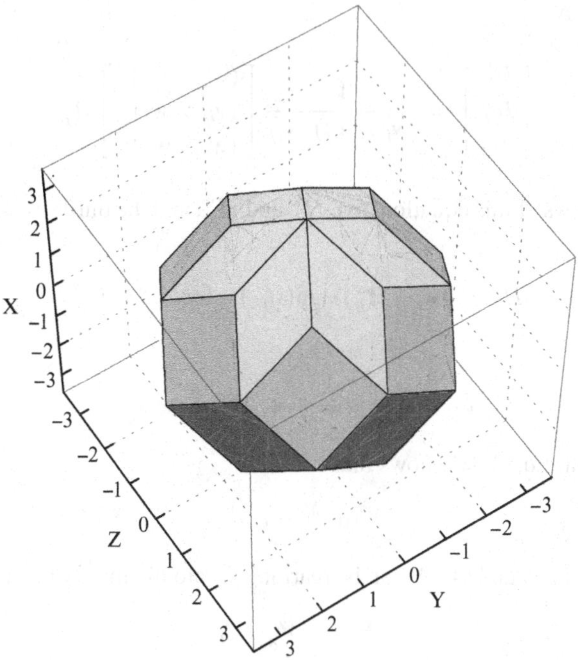

Fig. 4.10 Momentum polyhedron for six-wheel pyramid

magnitude H_0, and $n-1$ of the momenta have this magnitude along the edges. The polyhedron for H_0 equal to the maximum storage capacity of a single wheel bounds the angular momentum storage capacity of the wheel configuration. Figure 4.10 shows the polyhedron for a six-wheel pyramid configuration with the parameters $b = c = \sqrt{1/3}$, $a = d = \sqrt{2/3}$, and axes labeled in units of H_0.[10]

We now want to invert Eq. (4.84) in order to transform the body frame angular momentum into the wheel frame. Writing this as a matrix equation,

$$\mathbf{H}_B^w = \begin{bmatrix} \mathbf{w}_i & \mathbf{w}_j & \mathbf{v}_{ij} \end{bmatrix} \begin{bmatrix} H_i^w \\ H_j^w \\ H_0 \end{bmatrix} \tag{4.87}$$

[10]The momentum polyhedron of the dodecahedron reaction wheel configuration takes the form of a rhombic tricontahedron, the shape of a 30-sided die.

and inverting gives

$$\begin{bmatrix} H_i^w \\ H_j^w \\ H_0 \end{bmatrix} = \frac{1}{(\mathbf{w}_i \times \mathbf{w}_j) \cdot \mathbf{v}_{ij}} \begin{bmatrix} (\mathbf{w}_j \times \mathbf{v}_{ij})^T \\ (\mathbf{v}_{ij} \times \mathbf{w}_i)^T \\ (\mathbf{w}_i \times \mathbf{w}_j)^T \end{bmatrix} \mathbf{H}_B^w \tag{4.88}$$

The first two rows of this equation give H_i^w and H_j^w, and the third row and Eq. (4.82) give

$$H_k^w = (\mathbf{w}_{ij} \cdot \mathbf{H}_B^w)\,\text{sign}(s_{ijk}) \quad \text{for} \quad k \neq i, j \tag{4.89}$$

where

$$\mathbf{w}_{ij} \equiv [(\mathbf{w}_i \times \mathbf{w}_j) \cdot \mathbf{v}_{ij}]^{-1}(\mathbf{w}_i \times \mathbf{w}_j) \tag{4.90}$$

Comparing with Eq. (4.86) shows that

$$\|\mathbf{w}_{ij}\| = d_{ij}^{-1} \tag{4.91}$$

Equations (4.88) and (4.89) can be rearranged into the matrix form

$$\mathbf{H}_W^w = \mathscr{W}_n^{ij} \mathbf{H}_B^w \tag{4.92}$$

where the elements of the matrix $\mathscr{W}_n^{ij}$ are specified by Eqs. (4.88) and (4.89). This matrix was explicitly constructed to satisfy Eq. (4.65b), so it follows that

$$\mathscr{W}_n \mathscr{W}_n^{ij} = I_3 \tag{4.93}$$

which is the same as Eq. (4.72) for the pseudoinverse. Thus $\mathscr{W}_n^{ij}$ satisfies all the requirements for being a pseudoinverse except that $\mathscr{W}_n^{ij}\mathscr{W}_n$ is not symmetric in general. It appears that we have $n(n-1)$ matrices for computing the minimax momentum distribution, and we need a criterion to choose among them. Before proceeding, however, we note from the symmetry of the equations under interchange of the indices i and j that $\mathscr{W}_n^{ij} = \mathscr{W}_n^{ji}$, so we actually have only $n(n-1)/2$ independent minimax distribution matrices.

We will show the minimax distribution matrices for the four-wheel pyramid as an illustration. With $\sigma \equiv \text{sign}(bc)$, these matrices are found to be

$$\mathscr{W}_4^{12} = \frac{1}{2}\begin{bmatrix} 1/a & 1/b & 0 \\ -1/a & 1/b & 0 \\ 0 & 0 & 1/d \\ 0 & 0 & -1/d \end{bmatrix}, \quad \mathscr{W}_4^{34} = \frac{1}{2}\begin{bmatrix} 1/a & 0 & 0 \\ -1/a & 0 & 0 \\ 0 & 1/c & 1/d \\ 0 & 1/c & -1/d \end{bmatrix}$$

$$\mathscr{W}_4^{13} = \frac{1}{2(b+\sigma c)}\begin{bmatrix} (b+2\sigma c)/a & 1 & -c/d \\ -b/a & 1 & -c/d \\ -\sigma b/a & \sigma & (2b+\sigma c)/d \\ -\sigma b/a & \sigma & -\sigma c/d \end{bmatrix}$$

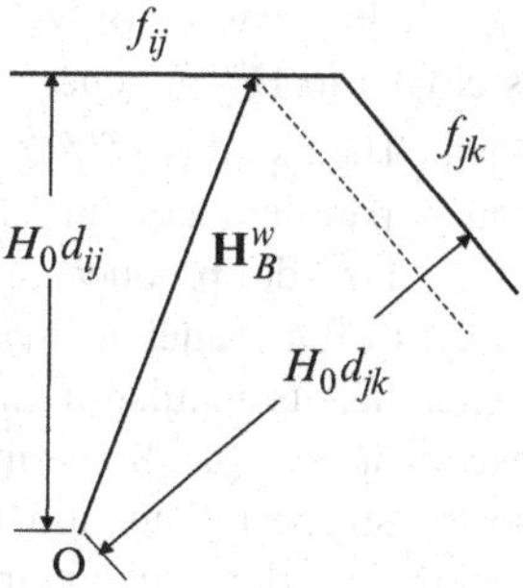

Fig. 4.11 Identifying the correct face. See text for explanation

$$\mathscr{W}_4^{14} = \frac{1}{2(b+\sigma c)} \begin{bmatrix} (b+2\sigma c)/a & 1 & c/d \\ -b/a & 1 & c/d \\ -\sigma b/a & \sigma & \sigma c/d \\ -\sigma b/a & \sigma & -(2b+\sigma c)/d \end{bmatrix}$$

$$\mathscr{W}_4^{23} = \frac{1}{2(b+\sigma c)} \begin{bmatrix} b/a & 1 & -c/d \\ -(b+2\sigma c)/a & 1 & -c/d \\ \sigma b/a & \sigma & (2b+\sigma c)/d \\ \sigma b/a & \sigma & -\sigma c/d \end{bmatrix}$$

$$\mathscr{W}_4^{24} = \frac{1}{2(b+\sigma c)} \begin{bmatrix} b/a & 1 & c/d \\ -(b+2\sigma c)/a & 1 & c/d \\ \sigma b/a & \sigma & \sigma c/d \\ \sigma b/a & \sigma & -(2b+\sigma c)/d \end{bmatrix} \tag{4.94}$$

Finding the correct matrix to compute the minimax distribution requires identifying the face of the polyhedron that the body momentum vector falls on. Figure 4.11 is an edge-on view of two adjoining faces, f_{ij} and f_{jk}, which means that $\mathbf{w}_j$ is normal to the plane of the paper. The center of the polyhedron, the origin of body momentum space, is labeled O, and the normal distances from the origin to the faces are specified. The projection of the body momentum vector $\mathbf{H}_B^w$ into the plane of the paper is also shown. We assume that the momentum is on face f_{ij}, as indicated by the tip of the arrow. Equations (4.90) and (4.91) tell us that $\mathbf{w}_{ij} = d_{ij}^{-1}\mathbf{e}_{ij}$, where $\mathbf{e}_{ij}$ is the unit vector normal to f_{ij}, so we see from the figure that

$$\frac{\mathbf{w}_{ij} \cdot \mathbf{H}_B^w}{\mathbf{w}_{jk} \cdot \mathbf{H}_B^w} = \frac{d_{ij}^{-1}\mathbf{e}_{ij} \cdot \mathbf{H}_B^w}{d_{jk}^{-1}\mathbf{e}_{jk} \cdot \mathbf{H}_B^w} > 1 \tag{4.95}$$

The inequality holds because $\mathbf{e}_{ij} \cdot \mathbf{H}_B^w = H_0 d_{ij}$ and $\mathbf{e}_{jk} \cdot \mathbf{H}_B^w$, the distance from the origin to the dashed line in the figure, is less than $H_0 d_{jk}$, the distance to f_{jk}. Thus the angular momentum will be on face f_{ij} if $\mathbf{w}_{ij} \cdot \mathbf{H}_B^w > \mathbf{w}_{jk} \cdot \mathbf{H}_B^w$. Generalizing this to the entire polyhedron establishes that the correct face is the one producing the maximum value of $\mathbf{w}_{ij} \cdot \mathbf{H}_B^w$.[11] Note that f_{ij} is *not* the face whose normal vector is closest to $\mathbf{H}_B^w$ in general, because the distances d_{ij} to the different faces are not equal. The equality of $\mathscr{W}_n^{ij}$ and $\mathscr{W}_n^{ji}$ means the we only have to identify the correct pair of parallel faces, and the relation $\mathbf{w}_{ij} = -\mathbf{w}_{ji}$ says that this requires evaluating $n(n-1)/2$ dot products and finding the maximum value of $|\mathbf{w}_{ij} \cdot \mathbf{H}_B^w|$.

It is often better to maximize the torque provided by a reaction wheel array, rather than its angular momentum storage capacity. The Swift spacecraft provides an excellent example. Being in near-Earth orbit, it can unload momentum continuously, so the stored momentum is always small; and high torques are required to slew rapidly in order to image transient gamma-ray bursts. Thus we command the wheel torque vector $\mathbf{L}_W^w = \mathscr{W}_n^{jk}\mathbf{L}_B^w$, where f_{jk} is the face that maximizes $|\mathbf{w}_{jk} \cdot \mathbf{L}_B^w|$. This torque will not necessarily maintain a minimax distribution of reaction wheel momenta, because the face on which $|\mathbf{w}_{jk} \cdot \mathbf{L}_B^w|$ is maximized is generally not the same as the face on which $|\mathbf{w}_{ij} \cdot \mathbf{H}_B^w|$ is maximized. This can be dealt with by setting up a slower control loop to asymptotically drive the wheels to their minimax momentum distribution. The controller can use wheel tachometer data to compute $\mathbf{H}_W^w$ and Eq. (4.65b) to compute the body angular momentum. Maximizing $|\mathbf{w}_{ij} \cdot \mathbf{H}_B^w|$ identifies the face providing the minimax momentum distribution and thus provides the desired wheel momentum $\mathscr{W}_n^{ij}\mathbf{H}_B^w = \mathscr{W}_n^{ij}\mathscr{W}_n\mathbf{H}_W^w$. Then the torque command is

$$\mathbf{L}_W^w = \mathscr{W}_n^{jk}\mathbf{L}_B^w + \kappa(\mathscr{W}_n^{ij}\mathscr{W}_n\mathbf{H}_W^w - \mathbf{H}_W^w) = \mathscr{W}_n^{jk}\mathbf{L}_B^w + \kappa(\mathscr{W}_n^{ij}\mathscr{W}_n - I_n)\mathbf{H}_W^w \quad (4.96)$$

where $\mathbf{L}_B^w$ is the desired torque in the body frame and κ is a feedback gain. Equation (4.93) shows that the feedback term is in the null space of $\mathscr{W}_n$, so it does not affect the resultant torque in the body frame.

Another observation about minimax distribution of momentum is that unacceptable jitter can result if $n-2$ wheel speeds coincide with the frequency of a flexible mode of the spacecraft structure. It is also desirable to avoid sustained operation of reaction wheels near zero speed. These problems can be avoided without perturbing the spacecraft pointing by using torque or momentum in the $(n-3)$-dimensional null space of $\mathscr{W}_n$ to separate equal wheel speeds or to drive a wheel's speed rapidly through zero or past a resonant frequency [41]. Minimizing jitter is most important during stable pointing of the spacecraft, especially for a precision pointing mission like JWST with its milli-arcsecond level stability requirements. Avoiding wheel speeds near zero or a resonant frequency is less important and sometimes unavoidable while executing slews, as will be seen in the examples in the next section.

---

[11]This relation, which is crucial for real-time implementation of the minimax algorithm, was discovered by Frank X. Liu.

Table 4.2 Multiple wheel capability as a multiple of the capacity of a single wheel for pseudoinverse and minimax distribution

Configuration	Wheel Failure	Pseudoinverse	Minimax
NASA Standard	None	1.155	1.414
	Any Orthogonal	0.577	0.577
	Skew	1.000	1.000
Four-Wheel Pyramid	None	1.333	1.633
	Any One	0.816	0.816
Six-Wheel Pyramid	None	2.000	2.667
	Any One	1.309	1.667
Dodecahedron	None	2.000	2.753
	Any One	1.581	1.902

4.8.3.3 Comparison of Pseudoinverse and Minimax Distribution Laws

We will now examine the maximum momentum storage capacity in the least favorable direction of several wheel configurations, using the minimax or pseudoinverse method to distribute the angular momentum. Recall that the same relations will hold for the torque capability of the array. We denote the momentum storage capacity of a single wheel by $H_{\max}$. The momentum storage capacity is determined by considering the maximum momentum that can be assigned to any wheel for a given magnitude of body momentum. This occurs for the pseudoinverse method when $\mathbf{H}_B^w$ is parallel to the row of the pseudoinverse matrix corresponding to that wheel, and the magnitude of the momentum assigned to the wheel will be the product of the magnitude of the body momentum and the Euclidean norm of that row vector. Thus the momentum capacity in the least favorable direction is $H_{\max}$ divided by the maximum of the norms of the rows of the pseudoinverse matrix. Analysis of the minimax method indicates that the maximum storage on face f_{ij} is at least as large as d_{ij}.[12] Thus the minimax storage capacity in the least favorable direction is $H_{\max}$ times the minimum of d_{ij} over all the faces.

Table 4.2 shows the capabilities of four wheel configurations: the NASA Standard configuration with $\alpha = \beta = \gamma = 1/\sqrt{3}$, the four-wheel and six-wheel pyramids with $b = c = \sqrt{1/3}$ and $a = d = \sqrt{2/3}$, and the six-wheel dodecahedron. Reference [41] proves that these parameters for the pyramid configurations make their storage capability as isotropic as possible.[13] It must be emphasized that these choices will rarely be optimal for any specific mission, since moments of inertia, slewing requirements, and environmental torques are almost never isotropic. However, they provide a baseline for comparing the different

[12]This is true for all the configurations considered here, but may not hold for some pathological configurations [41].

[13]The four-wheel and six-wheel pyramids with these parameters, and also the dodecahedron, have $\mathscr{W}_n \mathscr{W}_n^T = (n/3) I_3$. Then Eq. (4.75) shows that the pseudoinverse distribution method gives $\|\mathbf{H}_W^w\|^2 = (3/n)\|\mathbf{H}_B^w\|^2$ for these configurations.

configurations and distribution methods. We also present the storage capacities of the four configurations with one wheel failure.

The minimax method provides more momentum storage or torque capability than the pseudoinverse method for almost all cases in Table 4.2. The only exceptions are the four-wheel configurations with a failed wheel, which are three-wheel configurations for which $\mathscr{W}_3$ has a unique inverse. The NASA Standard configuration provides less capability than the four-wheel pyramid with all wheels active, and is quite a bit less capable in the single-point failure case if one of its orthogonal wheels fails. It is better only in the statistically less likely case that the failed wheel is the skew wheel. The dodecahedron provides little advantage over the six-wheel pyramid configuration if all six wheels are functioning, but it performs better in the single-point failure case.

We also compare simulations of a rest-to-rest 90° slew using the minimax and pseudoinverse methods to distribute angular momentum among six wheels in the same pyramid configuration as was used for the capacity comparisons. This example is motivated by JWST, but differs in several details. The slew is about the body $\mathbf{e}_1$ axis, the cone axis of the wheel configuration, and starts with system momentum of $2H_{\max}$ in the body $+\mathbf{e}_2$ direction. We ignore the very small changes in the system momentum during the slew arising from external torques. The slew begins with a constant angular acceleration of 1.6×10^{-4} deg/s^2 or 90 deg/(750 s)^2 for a time $t_{\text{accel}} \leq 750$ s. This is followed by a coast at a constant angular rate between t_{accel} and $(750\text{ s})^2/t_{\text{accel}}$. The slew ends with a deceleration at -1.6×10^{-4} deg/s^2 for a time t_{accel}. There is no coast period if $t_{\text{accel}} = 750$ s, so this limit represents an acceleration-limited slew in the minimum time of 25 min. The system momentum in the spacecraft body frame is

$$\mathbf{H}_B = J\omega + \mathbf{H}_B^w = A(\mathbf{e}_1, \theta) \begin{bmatrix} 0 \\ 2H_{\max} \\ 0 \end{bmatrix} \tag{4.97}$$

which gives, assuming a diagonal inertia tensor,

$$\mathbf{H}_B^w = \begin{bmatrix} -J_1\dot{\theta} \\ 2H_{\max}\cos\theta \\ 2H_{\max}\sin\theta \end{bmatrix} \tag{4.98}$$

The simulation uses $J_1 = H_{\max} \times 750$ s, consistent with values of 60 Nms for $H_{\max}$ and 45,000 kg-m^2 for J_1, the approximate values for JWST. It follows from Eq. (4.98) that $\|\mathbf{H}_B^w\|^2 = 4H_{\max}^2 + (J_1\dot{\theta})^2$. With the parameters used in the simulations, this has the maximum value

$$\|\mathbf{H}_B^w\|_{\max} = 2H_{\max}\sqrt{1 + (\pi t_{\text{accel}}/3{,}000\text{ s})^2} \tag{4.99}$$

Figure 4.12 shows the wheel momenta that result from using the pseudoinverse algorithm to distribute the momentum of Eq. (4.98) among the six wheels, identified

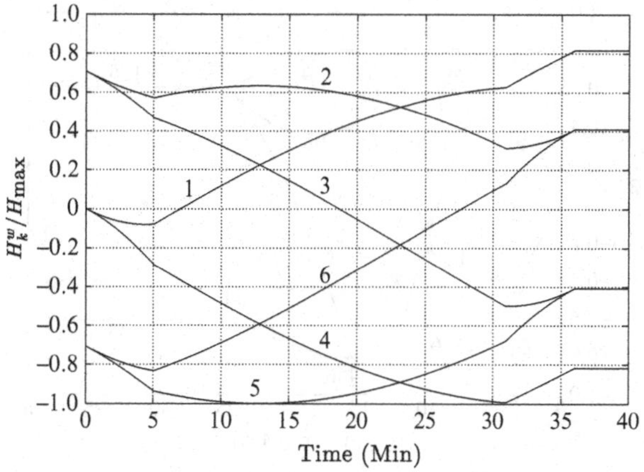

Fig. 4.12 Wheel momenta for 36-min slew using pseudoinverse distribution

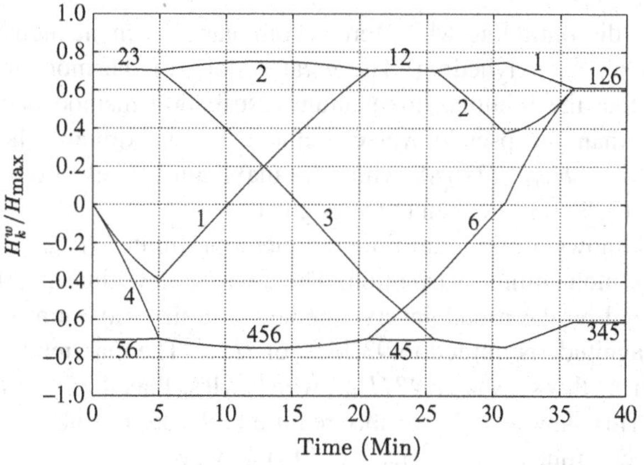

Fig. 4.13 Wheel momenta for 36-min slew using minimax distribution

by the labels on the curves. The acceleration time is $t_{accel} = 303$ s, which gives a slew time of 36 min. The magnitude of the angular momentum of wheel 5 reaches 99.97 % of H_{max}, so this 90° slew cannot be performed in less than 36 min using the pseudoinverse momentum distribution method. By Eq. (4.99), the maximum angular momentum in the body frame during this slew is $2.098H_{max}$, greater than the maximum value of $2H_{max}$ allowed by Table 4.2. That is not a contradiction, because the angular momentum in this simulation is not in the least favorable direction.

The minimax momentum distribution method can be used to either reduce the maximum wheel angular momentum for a given slew time or to slew more rapidly. Figure 4.13 shows the first option, keeping $t_{accel} = 303$ s and reducing the maximum wheel angular momentum magnitude to less than 75 % of H_{max}, a reduction of 25 % compared to the pseudoinverse method. The plots show that there are always four wheels with the same angular momentum magnitude, and also that the minimax

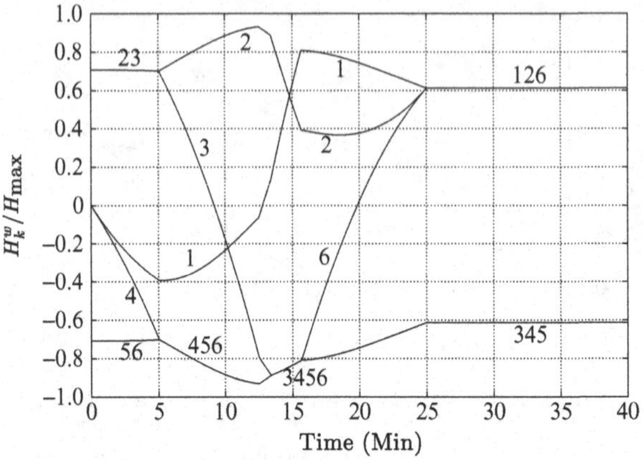

Fig. 4.14 Wheel momenta for 25-min slew using minimax distribution

method leads to discontinuous wheel torque commands when the momentum moves between faces of the polyhedron. The greater slope of the momentum vs. time curves shows that the minimax momentum distribution method demands higher wheel torques than the pseudoinverse method. The maximum wheel torque in Fig. 4.12 is $L_1 = H_{max}/(776\text{ s})$, while the maximum wheel torque in Fig. 4.13 is $L_1 = H_{max}/(353\text{ s})$, more than twice as great.

Figure 4.14 shows a minimum-time, acceleration-limited ($t_{accel} = 750\text{ s}$) slew using minimax momentum distribution. The 25 min slew time is 30 % less than the time required by the pseudoinverse method, and the maximum wheel angular momentum magnitude is limited to 93.16 % of H_{max}. The maximum body angular momentum during this slew is $2.543H_{max}$, which is less than the $2.667H_{max}$ allowed by Table 4.2. This slew would probably require high-torque wheels to provide the maximum wheel torque of $L_1 = H_{max}/(179\text{ s})$, however.

4.9 Control Moment Gyros

This section expands on the brief discussion of CMGs for attitude control in Sect. 3.3.5.1. Although double-gimbal control moment gyros have been developed, their mechanical complexity has led to their being supplanted by single-gimbal control moment gyros (SGCMGs) in most applications. A SGCMG has a wheel spinning at constant speed about a spin axis $\mathbf{w}_j$ with an angular momentum of magnitude H_j^w.[14] Rotating the angular momentum about a gimbal axis $\mathbf{g}_j$ perpendicular to the spin axis requires a torque

[14] Variable-speed CMGs, combining the properties of CMGs and reaction wheels, have been proposed but are not widely employed.

$$\mathbf{L}_j^w = \frac{d(H_j^w \mathbf{w}_j)}{dt} = H_j^w \dot{\mathbf{w}}_j = H_j^w \dot{\theta}_j \mathbf{g}_j \times \mathbf{w}_j \tag{4.100}$$

where θ_j is the gimbal angle in radians. The reaction torque exerted on the spacecraft is the negative of this. CMGs are capable of providing higher torques than reaction wheels because it is easier to apply a torque to a slowly-moving gimbal than to a rapidly-rotating wheel. CMGs are also advantageous from energy considerations, because the rotational kinetic energy stored in a CMG rotor moving at constant speed does not change, whereas energy must be added to or subtracted from a reaction wheel to change its speed. The angular momentum of CMGs covers a wide range from less than 1 Nms to greater than 2,000 Nms. Maximum gimbal rates are typically in the range from 0.2 to 2 rad/s, which gives an idea of the available torques.

It is clear that at least three SGCMGs are needed to provide three-axis control, and a system of n SGCMGs gives

$$\mathbf{L}^w = \sum_{j=1}^{n} H_j^w \dot{\theta}_j \mathbf{g}_j \times \mathbf{w}_j = \mathscr{W}_n^{CMG} \dot{\boldsymbol{\theta}} \tag{4.101}$$

where

$$\mathscr{W}_n^{CMG} \equiv \left[H_1^w \mathbf{g}_1 \times \mathbf{w}_1 \;\; H_2^w \mathbf{g}_2 \times \mathbf{w}_2 \;\; \cdots \;\; H_n^w \mathbf{g}_n \times \mathbf{w}_n \right] \tag{4.102}$$

and

$$\dot{\boldsymbol{\theta}} \equiv \left[\dot{\theta}_1 \; \dot{\theta}_2 \cdots \dot{\theta}_n \right]^T \tag{4.103}$$

The methods developed for reaction wheels in Sect. 4.8.3 can be used to invert these equations in order to find $\dot{\boldsymbol{\theta}}$ in terms of $\mathbf{L}^w$, but there are additional complications for SGCMGs. The directions of the gimbal axes are fixed, but the directions of the spin axes vary as the gimbal angles change, so the matrix $\mathscr{W}_n^{CMG}$ is not constant. A greater problem is that a system of SGCMGs can reach a singular configuration in which all the available torques lie in a plane, resulting in a loss of control authority about the axis perpendicular to the plane. These singular configurations, and methods for avoiding them, have been studied extensively [30, 40, 67]. In a system with only three SGCMGs, the commanded gimbal rates are determined uniquely by the inverse of $\mathscr{W}_n^{CMG}$, so there are no extra degrees of freedom for steering around singular configurations. For this reason, systems of SGCMGs invariably employ four or more CMGs.

4.10 Magnetic Torquers

Magnetic torquers create a magnetic dipole moment, **m**, which in turn creates a torque given by Eq. (3.159). Magnetic torques can be used either directly for attitude control or to unload momentum accumulated by reaction wheels or CMGs. The simplest torquers are made of N turns of wire in a loop of area A; sending a current I through such a coil produces a magnetic dipole moment of magnitude $m = NIA$ in a direction perpendicular to the plane of the coil. This relation has the advantage of being a perfectly linear function of the current, but large areas or many turns are required to produce the dipole moments needed by many spacecraft. For this reason *torque rods*, which are coils of wire wrapped around ferromagnetic cores, are employed to obtain greater dipole moments for a given amount of current.

Ferromagnetic materials exhibit hysteresis, as shown in Fig. 4.15. The dipole moment has a maximum value known as the saturation value, m_{sat}, when all the magnetic dipoles in the ferromagnetic material line up in the same direction. There is also a residual dipole, m_{res}, when no current is applied. The sign of the residual dipole depends on the direction from which the zero-current condition is approached, as the hysteresis curve is traversed in a counterclockwise direction. For most commercially available torque rods, the residual dipole moment is less than 1 % of the saturation moment, and the dipole moment is an approximately linear function of the current for dipole moments of at least 80 % of the saturation value, as indicated in the figure. Torque rods are available with saturation moments ranging from about 1 to 1,000 Am^2 and scale factors ranging from less than 0.1 Am^2/mA to almost 3 Am^2/mA, with the larger scale factors corresponding to the larger moments. Note that attaining a scale factor of 1 Am^2/mA without a ferromagnetic core would require a number of turns and coil area satisfying $NA = 1{,}000\,\text{m}^2$.

Most applications of magnetic torquers use three torquers producing moments on orthogonal axes. It is generally not necessary to employ extra torque rods for redundancy, because they usually have dual windings to provide internal redundancy. Sometimes more than three torque rods are used to provide additional capability, giving a total magnetic dipole moment equal to the vector sum of the moments provided by the individual rods. In this case, the desired net magnetic

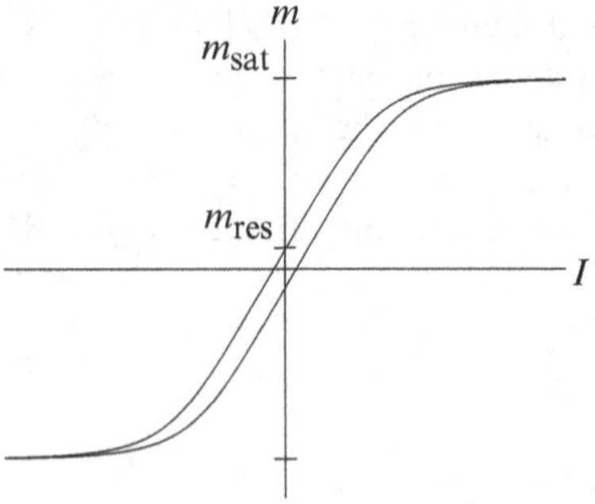

Fig. 4.15 Typical hysteresis loop; magnetic dipole moment as a function of applied current

moment can be distributed among the torquers by using one of the algorithms developed in Sect. 4.8.3 to distribute torque or angular momentum among reaction wheels.

4.11 Thrusters

Thrusters generate both forces and torques, so they can be used for both trajectory control and attitude control, as was briefly discussed in Sect. 3.3.6.5. Reference [33] provides an excellent overview of spacecraft propulsion systems. Thrusters have the advantage of not being dependent on an ambient magnetic or gravitational field, so they can be used for control or momentum unloading in any orbit. They have the disadvantage of requiring an expendable propellant, so that the life of a thruster-dependent mission is limited by the propellant supply. The force and torque provided by a thruster are

$$\mathbf{F}_{\text{mexp}} = -\dot{m}\mathbf{v}_{\text{rel}} \tag{4.104a}$$

$$\mathbf{L}_{\text{mexp}} = \mathbf{r} \times \mathbf{F}_{\text{mexp}} \tag{4.104b}$$

where $\dot{m}$ is the rate at which mass is expelled, $\mathbf{v}_{\text{rel}}$ is the velocity of the expelled mass relative to the spacecraft, and $\mathbf{r}$ is the vector from the spacecraft center of mass to the thruster. These equations show that the thrust and torque for a given mass expenditure are linearly proportional to the velocity v_{rel}. Writing Eq. (4.104a) as $v_{\text{rel}} = \Delta p/\Delta m$ shows that this parameter has the units of impulse per unit mass of propellant, so it is referred to as the *specific impulse* and can be expressed in units of Ns/kg. It is more common to normalize the specific impulse by the acceleration of gravity at the surface of the Earth, $g_0 = 9.80665$ m/s^2, to express it in time units:

$$I_{\text{sp}} \equiv v_{\text{rel}}/g_0 \tag{4.105}$$

Three types of thrusters are in common use for attitude control. Cold gas thrusters use a non-reacting gas stored at high pressure, on the order of 30 MPa. The most common gases are nitrogen, with a specific impulse of about 70 s, and helium, with a specific impulse of about 175 s. Using helium saves mass, but it is more expensive and more prone to leakage because of the small size of the helium molecule. Helium also requires a propellant tank with seven times as much volume or seven times higher pressure than the equivalent amount of nitrogen. Other propellants are used, but more rarely, in cold gas systems.

Monopropellant thrusters use a propellant that is decomposed catalytically. Hydrazine, with $I_{\text{sp}} = 242$ s, is by far the most common monopropellant. Monopropellant systems do not require the high pressures needed for cold gas, but hydrazine is a highly toxic material. This has led to interest in less toxic "green" propellants, such as hydroxyl ammonium nitrate (HAN), with $I_{\text{sp}} = 266$ s, and ammonium dinitramide (ADN), with $I_{sp} = 253$ s. These new propellants are less harmful to the environment, diminish operational hazards, and actually increase fuel efficiency due to their higher specific impulse [68].

The third class of thrusters uses electric fields to accelerate the expelled mass to a high velocity. The specific impulse of electric thrusters ranges from 500 to 3,000 s, so the saving in propellant mass is apparent. The kinetic energy that must be provided to the expelled mass is $mv_{rel}^2/2$, so the power required by the thruster, assuming 100 % efficiency, is

$$P_{thruster} = \frac{1}{2}\dot{m}v_{rel}^2 = \frac{1}{2}F_{mexp}v_{rel} \tag{4.106}$$

This energy is stored in the pressure vessel for cold gas thrusters and is provided by chemical decomposition for monopropellant thrusters, but it must be provided by the spacecraft power system for electric propulsion. Thus we see that an electric propulsion system with a higher specific impulse requires less propellant but more power to generate a given level of thrust.

An attitude torque provided by thrusters will perturb the spacecraft's orbit unless it is provided by a pair of equal and opposite forces comprising a perfect couple

$$\mathbf{L}_{couple} = \mathbf{r}_1 \times \mathbf{F}_{mexp} + \mathbf{r}_2 \times (-\mathbf{F}_{mexp}) = (\mathbf{r}_1 - \mathbf{r}_2) \times \mathbf{F}_{mexp} \tag{4.107}$$

Couples also have the advantage of being insensitive to a shift in the spacecraft's center of mass, which would give $\mathbf{r}_1 \to \mathbf{r}_1 - \Delta\mathbf{r}$ and $\mathbf{r}_2 \to \mathbf{r}_2 - \Delta\mathbf{r}$, and thus would leave $\mathbf{L}_{couple}$ unchanged. Several factors make it impractical to provide perfect couples, though. It is often desirable to have all the thrusters on one face of a spacecraft to facilitate modular integration of the propulsion system, to minimize the amount of propellant plumbing required, or to protect sensitive instruments from the deleterious effects of thruster plume impingement. The difficulty of providing perfect couples makes it important to minimize the migration of the spacecraft's center of mass, either by locating a single propellant tank at the center of mass or by distributing tanks symmetrically around the center of mass and drawing equal masses of propellant from them.

4.12 Nutation Dampers

Nutation dampers are passive devices designed to dissipate energy, thereby producing steady rotation of a spinning spacecraft about its axis of maximum principal moment of inertia, as described in Sect. 3.3.3.2. Libration dampers are similar energy-dissipation mechanisms used to eliminate libration on spacecraft employing gravity gradient stabilization, as discussed in Sect. 3.3.8. The energy-dissipating medium in either case can be a viscous fluid, a flexible spring with high structural damping, or electrical eddy currents [66].

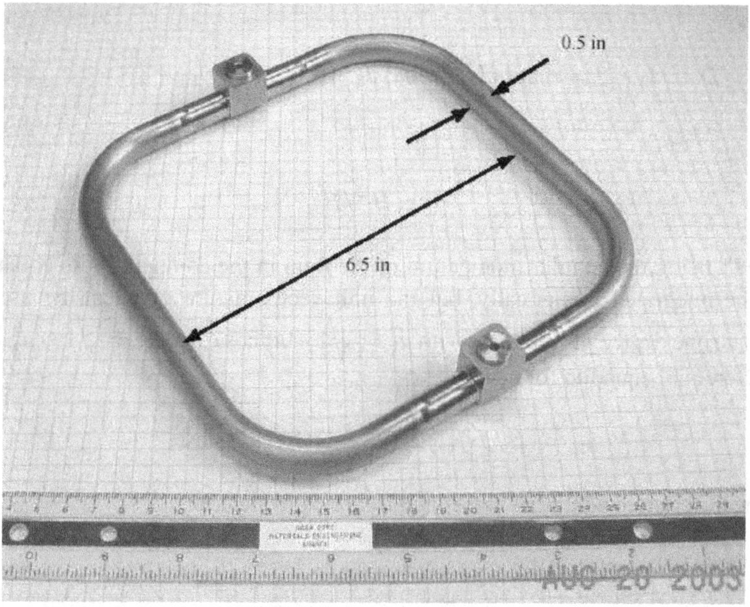

Fig. 4.16 ST5 passive nutation damper. Source: NASA

A very common type of nutation damper is the *viscous ring nutation damper*, a tube bent into the shape of a ring and filled with a viscous fluid [7, 27]. If the fluid does not fill the ring completely, surface tension at the interfaces of the fluid and gas (or vacuum) can cause the fluid to be locked in place, with a resultant loss of damping action. For this reason, viscous ring dampers are commonly fully filled and equipped with bellows to accommodate thermal expansion and contraction of the fluid. Some dampers, however, have omitted bellows because of space limitations, tolerating vacuum bubbles at non-operational low temperatures and using tubing strong enough to handle high pressures at the upper end of the operational temperature range. The viscous ring nutation damper of NASA's Space Technology-5 (ST5) mission, illustrated in Fig. 4.16 provides a good example [47].

We will use the methods of Sect. 3.3.5.1 to simulate the effect of a single fully-filled viscous ring nutation damper on a triaxial spacecraft with principal moments $J_1 < J_2 < J_3$. The damper is modeled as a reaction wheel with its axis perpendicular to the plane of the ring, along the axis of intermediate inertia, the orientation providing the most effective damping. We will neglect external torques, so the system angular momentum will have a constant magnitude H. Letting $\mathbf{H}_B = H\mathbf{u}$, where $\mathbf{u}$ is a unit vector, Eq. (3.80) can be written

$$\dot{\mathbf{u}} = -\boldsymbol{\omega}_B^{BI} \times \mathbf{u} = -(H/J_3)[D(\mathbf{u} - \chi\mathbf{e}_2)] \times \mathbf{u} \tag{4.108}$$

where

$$D \equiv J_3 J_B^{-1} = \mathrm{diag}([\, J_3/J_1 \;\; J_3/J_2 \;\; 1]) \equiv \mathrm{diag}([\gamma_1 \; \gamma_2 \; 1]) \tag{4.109}$$

and

$$\chi \equiv H^w/H \tag{4.110}$$

with H^w being the angular momentum of the fluid in the damper. These equations must be integrated numerically, but the unit vector nature of $\mathbf{u}$ can be used to eliminate one degree of freedom. With $u_1 = \sqrt{1-u_2^2}\cos\zeta$ and $u_3 = \sqrt{1-u_2^2}\sin\zeta$, the differential equation for u_2 is

$$\dot{u}_2 = (H/J_3)(1/2)(\gamma_1 - 1)(1 - u_2^2)\sin 2\zeta \tag{4.111}$$

and the differential equations for u_1 and u_3 are equivalent to

$$\dot{\zeta} = (H/J_3)\{(1/2)[2\gamma_2 - \gamma_1 - 1 - (\gamma_1 - 1)\cos 2\zeta\,]u_2 - \gamma_2\chi\} \tag{4.112}$$

We assume that the torque acting on the damper fluid is linearly proportional to and in the opposite direction from the fluid's angular velocity ω^w:

$$L^w = -\alpha(J^{\|}/J_3)H\omega^w = -\alpha\beta H\omega^w \tag{4.113}$$

where α is a dimensionless damping coefficient, $J^{\|}$ is the MOI of the moving fluid, and β is defined by the last equality. The actual value of the damping coefficient depends on the diameter of the tubing and the density and viscosity of the fluid, incorporated in a "wobble Reynolds number," but we will treat it simply as a tunable parameter [7,27]. Then Eqs. (3.145) and (3.141) give the equation of motion of the dimensionless damper angular momentum:

$$\begin{aligned}\dot{\chi} = L^w/H &= -\alpha\beta\omega^w = -\alpha\beta[(H^w/J^{\|}) - \mathbf{e}_2\cdot\boldsymbol{\omega}_B^{BI}]\\ &= -\alpha\beta[(\chi H/J^{\|}) - (u_2-\chi)H/J_2] = \alpha(H/J_3)[\beta\gamma_2 u_2 - (1+\beta\gamma_2)\chi]\end{aligned} \tag{4.114}$$

The MATLAB integrator ode45 was used to integrate Eqs. (4.111), (4.112), and (4.114) for a spacecraft with the inertia ratios of Fig. 3.2b, which give $\gamma_1 = 2$ and $\gamma_2 = 3/2$. The damper MOI was assumed to give $\beta = 0.005$, which about an order of magnitude larger than a reasonable damper could provide, but was chosen to give readable plots. The damping coefficient was assigned the value $\alpha = 1/\sqrt{2}$, which we will show to be the optimal value. The initial values of $\zeta = -0.01$ rad and $u_2 = \chi = 0$ represent very nearly pure spin about $\mathbf{e}_1$, the spacecraft axis of minimum MOI. The results of the simulation, displayed in Fig. 4.17, illustrate quantitatively in non-dimensional units the qualitative behavior discussed

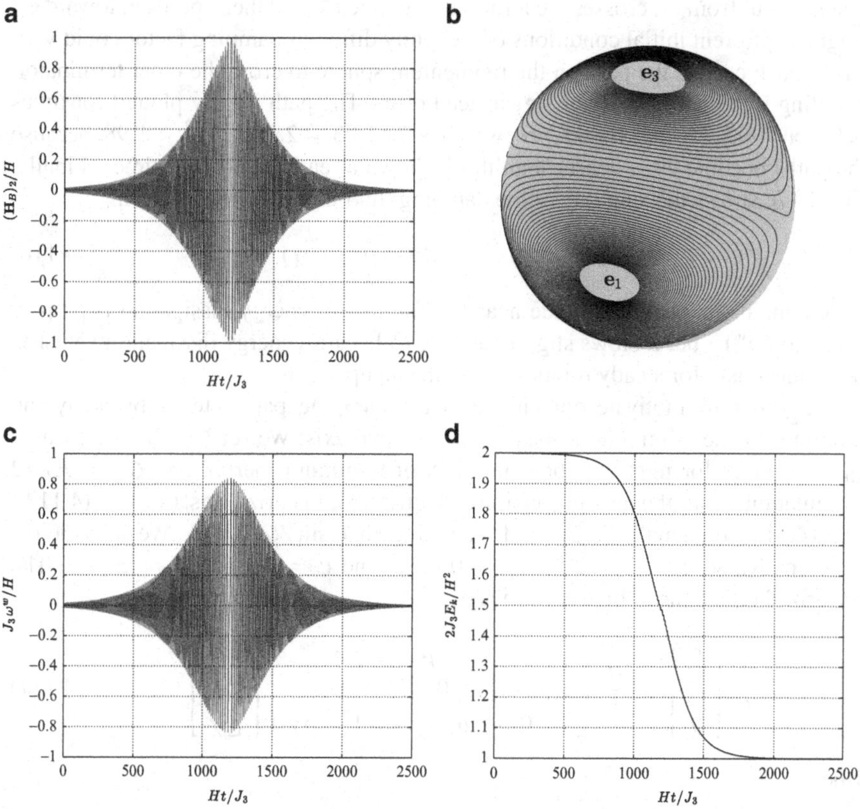

Fig. 4.17 Simulation of spinning spacecraft with nutation damper. (**a**) Intermediate axis angular momentum. (**b**) Path on momentum sphere. (**c**) Damping fluid spin rate. (**d**) Rotational kinetic energy

in Sect. 3.3.3. Figure 4.17a shows the component of the angular momentum along the axis of intermediate inertia; it begins at zero, then acquires an oscillatory behavior that grows to an amplitude close to the total system angular momentum, and finally decays to zero. The rotational kinetic energy in dimensionless form,

$$2J_3E_k/H^2 = (1-u_2^2)(\gamma_1\cos^2\zeta + \sin^2\zeta) + \gamma_2(u_2-\chi)^2 + \chi^2/\beta \qquad (4.115)$$

is plotted in Fig. 4.17d. It starts at the value $\gamma_1 = 2$, the value for pure rotation around the axis of minimum inertia, and ends at 1, the value for pure rotation around the axis of maximum inertia. It crosses the value $\gamma_2 = 3/2$, the value for pure rotation around the axis of intermediate inertia, at time $t^* = 1184J_3/H$, when the intermediate axis momentum reaches its maximum amplitude. Figure 4.17b shows the path followed by the angular momentum on the angular momentum sphere;

it spirals out from $\mathbf{e}_1$, crosses the terminator at time t^*, and then spirals in toward $\mathbf{e}_3$. Slightly different initial conditions or a slightly different damping factor could very easily have caused the path on the momentum sphere to cross the other terminator, resulting in a final spin about $-\mathbf{e}_3$ instead of $\mathbf{e}_3$. The path is only plotted for times between $700J_3/H$ and $1600J_3/H$, which is for $1.03 < 2J_3E_k/H^2 < 1.98$, because the paths become even denser than those shown at earlier and later times. Finally, Fig. 4.17c shows the spin rate of the damping fluid in dimensionless form,

$$J_3\omega^w/H = \chi/\beta - \gamma_2(u_2 - \chi) \tag{4.116}$$

which attains its peak amplitude near t^*. The rate of energy dissipation is proportional to $(\omega^w)^2$, but it slows slightly at time t^* because energy dissipation vanishes instantaneously for steady rotation about the damper axis.

We will now motivate our choice of the damping parameter α by analyzing motion near the minimum or maximum principal axis. We set $\zeta = \zeta_0 + \zeta'$, where $\zeta_0 = 0$ or π for nutation about the axis of minimum inertia, and $\zeta_0 = \pm\pi/2$ for nutation about the axis of maximum inertia. We linearize Eqs. (4.111), (4.112), and (4.114) for small ζ', u_2, and χ, noting that $\sin 2\zeta_0 = 0$. We also define $c \equiv \cos 2\zeta_0$ so that $c = 1$ for $\zeta_0 = 0$ or π, and $c = -1$ for $\zeta_0 = \pm\pi/2$. The linearized equations in matrix form are

$$\frac{d}{dt}\begin{bmatrix}\zeta' \\ u_2 \\ \chi\end{bmatrix} = \frac{H}{J_3}\begin{bmatrix} 0 & -c\mu & -\gamma_2 \\ c(\gamma_1 - 1) & 0 & 0 \\ 0 & \alpha\beta\gamma_2 & -\alpha(1+\beta\gamma_2)\end{bmatrix}\begin{bmatrix}\zeta' \\ u_2 \\ \chi\end{bmatrix} \tag{4.117}$$

where

$$\mu = \begin{cases} \gamma_1 - \gamma_2 & \text{for} \quad c = 1 \\ \gamma_2 - 1 & \text{for} \quad c = -1 \end{cases} \tag{4.118}$$

The general solution of these coupled linear differential equations is a superposition of three components of the form

$$\begin{bmatrix}\zeta'(t) \\ u_2(t) \\ \chi(t)\end{bmatrix} = \begin{bmatrix}\zeta'(0) \\ u_2(0) \\ \chi(0)\end{bmatrix} e^{\lambda t} \tag{4.119}$$

with coefficients satisfying initial conditions. A nontrivial solution exists only if λ is one of the three roots of the eigenvalue equation

$$\begin{aligned} 0 &= \det\left\{\lambda I_3 - \frac{H}{J_3}\begin{bmatrix} 0 & -c\mu & -\gamma_2 \\ c(\gamma_1 - 1) & 0 & 0 \\ 0 & \alpha\beta\gamma_2 & -\alpha(1+\beta\gamma_2)\end{bmatrix}\right\} \\ &= (\lambda^2 + \tilde{\omega}_p^2)[\lambda + H\alpha(1+\beta\gamma_2)/J_3] + c\alpha\beta(H/J_3)^3\gamma_2^2(\gamma_1 - 1) \end{aligned} \tag{4.120}$$

where

$$\tilde{\omega}_p = \frac{H}{J_3}\sqrt{\mu(\gamma_1 - 1)} = \begin{cases} (H/J_3)\sqrt{(J_3 - J_1)(J_3 - J_2)/J_1 J_2} & \text{for} \quad c = 1 \\ (H/J_1)\sqrt{(J_2 - J_1)(J_3 - J_1)/J_2 J_3} & \text{for} \quad c = -1 \end{cases} \tag{4.121}$$

The cubic eigenvalue equation does not have nice analytic solutions, but its solutions are close to those obtained by ignoring the last term on the right side because β is very small. Those approximate solutions are

$$\lambda_0 = -H\alpha(1 + \beta\gamma_2)/J_3 \quad \text{and} \quad \lambda_\pm = \pm i\tilde{\omega}_p \tag{4.122}$$

The real eigenvalue λ_0 has the eigenvector $[0\ \ 0\ \ 1]^T$ in the $\beta = 0$ limit. It describes only uninteresting transient behavior, the rapid decay of an initial value of H^w.

The purely imaginary eigenvalues are more interesting, describing sinusoidal motion with frequency $\tilde{\omega}_p$. In the limit of small nutation about the $\mathbf{e}_3$ axis, i.e. for $c = -1$ and $2E_k = H^2/J_3$, this frequency agrees up to a sign with the nutation frequency given by Eq. (3.120). The expression for $c = 1$, i.e. for nutation about the $\mathbf{e}_1$ axis, has the same form with J_1 and J_3 interchanged, as would be expected. Now we need to consider the effect of the term in Eq. (4.120) that we have ignored up to this point. This will cause a small change in $\lambda_\pm$, so we substitute $\lambda_\pm = \pm i\tilde{\omega}_p + \epsilon_\pm$ into Eq. (4.120), giving

$$\begin{aligned} 0 &= (\pm 2i\tilde{\omega}_p\epsilon_\pm + \epsilon_\pm^2)[\pm i\tilde{\omega}_p + \epsilon_\pm + H\alpha(1 + \beta\gamma_2)/J_3] + c\alpha\beta(H/J_3)^3\gamma_2^2(\gamma_1 - 1) \\ &\approx \pm 2i\tilde{\omega}_p\epsilon_\pm(\pm i\tilde{\omega}_p + H\alpha/J_3) + c\alpha\beta(H/J_3)^3\gamma_2^2(\gamma_1 - 1) \end{aligned} \tag{4.123}$$

The approximation ignores all terms of higher than first order in $\epsilon_\pm$ and β. Solving for $\epsilon_\pm$ gives

$$\epsilon_\pm = \frac{c\alpha\beta(H/J_3)^3\gamma_2^2(\gamma_1 - 1)}{2\tilde{\omega}_p(\tilde{\omega}_p \mp iH\alpha/J_3)} = \frac{c\alpha\beta(H/J_3)^3\gamma_2^2(\gamma_1 - 1)(\tilde{\omega}_p \pm iH\alpha/J_3)}{2\tilde{\omega}_p[\tilde{\omega}_p^2 + (H\alpha/J_3)^2]} \tag{4.124}$$

and thus

$$\lambda_\pm = \pm i\tilde{\omega}_p\left\{1 + \frac{c\alpha^2\beta\gamma_2^2}{2\mu[(J_3\tilde{\omega}_p/H)^2 + \alpha^2]}\right\} + \frac{cH}{2J_3}\frac{\alpha\beta\gamma_2^2(\gamma_1 - 1)}{(J_3\tilde{\omega}_p/H)^2 + \alpha^2} \tag{4.125}$$

The quantity in the curly brackets gives a small correction to the nutation frequency, less than 1 % for the parameters of our simulation. The second term is much more important; it gives an exponentially growing nutation amplitude for $c = 1$ and an exponentially decaying amplitude for $c = -1$, exactly what we expect for motion near the axis of minimum or maximum principal MOI, respectively. The time constant of the growth or decay is

$$\tau = \frac{2J_3}{H}\frac{(J_3\tilde{\omega}_p/H)^2 + \alpha^2}{\alpha\beta\gamma_2^2(\gamma_1 - 1)} = \frac{2\mu}{\beta\gamma_2^2\tilde{\omega}_p}\frac{\alpha^2 + \mu(\gamma_1 - 1)}{\alpha\sqrt{\mu(\gamma_1 - 1)}} \tag{4.126}$$

It is easy to see that the time constant is minimized for $\alpha = \sqrt{\mu(\gamma_1 - 1)}$. For $\gamma_1 = 2$ and $\gamma_2 = 3/2$, $\alpha = 1/\sqrt{2}$ provides the minimum time constant near both $\mathbf{e}_1$ and $\mathbf{e}_3$, which is why this value was chosen for the simulation. As a check, simulations with $\alpha = 0.6$ and $\alpha = 0.8$ both gave slower damping. Choosing the optimal damping coefficient gives

$$\tau_{min} = \frac{4\mu}{\beta\gamma_2^2\tilde{\omega}_p} \tag{4.127}$$

which is on the order of $1/\beta$ times the nutation period. A more realistic value of β than was used in the simulations would result in damping times about ten times as great.

Problems

4.1. Use the properties of the Poisson distribution to verify the probabilities for star availability presented in Sect. 4.2.4, i.e. $\bar{N} = 6.75$ gives a 90 % probability of having four or more stars in the FOV, $\bar{N} = 8$ gives a 90 % probability of having five or more stars in the FOV, $\bar{N} = 10$ gives a 99 % probability of having four or more stars in the FOV, and $\bar{N} = 11.7$ gives a 99 % probability of having five or more stars in the FOV.

4.2. The discrete-time gyro simulation model in Eq. (4.54) can also be derived from Farrenkopf's model given in Eq. (6.116), which is summarized by

$$\theta_{k+1}^{\text{true}} = \theta_k^{\text{true}} - \Delta t\, \beta_k^{\text{true}} + \Delta t\, \omega_k + w_{1_k}$$
$$\beta_{k+1}^{\text{true}} = \beta_k^{\text{true}} + w_{2_k}$$

where $E\left\{\mathbf{w}_k \mathbf{w}_k^T\right\} = Q$, with $\mathbf{w}_k = [w_{1_k}\ \ w_{2_k}]^T$ and

$$Q = \begin{bmatrix} \sigma_v^2 \Delta t + \frac{1}{3}\sigma_u^2 \Delta t^3 & -\frac{1}{2}\sigma_u^2 \Delta t^2 \\ -\frac{1}{2}\sigma_u^2 \Delta t^2 & \sigma_u^2 \Delta t \end{bmatrix}$$

To simulate the correlated process noise process a Cholesky decomposition of Q can be used, with $L^T L = Q$. First, prove that this can be done. Consider a zero-mean Gaussian white-noise variable $\mathbf{w}_k$ with covariance Q. Show that $\mathbf{w}_k = L^T \mathbf{p}_k$ can be used to simulate the correlated process noise, where $\mathbf{p}_k$ is zero-mean Gaussian white-noise process with covariance given by the identity matrix.

Assume the following form for L:

$$L = \begin{bmatrix} \alpha & \gamma \\ \beta & 0 \end{bmatrix}$$

Determine α, β and γ from Q in terms of σ_u, σ_v and Δt. Next, using $(\theta_{k+1}^{\text{true}} - \theta_k^{\text{true}})/\Delta t = \omega_k^{\text{true}}$ prove that the scalar version of Eq. (4.54) can be derived from Farrenkopf's model.

4.3. Show that the discrete-time gyro simulation model in Eq. (4.54) can effectively be executed using the following model for each axis:

$$\omega_{k+1} = \omega_{k+1}^{\text{true}} + y_k + \left(\frac{\sigma_v^2}{\Delta t} + \frac{1}{12}\sigma_u^2\,\Delta t\right)^{1/2} N_{v_k}$$

$$\beta_{k+1}^{\text{true}} = \beta_k^{\text{true}} + \sigma_u\,\Delta t^{1/2} N_{u_k}$$

$$y_k = \beta_k^{\text{true}} + \frac{1}{2}\sigma_u\,\Delta t^{1/2} N_{u_k}$$

Note that the last two equations give the following discrete-time system matrices as defined in Eq. (12.49): $\Phi = 1$, $\Gamma = 1$, $H = 1$, and $D = 1/2$ with input given by $\sigma_u\,\Delta t^{1/2} N_{u_k}$. Therefore, y_k is first obtained using the aforementioned discrete-time state model and then substituted into the ω_{k+1} equation to provide simulated discrete-time gyro measurements.

4.4. Find, as a function of γ and ω_0, the wheel speed at which the upper whirl frequency ω^+ of Eq. (4.63) coincides with the first harmonic of the wheel speed $\omega = 2\omega^w$. Find the numerical value of this in rpm for $\gamma = 0.91$ and $\omega_0 = 60$ Hz. Discuss.

4.5. This problem has two parts:

a) Derive the wheel capabilities in Table 4.2.
b) Derive the wheel capabilities in the two-failure cases for the six-wheel configurations in Table 4.2. There are three different possibilities for the six-wheel pyramid: failure of adjacent wheels (e.g. 1&2), failure of opposite wheels (e.g. 1&4), and the intermediate case (e.g. 1&3). In contrast to this, all two-wheel failures in the dodecahedron configuration are equivalent.

4.6. Repeat the analysis leading to Figs. 4.12, 4.13, and 4.14 using the four-wheel configuration $\mathscr{W}_4$ of Eq. (4.66) with $b = c = \sqrt{1/3}$ and $a = d = \sqrt{2/3}$. As in the example in the text, assume a rest-to-rest 90° slew about the body $\mathbf{e}_1$ axis, beginning with a constant angular acceleration of 1.6×10^{-4} deg/s^2 for a time $t_{\text{accel}} \leq 750$ s, followed by a coast at a constant angular rate between t_{accel} and $(750\ \text{s})^2/t_{\text{accel}}$, and ending with a deceleration at -1.6×10^{-4} deg/s^2 for a time t_{accel}. Use $J_1 = H_{\max} \times 750$ s, but assume that the system momentum for the four-wheel case is only $H_{\max}$ in the body $+\mathbf{e}_2$ direction rather than $2H_{\max}$, as in the six-wheel case.

Find the minimum slew time for the pseudoinverse distribution law. Find the maximum value of any reaction wheel momentum for the minimax distribution law if the slew is performed in this time. Find the minimum slew time for the minimax distribution law.

4.7. A common thruster configuration has four thrusters at the corners of the $-x$ face of a spacecraft, which is a distance d from the center of mass, so the vectors from the center of mass to the thrusters are

$$\mathbf{r}_1 = \begin{bmatrix} -d \\ a \\ a \end{bmatrix}, \quad \mathbf{r}_2 = \begin{bmatrix} -d \\ -a \\ a \end{bmatrix}, \quad \mathbf{r}_3 = \begin{bmatrix} -d \\ -a \\ -a \end{bmatrix}, \quad \mathbf{r}_4 = \begin{bmatrix} -d \\ a \\ -a \end{bmatrix}$$

The thrusters are canted at an angle θ, with $\tan\theta < a/d$, so that

$$\mathbf{F}_1 = \mathbf{F}_2 = F \begin{bmatrix} \cos\theta \\ 0 \\ -\sin\theta \end{bmatrix}, \quad \mathbf{F}_3 = \mathbf{F}_4 = F \begin{bmatrix} \cos\theta \\ 0 \\ \sin\theta \end{bmatrix}$$

a) Find the torques due to firing the thrusters individually.
b) Show that firing all four thrusters simultaneously gives a net thrust but no net torque.
c) Show that firing the thrusters in pairs can give a positive or negative torque about any axis. Thus attitude control is provided by on-pulsing pairs of thrusters when not performing an orbit adjustment, or by off-pulsing pairs of thrusters when the orbit is being adjusted.

References

1. Al-Bender, F., Swevers, J.: Characterization of friction force dynamics. IEEE Contr. Syst. Mag. **28**(6), 64–81 (2008)
2. Åström, K.J., Canudas de Wit, C.: Revisiting the LuGre model. IEEE Contr. Syst. Mag. **28**(6), 101–114 (2008)
3. Axelrad, P., Behre, C.P.: Satellite attitude determination based on GPS signal-to-noise ratio. Proc. IEEE **87**(1), 133–144 (1999)
4. Bahcall, J.N.: Star counts and galactic structure. Annu. Rev. Astron. Astrophys. **24**, 577–611 (1986)
5. Bhanderi, D.D.V.: Modeling Earth albedo currents on Sun sensors for improved vector observation. In: AIAA Guidance, Navigation and Control Conference. Keystone (2006). AIAA 2006-6592
6. Bhanderi, D.D.V., Bak, T.: Modeling Earth albedo for satellites in Earth orbit. In: AIAA Guidance, Navigation and Control Conference. San Francisco (2005). AIAA 2005-6465
7. Bhuta, P.G., Koval, L.R.: A viscous ring damper for a freely precessing satellite. Int. J. Mech. Sci. **8**(5), 383–395 (1966)

8. Bialke, B.: High fidelity mathematical modeling of reaction wheel performance. In: Culp, R.D., Igli, D. (eds.) Guidance and Control 1998, Advances in the Astronautical Sciences, vol. 98, pp. 483–496. Univelt, San Diego (2007)
9. Bialke, B.: Microvibration disturbance fundamentals for rotating mechanisms. In: Miller, K.B. (ed.) Guidance and Control 2011, Advances in the Astronautical Sciences, vol. 141, pp. 417–432. Univelt, San Diego (2007)
10. Braasch, M.S.: Multipath effects. In: Parkinson, B., Spilker, J. (eds.) Global Positioning System: Theory and Applications, Progress in Astronautics and Aeronautics, vol. 64, chap. 14. American Institute of Aeronautics and Astronautics, Washington, DC (1996)
11. Bronowicki, A.J.: Forensic investigation of reaction wheel nutation on isolator. In: AIAA/ASME/AHS/ASC Structures, Structural Dynamics, and Materials Conference. Schaumberg (2008)
12. Canudas de Wit, C., Olsson, H., Åström, K.J., Lischinsky, P.: A new model for control of systems with friction. IEEE Trans. Automat. Contr. **40**(3), 419–425 (1995)
13. Chen, X., Steyn, W.H.: Robust combined eigenaxis slew maneuver. In: AIAA Guidance, Navigation and Control Conference, pp. 521–529. Portland (1999). AIAA 1999-4048
14. Chow, W.W., Sanders, V.E., Schleich, W., Scully, M.O.: The ring laser gyro. Rev. Mod. Phys. **57**(1), 61–103 (1985)
15. Cohen, C.E.: Attitude determination using GPS. Ph.D. thesis, Stanford University (1993)
16. Cohen, C.E.: Attitude determination. In: B. Parkinson, J. Spilker (eds.) Global Positioning System: Theory and Applications, Progress in Astronautics and Aeronautics, vol. 64, chap. 19. American Institute of Aeronautics and Astronautics, Washington, DC (1996)
17. Cowen, R.: The wheels come off Kepler. Nature **497**(7450), 417–418 (2013)
18. Dahl, P.R.: A solid friction model. Contractor Report TOR-0158(3107-18)-1, The Aerospace Corporation, Los Angeles (1968)
19. Fallon III, L.: Gyroscopes. In: J.R. Wertz (ed.) Spacecraft Attitude Determination and Control, chap. 6.5. Kluwer Academic, Dordrecht (1978)
20. Farrenkopf, R.L.: Analytic steady-state accuracy solutions for two common spacecraft attitude estimators. J. Guid. Contr. **1**(4), 282–284 (1978)
21. Golub, G.H., Van Loan, C.F.: Matrix Computations, 3rd edn. The Johns Hopkins University Press, Baltimore (1996)
22. Hablani, H.B.: Momentum accumulation due to solar radiation torque, and reaction wheel sizing, with configuration optimization. In: Proceedings of the Flight Mechanics/Estimation Theory Symposium, pp. 3–22. NASA-Goddard Space Flight Center, Greenbelt (1992)
23. Hablani, H.B.: Sun-tracking commands and reaction wheel sizing with configuration optimization. J. Guid. Contr. Dynam. **17**(4), 805–814 (1994)
24. Haworth, D.: How many stars can you observe (2013). http://www.stargazing.net/David/constel/howmanystars.html
25. Horn, R.A., Johnson, C.R.: Matrix Analysis. Cambridge University Press, Cambridge (1985)
26. Horri, N.M., Palmer, P., Giffen, A.: Active attitude control mechanisms. In: Blockley, R., Shyy, W. (eds.) Encyclopedia of Aerospace Engineering. Wiley, Chichester (2010)
27. Hubert, C.: Modeling completely filled viscous ring nutation dampers. Technical Report B2027, NASA Goddard Space Flight Center (2002)
28. Kawaguchi, J., Maeda, K., Matsuo, H., Ninomiya, K.: Closed loop momentum transfer maneuvers using multiwheels. J. Guid. Contr. Dynam. **18**(4), 867–874 (1995)
29. Kumar, K.: A microstructure study of corrosion in Ag-Cu flex leads, Journal of the Electrochemical Society, **127**(4), 906–910 (1980)
30. Kurokawa, H.: Survey of theory and steering laws of single-gimbal control moment gyros. J. Guid. Contr. Dynam. **30**(5), 1331–1340 (2007)
31. Lee, F.C., Werner, M.: Reaction wheel jitter analysis including rocking dynamics & bearing harmonic disturbances. In: Hollowell, H.E., Culp, R.D. (eds.) Guidance and Control 2007, Advances in the Astronautical Sciences, vol. 128, pp. 93–110. Univelt, San Diego (2007)
32. Leick, A.: GPS Satellite Surveying, 3rd edn. Wiley, Chichester (2004)

33. Leyva, I.A.: Spacecraft subsystems I – propulsion. In: Wertz, J.R., Everett, D.F., Puschell, J.J. (eds.) Space Mission Engineering: The New SMAD, Space Technology Library. Microcosm Press, Hawthorne (2011)
34. Liebe, C.C.: Accuracy performance of star trackers – a tutorial. IEEE Trans. Aero. Electron. Syst. **38**(2), 587–599 (2002)
35. Lightsey, E.G.: Development and flight demonstration of a GPS receiver for space. Ph.D. thesis, Stanford University (1997)
36. Lightsey, E.G., Madsen, J.: Three-axis attitude determination using Global Positioning System signal strength measurements. J. Guid. Contr. Dynam. **26**(2), 304–310 (2003)
37. Liu, K.C., Kenney, T., Maghami, P., Mulé, P., Blaurock, C., Haile, W.B.: Jitter test program and on-orbit mitigation strategies for Solar Dynamics Observatory. In: 20th International Symposium on Space Flight Dynamics. NASA Goddard Space Flight Center, Annapolis (2007)
38. Liu, K.C., Maghami, P., Blaurock, C.: Reaction wheel disturbance modeling, jitter analysis, and validation tests for Solar Dynamics Observatory. In: AIAA Guidance, Navigation and Control Conference. Honolulu (2008). AIAA 2008-7232
39. Madsen, J., Lightsey, E.G.: Robust spacecraft attitude determination using Global Positioning System receivers. J. Spacecraft Rockets **41**(4), 635–643 (2004)
40. Margulies, G., Aubrun, J.N.: Geometric theory of single-gimbal control moment gyro systems. J. Astronaut. Sci. **26**(2), 221–238 (1978)
41. Markley, F.L., Reynolds, R.G., Liu, F.X., Lebsock, K.L.: Maximum torque and momentum envelopes for reaction-wheel arrays. J. Guid. Contr. Dynam. **33**(5), 1606–1614 (2010)
42. Masterson, R.A., Miller, D.W., Grogan, R.L.: Development and validation of reaction wheel disturbance models: Empirical model. J. Sound Vib. **249**(3), 575–598 (2002)
43. McQuerry, J.P., Radovich Jr., M.A., Deters, R.A.: A precision star tracker for the nineties: A system guide to applications. In: Culp, R.D., Gravseth, A.D. (eds.) Guidance and Control 1990, Advances in the Astronautical Sciences, vol. 72, pp. 83–104. Univelt, San Diego (1990). AAS 90–014
44. Merhav, S.: Aerospace Sensor Systems and Applications. Springer, New York (1962)
45. Merlo, S., Norgia, M., Donati, S.: Fiber gyroscope principles. In: Lòpez-Higuera, J.M. (ed.) Handbook of Optical Fibre Sensing Technology, chap. 16, pp. 331–347. Wiley, Chichester (2002)
46. Nicolle, M., Fusco, T., Rousset, G., Michau, V.: Improvement of Shack-Hartmann wave-front sensor measurement for extreme adaptive optics. Optic Lett. **29**(3), 2743–2745 (2004)
47. O'Donnell Jr., J.R., Concha, M., Tsai, D.C., Placanica, S.J., Morrissey, J.R., Russo, A.M.: Space Technology 5 launch and operations. In: Hallowell, H.E., Culp, R.D. (eds.) Guidance and Control 2007, Advances in the Astronautical Sciences, vol. 128, pp. 735–753. Univelt, San Diego (1999). AAS 07–091
48. Pandiyan, R., Solaiappan, A., Malik, N.: A one step batch filter for estimating gyroscope calibration parameters using star vectors. In: AIAA/AAS Astrodynamics Specialist Conference and Exhibit. Providence (2004). AIAA 04–4858
49. Park, K., Crassidis, J.L.: Attitude determination methods using pseudolite signal phase measurements. J. Inst. Navigation **53**(2), 121–134 (2006)
50. Park, K., Crassidis, J.L.: A robust GPS receiver self survey algorithm. J. Inst. Navigation **53**(4), 259–268 (2006)
51. Parkinson, B., Spilker, J. (eds.): Global Positioning System: Theory and Applications, Progress in Astronautics and Aeronautics, vol. 64. American Institute of Aeronautics and Astronautics, Washington, DC (1996)
52. Parkinson, B.W.: GPS error analysis. In: Parkinson, B., Spilker, J. (eds.) Global Positioning System: Theory and Applications, Progress in Astronautics and Aeronautics, vol. 64, chap. 11. American Institute of Aeronautics and Astronautics, Washington, DC (1996)
53. Pittelkau, M.E.: Kalman filtering for spacecraft system alignment calibration. J. Guid. Contr. Dynam. **24**(6) (2001)
54. Pittelkau, M.E.: Sensors for attitude determination. In: Blockley, R., Shyy, W. (eds.) Encyclopedia of Aerospace Engineering. Wiley, Chichester (2010)

55. Post, E.J.: Sagnac effect. Rev. Mod. Phys. **39**(2), 475–493 (1967)
56. Salomon, P.M., Glavich, T.A.: Image signal processing in sub-pixel accuracy star trackers. In: Barbe, D.F. (ed.) Smart Sensors II, SPIE Proceedings, vol. 252, pp. 64–74 (1980)
57. Secroun, A., Lampton, M., Levi, M.: A high-accuracy, small field of view star guider with application to SMAP. Exp. Astron. **12**(2), 69–85 (2000)
58. Serrano, J., Potti, J., Bernedo, P., Silvestrin, P.: A new spacecraft attitude determination scheme based on the use of GPS line-of-sight vectors. In: Proceedings of the ION GPS-95, pp. 1797–1806. Institute of Navigation, Fairfax (1995)
59. Shuster, M.D.: Stellar aberration and parallax: A tutorial. J. Astronaut. Sci. **51**(4), 477–494 (2003)
60. Sinnott, R., Perryman, M.: Millennium Star Atlas, vol. 1. Sky Publishing Corporation & European Space Agency, Cambridge (1997)
61. Spinney, V.W.: Applications of the Global Positioning System as an attitude reference for near Earth users. In: ION National Aerospace Meeting, Naval Air Development Center. Warminster (1976)
62. Spratling IV, B.B., Mortari, D.: A survey on star identification algorithms. Algorithms **2**(1), 93–107 (2009)
63. Steyn, W.H.: Near-minimum-time eigenaxis rotation maneuvers using reaction wheels. J. Guid. Contr. Dynam. **18**(5), 1184–1189 (1995)
64. van Bezooijen, R.W.H., Anderson, K.A., Ward, D.K.: Performance of the AST-201 star tracker for the Microwave Anisotropy Probe. In: AIAA Guidance, Navigation and Control Conference. Monterey (2002). AIAA 2002-4582
65. Wertheimer, J.G., Laughlin, G.: Are Proxima Centauri and α Centauri gravitationally bound? Astron. J. **132**(5), 1995–1997 (2007)
66. Wertz, J.R. (ed.): Spacecraft Attitude Determination and Control. Kluwer Academic, Dordrecht (1978)
67. Wie, B.: Singularity analysis and visualization for single-gimbal control moment gyro systems. J. Guid. Contr. Dynam. **27**(2), 271–282 (2004)
68. Witze, A.: Green fuels blast off. Nature **500**(7464), 509–510 (2013)

Chapter 5
Static Attitude Determination Methods

Attitude determination typically requires finding three independent quantities, such as any minimal parameterization of the attitude matrix. The mathematics behind attitude determination can be broadly characterized into approaches that use stochastic analysis and approaches that do not. We restrict the term "estimation" to approaches that explicitly account for stochastic variables in the mathematical formulation, such as a Kalman filter or a maximum likelihood approach [29]. Black's 1964 TRIAD algorithm was the first published method for determining the attitude of a spacecraft using body and reference observations, but his method could only combine the information from two measurements [2]. One year later, Wahba formulated a general criterion for attitude determination using two or more vector measurements [36]. However, explicit relations to stochastic errors in the body measurements are not shown in these formulations. The connection to the stochastic nature associated with random measurement noise was first made by Farrell in a Kalman filtering application that appeared in a NASA report in 1964 [11], but was not published in the archival literature until 1970 [12]. Farrell's filter did not account for errors in the system dynamics, which were first accounted for in a Kalman filter developed by Potter and Vander Velde in 1968 [27].

It is useful to divide attitude determination approaches into two other categories. The first category comprises static determination approaches that depend on measurements taken at the same time, or close enough in time that spacecraft motion between the measurements can be ignored or easily compensated for. The second category comprises filtering approaches that make explicit use of knowledge of the motion of the spacecraft to accumulate a "memory" of past measurements. Static approaches have the requirement that at each time there are enough observations available to fully compute the attitude, but they typically require no a priori attitude estimate. A purely deterministic approach incorporates just enough observation information to uniquely determine the attitude. For example, the TRIAD algorithm uses only three out of the four pieces of information obtainable from two measured vectors to compute an attitude solution. Another deterministic approach involves a single vector measurement and a single angle measurement [30]. Over-determined

F.L. Markley and J.L. Crassidis, *Fundamentals of Spacecraft Attitude Determination and Control*, Space Technology Library 33, DOI 10.1007/978-1-4939-0802-8_5,
© Springer Science+Business Media New York 2014

approaches use more observation information than is required to compute a full attitude solution. For example, a star tracker onboard a space vehicle observes line-of-sight vectors to several stars, which are compared with known inertial line-of-sight vectors from a star catalog to determine the attitude of the vehicle. Statistical factors, such as an attitude error covariance, can be developed from a static solution.

In many spacecraft attitude systems, the attitude observations are naturally represented as unit vectors. Typical examples are the unit vectors giving the direction to the Sun or a star and the unit vector in the direction of the Earth's magnetic field. This chapter will consider the static approaches to attitude determination using vector measurements, beginning with TRIAD and then going on to consideration of methods accommodating more than two measurements. We will also consider the error bounds of the various estimators.

5.1 The TRIAD Algorithm

Some spacecraft attitude determination methods use exactly two vector measurements. The two vectors are typically the unit vector to the Sun and the Earth's magnetic field vector for coarse Sun-magnetic attitude determination or unit vectors to two stars tracked by two star trackers for fine attitude determination. TRIAD, the earliest published algorithm for determining spacecraft attitude from two vector measurements, has been widely used in both ground-based and onboard attitude determination. The name "TRIAD" can be considered either as the word "triad" or as an acronym for TRIaxial Attitude Determination.

Suppose that we have measured the components of two unit vectors in the spacecraft body frame and also know these components in some reference frame. We consider only unit vectors because the length of the vector has no information relevant to attitude determination. Owing to the norm constraint, each of these unit vectors contains only two independent scalar pieces of attitude information. We have seen that three parameters are required to specify the attitude matrix A, so two unit vector measurements are required to determine the attitude matrix, in general. In fact two vectors overdetermine the attitude. The notation we have been using up to now would denote the representation of a vector in the body frame as $\mathbf{x}_B$ and the representation in the reference frame by $\mathbf{x}_R$. To avoid the proliferation of subscripts, we shall denote the representation of the two vectors in the body frame as $\mathbf{b}_1$ and $\mathbf{b}_2$ and the representations of the corresponding vectors in the reference frame by $\mathbf{r}_1$ and $\mathbf{r}_2$. It is clear that the attitude matrix is not uniquely determined if either the pair $\mathbf{b}_1$ and $\mathbf{b}_2$ or the pair $\mathbf{r}_1$ and $\mathbf{r}_2$ are parallel or antiparallel, because the attitude can only be determined up to a rotation by angle about the pair of vectors in that case.

The attitude matrix to be determined is the matrix that rotates vectors from the reference frame to the spacecraft body frame, so we would like to find an attitude matrix such that

$$A\mathbf{r}_i = \mathbf{b}_i, \quad \text{for} \quad i = 1, 2 \tag{5.1}$$

This is not possible in general, however, for Eq. (5.1) implies that

$$\mathbf{b}_1 \cdot \mathbf{b}_2 = (A\mathbf{r}_1) \cdot (A\mathbf{r}_2) = \mathbf{r}_1^T A^T A \mathbf{r}_2 = \mathbf{r}_1 \cdot \mathbf{r}_2 \tag{5.2}$$

This equality is true for error-free measurements, but is not generally true in the presence of measurement errors. Thus it is impossible to satisfy both of Eq. (5.1) simultaneously if Eq. (5.2) is not obeyed. The classical TRIAD algorithm is based on the assumption that one of the unit vectors, which is conventionally denoted by $\mathbf{b}_1$, is much more accurately determined than the other, so the estimate satisfies $A\mathbf{r}_1 = \mathbf{b}_1$ exactly, but $A\mathbf{r}_2 = \mathbf{b}_2$ only approximately.

TRIAD is based on the following idea. If we have an orthonormal right-handed triad of vectors $\{\mathbf{v}_1, \mathbf{v}_2, \mathbf{v}_3\}$ in the reference frame, and a corresponding orthonormal right-handed triad $\{\mathbf{w}_1, \mathbf{w}_2, \mathbf{w}_3\}$ in the spacecraft body frame, then the attitude matrix

$$A \equiv [\mathbf{w}_1\, \mathbf{w}_2\, \mathbf{w}_3][\mathbf{v}_1\, \mathbf{v}_2\, \mathbf{v}_3]^T = \sum_{i=1}^{3} \mathbf{w}_i \mathbf{v}_i^T \tag{5.3}$$

will transform the $\mathbf{v}_i$ to the $\mathbf{w}_i$ by

$$A\mathbf{v}_i = \mathbf{w}_i, \quad \text{for} \quad i = 1, 2, 3 \tag{5.4}$$

TRIAD forms the triad $\{\mathbf{v}_1, \mathbf{v}_2, \mathbf{v}_3\}$ from $\mathbf{r}_1$ and $\mathbf{r}_2$, and the triad $\{\mathbf{w}_1, \mathbf{w}_2, \mathbf{w}_3\}$ from $\mathbf{b}_1$ and $\mathbf{b}_2$ by means of

$$\mathbf{v}_1 = \mathbf{r}_1, \quad \mathbf{v}_2 = \mathbf{r}_\times \equiv \frac{\mathbf{r}_1 \times \mathbf{r}_2}{\|\mathbf{r}_1 \times \mathbf{r}_2\|}, \quad \mathbf{v}_3 = \mathbf{r}_1 \times \mathbf{r}_\times \tag{5.5a}$$

$$\mathbf{w}_1 = \mathbf{b}_1, \quad \mathbf{w}_2 = \mathbf{b}_\times \equiv \frac{\mathbf{b}_1 \times \mathbf{b}_2}{\|\mathbf{b}_1 \times \mathbf{b}_2\|}, \quad \mathbf{w}_3 = \mathbf{b}_1 \times \mathbf{b}_\times \tag{5.5b}$$

Then the estimate of the attitude matrix, indicated by a caret, is given by Eq. (5.3) as

$$\hat{A}_{\mathrm{TRIAD}} = \mathbf{b}_1 \mathbf{r}_1^T + (\mathbf{b}_1 \times \mathbf{b}_\times)(\mathbf{r}_1 \times \mathbf{r}_\times)^T + \mathbf{b}_\times \mathbf{r}_\times^T \tag{5.6}$$

It can easily be seen that this solution satisfies the first equality of Eq. (5.1), and some manipulation of dot and cross products shows that it satisfies the second equality if and only if Eq. (5.2) is obeyed. We note that the attitude matrix is undefined if either the reference vectors or the observed vectors are parallel or antiparallel. This is the case noted above in which there is insufficient information to determine the attitude uniquely.

This TRIAD attitude estimate is completely independent of the component of $\mathbf{b}_2$ along $\mathbf{b}_1$ and the component of $\mathbf{r}_2$ along $\mathbf{r}_1$. This is the information that is discarded to give the three pieces of information needed to specify the attitude matrix. Variants of TRIAD have been proposed that use other combinations of the reference and body vectors to form the orthonormal triads. These have been superseded by an algorithm to be derived later in this chapter that uses an arbitrary scalar weighting of the two vectors.

5.2 Wahba's Problem

We can improve on the TRIAD method in two ways: by allowing arbitrary weighting of the measurements and by allowing the use of more than two measurements. The latter is especially important for use with star trackers that can track many stars simultaneously. Almost all of these improvements are based on a problem posed in 1965 by Grace Wahba [36]. Wahba's problem is to find the orthogonal matrix A with determinant $+1$ that minimizes the loss function

$$L(A) \equiv \frac{1}{2}\sum_{i=1}^{N} a_i \|\mathbf{b}_i - A\mathbf{r}_i\|^2 \tag{5.7}$$

where $\{\mathbf{b}_i\}$ is a set of N unit vectors measured in a spacecraft's body frame, $\{\mathbf{r}_i\}$ are the corresponding unit vectors in a reference frame, and $\{a_i\}$ are non-negative weights. We will discuss the choice of weights in Sect. 5.5.

Using the orthogonality of A, the unit norm of the unit vectors, and the cyclic invariance of the trace, Eq. (2.11c), gives

$$\|\mathbf{b}_i - A\mathbf{r}_i\|^2 = \|\mathbf{b}_i\|^2 + \|A\mathbf{r}_i\|^2 - 2\mathbf{b}_i \cdot (A\mathbf{r}_i) = 2 - 2\,\mathrm{tr}(A\mathbf{r}_i\mathbf{b}_i^T) \tag{5.8}$$

Thus we can write the loss function in the very convenient form

$$L(A) = \lambda_0 - \mathrm{tr}(AB^T) \tag{5.9}$$

with

$$\lambda_0 \equiv \sum_{i=1}^{N} a_i \tag{5.10}$$

and the "attitude profile matrix" B defined by

$$B \equiv \sum_{i=1}^{N} a_i \mathbf{b}_i \mathbf{r}_i^T \tag{5.11}$$

Now it is clear the loss function is minimized when $\mathrm{tr}(AB^T)$ is maximized. This has a close relation to the orthogonal Procrustes problem, which is equivalent to finding the orthogonal matrix A that is closest to B in the Frobenius norm (also known as the Euclidean, Schur, or Hilbert-Schmidt norm) [17]

$$\|M\|_F^2 \equiv \mathrm{tr}(MM^T) \tag{5.12}$$

It follows from

$$\|A - B\|_F^2 = \|A\|_F^2 + \|B\|_F^2 - 2\text{tr}(AB^T) = 3 + \|B\|_F^2 - 2\text{tr}(AB^T) \qquad (5.13)$$

that Wahba's problem is equivalent to the orthogonal Procrustes problem with the proviso that the determinant of A must be $+1$.

Algorithms for solving Wahba's problem fall into two classes. The first solves for the attitude matrix directly, and the second solves for the quaternion representation of the attitude matrix. With error-free mathematics, all algorithms should lead to the same attitude, and testing shows this to be the case [5, 24]. Some algorithms are faster than others, but execution speeds with modern processors make speed differences less important [4, 24]. Quaternion solutions have proven to be much more useful in practice, so we will consider them first.

5.3 Quaternion Solutions of Wahba's Problem

5.3.1 Davenport's q Method

Paul Davenport provided the first useful solution of Wahba's problem for spacecraft attitude determination [18]. He substituted Eq. (2.129) into Wahba's loss function to get

$$L(A(\mathbf{q})) = \lambda_0 - \sum_{i=1}^{N} a_i \mathbf{b}_i^T A(\mathbf{q})\mathbf{r}_i = \lambda_0 - \sum_{i=1}^{N} a_i \mathbf{b}_i^T \Xi^T(\mathbf{q})\Psi(\mathbf{q})\mathbf{r}_i \qquad (5.14)$$

Using several relations from Sect. 2.7 gives

$$\begin{aligned} \mathbf{b}_i^T \Xi^T(\mathbf{q})\Psi(\mathbf{q})\mathbf{r}_i &= \begin{bmatrix} \mathbf{b}_i \\ 0 \end{bmatrix}^T [\mathbf{q}\odot]^T[\mathbf{q}\otimes] \begin{bmatrix} \mathbf{r}_i \\ 0 \end{bmatrix} = (\mathbf{q} \odot \mathbf{b}_i)^T (\mathbf{q} \otimes \mathbf{r}_i) \\ &= (\mathbf{b}_i \otimes \mathbf{q})^T (\mathbf{r}_i \odot \mathbf{q}) = \mathbf{q}^T [\mathbf{b}_i \otimes]^T [\mathbf{r}_i \odot]\mathbf{q} \end{aligned} \qquad (5.15)$$

This allows us to express the loss function as

$$L(A(\mathbf{q})) = \lambda_0 - \mathbf{q}^T K(B)\mathbf{q} \qquad (5.16)$$

where $K(B)$ is the symmetric traceless matrix

$$K(B) = \sum_{i=1}^{N} a_i [\mathbf{b}_i \otimes]^T [\mathbf{r}_i \odot] = \begin{bmatrix} B + B^T - (\text{tr}B)I_3 & \mathbf{z} \\ \mathbf{z}^T & \text{tr}B \end{bmatrix} \qquad (5.17)$$

with

$$\mathbf{z} \equiv \begin{bmatrix} B_{23} - B_{32} \\ B_{31} - B_{13} \\ B_{12} - B_{21} \end{bmatrix} = \sum_{i=1}^{N} a_i (\mathbf{b}_i \times \mathbf{r}_i) \tag{5.18}$$

It is known that a real symmetric $n \times n$ matrix has n real eigenvalues and n real eigenvectors that can be chosen to form an orthonormal basis [15, 17]. The eigenvalue/eigenvector decomposition of K can be written as

$$K(B) = \sum_{i=1}^{4} \lambda_i \mathbf{q}_i \mathbf{q}_i^T \tag{5.19}$$

where $\mathbf{q}_i$ is the eigenvector with eigenvalue λ_i, and the eigenvalues are labeled so that $\lambda_{\max} \equiv \lambda_1 \geq \lambda_2 \geq \lambda_3 \geq \lambda_4$. Taking the trace of this equation and using the cyclic invariance of the trace and the orthonormality of the eigenvectors gives the well known result that the trace of a diagonalizable matrix is equal to the sum of its eigenvalues

$$\text{tr}(K(B)) = \sum_{i=1}^{4} \lambda_i \tag{5.20}$$

Equation (5.17) shows that the trace of $K(B)$ is zero, so this means that the sum of the eigenvalues is zero:

$$\sum_{i=1}^{4} \lambda_i = 0 \tag{5.21}$$

Substituting Eq. (5.19) into Eq. (5.16) gives

$$\begin{aligned} L(A(\mathbf{q})) &= \lambda_0 - \sum_{i=1}^{4} \lambda_i (\mathbf{q}^T \mathbf{q}_i)^2 = \lambda_0 - \lambda_1 \sum_{i=1}^{4} (\mathbf{q}^T \mathbf{q}_i)^2 + \sum_{i=1}^{4} (\lambda_1 - \lambda_i)(\mathbf{q}^T \mathbf{q}_i)^2 \\ &= \lambda_0 - \lambda_1 + \sum_{i=2}^{4} (\lambda_1 - \lambda_i)(\mathbf{q}^T \mathbf{q}_i)^2 \end{aligned} \tag{5.22}$$

where the third equality holds because the four eigenvectors constitute an orthonormal basis and because the $i = 1$ term in the second sum vanishes identically. The loss function is minimized if the quaternion is orthogonal to $\mathbf{q}_2$, $\mathbf{q}_3$, and $\mathbf{q}_4$, i.e. the optimal quaternion is the normalized eigenvector of K corresponding to the largest eigenvalue

$$\hat{\mathbf{q}} = \mathbf{q}_1 \tag{5.23}$$

This result can also be derived by using a Lagrange multiplier to append the quaternion norm constraint to the loss function [8].

The optimized loss function is easily seen to be equal to

$$L(A(\hat{\mathbf{q}})) = \lambda_0 - \lambda_{\max} \tag{5.24}$$

which explains our choice of the notation λ_0 for the sum of the weights. The loss function is non-negative by definition, so it must be true that $\lambda_{\max} \leq \lambda_0$. Since we expect the optimal value of the loss function to be small, the value of $\lambda_0 - \lambda_{\max}$ provides a very useful data quality check. Shuster has shown that if the measurement errors are Gaussian and the weights a_i are chosen to be the inverse measurement variances, then TASTE $\equiv 2(\lambda_0 - \lambda_{\max})$ will obey a χ^2 probability distribution with $2N - 3$ degrees of freedom to a very good approximation [33].

Davenport's algorithm does not have a unique solution if the two largest eigenvalues of $K(B)$ are equal. This is not a failure of the q method; it means that the data are not sufficient to determine the attitude uniquely. Very robust algorithms exist to solve the symmetric eigenvalue problem, and Davenport's method remains the best method for solving Wahba's problem if one has access to one of these eigenvalue decomposition algorithms.

5.3.2 *Quaternion Estimator (QUEST)*

Davenport's q method was used to compute attitude estimates for the High Energy Astronomy Observatory (HEAO-B) in 1978; but it could not provide the more frequent attitude computations required by the MAGSAT spacecraft, launched 1 year later, using the computers of the time. The QUEST algorithm was devised to answer this need, and has become the most widely used algorithm for solving Wahba's problem [24, 31, 35].

We can express Davenport's eigenvalue condition as

$$\mathbf{0}_4 = H(\lambda_{\max})\,\hat{\mathbf{q}} \tag{5.25}$$

where

$$H(\lambda) \equiv \lambda I_4 - K(B) = \begin{bmatrix} (\lambda + \operatorname{tr}B)I_3 - S & -\mathbf{z} \\ -\mathbf{z}^T & \lambda - \operatorname{tr}B \end{bmatrix} \tag{5.26}$$

and

$$S \equiv B + B^T \tag{5.27}$$

Equation (5.25) is equivalent to the two equations

$$(\rho I_3 - S)\hat{\mathbf{q}}_{1:3} = \hat{q}_4 \mathbf{z} \tag{5.28a}$$

$$(\lambda_{\max} - \operatorname{tr}B)\hat{q}_4 - \mathbf{z}^T \hat{\mathbf{q}}_{1:3} = 0 \tag{5.28b}$$

where

$$\rho \equiv \lambda_{\max} + \mathrm{tr}B \tag{5.29}$$

If we knew $\lambda_{\max}$, Eqs. (5.28a) and (2.16) would give the optimal quaternion as

$$\hat{\mathbf{q}} = \alpha \begin{bmatrix} \mathrm{adj}(\rho I_3 - S)\mathbf{z} \\ \det(\rho I_3 - S) \end{bmatrix} \tag{5.30}$$

where α is determined by normalization of $\hat{\mathbf{q}}$. Substituting Eq. (5.30) into Eq. (5.28b) gives an implicit equation for the maximum eigenvalue

$$(\lambda_{\max} - \mathrm{tr}B)\det(\rho I_3 - S) - \mathbf{z}^T \mathrm{adj}(\rho I_3 - S)\mathbf{z} = 0 \tag{5.31}$$

This is just the characteristic equation of $K(B)$. We convert it to an explicit equation for $\lambda_{\max}$ by using the definitions of the adjoint and the determinant to write

$$\mathrm{adj}(\rho I_3 - S) = \mathrm{adj}\, S + \rho S + \rho(\rho - \mathrm{tr}S)I_3 \tag{5.32}$$

and

$$\det(\rho I_3 - S) = \rho^3 - \rho^2 \mathrm{tr}S + \kappa\rho - \det S \tag{5.33}$$

where

$$\kappa \equiv \mathrm{tr}(\mathrm{adj}S) \tag{5.34}$$

The expression for the adjoint is further simplified by applying the Cayley-Hamilton Theorem, which says that a matrix obeys its own characteristic equation [17], to the matrix S. With Eq. (2.103) this means that

$$S^3 - S^2 \mathrm{tr}S + S\,\mathrm{tr}(\mathrm{adj}S) - I_3 \det S = 0_{3\times 3} \tag{5.35}$$

Multiplying through by $\mathrm{adj}S$ and using Eq. (2.15) and a little algebra gives

$$\mathrm{adj}S = S^2 - S\mathrm{tr}S + \kappa I_3 \tag{5.36}$$

With these substitutions Eq. (5.31) can be written as a quartic equation for λ:

$$\begin{aligned} 0 = \psi_{\mathrm{QUEST}}(\lambda) \equiv {} & \left[\lambda^2 - (\mathrm{tr}B)^2 + \kappa\right]\left[\lambda^2 - (\mathrm{tr}B)^2 - \|\mathbf{z}\|^2\right] \\ & - (\lambda - \mathrm{tr}B)(\mathbf{z}^T S\mathbf{z} + \det S) - \mathbf{z}^T S^2 \mathbf{z} = \prod_{i=1}^{4}(\lambda - \lambda_i) \end{aligned} \tag{5.37}$$

We have written λ instead of $\lambda_{\max}$ because this equation has four roots. We are only interested in the largest root, though. Shuster's important observation was that $\lambda_{\max}$ can be easily obtained by Newton-Raphson iteration of Eq. (5.37) starting from λ_0 as the initial estimate, since these two values will be very nearly equal if the optimized loss function is small, as we have previously observed. In fact, a single iteration is generally sufficient, and the approximation $\lambda_{\max} = \lambda_0$ is adequate in many cases. It is useful to perform at least one iteration, however, in order to compute the quantity TASTE for data validation.

The efficiency of the QUEST algorithm results from its replacement of the iterative operations on 4×4 matrices required by Davenport's q method with iterative scalar computations followed by straightforward matrix multiplications. Numerical analysts know that solving the characteristic equation is not the best way to find eigenvalues, in general, so QUEST is in principle less robust than Davenport's method. QUEST has proven to be extremely robust in practice, though, as long as the characteristic equation is evaluated in the partially-factored form shown in Eq. (5.37) [5]. The quartic characteristic equation has an analytic solution, but this solution is slower, no more accurate, and sometimes less reliable than Newton-Raphson iteration.

5.3.3 Another View of QUEST

The QUEST solution can be related to the adjoint of the matrix H defined in Eq. (5.26). With Eq. (5.19) and $I_4 = \sum_{i=1}^{4} \mathbf{q}_i \mathbf{q}_i^T$, we can express H in the form

$$H(\lambda) = \sum_{i=1}^{4} (\lambda - \lambda_i) \mathbf{q}_i \mathbf{q}_i^T \tag{5.38}$$

The adjoint of this matrix is

$$\text{adj}\, H(\lambda) = \sum_{i=1}^{4} (\lambda - \lambda_j)(\lambda - \lambda_k)(\lambda - \lambda_\ell) \mathbf{q}_i \mathbf{q}_i^T \tag{5.39}$$

where $\{i, j, k, \ell\}$ is a permutation of $\{1, 2, 3, 4\}$. Setting $\lambda = \lambda_{\max} = \lambda_1$ gives

$$\text{adj}\, H(\lambda_{\max}) = (\lambda_{\max} - \lambda_2)(\lambda_{\max} - \lambda_3)(\lambda_{\max} - \lambda_4) \mathbf{q}_1 \mathbf{q}_1^T = \gamma\, \hat{\mathbf{q}} \hat{\mathbf{q}}^T \tag{5.40}$$

where γ is positive if $\lambda_1 \neq \lambda_2$, which we have seen to be the condition for uniqueness of the attitude solution. In fact, γ is just $d\psi_{\text{QUEST}}/d\lambda$ evaluated at $\lambda = \lambda_{\max}$. Daniele Mortari was the first to recognize the significance of the matrix adj H [25].

The 4−4 component of Eq. (5.40) is[1]

$$[\text{adj}\, H(\lambda_{\max})]_{44} = \det(\rho I_3 - S) = \gamma\, \hat{q}_4^2 \tag{5.41}$$

and the other three elements of the fourth column are

$$[\text{adj}\, H(\lambda_{\max})]_{k4} = [\text{adj}(\rho I_3 - S)\mathbf{z}]_k = \gamma\, \hat{q}_4 \hat{q}_k \quad \text{for} \quad k = 1, 2, 3 \tag{5.42}$$

Comparison with Eq. (5.30) establishes that the QUEST quaternion estimate is just the normalized fourth column of adj $H(\lambda_{\max})$ and that

$$\alpha = \pm(\gamma \hat{q}_4)^{-1} = \pm[\gamma \det(\rho I_3 - S)]^{-1/2} \tag{5.43}$$

Equation (5.41) clearly shows that q_4 is zero if $\rho I_3 - S$ is singular, or equivalently that the estimate in this case is a 180° rotation. We can also see that α is infinite in this limit. Quaternion normalization requires that $\|\text{adj}(\rho I_3 - S)\mathbf{z}\|$ tends to $[\gamma \det(\rho I_3 - S)]^{1/2}$ in the singular limit, so it is clear that the QUEST computation of the quaternion loses all numerical significance. This is quite different from the failure of Wahba's problem to have a unique solution if the two largest eigenvalues of $K(B)$ are equal, and it can be avoided by a method discussed in the following section.

5.3.4 *Method of Sequential Rotations*

Shuster discovered the indeterminacy of the QUEST solution when $\rho I_3 - S$ is singular, and he introduced the method of sequential rotations to deal with the problem [31, 34, 35]. The idea behind this method is to solve for a quaternion $\mathbf{q}^k \equiv \mathbf{q}_{BR_k}$ representing the attitude with respect to a rotated reference frame R_k. The attitude quaternion $\mathbf{q} \equiv \mathbf{q}_{BR}$ with respect to the original frame R is the product of $\mathbf{q}^k$ and the quaternion $\mathbf{q}_{R_k R}$ representing the rotation between frames R and R_k. For simplicity, the frames are related by a 180° rotation about one of the coordinate axes, i.e.

$$\begin{aligned}\mathbf{q} \equiv \mathbf{q}_{BR} &= \mathbf{q}_{BR_k} \otimes \mathbf{q}_{R_k R} = \mathbf{q}^k \otimes \mathbf{q}(\mathbf{e}_k, \pi) \\ &= \begin{bmatrix} \mathbf{q}^k_{1:3} \\ q^k_4 \end{bmatrix} \otimes \begin{bmatrix} \mathbf{e}_k \\ 0 \end{bmatrix} = \begin{bmatrix} q^k_4 \mathbf{e}_k + \mathbf{e}_k \times \mathbf{q}^k_{1:3} \\ -\mathbf{e}_k \cdot \mathbf{q}^k_{1:3} \end{bmatrix} = E_k \mathbf{q}^k\end{aligned} \tag{5.44}$$

where $\mathbf{e}_k$ is the unit vector along the kth coordinate axis and where

$$E_k \equiv \begin{bmatrix} [\mathbf{e}_k \times] & \mathbf{e}_k \\ -\mathbf{e}_k^T & 0 \end{bmatrix} \tag{5.45}$$

[1]We are indebted to Yang Cheng for providing the basis of the discussion in this paragraph.

The matrix E_k is skew-symmetric and orthogonal. It has exactly one element with the value ± 1 in each column and exactly one element with the value ± 1 in each row; all its other elements are zero. Thus E_k is like a permutation matrix except for the minus signs. This means that the transformation from $\mathbf{q}^k$ to $\mathbf{q}$ requires no multiplications; it merely permutes the components of the quaternion with some sign changes. The inverse transformation is

$$\mathbf{q}^k = \mathbf{q} \otimes \mathbf{q}(\mathbf{e}_k, -\pi) = E_k^T \mathbf{q} \tag{5.46}$$

We see that the kth component of $\mathbf{q}$ ends up as the fourth component of $\mathbf{q}^k$. Because $\mathbf{q}$ is a unit quaternion, at least one of its components must have magnitude greater than or equal to $1/2$, so it is always possible to find a reference frame rotation that results in q_4^k having magnitude of at least $1/2$. Since QUEST prefers to keep the magnitude of this component away from zero, it follows that the optimal reference frame rotation is a rotation about the axis corresponding to the component of $\mathbf{q}$ having the largest magnitude. If the fourth component of $\mathbf{q}$ has the largest magnitude, no reference frame rotation is required.

These rotations are easy to implement on the input data, since a 180° rotation about the kth coordinate axis simply changes the signs of the ith and jth components of each reference vector, where $\{i, j, k\}$ is a permutation of $\{1, 2, 3\}$. This is equivalent to changing the signs of the ith and jth columns of the B matrix. The reference system rotation is easily undone by Eq. (5.44) after the optimal quaternion in the rotated frame has been computed.

The original QUEST implementation performed sequential rotations one axis at a time, until an acceptable reference coordinate system was found. It is not necessary to find a rotation resulting in $q_4^k \geq 1/2$, it is only necessary for q_4^k to be larger than some chosen value. It is clearly preferable to save computations by choosing a single desirable rotation as early in the computation as possible. This can be accomplished by considering the components of an a priori quaternion if one is available. An a priori quaternion is generally available before computing the final attitude estimate in a star tracker application since an approximate attitude estimate is needed to identify the stars in the tracker's field of view. This is either available from a previous estimate or is produced by a "lost-in-space" algorithm using fewer (generally three or four) stars than are employed for the final attitude estimate.

We now want to show the effect of reference frame rotations on Davenport's K matrix. This is not part of the implementation of QUEST, but serves to show the relation to the next algorithm we will consider. Putting $A(\mathbf{q})$ instead of B into Davenport's definition of the K matrix gives

$$K(A(\mathbf{q})) = 4\mathbf{q}\,\mathbf{q}^T - I_4 \tag{5.47}$$

Equation (5.21) allows us to rewrite Eq. (5.19) as

$$K(B) = K(B) - \frac{1}{4}\left(\sum_{i=1}^{4} \lambda_i\right) I_4 = \frac{1}{4}\sum_{i=1}^{4} \lambda_i (4\mathbf{q}_i \mathbf{q}_i^T - I_4) = \frac{1}{4}\sum_{i=1}^{4} \lambda_i K(A(\mathbf{q}_i)) \tag{5.48}$$

Because $K(\cdot)$ is a linear function of its argument, it follows from Eq. (5.17) that[2]

$$B = \frac{1}{4}\sum_{i=1}^{4} \lambda_i A(\mathbf{q}_i) \tag{5.49}$$

In the rotated reference frame, we have

$$\begin{aligned} B_k = BA(\mathbf{e}_k, -\pi) &= \frac{1}{4}\sum_{i=1}^{4} \lambda_i A(\mathbf{q}_i)A(\mathbf{e}_k, -\pi) \\ &= \frac{1}{4}\sum_{i=1}^{4} \lambda_i A(\mathbf{q}_i \otimes \mathbf{q}(\mathbf{e}_k, -\pi)) = \frac{1}{4}\sum_{i=1}^{4} \lambda_i A(E_k^T \mathbf{q}_i) \end{aligned} \tag{5.50}$$

But then we have

$$K(B_k) = \frac{1}{4}\sum_{i=1}^{4} \lambda_i K(A(E_k^T \mathbf{q}_i)) = \frac{1}{4}\sum_{i=1}^{4} \lambda_i E_k^T (4\mathbf{q}_i \mathbf{q}_i^T - I_4)E_k = E_k^T K(B) E_k \tag{5.51}$$

This result would be achieved much more directly if we could simply assert that all the eigenvectors of K transform according to Eq. (5.46) under reference frame rotations, but that is not necessarily true unless the four eigenvalues of $K(B)$ are distinct.

We have shown that the effect of a reference frame rotation on $K(B)$ is to permute its rows and columns, with some sign changes. The form of E_k shows that rows and columns 4 and k are interchanged, as are the rows and columns labeled i and j. From this viewpoint, the purpose of reference frame rotations is to move the kth row and column of $K(B)$ to the lower-right corner of $K(B_k)$.

5.3.5 *Estimator of the Optimal Quaternion (ESOQ)*

ESOQ avoids the need for explicit reference frame rotations by treating the four components of the quaternion more symmetrically than QUEST [24, 25]. It is similar in finding $\lambda_{\max}$ by Newton-Raphson iteration of the characteristic equation of Davenport's K matrix and in locating the component of $\mathbf{q}$ with the maximum magnitude. The algorithm is based on Mortari's observation that the optimal

---

[2]Let M denote the 3×3 matrix defined by the right side of Eq. (5.49). The $4-4$ component of Eq. (5.48) means that $\text{tr}B = \text{tr}M$. Then the upper left 3×3 submatrix of Eq. (5.48) says that $B + B^T = M + M^T$. Finally, the remaining 3×1 and 1×3 submatrices tell us that $B - B^T = M - M^T$, establishing Eq. (5.49).

quaternion can be computed by normalizing any column of $\text{adj}H(\lambda_{\max})$, not just the fourth column. The implementation of this idea is straightforward.

Let H_k denote the symmetric 3×3 matrix obtained by deleting the kth row and kth column from $H(\lambda_{\max})$, where k can be any index between 1 and 4, and let $\mathbf{h}_k$ denote the three-component column vector obtained by deleting the kth element from the kth column of $H(\lambda_{\max})$. Then the kth component of the optimal quaternion is given by

$$\hat{q}_k = -\alpha' \det H_k \tag{5.52}$$

and the other three components are given by[3]

$$\begin{bmatrix} \hat{\mathbf{q}}_{1:k-1} \\ \hat{\mathbf{q}}_{k+1:4} \end{bmatrix} = \alpha'(\text{adj } H_k)\mathbf{h}_k \tag{5.53}$$

where α' can be determined by quaternion normalization. The index k is chosen by finding the largest diagonal element $[\text{adj}H(\lambda_{\max})]_{kk} = \det H_k$ of the adjoint matrix, or as the index of the largest component of an a priori quaternion.

Assuming that QUEST and ESOQ use the same value for $\lambda_{\max}$, it is obvious that their equations for the optimal quaternion are identical for $k = 4$. The results of the previous section show that the only difference for other values of k is in some intermediate multiplications, for which any reasonable computer gives $(-a)b = a(-b)$ and $(-a)(-b) = ab$ exactly. Variations in implementation details may give rise to some differences, of course.

Between these two algorithms, ESOQ may be preferred for replacing some computations with simple indexing operations. However, the computational savings are not large, and implementations of QUEST incorporating extensive error checks have a long history of successful application.

5.3.6 *Second Estimator of the Optimal Quaternion (ESOQ2)*

The following presentation differs somewhat from Mortari's original derivation, but is equivalent [24, 26]. Multiplying Eq. (5.28a) by $(\lambda_{\max} - \text{tr}B)$, substituting Eq. (5.28b) for $(\lambda_{\max} - \text{tr}B)\hat{q}_4$ on the right side, and rearranging gives

$$M\hat{\mathbf{q}}_{1:3} = \mathbf{0}_3 \tag{5.54}$$

where

$$M \equiv (\lambda_{\max} - \text{tr}B)(\rho I_3 - S) - \mathbf{z}\mathbf{z}^T = [\mathbf{m}_1 \;\; \mathbf{m}_2 \;\; \mathbf{m}_3] \tag{5.55}$$

Computing the determinant of M and comparing with Eq. (5.31) gives

$$\det M = (\lambda_{\max} - \text{tr}B)^2 \det(\lambda_{\max} I_4 - K) \tag{5.56}$$

---

[3]We employ the convention that $\mathbf{x}_{1:0}$ or $\mathbf{x}_{5:4}$ is an empty vector, with no components.

It follows from $\lambda_{\max}$ being an eigenvalue of $K(B)$ that M is singular, as it must be for Eq. (5.54) to have a non-trivial solution. Thus the three columns of M lie in a plane to which $\hat{\mathbf{q}}_{1:3}$ must be orthogonal, so we can write

$$\hat{\mathbf{q}}_{1:3} = \alpha''(\lambda_{\max} - \mathrm{tr}B)(\mathbf{m}_i \times \mathbf{m}_j) \tag{5.57}$$

for some scalar α'' and for any pair of non-equal indices i and j. It is best to choose the indices to select the cross product with maximum norm.[4] Then $\hat{q}_4$ is determined by Eq. (5.28b), giving

$$\hat{\mathbf{q}} = \alpha'' \begin{bmatrix} (\lambda_{\max} - \mathrm{tr}B)(\mathbf{m}_i \times \mathbf{m}_j) \\ \mathbf{z} \cdot (\mathbf{m}_i \times \mathbf{m}_j) \end{bmatrix} \tag{5.58}$$

where α'' can be determined by quaternion normalization. These computations lose numerical significance if $(\lambda_{\max} - \mathrm{tr}B)$ is close to zero, but we can avoid this singular condition by using a reference system rotation to minimize $\mathrm{tr}B$. No rotation is performed if $\mathrm{tr}B$ is the minimum of $\{B_{11}, B_{22}, B_{33}, \mathrm{tr}B\}$ while a rotation about the ith axis is performed if B_{ii} is the minimum. As in the QUEST case, the rotation is easily undone by Eq. (5.44) after the quaternion has been computed. Note that either an optimal or a merely acceptable rotated frame for ESOQ2 can be found by inspection of the B matrix, requiring neither an a priori attitude estimate nor an iterative search.

5.4 Matrix Solutions of Wahba's Problem

5.4.1 *Singular Value Decomposition (SVD) Method*

J. L. Farrell and J. C. Stuelpnagel presented one of the first solutions of Wahba's problem [13]. They performed a polar decomposition of B into the product of an orthogonal matrix and a symmetric positive semidefinite matrix, followed by a diagonalization of the symmetric matrix. This two-step process is equivalent to the single-step procedure known as the singular value decomposition, for which very robust algorithms have been developed, so we will only discuss the latter method [15, 17, 19].

The singular value decomposition of the attitude profile matrix is given by

$$B = U\Sigma V^T = U\mathrm{diag}([s_1\ s_2\ s_3])V^T \tag{5.59}$$

[4]One of the eigenvalues of the singular matrix M is zero, so its eigenvalue decomposition is $M = \mu_1 \mathbf{v}\mathbf{v}^T + \mu_2 \mathbf{w}\mathbf{w}^T$. Then $\mathbf{m}_i \times \mathbf{m}_j = \mu_1\mu_2(\mathbf{v}\times\mathbf{w})_k(\mathbf{v}\times\mathbf{w})$, where $\{i, j, k\}$ is a cyclic permutation of $\{1, 2, 3\}$. The optimal indices are thus those with maximum $|(\mathbf{m}_i \times \mathbf{m}_j)_k|$.

where U and V are orthogonal, and $s_1 \geq s_2 \geq s_3 \geq 0$ [15]. The matrices U and V are not guaranteed to have determinant $+1$, and we are only interested in proper orthogonal matrices, so it is convenient to define the rotation matrices

$$U_+ \equiv U \mathrm{diag}([1\ 1\ \det U]) \quad \text{and} \quad V_+ \equiv V \mathrm{diag}([1\ 1\ \det V]) \tag{5.60}$$

Then

$$B = U_+ \Sigma' V_+^T = U_+ \mathrm{diag}([s_1\ s_2\ s_3'])V_+^T \tag{5.61}$$

where

$$s_3' \equiv s_3 \det U \det V \tag{5.62}$$

and $s_1 \geq s_2 \geq |s_3'|$. We now define a rotation matrix and its Euler axis/angle representation by

$$W \equiv U_+^T A V_+ = (\cos\vartheta)I_3 - \sin\vartheta[\mathbf{e}\times] + (1-\cos\vartheta)\mathbf{e}\mathbf{e}^T \tag{5.63}$$

Using the cyclic invariance of the trace, Eq. (2.11c), gives

$$\begin{aligned} \mathrm{tr}(AB^T) &= \mathrm{tr}(W\Sigma') = \mathbf{e}^T\Sigma'\mathbf{e} + \cos\vartheta\left(\mathrm{tr}\Sigma' - \mathbf{e}^T\Sigma'\mathbf{e}\right) \\ &= \mathbf{e}^T\Sigma'\mathbf{e} + \cos\vartheta[s_2 + s_3' + e_2^2(s_1 - s_2) + e_3^2(s_1 - s_3')] \end{aligned} \tag{5.64}$$

The trace is maximized for $\vartheta = 0$, which gives $W = I_3$ and thus the optimal attitude matrix

$$\hat{A} = U_+ V_+^T = U \mathrm{diag}([1\ 1\ \det U \det V])V^T \tag{5.65}$$

The optimized loss function is easily seen to be equal to

$$L(\hat{A}) = \lambda_0 - \mathrm{tr}(\hat{A}B^T) = \lambda_0 - \mathrm{tr}\Sigma' = \lambda_0 - (s_1 + s_2 + s_3') \tag{5.66}$$

and comparison with Eq. (5.24) shows that

$$\lambda_{\max} = s_1 + s_2 + s_3' \tag{5.67}$$

Equation (5.64) reduces to $\mathrm{tr}(AB^T) = s_1 - (1-\cos\vartheta)[e_2^2(s_1-s_2)+e_3^2(s_1-s_3')]$ if $s_2 + s_3' = 0$. The loss function does not have a unique minimum in this case because it is invariant under a rotation by any angle about the axis $\mathbf{e}_1 \equiv [1\ 0\ 0]^T$. We will see shortly that this non-uniqueness is the same as the one found in Davenport's method when the two largest eigenvalues of $K(B)$ are not distinct.

5.4.2 Fast Optimal Attitude Matrix (FOAM)

This method bears the same relationship to the SVD method as QUEST bears to Davenport's q method. It computes λ_{max} iteratively and uses simple matrix operations to avoid the need to perform the singular value decomposition of the attitude profile matrix [20]. Equation (5.61) allows us to write

$$\det B = s_1 s_2 s'_3 \tag{5.68a}$$

$$\|B\|_F^2 = s_1^2 + s_2^2 + (s'_3)^2 \tag{5.68b}$$

$$\text{adj}\, B^T = U_+ \text{diag}([s_2 s'_3 \;\; s'_3 s_1 \;\; s_1 s_2]) V_+^T \tag{5.68c}$$

$$B B^T B = U_+ \text{diag}([s_1^3 \;\; s_2^3 \;\; (s'_3)^3]) V_+^T \tag{5.68d}$$

Then the optimal attitude matrix can be represented as

$$\hat{A} = \frac{\left(\lambda_{max}^2 + \|B\|_F^2\right) B + 2\lambda_{max}\, \text{adj}\, B^T - 2 B B^T B}{\lambda_{max}\left(\lambda_{max}^2 - \|B\|_F^2\right) - 2 \det B} \tag{5.69}$$

Substituting Eqs. (5.61), (5.67) and (5.68) into this representation, performing some straightforward algebra, and comparing the result with Eq. (5.65) establishes its validity. With the exception of λ_{max}, all the quantities in Eq. (5.69) can be computed by straightforward algebraic operations. We find the maximum eigenvalue from

$$\lambda_{max} = \lambda_0 - L(\hat{A}) = \text{tr}(\hat{A} B^T) \tag{5.70}$$

Substituting Eq. (5.69) and taking the trace, with λ in place of λ_{max}, gives

$$0 = \psi_{FOAM}(\lambda) \equiv \left(\lambda^2 - \|B\|_F^2\right)^2 - 8\lambda \det B - 4\|\text{adj} B\|_F^2 = \prod_{i=1}^{4} (\lambda - \lambda_i) \tag{5.71}$$

where we have used the identity

$$\|B\|_F^4 - \text{tr}(B B^T B B^T) = 2\|\text{adj} B\|_F^2 \tag{5.72}$$

which is easily verified using the singular value decomposition. The numerical coefficients in $\psi_{QUEST}(\lambda)$ and $\psi_{FOAM}(\lambda)$ must be identical with error-free mathematics, since both functions are forms of the characteristic equation of Davenport's K matrix. The FOAM form is somewhat simpler, though, and has also been employed in ESOQ and ESOQ2 implementations.

Substituting Eqs. (5.68a)–(5.68c) into Eq. (5.71) gives the eigenvalues of the K matrix in terms of the singular values[5]:

$$\lambda_1 = s_1 + s_2 + s_3' \tag{5.73a}$$

$$\lambda_2 = s_1 - s_2 - s_3' \tag{5.73b}$$

$$\lambda_3 = -s_1 + s_2 - s_3' \tag{5.73c}$$

$$\lambda_4 = -s_1 - s_2 + s_3' \tag{5.73d}$$

The denominator of Eq. (5.69) is equal to

$$2(s_2 + s_3')(s_3' + s_1)(s_1 + s_2) = (\lambda_1 - \lambda_2)(\lambda_1 - \lambda_3)(\lambda_1 - \lambda_4)/4 \tag{5.74}$$

This clearly shows that the FOAM algorithm fails if $s_2 + s_3' = \lambda_1 - \lambda_2 = 0$, which is the same indication of unobservability that we noted with Davenport's q method and the SVD method.

5.4.3 *Wahba's Problem with Two Observations*

The two-observation case is of special interest for two reasons. The first is that the characteristic equation for $K(B)$ has a simple closed-form solution in this case, as was noted very early [35]. The second reason is that the solution to Wahba's problem with two observations gives a generalization of the TRIAD method for arbitrary measurement weights [21,22].

Several simplifications follow from having only two observations:

$$\det B = 0 \tag{5.75a}$$

$$BB^T = a_1^2 \mathbf{b}_1 \mathbf{b}_1^T + a_2^2 \mathbf{b}_2 \mathbf{b}_2^T + a_1 a_2 (\mathbf{r}_1 \cdot \mathbf{r}_2)(\mathbf{b}_1 \mathbf{b}_2^T + \mathbf{b}_2 \mathbf{b}_1^T) \tag{5.75b}$$

$$\|B\|_F^2 = a_1^2 + a_2^2 + 2a_1 a_2 (\mathbf{b}_1 \cdot \mathbf{b}_2)(\mathbf{r}_1 \cdot \mathbf{r}_2) \tag{5.75c}$$

$$\text{adj}\, B^T = a_1 a_2 (\mathbf{b}_1 \times \mathbf{b}_2)(\mathbf{r}_1 \times \mathbf{r}_2)^T \tag{5.75d}$$

$$\|\text{adj}\, B\|_F = a_1 a_2 \|\mathbf{b}_1 \times \mathbf{b}_2\| \, \|\mathbf{r}_1 \times \mathbf{r}_2\| \tag{5.75e}$$

It then follows from Eq. (5.71) that

$$\lambda_{\max} = \left\{a_1^2 + a_2^2 + 2a_1 a_2 [(\mathbf{b}_1 \cdot \mathbf{b}_2)(\mathbf{r}_1 \cdot \mathbf{r}_2) + \|\mathbf{b}_1 \times \mathbf{b}_2\| \, \|\mathbf{r}_1 \times \mathbf{r}_2\|]\right\}^{1/2} \tag{5.76}$$

[5] Paul Davenport discovered these relations, but did not publish them.

This can be used in QUEST, ESOQ, ESOQ2, or FOAM, avoiding the need for Newton-Raphson iteration. If we define θ_b to be the angle between $\mathbf{b}_1$ and $\mathbf{b}_2$, and θ_r to be the angle between $\mathbf{r}_1$ and $\mathbf{r}_2$, then

$$\lambda_{\max} = \left[\lambda_0^2 - 4a_1a_2\sin^2\left(\frac{\theta_b - \theta_r}{2}\right)\right]^{1/2} \tag{5.77}$$

This form provides some insight, but is less useful for computation than Eq. (5.76).

Substituting Eqs. (5.75a)–(5.75e) into (5.69) gives, after some algebra

$$\begin{aligned}\hat{A} = &(a_1/\lambda_{\max})\left[\mathbf{b}_1\mathbf{r}_1^T + (\mathbf{b}_1 \times \mathbf{b}_\times)(\mathbf{r}_1 \times \mathbf{r}_\times)^T\right] \\ &+(a_2/\lambda_{\max})\left[\mathbf{b}_2\mathbf{r}_2^T + (\mathbf{b}_2 \times \mathbf{b}_\times)(\mathbf{r}_2 \times \mathbf{r}_\times)^T\right] + \mathbf{b}_\times\mathbf{r}_\times^T\end{aligned} \tag{5.78}$$

using the notation of Sect. 5.1. It is obvious from this equation that the optimal attitude estimate maps $\mathbf{r}_1$ and $\mathbf{r}_2$ into the plane spanned by $\mathbf{b}_1$ and $\mathbf{b}_2$. The TRIAD solution also has this property, so the difference between the Wahba and TRIAD estimates can be represented as a rotation about the axis $\mathbf{b}_\times$, which is the normal vector to the plane containing $\mathbf{b}_1$ and $\mathbf{b}_2$.

It is also easy to see that the solution has a unique limit as either observation weight goes to zero. This limit is the TRIAD solution of Eq. (5.3) if $a_2 = 0$, and it is the TRIAD solution with the roles of the two observations interchanged if $a_1 = 0$. This limit was already noted in [35]. For general weights, the optimal solution is some sort of an average of the two TRIAD solutions. It is not a simple average, though, because a simple average would not give an orthogonal attitude matrix. It is interesting that Eq. (5.78) has a unique limit as either observation weight goes to zero even though Wahba's loss function does not have a unique minimum in either limit, because it effectively includes only one observation.

There is an equivalent closed-form solution for the quaternion [21]:

$$\hat{\mathbf{q}} = \begin{cases} \dfrac{1}{2\sqrt{\gamma(\gamma+\alpha)(1+\mathbf{b}_\times\cdot\mathbf{r}_\times)}}\begin{bmatrix}(\gamma+\alpha)(\mathbf{b}_\times\times\mathbf{r}_\times)+\beta(\mathbf{b}_\times+\mathbf{r}_\times)\\ (\gamma+\alpha)(1+\mathbf{b}_\times\cdot\mathbf{r}_\times)\end{bmatrix} & \text{for} \quad \alpha \geq 0 \\[2ex] \dfrac{1}{2\sqrt{\gamma(\gamma-\alpha)(1+\mathbf{b}_\times\cdot\mathbf{r}_\times)}}\begin{bmatrix}\beta(\mathbf{b}_\times\times\mathbf{r}_\times)+(\gamma-\alpha)(\mathbf{b}_\times+\mathbf{r}_\times)\\ \beta(1+\mathbf{b}_\times\cdot\mathbf{r}_\times)\end{bmatrix} & \text{for} \quad \alpha < 0 \end{cases} \tag{5.79}$$

where

$$\alpha \equiv (1+\mathbf{b}_\times\cdot\mathbf{r}_\times)(a_1\mathbf{b}_1\cdot\mathbf{r}_1 + a_2\mathbf{b}_2\cdot\mathbf{r}_2) + (\mathbf{b}_\times\times\mathbf{r}_\times)\cdot(a_1\mathbf{b}_1\times\mathbf{r}_1 + a_2\mathbf{b}_2\times\mathbf{r}_2) \tag{5.80a}$$

$$\beta \equiv (\mathbf{b}_\times + \mathbf{r}_\times)\cdot(a_1\mathbf{b}_1\times\mathbf{r}_1 + a_2\mathbf{b}_2\times\mathbf{r}_2) \tag{5.80b}$$

$$\gamma \equiv \sqrt{\alpha^2+\beta^2} \tag{5.80c}$$

The above expressions are indeterminate when $\mathbf{b}_\times = -\mathbf{r}_\times$, and are unreliable near this singular condition. This problem can be avoided by applying one of the reference frame rotations of Sect. 5.3.4. The effect of a 180° rotation about the ith axis on the inner product $\mathbf{b}_\times \cdot \mathbf{r}_\times$ is

$$\begin{aligned}(\mathbf{b}_\times \cdot \mathbf{r}_\times)_{\text{rotated}} &= [(\mathbf{b}_\times)_i \cdot (\mathbf{r}_\times)_i - (\mathbf{b}_\times)_j \cdot (\mathbf{r}_\times)_j - (\mathbf{b}_\times)_k \cdot (\mathbf{r}_\times)_k]_{\text{unrotated}} \\ &= [2(\mathbf{b}_\times)_i \cdot (\mathbf{r}_\times)_i - \mathbf{b}_\times \cdot \mathbf{r}_\times]_{\text{unrotated}} \end{aligned} \quad (5.81)$$

We ensure the largest value for $(\mathbf{b}_\times \cdot \mathbf{r}_\times)_{\text{rotated}}$ by performing the estimation in the original reference frame if $\mathbf{b}_\times \cdot \mathbf{r}_\times$ is larger than any of the products $(\mathbf{b}_\times)_i \cdot (\mathbf{r}_\times)_i$ in this frame, or by rotating about the axis with maximum $(\mathbf{b}_\times)_i \cdot (\mathbf{r}_\times)_i$ if this is greater than $\mathbf{b}_\times \cdot \mathbf{r}_\times$. More details can be found in [21].

5.5 Error Analysis of Wahba's Problem

The value of an estimate is greatly enhanced by knowledge of its accuracy. In fact, Malcolm Shuster's covariance analysis of the QUEST algorithm was at least as important as his invention of the algorithm itself.[6] This section will provide an estimate of the accuracy of a solution to Wahba's problem. Because all algorithms yield the same estimate, assuming that it is unique, we only need to analyze the errors exhibited by one algorithm. We will use Davenport's q method, which leads most directly to the desired result.

5.5.1 *Attitude Error Vector*

Attitude errors are most usefully expressed in the spacecraft's body frame, so we parameterize the attitude errors by the first-order rotation vector $\boldsymbol{\delta\vartheta}$ representing the rotation between the estimated body frame $\hat{B}$ and the true body frame B. The goal of this section is to find an expression for $\boldsymbol{\delta\vartheta}$ in terms of the errors of the body and reference frame vectors. Attitude errors are expected to be small, so we do not have to worry about singularities of the rotation vector representation. We represent the attitude by a quaternion, so

$$\hat{\mathbf{q}} \equiv \mathbf{q}_{\hat{B}R} = \mathbf{q}_{\hat{B}B} \otimes \mathbf{q}_{BR} = [\mathbf{q}_{\hat{B}B} \otimes]\mathbf{q}_{BR} = \exp[(\boldsymbol{\delta\vartheta}/2)\otimes]\mathbf{q}_{BR} \quad (5.82)$$

where we have used Eq. (2.127). All our error analysis will be carried out only to first order in small quantities, so we can approximate

$$\exp[(\boldsymbol{\delta\vartheta}/2)\otimes] = I_4 + [(\boldsymbol{\delta\vartheta}/2)\otimes] \quad (5.83)$$

[6] Much more important, in Paul Davenport's opinion.

which gives

$$\hat{\mathbf{q}} = \mathbf{q}_{BR} + (1/2)\,\boldsymbol{\delta\vartheta} \otimes \mathbf{q}_{BR} \tag{5.84}$$

The analysis of errors in the spacecraft's body frame does not depend on the specification of the reference frame R, so we are free to choose this to coincide with the body frame. With this choice, $\mathbf{q}_{BR}$ is the identity quaternion, and Eq. (5.84) reduces to

$$\hat{\mathbf{q}} = \begin{bmatrix} \mathbf{0}_3 \\ 1 \end{bmatrix} + \frac{1}{2}\begin{bmatrix} \boldsymbol{\delta\vartheta} \\ 0 \end{bmatrix} \tag{5.85}$$

The body frame vectors $\mathbf{b}_i$ and reference frame vectors $\mathbf{r}_i$ are equal to their true, error-free, values plus some errors. Because we take the reference frame to be the body frame, the true values of the reference vectors and body vectors are identical. Thus the attitude profile matrix is

$$B = \sum_{i=1}^{N} a_i(\mathbf{b}_i^{\text{true}} + \Delta\mathbf{b}_i)(\mathbf{b}_i^{\text{true}} + \Delta\mathbf{r}_i)^T = B_0 + \Delta B \tag{5.86}$$

where

$$B_0 \equiv \sum_{i=1}^{N} a_i \mathbf{b}_i^{\text{true}}(\mathbf{b}_i^{\text{true}})^T \tag{5.87}$$

$$\Delta B \equiv \sum_{i=1}^{N} a_i[\mathbf{b}_i^{\text{true}}\Delta\mathbf{r}_i^T + \Delta\mathbf{b}_i(\mathbf{b}_i^{\text{true}})^T] \tag{5.88}$$

and we have ignored terms of higher than first order. The matrix B_0 is symmetric, and its trace is λ_0, so

$$K(B_0) = \begin{bmatrix} 2B_0 - \lambda_0 I_3 & \mathbf{0}_3 \\ \mathbf{0}_3^T & \lambda_0 \end{bmatrix} \tag{5.89}$$

We see that $K(B_0)\mathbf{q}_{BR} = K(B_0)\mathbf{I}_q = \lambda_0\mathbf{I}_q$, so the maximum eigenvalue is equal to λ_0 for error-free measurements. This is expected because the loss function must be zero if the measurement errors are zero. The other three eigenvectors of K_0 have three-vector part $\mathbf{q}_{1:3} = \mathbf{x}_i$ and scalar part $q_4 = 0$, where $\mathbf{x}_i$ is an eigenvector of $(2B_0 - \lambda_0 I_3)$ with eigenvalue λ_{i+1}, for $i = 1, 2, 3$.

Now let us turn to ΔB. Both $\mathbf{b}_i^{\text{true}} + \Delta\mathbf{b}_i$ and $\mathbf{b}_i^{\text{true}}$ are unit vectors, so

$$1 = \|\mathbf{b}_i^{\text{true}} + \Delta\mathbf{b}_i\|^2 = 1 + 2\mathbf{b}_i^{\text{true}} \cdot \Delta\mathbf{b}_i + \|\Delta\mathbf{b}_i\|^2 \tag{5.90}$$

The same relation holds for $\mathbf{r}_i = \mathbf{b}_i^{\text{true}} + \Delta\mathbf{r}_i$, so

$$\mathbf{b}_i^{\text{true}} \cdot \Delta\mathbf{b}_i = -\frac{1}{2}\|\Delta\mathbf{b}_i\|^2 \quad \text{and} \quad \mathbf{b}_i^{\text{true}} \cdot \Delta\mathbf{r}_i = -\frac{1}{2}\|\Delta\mathbf{r}_i\|^2 \tag{5.91}$$

It follows that $\text{tr}(\Delta B) = 0$, to first order in the vector errors. The **z** vector associated with ΔB is, also to first order

$$\Delta\mathbf{z} = \sum_{i=1}^{N} a_i(\mathbf{b}_i^{\text{true}} + \Delta\mathbf{b}_i) \times (\mathbf{b}_i^{\text{true}} + \Delta\mathbf{r}_i) = \sum_{i=1}^{N} a_i \mathbf{b}_i^{\text{true}} \times (\Delta\mathbf{r}_i - \Delta\mathbf{b}_i) \tag{5.92}$$

and so

$$K(\Delta B) = \begin{bmatrix} \Delta B + \Delta B^T & \Delta\mathbf{z} \\ \Delta\mathbf{z}^T & 0 \end{bmatrix} \tag{5.93}$$

Putting all this into Davenport's eigenvalue equation gives

$$K(B)\hat{\mathbf{q}} = [K(B_0) + K(\Delta B)]\left(\begin{bmatrix} \mathbf{0}_3 \\ 1 \end{bmatrix} + \frac{1}{2}\begin{bmatrix} \boldsymbol{\delta\vartheta} \\ 0 \end{bmatrix}\right)$$
$$= \lambda_{\max}\hat{\mathbf{q}} = (\lambda_0 + \Delta\lambda)\left(\begin{bmatrix} \mathbf{0}_3 \\ 1 \end{bmatrix} + \frac{1}{2}\begin{bmatrix} \boldsymbol{\delta\vartheta} \\ 0 \end{bmatrix}\right) \tag{5.94}$$

Multiplying this out, discarding terms of second order, and canceling the common zeroth-order term $K(B_0)\mathbf{I}_q = \lambda_0\mathbf{I}_q$ on the two sides gives

$$\frac{1}{2}K(B_0)\begin{bmatrix} \boldsymbol{\delta\vartheta} \\ 0 \end{bmatrix} + K(\Delta B)\begin{bmatrix} \mathbf{0}_3 \\ 1 \end{bmatrix} = \frac{1}{2}\lambda_0\begin{bmatrix} \boldsymbol{\delta\vartheta} \\ 0 \end{bmatrix} + \Delta\lambda\begin{bmatrix} \mathbf{0}_3 \\ 1 \end{bmatrix} \tag{5.95}$$

Inserting the explicit forms of $K(B_0)$ and $K(\Delta B)$ gives

$$\begin{bmatrix} (1/2)(2B_0 - \lambda_0 I_3)\boldsymbol{\delta\vartheta} + \Delta\mathbf{z} \\ 0 \end{bmatrix} = \begin{bmatrix} (1/2)\lambda_0\,\boldsymbol{\delta\vartheta} \\ \Delta\lambda \end{bmatrix} \tag{5.96}$$

The fourth component of this equation says that $\Delta\lambda$ is zero to first order in the errors in the vectors. This means that the loss function is of second order in these errors, which is in accordance with its definition. The first three components of Eq. (5.96) are more interesting; they give

$$\boldsymbol{\delta\vartheta} = (\lambda_0 I_3 - B_0)^{-1}\Delta\mathbf{z} = (\lambda_0 I_3 - B_0)^{-1}\sum_{i=1}^{N} a_i \mathbf{b}_i^{\text{true}} \times (\Delta\mathbf{r}_i - \Delta\mathbf{b}_i) \tag{5.97}$$

This is the key result; it expresses the attitude error rotation vector in terms of the errors in the body frame vectors and the reference vectors, with the latter mapped into the true body frame. Is is also interesting to note that

$$\lambda_0 I_3 - B_0 = \sum_{i=1}^{N} a_i \left[I_3 - \mathbf{b}_i^{\text{true}} (\mathbf{b}_i^{\text{true}})^T \right] \tag{5.98}$$

The discussion following Eq. (5.89) shows that this matrix is singular if the largest two eigenvalues of K_0 are equal.

5.5.2 Covariance Analysis of Wahba's Problem

If we knew the errors of the vectors in the spacecraft body frame and the reference frame, we would correct the vectors before performing the estimation, and they would no longer be errors. The usual situation is that the errors are unknown, but they have some statistical distribution with known properties. These properties are given by *expected values* over the probability distribution. The details of the probability distribution are not important for this section; we only need to know the expected values of some quantities.

We assume that the vector errors have zero mean, except for components along $\mathbf{b}_i^{\text{true}}$ that are necessary to satisfy the norm constraint of Eq. (5.91). Thus their expected values are assumed to satisfy, for all i

$$E\{\Delta\mathbf{b}_i\} = -\frac{1}{2} E\{\|\Delta\mathbf{b}_i\|^2\}\mathbf{b}_i^{\text{true}} \quad \text{and} \quad E\{\Delta\mathbf{r}_i\} = -\frac{1}{2} E\{\|\Delta\mathbf{r}_i\|^2\}\mathbf{b}_i^{\text{true}} \tag{5.99}$$

The usual justification of this assumption is that we would estimate and correct for any other non-zero means of these quantities before estimating the attitude. Now the only stochastic quantities in Eq. (5.97) are $\boldsymbol{\delta\vartheta}$, $\Delta\mathbf{z}$, $\Delta\mathbf{r}_i$, and $\Delta\mathbf{b}_i$; the other quantities are deterministic. Computing expected values is a linear operation, and the cross product of a vector with itself is zero, so it follows that

$$E\{\boldsymbol{\delta\vartheta}\} = (\lambda_0 I_3 - B_0)^{-1} \sum_{i=1}^{N} a_i \mathbf{b}_i^{\text{true}} \times (E\{\Delta\mathbf{r}_i\} - E\{\Delta\mathbf{b}_i\}) = \mathbf{0}_3 \tag{5.100}$$

This shows that the expected value of the attitude error $\boldsymbol{\delta\vartheta}$ is zero, which means that a solution of Wahba's problem is an unbiased estimator.

We are interested in the expected spread of the attitude estimates about the mean value. This is given by the covariance matrix

$$P_{\vartheta\vartheta} \equiv E\left\{ (\boldsymbol{\delta\vartheta} - E\{\boldsymbol{\delta\vartheta}\})(\boldsymbol{\delta\vartheta} - E\{\boldsymbol{\delta\vartheta}\})^T \right\} = E\{\boldsymbol{\delta\vartheta}\, \boldsymbol{\delta\vartheta}^T\} \tag{5.101}$$

Substituting Eq. (5.97) gives

$$P_{\vartheta\vartheta} = (\lambda_0 I_3 - B_0)^{-1} E\{\Delta\mathbf{z}\,\Delta\mathbf{z}^T\}(\lambda_0 I_3 - B_0)^{-1} \tag{5.102}$$

Further progress requires some assumptions about the covariance of the vector errors. We assume that the errors in the individual vectors are uncorrelated, meaning that

$$E\{\Delta\mathbf{b}_i\,\Delta\mathbf{b}_j^T\} = E\{\Delta\mathbf{r}_i\,\Delta\mathbf{r}_j^T\} = 0 \quad \text{for } i \neq j \tag{5.103a}$$

$$E\{\Delta\mathbf{b}_i\,\Delta\mathbf{r}_j^T\} = 0 \quad \text{for all } i, j \tag{5.103b}$$

We also define the measurement covariance matrices by

$$R_{b_i} \equiv E\{\Delta\mathbf{b}_i\,\Delta\mathbf{b}_i^T\} \tag{5.104a}$$

$$R_{r_i} \equiv E\{\Delta\mathbf{r}_i\,\Delta\mathbf{r}_i^T\} \tag{5.104b}$$

Then

$$\begin{aligned} E\{\Delta\mathbf{z}\,\Delta\mathbf{z}^T\} &= \sum_{i,j=1}^{N} a_i a_j [\mathbf{b}_i^{\text{true}}\times] E\{(\Delta\mathbf{r}_i - \Delta\mathbf{b}_i)(\Delta\mathbf{r}_j - \Delta\mathbf{b}_j)^T\}[\mathbf{b}_j^{\text{true}}\times]^T \\ &= \sum_{i=1}^{N} a_i^2 [\mathbf{b}_i^{\text{true}}\times](R_{r_i} + R_{b_i})[\mathbf{b}_i^{\text{true}}\times]^T \end{aligned} \tag{5.105}$$

and

$$P_{\vartheta\vartheta} = (\lambda_0 I_3 - B_0)^{-1} \left\{ \sum_{i=1}^{N} a_i^2 [\mathbf{b}_i^{\text{true}}\times](R_{r_i} + R_{b_i})[\mathbf{b}_i^{\text{true}}\times]^T \right\} (\lambda_0 I_3 - B_0)^{-1} \tag{5.106}$$

We now make a further simplifying assumption about the vector errors, namely that they are axially symmetric about the true vectors. We also ignore the components along the true vectors shown in Eq. (5.99), which are of higher order than the terms we retain. These assumptions are expressed in the equations

$$R_{r_i} = \sigma_{r_i}^2 \left[I_3 - \mathbf{r}_i^{\text{true}}(\mathbf{r}_i^{\text{true}})^T \right] = -\sigma_{r_i}^2 \left[\mathbf{r}_i^{\text{true}}\times \right]^2 \tag{5.107a}$$

$$R_{b_i} = \sigma_{b_i}^2 \left[I_3 - \mathbf{b}_i^{\text{true}}(\mathbf{b}_i^{\text{true}})^T \right] = -\sigma_{b_i}^2 \left[\mathbf{b}_i^{\text{true}}\times \right]^2 \tag{5.107b}$$

This has become known as the *QUEST Measurement Model* (QMM). If we think of the error-corrupted vectors as arrows with a common base, their points should all lie on a unit sphere. The QUEST measurement model is essentially the approximation that the points all lie on the plane that is tangent to the unit sphere at the location of the true unit vector. This is certainly a good approximation if the errors are small. We note that $E\{\|\Delta\mathbf{r}_i\|^2\} = \mathrm{tr}R_{r_i}$ and $E\{\|\Delta\mathbf{b}_i\|^2\} = \mathrm{tr}R_{b_i}$, so Eq. (5.99) becomes, for the QMM,

$$E\{\Delta\mathbf{r}_i\} = -\sigma_{r_i}^2\mathbf{r}_i^{\text{true}} \quad \text{and} \quad E\{\Delta\mathbf{b}_i\} = -\sigma_{b_i}^2\mathbf{b}_i^{\text{true}} \tag{5.108}$$

It is useful to combine the error variances of the body frame vectors and reference frame vectors into overall measurement variances

$$\sigma_i^2 \equiv \sigma_{r_i}^2 + \sigma_{b_i}^2 \tag{5.109}$$

Remembering that we have assumed $\mathbf{r}_i^{\text{true}} = \mathbf{b}_i^{\text{true}}$ for our error analysis, the error covariance with the QMM is

$$P_{\vartheta\vartheta} = (\lambda_0 I_3 - B_0)^{-1} M (\lambda_0 I_3 - B_0)^{-1} \tag{5.110}$$

where

$$M \equiv \sum_{i,=1}^{N} a_i^2\sigma_i^2[\mathbf{b}_i^{\text{true}}\times][\mathbf{b}_i^{\text{true}}\times]^T = \sum_{i,=1}^{N} a_i^2\sigma_i^2\left[I_3 - \mathbf{b}_i^{\text{true}}(\mathbf{b}_i^{\text{true}})^T\right] \tag{5.111}$$

It is interesting to note that using the nonsingular, but nonphysical, measurement covariance matrices $R_{r_i} = \sigma_{r_i}^2 I_3$ and $R_{b_i} = \sigma_{b_i}^2 I_3$ yields the same result.

If we choose the weights to be proportional to the inverses of the measurement variances,

$$a_i = c/\sigma_i^2 \tag{5.112}$$

for some constant c, we get the very simple result

$$M = c^2 F, \quad \lambda_0 I_3 - B_0 = cF, \quad \text{and} \quad P_{\vartheta\vartheta} = F^{-1} \tag{5.113}$$

where

$$F \equiv \sum_{i=1}^{N} \sigma_i^{-2}\left[I_3 - \mathbf{b}_i^{\text{true}}(\mathbf{b}_i^{\text{true}})^T\right] \tag{5.114}$$

We will show in Sect. 5.6.1 that this choice of weights in Wahba's loss function also results in the best estimate.[7] Note that the covariance is independent of the scaling

[7] We can show that Eq. (5.112) gives an extremum of $P_{\vartheta\vartheta}$ by differentiating Eq. (5.110) with respect to any a_k, then substituting Eqs. (5.112) and (5.113) after differentiating, which gives zero for the derivative.

parameter c. Most recent work sets $c = 1$, so the weights are equal to the inverse variances, but earlier treatments often set $c = \sigma_{tot}^2$, where

$$\sigma_{tot} \equiv \left(\sum_{i=1}^{N} \sigma_i^{-2} \right)^{-1/2} \tag{5.115}$$

so that the weights sum to unity.

Equation (5.114) has a nice physical interpretation. The covariance matrix is a measure of the uncertainty, or the lack of complete information, in computing the estimate. The inverse of the covariance matrix, F, is therefore a measure of the information contributing to the estimate. Each term in the sum in Eq. (5.114) can be interpreted as the information contributed by a single vector measurement. For an *efficient* estimator F is equivalent to the *Fisher Information Matrix*. See Sect. 12.3.4 for more details.

In practice, we would like to compute an estimate of the covariance without requiring knowledge of the true body frame vectors. We could simply use the measured body frame vectors $\mathbf{b}_i$ in Eq. (5.114), but it is better to use the predicted body frame vectors $\hat{A}\mathbf{r}_i$, which usually have smaller errors. We can avoid the need to perform the additional summation of Eq. (5.114) by using the B matrix, which must be computed as part of the attitude estimation. Dispensing with the assumption that the reference frame coincides with the body frame, we see from Eq. (5.113) that a good estimate of the covariance is

$$P_{\vartheta\vartheta} = c(\lambda_{max} I_3 - B\hat{A}^T)^{-1} \tag{5.116}$$

We use λ_{max} instead of λ_0 in this equation to preserve the important property that the covariance is infinite if the two largest eigenvalues of K are equal. Analysis parallel to that used to derive the FOAM attitude estimate gives the alternative form for the covariance

$$P_{\vartheta\vartheta} = c \frac{\left(\lambda_{max}^2 - \|B\|_F^2\right) I_3 + 2BB^T}{\lambda_{max}\left(\lambda_{max}^2 - \|B\|_F^2\right) - 2\det B} \tag{5.117}$$

which avoids a matrix inversion.

5.5.3 *Covariance with Two Observations*

The covariance for the case of two observations can be found by substituting Eqs. (5.75a)–(5.75c) and (5.76) into Eq. (5.117). We also assume for simplicity that $\mathbf{r}_1 \cdot \mathbf{r}_2 = \mathbf{b}_1 \cdot \mathbf{b}_2$, which leads to $\|\mathbf{b}_1 \times \mathbf{b}_2\| = \|\mathbf{r}_1 \times \mathbf{r}_2\|$ and $\lambda_{max} = a_1 + a_2$. Putting all this together gives

$$P_{\vartheta\vartheta} = \frac{c}{a_1 + a_2} \left[I_3 + \frac{a_1^2 \mathbf{b}_1 \mathbf{b}_1^T + a_2^2 \mathbf{b}_2 \mathbf{b}_2^T + a_1 a_2 (\mathbf{b}_1 \cdot \mathbf{b}_2)(\mathbf{b}_1 \mathbf{b}_2^T + \mathbf{b}_2 \mathbf{b}_1^T)}{a_1 a_2 \|\mathbf{b}_1 \times \mathbf{b}_2\|^2} \right] \tag{5.118}$$

We now use the expansion of the identity matrix in terms of the orthonormal triad of Sect. 5.1:

$$\begin{aligned} I_3 &= \mathbf{w}_1\mathbf{w}_1^T + \mathbf{w}_2\mathbf{w}_2^T + \mathbf{w}_3\mathbf{w}_3^T \\ &= \frac{\mathbf{b}_1\mathbf{b}_1^T + \mathbf{b}_2\mathbf{b}_2^T - (\mathbf{b}_1 \cdot \mathbf{b}_2)(\mathbf{b}_1\mathbf{b}_2^T + \mathbf{b}_2\mathbf{b}_1^T)}{\|\mathbf{b}_1 \times \mathbf{b}_2\|^2} + \mathbf{b}_\times\mathbf{b}_\times^T \end{aligned} \tag{5.119}$$

and substitute $a_i = c/\sigma_i^2$ to get the final result

$$P_{\vartheta\vartheta} = \frac{\sigma_2^2\mathbf{b}_1\mathbf{b}_1^T + \sigma_1^2\mathbf{b}_2\mathbf{b}_2^T}{\|\mathbf{b}_1 \times \mathbf{b}_2\|^2} + \frac{\sigma_1^2\sigma_2^2}{\sigma_1^2 + \sigma_2^2}\mathbf{b}_\times\mathbf{b}_\times^T \tag{5.120}$$

It is easy to see with the help of Eq. (5.119) that this is the inverse of Eq. (5.114) if measurement errors are negligible.

The covariance of the TRIAD estimate is found as a limiting case of the covariance of the Wahba estimate. We have seen that the TRIAD estimate is the limiting case of Eq. (5.78) for $a_1 \gg a_2$, or equivalently for $\sigma_2 \gg \sigma_1$. Taking this limit of Eq. (5.120) gives

$$P_{\text{TRIAD}} = \frac{\sigma_2^2\mathbf{b}_1\mathbf{b}_1^T + \sigma_1^2\mathbf{b}_2\mathbf{b}_2^T}{\|\mathbf{b}_1 \times \mathbf{b}_2\|^2} + \sigma_1^2\mathbf{b}_\times\mathbf{b}_\times^T \tag{5.121}$$

We noted that the Wahba and TRIAD attitude estimates only differ by a rotation around $\mathbf{b}_\times$, so it is not surprising that the only difference in the covariance of the two estimates is in that direction. The TRIAD information matrix is the inverse of the covariance, namely

$$F_{\text{TRIAD}} = \sigma_1^{-2}(I_3 - \mathbf{b}_1\mathbf{b}_1^T) + \sigma_2^{-2}(\mathbf{b}_2 \times \mathbf{b}_\times)(\mathbf{b}_2 \times \mathbf{b}_\times)^T \tag{5.122}$$

The first vector measurement contributes the same information to the TRIAD estimate as to the Wahba estimate, two components perpendicular to $\mathbf{b}_1$; but the second vector measurement contributes only one component of information, perpendicular to both $\mathbf{b}_2$ and $\mathbf{b}_\times$, to the TRIAD estimate. This derivation of the TRIAD covariance and information matrices is somewhat heuristic, but a more rigorous analysis gives the same result [32,35].

5.6 MLE for Attitude Determination

In this section maximum likelihood estimation (MLE) for the attitude determination problem is studied. The treatment in Sect. 12.3.4 requires the measurement covariance matrix R to be nonsingular, as will usually be the case if the measurement vector $\mathbf{y}$ contains actual measurements. Considering a star tracker, for example, the actual measurements are the horizontal and vertical positions of the centroids of the star images in the tracker's field of view.[8] The measurement covariance matrix for these measurements is nonsingular, but it is not particularly simple in the general case [3].

Here we want to focus on modeling the measurements as unit vectors, which leads to the elegant and useful estimation methods based on Wahba's loss function. An apparent difficulty with this approach is that the measurement covariance matrix of the QMM is singular, as shown in Eq. (5.107). Shuster has shown how to define a conditional pdf for vector measurements that overcomes this problem [29], and we will follow his approach.

We first assume that no errors exist in the reference vectors, and will relax this assumption later. In this case, the conditional pdf for a unit-vector measurement $\mathbf{b}_i$ is

$$p(\mathbf{b}_i|A^{\text{true}}, \mathbf{r}_i^{\text{true}}) = \mathcal{N}_{b_i} \exp\left(-\frac{1}{2\sigma_{b_i}^2}\|\mathbf{b}_i - A^{\text{true}}\mathbf{r}_i^{\text{true}}\|^2\right) \tag{5.123}$$

This pdf is defined over the unit sphere i.e. for $\mathbf{b}_i$ that satisfies $\|\mathbf{b}_i\| = 1$. In this pdf, the single parameter $\sigma_{b_i}^2$ takes the place of the measurement covariance matrix, so this models measurement errors with a high degree of symmetry. In fact, the errors in this model are as symmetric as possible consistent with the unit vector constraint on the measurements. We will show that this model is equivalent to the QMM to a very good approximation.

The normalization coefficient $\mathcal{N}_{b_i}$ must be specified so that the total probability is unity. To compute the normalization and other integrals of interest, we parameterize $\mathbf{b}_i$ as $\mathbf{b}_i = A^{\text{true}}\mathbf{r}_i^{\text{true}} \cos\theta + \mathbf{u}\sin\theta\cos\phi + \mathbf{v}\sin\theta\sin\phi$, where $\mathbf{u}$ and $\mathbf{v}$ are two unit vectors perpendicular to $A^{\text{true}}\mathbf{r}_i^{\text{true}}$ and to each other. The area element on the unit sphere is $\sin\theta\, d\theta\, d\phi = d\zeta\, d\phi$, where $\zeta \equiv 1 - \cos\theta$. Then the normalization constant is given by

$$\mathcal{N}_{b_i} = \left[\int_0^{2\pi}\int_0^2 \exp(-\zeta/\sigma_{b_i}^2)\, d\zeta\, d\phi\right]^{-1} = \left[2\pi\sigma_{b_i}^2\left(1 - e^{-2/\sigma_{b_i}^2}\right)\right]^{-1} \tag{5.124}$$

The exponential term can be neglected even for extremely large errors, reflecting the fact that the pdf becomes extremely small over most of the unit sphere. For example, if σ_i is 10°, or 0.175 rad, then $e^{-2/\sigma_{b_i}^2} = 3 \times 10^{-29}$. Neglecting the exponential term is equivalent to taking the upper limit of the ζ integral to be infinite, an approximation we will employ in the next paragraph.

[8]The really basic measurements are the electron counts in the individual pixels of the star tracker's focal plane, but these are invariably reduced to centroids before being communicated to the attitude control system.

It is interesting to see how the unit vector pdf of Eq. (5.123) is related to the QMM. First, with $\mathbf{b}_i^{\text{true}} \equiv A^{\text{true}}\mathbf{r}_i^{\text{true}}$ we see that

$$E\{\mathbf{b}_i - \mathbf{b}_i^{\text{true}}\} \approx \frac{1}{2\pi\sigma_{b_i}^2}\int_0^{2\pi}\int_0^{\infty}(-\zeta)\exp(-\zeta/\sigma_{b_i}^2)\,d\zeta\,d\phi\;\mathbf{b}_i^{\text{true}} \approx -\sigma_{b_i}^2\,\mathbf{b}_i^{\text{true}} \tag{5.125}$$

because the integral over ϕ gives zero components along $\mathbf{u}$ and $\mathbf{v}$. This shows that Eq. (5.108) holds in the approximation of ignoring $e^{-2/\sigma_{b_i}^2}$. Now we compute $E\{(\mathbf{b}_i - \mathbf{b}_i^{\text{true}})(\mathbf{b}_i - \mathbf{b}_i^{\text{true}})^T\}$. The integrals involving an odd power of $\cos\phi$ or $\sin\phi$ will vanish in the integration over ϕ, and the integrals over ϕ of $\cos^2\phi$ and $\sin^2\phi$ both give π. We also note that $\sin^2\theta = 2\zeta - \zeta^2$ and $\mathbf{u}\,\mathbf{u}^T + \mathbf{v}\,\mathbf{v}^T = I_3 - \mathbf{b}_i^{\text{true}}(\mathbf{b}_i^{\text{true}})^T$, so

$$\begin{aligned}
E\{(\mathbf{b}_i - \mathbf{b}_i^{\text{true}})(\mathbf{b}_i - \mathbf{b}_i^{\text{true}})^T\} &\approx \sigma_{b_i}^{-2}\int_0^{\infty}\zeta\exp(-\zeta/\sigma_{b_i}^2)\,d\zeta\,\left[I_3 - \mathbf{b}_i^{\text{true}}(\mathbf{b}_i^{\text{true}})^T\right] \\
&\quad - \frac{1}{2}\sigma_{b_i}^{-2}\int_0^{\infty}\zeta^2\exp(-\zeta/\sigma_{b_i}^2)\,d\zeta\,\left[I_3 - 3\,\mathbf{b}_i^{\text{true}}(\mathbf{b}_i^{\text{true}})^T\right] \\
&\approx \sigma_{b_i}^2\left[I_3 - \mathbf{b}_i^{\text{true}}(\mathbf{b}_i^{\text{true}})^T\right] - \sigma_{b_i}^4\left[I_3 - 3\,\mathbf{b}_i^{\text{true}}(\mathbf{b}_i^{\text{true}})^T\right]
\end{aligned} \tag{5.126}$$

The first term is the covariance of the QMM, Eq. (5.107b). The second term is a small correction, much larger than the neglected exponential terms but only 3 % for measurement errors as large as 10°. It is interesting to note that this term has zero trace.

The likelihood of a set of N statistically independent measurements is

$$\ell(\mathbf{b}_1,\ldots,\mathbf{b}_N|A^{\text{true}},\mathbf{r}_i^{\text{true}}) = \prod_{i=1}^{N}\mathcal{N}_{b_i}\exp\left(-\frac{1}{2\sigma_{b_i}^2}\|\mathbf{b}_i - A^{\text{true}}\mathbf{r}_i^{\text{true}}\|^2\right) \tag{5.127}$$

Maximizing this likelihood is equivalent to minimizing the negative log-likelihood, which is given by

$$-\ln\ell = \sum_{i=1}^{N}\left(\frac{1}{2\sigma_{b_i}^2}\|\mathbf{b}_i - A^{\text{true}}\mathbf{r}_i^{\text{true}}\|^2 - \ln\mathcal{N}_{b_i}\right) \tag{5.128}$$

The data-dependent portion of the negative log-likelihood, i.e. the part exclusive of the $\ln\mathcal{N}_{b_i}$ terms, is Wahba's loss function with the weights chosen according to Eq. (5.112), which establishes that Wahba's problem is equivalent to MLE with this choice of weights.

The MLE analysis up to this point has assumed that the reference vectors are perfectly known. This is never the case, although the reference vectors often have much smaller uncertainties than the body frame vectors. We can generalize the analysis by assuming that the reference vectors have a conditional pdf over the unit sphere of the form

$$p(\mathbf{r}_i|\mathbf{r}_i^{\text{true}}) = \mathcal{N}_{r_i} \exp\left(-\frac{1}{2\sigma_{r_i}^2}\|\mathbf{r}_i - \mathbf{r}_i^{\text{true}}\|^2\right) \tag{5.129}$$

With the reasonable assumption of statistical independence of the reference vector errors and the errors in the body frame measurements, the likelihood of the reference and body frame vectors is

$$\ell(\mathbf{b}_1, \ldots, \mathbf{b}_N, \mathbf{r}_1, \ldots, \mathbf{r}_N | A^{\text{true}}, \mathbf{r}_i^{\text{true}}) = \prod_{i=1}^{N} \mathcal{N}_{b_i}\mathcal{N}_{r_i}$$

$$\times \exp\left(-\frac{1}{2\sigma_{b_i}^2}\|\mathbf{b}_i - A^{\text{true}}\mathbf{r}_i^{\text{true}}\|^2 - \frac{1}{2\sigma_{r_i}^2}\|\mathbf{r}_i - \mathbf{r}_i^{\text{true}}\|^2\right) \tag{5.130}$$

and the negative log-likelihood function is

$$-\ln\ell = \sum_{i=1}^{N}\left[\frac{1}{2\sigma_{b_i}^2}\|\mathbf{b}_i - A^{\text{true}}\mathbf{r}_i^{\text{true}}\|^2 + \frac{1}{2\sigma_{r_i}^2}\|\mathbf{r}_i - \mathbf{r}_i^{\text{true}}\|^2 - \ln(\mathcal{N}_{b_i}\mathcal{N}_{r_i})\right]$$

$$= \sum_{i=1}^{N}\left\{\sigma_{b_i}^{-2} + \sigma_{r_i}^{-2} - \left[\sigma_{b_i}^{-2}(A^{\text{true}})^T\mathbf{b}_i + \sigma_{r_i}^{-2}\mathbf{r}_i\right]^T \mathbf{r}_i^{\text{true}} - \ln(\mathcal{N}_{b_i}\mathcal{N}_{r_i})\right\} \tag{5.131}$$

We have to use our error-corrupted reference vectors and body frame vectors to find a maximum likelihood estimate of both the attitude matrix A^{true} and each of the true reference vectors $\mathbf{r}_i^{\text{true}}$. It is clear that we minimize the negative log-likelihood function with respect to the true reference vectors by taking $\mathbf{r}_i^{\text{true}}$ to be the unit vector in the direction of $\sigma_{b_i}^{-2}(A^{\text{true}})^T\mathbf{b}_i + \sigma_{r_i}^{-2}\mathbf{r}_i$, i.e.

$$\mathbf{r}_i^{\text{true}} = \frac{\sigma_{b_i}^{-2}(A^{\text{true}})^T\mathbf{b}_i + \sigma_{r_i}^{-2}\mathbf{r}_i}{(\sigma_{b_i}^{-4} + \sigma_{r_i}^{-4} + 2\sigma_{b_i}^{-2}\sigma_{r_i}^{-2}\mathbf{b}_i^T A^{\text{true}}\mathbf{r}_i)^{1/2}} \tag{5.132}$$

This is exactly what we would expect, a weighted average of the estimates computed from the error-corrupted reference and body frame vectors. This estimate involves the as-yet-unknown attitude matrix. However, inserting this estimate of $\mathbf{r}_i^{\text{true}}$ into Eq. (5.131) gives the effective negative log-likelihood function for the attitude

$$-\ln\ell = \sum_{i=1}^{N}\left[\sigma_{b_i}^{-2} + \sigma_{r_i}^{-2} - (\sigma_{b_i}^{-4} + \sigma_{r_i}^{-4} + 2\sigma_{b_i}^{-2}\sigma_{r_i}^{-2}\mathbf{b}_i^T A^{\text{true}}\mathbf{r}_i)^{1/2} - \ln(\mathcal{N}_{b_i}\mathcal{N}_{r_i})\right] \tag{5.133}$$

This does not look like Wahba's loss function. However, we can write

$$\begin{aligned}\sigma_{b_i}^{-4} + \sigma_{r_i}^{-4} + 2\sigma_{b_i}^{-2}\sigma_{r_i}^{-2}\mathbf{b}_i^T A^{\text{true}}\mathbf{r}_i &= (\sigma_{b_i}^{-2} + \sigma_{r_i}^{-2})^2 - 2\sigma_{b_i}^{-2}\sigma_{r_i}^{-2}(1 - \mathbf{b}_i^T A^{\text{true}}\mathbf{r}_i)\\ &= (\sigma_{b_i}^{-2} + \sigma_{r_i}^{-2})^2 - \sigma_{b_i}^{-2}\sigma_{r_i}^{-2}\|\mathbf{b}_i - A^{\text{true}}\mathbf{r}_i\|^2\end{aligned} \tag{5.134}$$

It is always true that $\|\mathbf{b}_i - A^{\text{true}}\mathbf{r}_i\| \ll \sigma_{b_i}\sigma_{r_i}(\sigma_{b_i}^{-2} + \sigma_{r_i}^{-2}) = \sigma_{b_i}/\sigma_{r_i} + \sigma_{r_i}/\sigma_{b_i}$, because the right side of this inequality is never smaller than 2 and the left side is on the order of the σ^2 by Eq. (5.125). Thus expanding the square root in Eq. (5.133) as a Taylor series and retaining only the first two terms gives the useful expression

$$[(\sigma_{b_i}^{-2} + \sigma_{r_i}^{-2})^2 - \sigma_{b_i}^{-2}\sigma_{r_i}^{-2}\|\mathbf{b}_i - A^{\text{true}}\mathbf{r}_i\|^2]^{1/2}$$

$$\approx (\sigma_{b_i}^{-2} + \sigma_{r_i}^{-2})\left[1 - \frac{\|\mathbf{b}_i - A^{\text{true}}\mathbf{r}_i\|^2}{2\sigma_{b_i}^2\sigma_{r_i}^2\left(\sigma_{b_i}^{-2} + \sigma_{r_i}^{-2}\right)^2}\right] \tag{5.135}$$

and thus

$$-\ln \ell \approx \sum_{i=1}^{N}\left[\frac{1}{2\sigma_i^2}\|\mathbf{b}_i - A^{\text{true}}\mathbf{r}_i\|^2 - \ln(\mathcal{N}_{b_i}\mathcal{N}_{r_i})\right] \tag{5.136}$$

with $\sigma_i^2 = \sigma_{b_i}^2 + \sigma_{r_i}^2$, as in Eq. (5.109). The data-dependent portion of this is just Wahba's loss function with an effective measurement variance of $\sigma_{b_i}^2 + \sigma_{r_i}^2$.

5.6.1 Fisher Information Matrix for Attitude Determination

To determine the Fisher information matrix using the negative log-likelihood in Eq. (5.136), we must first decide what to use as the vector that we called $\mathbf{x}^{\text{true}}$ in Sect. 12.3.4. The components of the attitude matrix are not appropriate, because they are constrained by the orthogonality requirement. We know that three parameters provide a minimal representation of a rotation, so we use the three incremental error angles as defined in Eq. (2.123). Thus we replace A^{true} by

$$A^{\text{true}} \to \exp(-[\boldsymbol{\delta\vartheta}\times])A^{\text{true}} \approx (I_3 - [\boldsymbol{\delta\vartheta}\times])\,A^{\text{true}} \tag{5.137}$$

where $\boldsymbol{\delta\vartheta}$ is the vector of error angles. Substituting Eq. (5.137) into Eq. (5.136) and inserting the true values for $\mathbf{b}_i$ and $\mathbf{r}_i$ gives

$$-\ln \ell \approx \sum_{i=1}^{N}\left[\frac{1}{2\sigma_i^2}\|[\boldsymbol{\delta\vartheta}\times]\,A^{\text{true}}\mathbf{r}_i^{\text{true}}\|^2 - \ln(\mathcal{N}_{b_i}\mathcal{N}_{r_i})\right]$$

$$\approx \sum_{i=1}^{N}\left[\frac{1}{2\sigma_i^2}\boldsymbol{\delta\vartheta}^T[\mathbf{b}_i^{\text{true}}\times]^T[\mathbf{b}_i^{\text{true}}\times]\boldsymbol{\delta\vartheta} - \ln(\mathcal{N}_{b_i}\mathcal{N}_{r_i})\right] \tag{5.138}$$

because $\mathbf{b}_i^{\text{true}} = A^{\text{true}}\mathbf{r}_i^{\text{true}}$. This approximate expression can be considered as exact for the purpose of computing the Fisher information matrix, because the higher order terms neglected in Eq. (5.135) are of third and higher order in $\boldsymbol{\delta\vartheta}$ and vanish when computing the information matrix.

Now the Fisher information matrix is given by

$$F = -\partial^2(\ln \ell)/[(\partial \delta \boldsymbol{\vartheta})(\partial \delta \boldsymbol{\vartheta}^T)]\big|_{\delta\boldsymbol{\vartheta}=\mathbf{0}} = \sum_{i=1}^{N} \sigma_i^{-2}[\mathbf{b}_i^{\text{true}}\times]^T[\mathbf{b}_i^{\text{true}}\times]$$

$$= \sum_{i=1}^{N} \sigma_i^{-2}\left[I_3 - \mathbf{b}_i^{\text{true}}(\mathbf{b}_i^{\text{true}})^T\right] \tag{5.139}$$

remembering that $\mathbf{b}_i^{\text{true}}$ is a unit vector. Comparing this result to Eq. (5.114) shows that any solution that minimizes Wahba's problem leads to an efficient estimator, to within first-order, since $P_{\vartheta\vartheta} = F^{-1}$.

An analysis of the observable attitude axes using the Fisher information matrix is shown in [8], which is repeated here. This analysis is shown for one and two vector observations. For observation of a single vector the Fisher information matrix is given by

$$F = \sigma^{-2}\left[I_3 - \mathbf{b}^{\text{true}}(\mathbf{b}^{\text{true}})^T\right] \tag{5.140}$$

An eigenvalue/eigenvector decomposition can be useful to assess the observability of this system. Since F is a symmetric positive semi-definite matrix, then all of its eigenvalues are greater than or equal to zero. Furthermore, the matrix of eigenvectors is orthogonal, which can be used to define a coordinate system. The eigenvalues of this matrix are given by $\lambda_1 = 0$ and $\lambda_{2,3} = \sigma^{-2}$. This indicates that rotations about one of the eigenvectors are not observable. The eigenvector associated with the zero eigenvalue is along $\mathbf{b}^{\text{true}}$. Therefore, rotations about the body vector are unknown, which intuitively makes sense. The other observable axes are perpendicular to this unobservable axis, which also intuitively makes sense.

A more interesting case involves two vector observations. The information matrix for this case is given by

$$F = \sigma_1^{-2}\left[I_3 - \mathbf{b}_1^{\text{true}}(\mathbf{b}_1^{\text{true}})^T\right] + \sigma_2^{-2}\left[I_3 - \mathbf{b}_2^{\text{true}}(\mathbf{b}_2^{\text{true}})^T\right] \tag{5.141}$$

If two non-collinear vector observations exist, then the system is fully observable and no zero eigenvalues of F will exist. The maximum eigenvalue of F can be shown to be given by

$$\lambda_{\max} = \sigma_1^{-2} + \sigma_2^{-2} \tag{5.142}$$

Factoring this eigenvalue out of the characteristic equation, $\det(\lambda I_3 - F) = 0$, yields the following form for the remaining eigenvalues:

$$\lambda^2 - \lambda_{\max}\lambda + \sigma_1^{-2}\sigma_2^{-2}\|\mathbf{b}_1^{\text{true}} \times \mathbf{b}_2^{\text{true}}\|^2 = 0 \tag{5.143}$$

Therefore, the intermediate and minimum eigenvalues are given by

$$\lambda_{\text{int}} = \frac{\lambda_{\max}(1+\chi)}{2} \tag{5.144a}$$

$$\lambda_{\min} = \frac{\lambda_{\max}(1-\chi)}{2} \tag{5.144b}$$

where

$$\chi = \left[1 - \frac{4\|\mathbf{b}_1^{\text{true}} \times \mathbf{b}_2^{\text{true}}\|^2}{\sigma_1^2\sigma_2^2\lambda_{\max}^2}\right]^{1/2} \tag{5.145}$$

Note that $\lambda_{\max} = \lambda_{\min} + \lambda_{\text{int}}$. Also note that $\|\mathbf{b}_1^{\text{true}} \times \mathbf{b}_2^{\text{true}}\| = 0$ if $\mathbf{b}_1^{\text{true}}$ is collinear with $\mathbf{b}_2^{\text{true}}$. Equation (5.144) shows $\lambda_{\min} = 0$ and thus the inverse of the Fisher information matrix does not exist, as expected for this case. Also, $\lambda_{\text{int}} = \lambda_{\max}$ for this case.

The eigenvectors of F are computed by solving $\lambda\mathbf{v} = F\mathbf{v}$ for each eigenvalue. The eigenvector associated with the maximum eigenvalue can be shown to be given by

$$\mathbf{v}_{\max} = \pm\frac{\mathbf{b}_1^{\text{true}} \times \mathbf{b}_2^{\text{true}}}{\|\mathbf{b}_1^{\text{true}} \times \mathbf{b}_2^{\text{true}}\|} \tag{5.146}$$

The sign of this vector is not of consequence since we are only interested in rotations about this vector. This indicates that the most observable axis is perpendicular to the plane formed by $\mathbf{b}_1^{\text{true}}$ and $\mathbf{b}_2^{\text{true}}$, which intuitively makes sense. The remaining eigenvectors must surely lie in the $\mathbf{b}_1^{\text{true}}$-$\mathbf{b}_2^{\text{true}}$ plane. To determine the eigenvector associated with the minimum eigenvalue, we will perform a rotation about the $\mathbf{v}_{\max}$ axis and determine the angle from $\mathbf{b}_1$. Using the Euler axis and angle parameterization in Eq. (2.108) gives

$$\mathbf{v}_{\min} = \pm\left\{(\cos\vartheta)I_3 + (1-\cos\vartheta)\mathbf{v}_{\max}\mathbf{v}_{\max}^T - \sin\vartheta[\mathbf{v}_{\max}\times]\right\}\mathbf{b}_1^{\text{true}} \tag{5.147}$$

where ϑ is the angle used to rotate $\mathbf{b}_1^{\text{true}}$ to $\mathbf{v}_{\min}$. Using the fact that $\mathbf{v}_{\max}$ is perpendicular to $\mathbf{b}_1^{\text{true}}$ gives $\mathbf{v}_{\max}^T\mathbf{b}_1^{\text{true}} = 0$. Therefore, Eq. (5.147) reduces down to

$$\mathbf{v}_{\min} = \pm\{(\cos\vartheta)I_3 - \sin\vartheta[\mathbf{v}_{\max}\times]\}\,\mathbf{b}_1^{\text{true}} \tag{5.148}$$

Substituting Eq. (5.148) into $\lambda_{\min}\mathbf{v}_{\min} = F\mathbf{v}_{\min}$ and using the property of the cross product matrix leads to the following equation for ϑ:

$$\tan\vartheta = \frac{a+b}{c} \tag{5.149}$$

where

$$a \equiv \lambda_{\min}\sigma_1^{-2} \tag{5.150a}$$

$$b \equiv \sigma_1^{-2}\sigma_2^{-2}(\mathbf{b}_1^{\text{true}})^T[\mathbf{b}_2^{\text{true}}\times]^2\mathbf{b}_1^{\text{true}} \tag{5.150b}$$

$$c \equiv -\frac{\sigma_1^{-2}\sigma_2^{-2}(\mathbf{b}_1^{\text{true}})^T[\mathbf{b}_2^{\text{true}}\times]^2[\mathbf{b}_1^{\text{true}}\times]^2\mathbf{b}_2^{\text{true}}}{\|\mathbf{b}_1^{\text{true}} \times \mathbf{b}_2^{\text{true}}\|} \tag{5.150c}$$

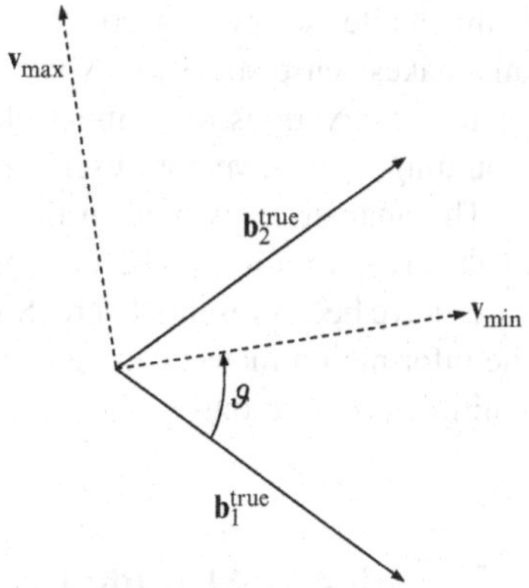

Fig. 5.1 Observable axes with two vector observations

Equation (5.149) can now be solved for ϑ, which can be used to determine $\mathbf{v}_{\rm min}$ from Eqs. (5.146) and (5.148). The intermediate axis is simply given by the cross product of $\mathbf{v}_{\rm max}$ and $\mathbf{v}_{\rm min}$:

$$\mathbf{v}_{\rm int} = \pm \mathbf{v}_{\rm max} \times \mathbf{v}_{\rm min} \tag{5.151}$$

A plot of the minimum and intermediate axes is shown in Fig. 5.1 for the case when the angle between $\mathbf{b}_1^{\rm true}$ and $\mathbf{b}_2^{\rm true}$ is less than 90°. Intuitively, this analysis makes sense since we expect that the least determined axis, $\mathbf{v}_{\rm min}$, is somewhere between $\mathbf{b}_1^{\rm true}$ and $\mathbf{b}_2^{\rm true}$ if these vector observations are less than 90° apart.

The previous analysis greatly simplifies if the reference vectors are unit vectors and the observation variances are equal, so that $\sigma_1^2 = \sigma_2^2 \equiv \sigma^2$. These assumptions are valid for a single field-of-view star camera. The eigenvalues are now given by

$$\lambda_{\rm max} = 2\sigma^{-2} \tag{5.152a}$$

$$\lambda_{\rm int} = \sigma^{-2}(1 + |(\mathbf{b}_1^{\rm true})^T \mathbf{b}_2|) \tag{5.152b}$$

$$\lambda_{\rm min} = \sigma^{-2}(1 - |(\mathbf{b}_1^{\rm true})^T \mathbf{b}_2|) \tag{5.152c}$$

The eigenvectors are now given by

$$\mathbf{v}_{\rm max} = \pm \frac{\mathbf{b}_1^{\rm true} \times \mathbf{b}_2^{\rm true}}{\|\mathbf{b}_1^{\rm true} \times \mathbf{b}_2^{\rm true}\|} \tag{5.153a}$$

$$\mathbf{v}_{\rm int} = \pm \frac{\mathbf{b}_1^{\rm true} - \text{sign}[(\mathbf{b}_1^{\rm true})^T \mathbf{b}_2^{\rm true}]\mathbf{b}_2^{\rm true}}{\|\mathbf{b}_1^{\rm true} - \text{sign}[(\mathbf{b}_1^{\rm true})^T \mathbf{b}_2^{\rm true}]\mathbf{b}_2^{\rm true}\|} \tag{5.153b}$$

$$\mathbf{v}_{\rm min} = \pm \frac{\mathbf{b}_1^{\rm true} + \text{sign}[(\mathbf{b}_1^{\rm true})^T \mathbf{b}_2^{\rm true}]\mathbf{b}_2^{\rm true}}{\|\mathbf{b}_1^{\rm true} + \text{sign}[(\mathbf{b}_1^{\rm true})^T \mathbf{b}_2^{\rm true}]\mathbf{b}_2^{\rm true}\|} \tag{5.153c}$$

where $\text{sign}[(\mathbf{b}_1^{\rm true})^T \mathbf{b}_2^{\rm true}]$ is used to ensure that the proper direction of the eigenvectors is determined when the angle between $\mathbf{b}_1^{\rm true}$ and $\mathbf{b}_2^{\rm true}$ is greater than 90°.

If this angle is less than 90° then $\mathbf{v}_{\min}$ is the *bisector* of $\mathbf{b}_1^{\text{true}}$ and $\mathbf{b}_2^{\text{true}}$. Intuitively this makes sense since we expect rotations perpendicular to the bisector of the two vector observations to be more observable than rotations about the bisector (again assuming that the vector observations are within 90° of each other).

The analysis presented in this section is extremely useful for the visualization of the observability of the determined attitude. Closed-form solutions for special cases have been presented here. Still, in general, the eigenvalues and eigenvectors of the information matrix can be used to analyze the observability for cases involving multiple observations.

5.7 Induced Attitude Errors from Orbit Errors

Missions sometime require accurate attitude knowledge for Earth-pointing spacecraft. As discussed in Chap. 1 this is true for the GOES series of spacecraft. A natural attitude sensor selection for Earth-pointing spacecraft is a horizon sensor because it provides an attitude estimate directly with respect to an Earth-centric frame. However, as is the case for GOES, a horizon sensor may not meet the desired attitude knowledge requirement. For GOES a star tracker was selected to provide higher attitude accuracy. But a star tracker provides an estimate with respect to an inertial frame. The conversion to an Earth-centric frame, such as the LVLH described in Sect. 2.6.4, requires an attitude rotation from an inertial frame to an Earth-centric frame. For the LVLH frame the required attitude matrix to accomplish this rotation is given by the transpose of the matrix in Eq. (2.79). Since this matrix depends on position and velocity, then these quantities will contribute to the overall attitude errors. This section derives the attitude error-covariance with additive position and velocity orbit errors in Eq. (2.79). Since the orbit determination errors are uncorrelated with the attitude sensor errors, then the overall attitude error is the sum of the sensor and orbit errors. In this way the additional errors induced by the coupling of position and velocity to attitude can be accounted for in the overall attitude knowledge budget.

First, the true variables must be defined from Eq. (2.78):

$$\mathbf{o}_{3I}^{\text{true}} = -\mathbf{r}_I^{\text{true}}/\|\mathbf{r}_I^{\text{true}}\| \equiv -g_3^{\text{true}}\mathbf{r}_I^{\text{true}} \tag{5.154a}$$

$$\mathbf{o}_{2I}^{\text{true}} = -(\mathbf{r}_I^{\text{true}} \times \mathbf{v}_I^{\text{true}})/\|\mathbf{r}_I^{\text{true}} \times \mathbf{v}_I^{\text{true}}\| \equiv -g_2^{\text{true}}(\mathbf{r}_I^{\text{true}} \times \mathbf{v}_I^{\text{true}}) \tag{5.154b}$$

$$\mathbf{o}_{1I}^{\text{true}} = g_2^{\text{true}} g_3^{\text{true}}[\,\|\mathbf{r}_I^{\text{true}}\|^2\mathbf{v}_I^{\text{true}} - (\mathbf{r}_I^{\text{true}} \cdot \mathbf{v}_I^{\text{true}})\mathbf{r}_I^{\text{true}}] \tag{5.154c}$$

Adding an error to position in $\mathbf{o}_{3I}$ simply gives

$$\mathbf{o}_{3I} = -\frac{\mathbf{r}_I^{\text{true}} + \Delta\mathbf{r}_I}{\|\mathbf{r}_I^{\text{true}} + \Delta\mathbf{r}_I\|} \tag{5.155}$$

where $\mathbf{r}_I^{\text{true}}$ is the true position and $\Delta\mathbf{r}_I$ is its associated error, which is assumed to be a zero-mean Gaussian noise process. In this section it is assumed that the signal-to-noise ratio is large enough so that second-order and higher error terms are negligible. The denominator in Eq. (5.155) is thus approximated by

$$\begin{aligned}\|\mathbf{r}_I^{\text{true}} + \Delta\mathbf{r}_I\| &= (\|\mathbf{r}_I^{\text{true}}\|^2 + 2\,\mathbf{r}_I^{\text{true}}\cdot\Delta\mathbf{r}_I + \|\Delta\mathbf{r}_I\|^2)^{1/2} \\ &\approx (\|\mathbf{r}_I^{\text{true}}\|^2 + 2\,\mathbf{r}_I^{\text{true}}\cdot\Delta\mathbf{r}_I)^{1/2}\end{aligned} \tag{5.156}$$

Next the following first-order binomial expansion is used:

$$(x+\Delta x)^n \approx x^n + n x^{n-1}\Delta x, \quad \Delta x << x \tag{5.157}$$

Using Eq. (5.157) with $x \equiv \|\mathbf{r}_I^{\text{true}}\|^2$, $\Delta x \equiv 2\,\mathbf{r}_I^{\text{true}}\cdot\Delta\mathbf{r}_I$ and $n = -1/2$ leads to

$$\mathbf{o}_{3I} \approx -(\mathbf{r}_I^{\text{true}} + \Delta\mathbf{r}_I)\left(\|\mathbf{r}_I^{\text{true}}\|^{-1} - \|\mathbf{r}_I^{\text{true}}\|^{-3}\mathbf{r}_I^{\text{true}}\cdot\Delta\mathbf{r}_I\right) \tag{5.158}$$

Ignoring second-order terms gives

$$\begin{aligned}\mathbf{o}_{3I} &\approx \mathbf{o}_{3I}^{\text{true}} + \frac{1}{\|\mathbf{r}_I^{\text{true}}\|^3}\left[(\mathbf{r}_I^{\text{true}})(\mathbf{r}_I^{\text{true}})^T - \|\mathbf{r}_I^{\text{true}}\|^2 I_3\right]\Delta\mathbf{r}_I \\ &= \mathbf{o}_{3I}^{\text{true}} + \frac{1}{\|\mathbf{r}_I^{\text{true}}\|^3}[\mathbf{r}_I^{\text{true}}\times]^2\Delta\mathbf{r}_I\end{aligned} \tag{5.159}$$

Therefore the error in $\mathbf{o}_{3I}$, denoted by $\Delta\mathbf{o}_{3I}$, is given by

$$\Delta\mathbf{o}_{3I} = \frac{1}{\|\mathbf{r}_I^{\text{true}}\|^3}[\mathbf{r}_I^{\text{true}}\times]^2\Delta\mathbf{r}_I \tag{5.160}$$

In a similar fashion the error in $\mathbf{o}_{2I}$, denoted by $\Delta\mathbf{o}_{2I}$, can by shown to be given by (which is left as an exercise to the reader)

$$\Delta\mathbf{o}_{2I} = \frac{1}{\|\mathbf{r}_I^{\text{true}}\times\mathbf{v}_I^{\text{true}}\|^3}[(\mathbf{r}_I^{\text{true}}\times\mathbf{v}_I^{\text{true}})\times]^2[(\mathbf{r}_I^{\text{true}}\times\Delta\mathbf{v}_I) - (\mathbf{v}_I^{\text{true}}\times\Delta\mathbf{r}_I)] \tag{5.161}$$

where $\mathbf{v}_I^{\text{true}}$ is the true velocity and $\Delta\mathbf{v}_I$ is its associated error, which is assumed to be a zero-mean Gaussian noise process. The error in $\mathbf{o}_{1I}$ is more complicated than the other ones, but fortunately this term is not required in the analysis.

The attitude matrix with errors can now be written as

$$A_{IO} = A_{IO}^{\text{true}} + \Delta A_{IO} \tag{5.162}$$

where

$$A_{IO} \equiv [\mathbf{o}_{1I}\ \ \mathbf{o}_{2I}\ \ \mathbf{o}_{3I}] \tag{5.163a}$$

$$A_{IO}^{\text{true}} \equiv [\mathbf{o}_{1I}^{\text{true}}\ \ \mathbf{o}_{2I}^{\text{true}}\ \ \mathbf{o}_{3I}^{\text{true}}] \tag{5.163b}$$

$$\Delta A_{IO} \equiv [\Delta\mathbf{o}_{1I}\ \ \Delta\mathbf{o}_{2I}\ \ \Delta\mathbf{o}_{3I}] \tag{5.163c}$$

We wish to use a multiplicative attitude error in the orbit frame, which is given by Eq. (2.180) as

$$\delta A_{IO} = (A_{IO}^{\text{true}})^T A_{IO} \approx I_3 - [\boldsymbol{\delta\vartheta} \times] \tag{5.164}$$

Substituting Eq. (5.162) into Eq. (5.164) and cancelling the I_3 terms gives

$$-[\boldsymbol{\delta\vartheta} \times] = (A_{IO}^{\text{true}})^T \Delta A_{IO} = \begin{bmatrix} 0 & \mathbf{o}_{1I}^{\text{true}} \cdot \Delta\mathbf{o}_{2I} & \mathbf{o}_{1I}^{\text{true}} \cdot \Delta\mathbf{o}_{3I} \\ \mathbf{o}_{2I}^{\text{true}} \cdot \Delta\mathbf{o}_{1I} & 0 & \mathbf{o}_{2I}^{\text{true}} \cdot \Delta\mathbf{o}_{3I} \\ \mathbf{o}_{3I}^{\text{true}} \cdot \Delta\mathbf{o}_{1I} & \mathbf{o}_{3I}^{\text{true}} \cdot \Delta\mathbf{o}_{2I} & 0 \end{bmatrix} \tag{5.165}$$

This is very similar to the derivation of Eq. (3.174), and as in that case we choose the easier dot products to evaluate. After some algebra that is left to the reader as an exercise, this leads to

$$\boldsymbol{\delta\vartheta} = \begin{bmatrix} \mathbf{o}_{2I}^{\text{true}} \cdot \Delta\mathbf{o}_{3I} \\ -\mathbf{o}_{1I}^{\text{true}} \cdot \Delta\mathbf{o}_{3I} \\ \mathbf{o}_{1I}^{\text{true}} \cdot \Delta\mathbf{o}_{2I} \end{bmatrix} = \mathscr{O} \begin{bmatrix} \Delta\mathbf{r}_I \\ \Delta\mathbf{v}_I \end{bmatrix} \tag{5.166}$$

where

$$\mathscr{O} \equiv \begin{bmatrix} g_3^{\text{true}} (\mathbf{o}_{2I}^{\text{true}})^T & \mathbf{0}_3^T \\ -g_3^{\text{true}} (\mathbf{o}_{1I}^{\text{true}})^T & \mathbf{0}_3^T \\ -g_2^{\text{true}} g_3^{\text{true}} (\mathbf{r}_I^{\text{true}} \cdot \mathbf{v}_I^{\text{true}}) (\mathbf{o}_{2I}^{\text{true}})^T & g_2^{\text{true}} g_3^{\text{true}} \|\mathbf{r}_I^{\text{true}}\|^2 (\mathbf{o}_{2I}^{\text{true}})^T \end{bmatrix} \tag{5.167}$$

The position and velocity error-covariance matrix is defined by

$$\mathscr{P}_{\text{orbit}} = E\left\{ \begin{bmatrix} \Delta\mathbf{r}_I \\ \Delta\mathbf{v}_I \end{bmatrix} \begin{bmatrix} \Delta\mathbf{r}_I^T & \Delta\mathbf{v}_I^T \end{bmatrix} \right\} \tag{5.168}$$

which is given by the orbit determination system. Therefore, the induced attitude error-covariance is given by

$$P_{\text{orbit}} = E\left\{ (\boldsymbol{\delta\vartheta})\, (\boldsymbol{\delta\vartheta})^T \right\} = \mathscr{O}\, \mathscr{P}_{\text{orbit}}\, \mathscr{O}^T \tag{5.169}$$

Note that P_{orbit} is a function of the true position and velocity variables in $\mathscr{O}$, but these can be replaced by their respective estimates from the orbit determination system in practice.

5.8 TRMM Attitude Determination

The Tropical Rainfall Measuring Mission (TRMM) spacecraft was launched on November 27, 1997 from the Tanegashima Space Center in Tanegashima, Japan. The main objectives of this mission include: (1) to obtain multi-year measurements

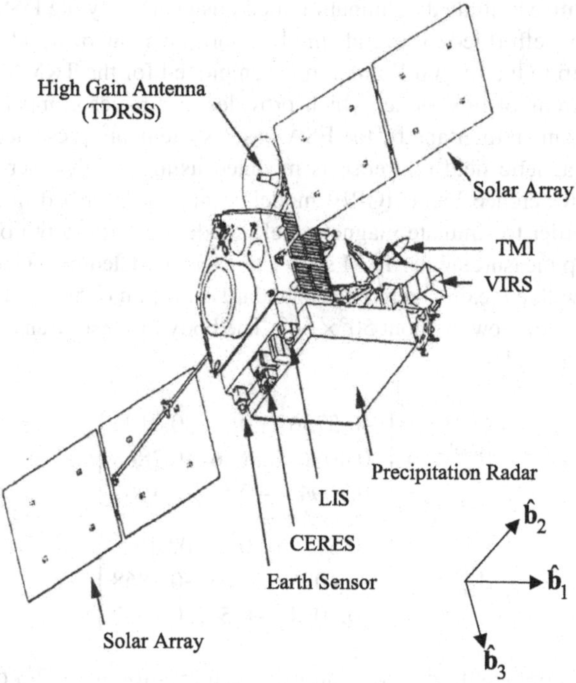

Fig. 5.2 TRMM spacecraft

of tropical and subtropical rainfall, (2) to understand how interactions between the sea, air, and land masses produce changes in global rainfall and climate, and (3) to help improve the modeling of tropical rainfall processes and their influence on global circulation.

TRMM's nominal orbit altitude was 350 km, until raised to 402 km to prolong mission life by reducing the fuel expenditure used to compensate for atmospheric drag [1]. The spacecraft is three-axis stabilized with an orbit inclination of 35°. A diagram of the TRMM spacecraft is shown in Fig. 5.2. The $\hat{\mathbf{b}}_1$ axis is parallel to the orbital velocity direction, the $\hat{\mathbf{b}}_2$ axis is parallel to the negative of the orbital momentum vector, and the $\hat{\mathbf{b}}_3$ axis points in the nadir direction. The nominal mission mode requires a rotation once per orbit about the spacecraft's $\hat{\mathbf{b}}_2$ axis while holding the remaining axis rotations near zero. Thus, the LVLH frame described in Sect. 2.6.4 can be used to compute the desired attitude matrix from the spacecraft's position/velocity vector in order to maintain an Earth-pointing configuration.

The attitude determination hardware consists of an Earth Sensor Assembly (ESA), Digital Sun Sensors (DSSs), Coarse Sun Sensors (CSSs), a Three-Axis Magnetometer (TAM), and rate-integrating gyroscopes (RIGs). The allotted attitude knowledge accuracy of 0.18° per axis was achieved by using the ESA for pitch and roll and the RIGs, updated twice per orbit by DSS measurements, for yaw. As was discussed in Chap. 1, potential problems with the ESA resulted in an effort to develop alternative attitude determination methods using only the DSSs, TAM, and RIGs [1, 7]. This effort led to several simple algorithms, one of which is used as an onboard backup to the full-up Kalman filter employed for the TRMM contingency mode. Simulations of this mode, which provides accuracies comparable to those given by the Kalman filter and by the ESA-based system, are presented here.

The true magnetic field reference is modeled using a 10th-order International Geomagnetic Reference Field (IGRF) model, which is described in Sect. 11.1 of Chap. 11. In order to simulate magnetic field modeling error, a 6th order IGRF is used to develop measurements. TAM sensor noise is modeled by a Gaussian white-noise process with a mean of zero and a standard deviation of 50 nT. The two DSSs each have a field of view of about 50°× 50°. The body to sensor transformations for each sensor is given by

$$A_{\text{body}}^{\text{DSS}_1} = \begin{bmatrix} -0.5736 & 0 & -0.8192 \\ 0.4096 & 0.8666 & -0.2868 \\ 0.7094 & -0.5 & -0.4967 \end{bmatrix} \tag{5.170a}$$

$$A_{\text{body}}^{\text{DSS}_2} = \begin{bmatrix} -0.5736 & 0 & 0.8192 \\ -0.4096 & 0.8666 & -0.2868 \\ -0.7094 & -0.5 & -0.4967 \end{bmatrix} \tag{5.170b}$$

The two DSSs combine to provide Sun measurements for about 2/3 of a complete orbit. The DSS sensor noise is also modeled by a Gaussian white-noise process with a mean of zero and a standard deviation of 0.05°. The gyro measurements are simulated using Eqs. (4.31) and (4.32) of Sect. 4.7.1, with $\sigma_u = \sqrt{10}\times 10^{-10}$ rad/s$^{3/2}$ and $\sigma_v = \sqrt{10} \times 10^{-7}$ rad/s$^{1/2}$. The initial bias for each axis is given by 0.1 deg/h. All sensors are sampled at 10-s intervals. Note that in general gyros are typically sampled at higher frequencies than other attitude sensors in order to provide rate information for other control aspects, such as spacecraft jitter.

Attitude results using the uncorrupted TAM, i.e. using the 10th-order model to generate the truth and the measurements, with the two DSSs are shown in Fig. 5.3a. The actual errors are plotted with their 3σ bounds, three times the square roots of the diagonal elements of the covariance matrix given by Eq. (5.120). Gaps are given when both of the DSSs are unavailable. The large values in the 3σ bounds are due to the vectors becoming more co-aligned, i.e. the angle between the magnetic field vector and Sun vector is small, thereby reducing usable information in the attitude determination solution. The computed 3σ bounds provide an accurate measure to quantify the actual errors. Results using the corrupted TAM, i.e. using the 10th-order model to generate the truth and the 6th-order model to generate the measurements, with the two DSSs are shown in Fig. 5.3b. Clearly the errors are now larger but they are within the typical accuracy provided by a TAM/DSS sensor suite for attitude determination on actual spacecraft.

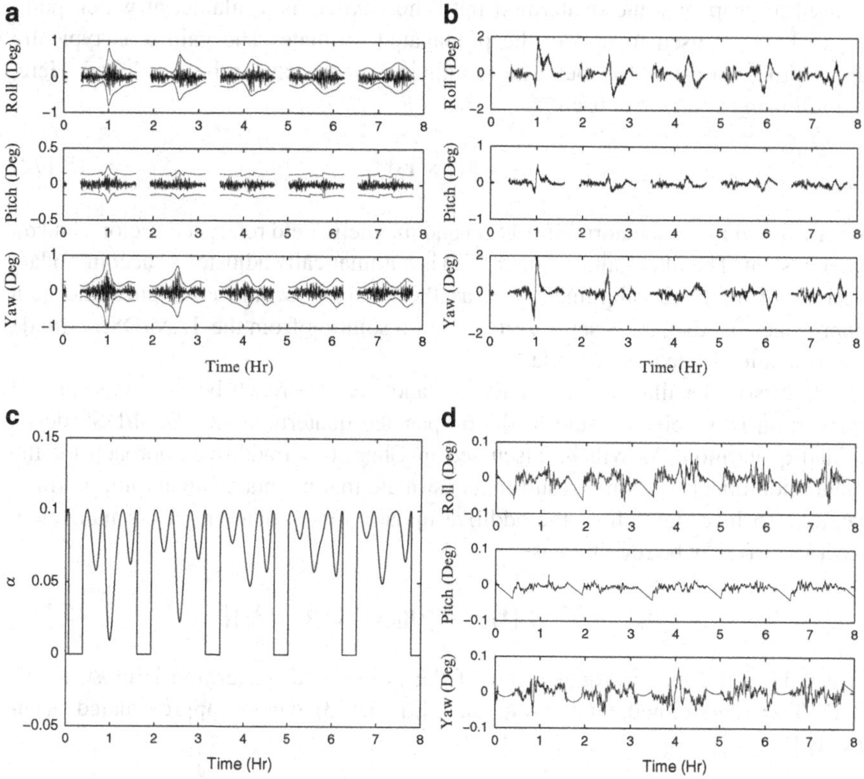

Fig. 5.3 TRMM attitude determination results. (**a**) Uncorrupted TAM results. (**b**) Corrupted TAM results. (**c**) Filter gain. (**d**) Filtered results

The accuracy can be improved by employing a simple first-order filter approach using the estimated quaternion together with gyro measurements [7]. Furthermore, the gyro can be used to propagate an attitude solution when no DSS measurements are available. The first-order filter is given by

$$\hat{\mathbf{q}}_k^+ = (1-\alpha)\hat{\mathbf{q}}_k^- + \alpha\tilde{\mathbf{q}}_k \tag{5.171a}$$

$$\hat{\mathbf{q}}_{k+1}^- = \exp\left(\frac{1}{2}[\tilde{\omega}_k\otimes]\Delta t\right)\hat{\mathbf{q}}_k^+ \tag{5.171b}$$

$$= \left[\cos\left(\frac{1}{2}\|\tilde{\omega}_k\|\Delta t\right)I_4 + \frac{1}{\|\tilde{\omega}_k\|}\sin\left(\frac{1}{2}\|\tilde{\omega}_k\|\Delta t\right)\Omega(\tilde{\omega}_k)\right]\hat{\mathbf{q}}_k^+ \tag{5.171c}$$

where $\tilde{\mathbf{q}}_k$ is the quaternion determined from the TAM/DSS attitude determination system, $\hat{\mathbf{q}}_k^-$ is the propagated estimate, $\hat{\mathbf{q}}_k^+$ is the updated estimated, $\tilde{\omega}_k$ is the vector of gyro measurements, α is a scalar gain between 0 and 1, and Δt is the sampling interval in the gyro. Using an initial estimate, which is typically $\tilde{\mathbf{q}}_k$, Eq. (5.171c) is used to propagate the quaternion until the next $\tilde{\mathbf{q}}_k$ is available, at which point Eq. (5.171a) is used to update the propagated estimate. The gain α is typically chosen to be constant [8], but a more judicious approach can be developed. Here, the following equation is used:

$$\alpha = \|\mathbf{r}_1 \times \mathbf{r}_2\|^2 \alpha_0 \tag{5.172}$$

where $\mathbf{r}_1$ and $\mathbf{r}_2$ are the normalized Sun and magnetic field reference vectors, and α_0 is a constant. The filter gain in Eq. (5.172) is automatically adjusted to accommodate periods of vector co-alignment, i.e. as the vectors become co-aligned the gain approaches 0. Also, α_0 is set to zero when no solution from the TAM/DSS attitude determination system is possible.

The first-order filter is essentially an "additive" approach because the updated quaternion is a weighted sum of the propagated quaternion and TAM/DSS determined quaternion. As will be discussed in Chap. 6, an additive approach for the quaternion update does not result in an estimate that maintains quaternion normalization. To investigate how the additive update affects quaternion normalization, Eq. (5.171a) may be rewritten as

$$\hat{\mathbf{q}}_k^+ = \hat{\mathbf{q}}_k^- \otimes \left\{\mathbf{I}_q + \alpha \left[(\hat{\mathbf{q}}_k^-)^{-1} \otimes \tilde{\mathbf{q}}_k - \mathbf{I}_q\right]\right\} \tag{5.173}$$

where $\mathbf{I}_q$ is the identity quaternion. If the propagated quaternion is close to the TAM/DSS determined quaternion, then Eq. (5.173) can be approximated accurately by

$$\hat{\mathbf{q}}_k^+ \approx \hat{\mathbf{q}}_k^- \otimes \begin{bmatrix} \frac{1}{2}\alpha\, \delta\boldsymbol{\vartheta} \\ 1 \end{bmatrix} \tag{5.174}$$

where $\delta\boldsymbol{\vartheta}$ is the angle vector used to form the rotation matrix between $\hat{\mathbf{q}}_k^-$ and $\tilde{\mathbf{q}}_k$. Therefore, since α is between 0 and 1, then normalization is maintained to within first order. For numerical precision, the quaternion estimates are explicitly normalized after the update process.

The simple filter is now applied to the TRMM example. Trial and error found that an initial gain of $\alpha_0 = 0.1$ provides good filtered estimates. A higher gain will produce estimates that are noisier and lower gain will result in estimates that suffer from greater gyro bias propagation errors. A plot of the filter gain is shown in Fig. 5.3c. Note that at the 1-h mark the gain becomes small, which is due to the fact that the Sun and magnetic field vectors are nearly co-aligned at that time. This forces the estimate to follow the gyro propagation more closely, which results in smaller estimate errors, as shown in Fig. 5.3d. Also, note that when no TAM/DSS quaternion estimate is given, the errors are dictated by the gyro bias, which is given as 0.1 deg/h. For the TRMM mission the attitude drift during these periods is within the mission specifications.

5.9 GPS Attitude Determination

GPS attitude determination uses phase differences of GPS signals received by antennas located at several locations on the spacecraft. Pairs of GPS antennas can be used to form a set of $n \geq 2$ baselines $\mathbf{b}_i$ in the spacecraft body frame, as illustrated in Fig. 4.4. The $m \geq 4$ unit vector sightlines $\mathbf{s}_j$ from the user spacecraft to the GPS satellites can be computed from the GPS position solution, as discussed in Sect. 4.6. The different path lengths from each GPS satellite to the antennas at the two ends of each baseline create mn phase differences of the received signals

$$\phi_{ij}^{\text{true}} = 2\pi(n_{ij}^{\phi} + \lambda^{-1}\mathbf{b}_i^T A\mathbf{s}_j) \tag{5.175}$$

where A is the attitude matrix, λ is the wavelength of the GPS signal, and the integers n_{ij}^{ϕ} are the number of full wavelength differences of the paths from GPS satellite j to the two antennas at the ends of baseline i. The latter are referred to as the *integer phase ambiguities*, and several algorithms have been proposed to compute them [6, 10, 28]. We will not discuss these, but assume that they have successfully solved for the integer ambiguities. We then use the measured phase differences to compute the normalized measurements

$$z_{ij} \equiv \lambda(\phi_{ij}/2\pi - n_{ij}^{\phi}) \tag{5.176}$$

The optimal attitude solution is the attitude matrix minimizing the loss function

$$L_{\text{GPS}}(A) \equiv \frac{1}{2}\sum_{i=1}^{n}\sum_{j=1}^{m} a_{ij}(z_{ij} - \mathbf{b}_i^T A\mathbf{s}_j)^2 \tag{5.177}$$

for some weights a_{ij}. This is similar to Wahba's loss function, but is not as easy to minimize. Solutions can be found with difficulty, however [6].

If we knew the representations of the baseline vectors in the reference frame, which we denote by $\mathbf{r}_i$ as in Sect. 5.2, we could use one of the algorithms of that section to compute the attitude matrix. In view of Eq. (5.1), we compute these as the $\mathbf{r}_i$ minimizing the loss function [9]

$$L_i(\mathbf{r}_i) \equiv \frac{1}{2}\sum_{j=1}^{m} a_{ij}(z_{ij} - \mathbf{r}_i^T\mathbf{s}_j)^2 \tag{5.178}$$

for $i = 1, 2, \ldots, n$. The minimization gives

$$\mathbf{r}_i = S_i^{-1}\sum_{j=1}^{m} a_{ij}z_{ij}\mathbf{s}_j \tag{5.179}$$

where

$$S_i \equiv \sum_{j=1}^{m} a_{ij}\mathbf{s}_j\mathbf{s}_j^T \tag{5.180}$$

These solutions only exist if the matrices S_i all have rank three, which requires that the sightlines $\mathbf{s}_j$ not be coplanar, as will generally be the case. The computational burden is reduced if the weights a_{ij} are independent of the baseline label i, since then all the S_i will be equal, and only one matrix must to be computed and inverted for each set of sightlines.

The computational burden would also be reduced if we reversed the roles of the baselines and sightlines, since the matrix corresponding to S_i would depend on the baselines in the body frame. Since these are constant, the matrix inverse in the analog of Eq. (5.179) would only need to be computed once, rather than at each measurement time. This would require at least three non-coplanar baselines, however, requiring more GPS antennas with unobstructed sightlines.

The final step of this method is to solve Wahba's problem for the vector sets $\{\mathbf{r}_i\}$ and $\{\mathbf{b}_i\}$; the fact that they are not unit vectors is not important. There are several options for choosing the weights a_i in Wahba's loss function, but no choice produces an optimal minimum of the loss function of Eq. (5.177), in general. The estimates are nearly optimal unless the sightlines are nearly coplanar, though; and simulations show that the computational advantages of this method do not entail a significant loss of accuracy [9].

Problems

5.1. A measure for the error between the estimated attitude, A, and the true attitude, A^{true}, is given by

$$A\,(A^{\text{true}})^T = (\cos\vartheta_{\text{err}})\,I_3 - \sin\vartheta_{\text{err}}[\mathbf{e}\times] + (1-\cos\vartheta_{\text{err}})\mathbf{e}\,\mathbf{e}^T$$

where ϑ_{err} is the error angle of rotation. Show that ϑ_{err} is given by the following equation:

$$\vartheta_{\text{err}} = 2\sin^{-1}\left(\|A\,(A^{\text{true}})^T - I_3\|_F/\sqrt{8}\right) = 2\sin^{-1}\left(\|A - A^{\text{true}}\|_F/\sqrt{8}\right)$$

where the Frobenius norm is given by Eq. (5.12).

5.2. Consider the following four reference vectors:

$$\mathbf{r}_1 = \begin{bmatrix} 0 \\ 0 \\ 1 \end{bmatrix}, \quad \mathbf{r}_2 = \frac{1}{\sqrt{0.1^2+1}}\begin{bmatrix} 0 \\ 0.1 \\ 1 \end{bmatrix}$$

$$\mathbf{r}_3 = \frac{1}{\sqrt{0.1^2+1}}\begin{bmatrix} 0 \\ -0.1 \\ 1 \end{bmatrix}, \quad \mathbf{r}_4 = \frac{1}{\sqrt{0.1^2+1}}\begin{bmatrix} 0.1 \\ 0 \\ 1 \end{bmatrix}$$

Assume that the true attitude matrix is given by the identity matrix. To generate body measurements and measurement errors use the QMM in Eq. (5.107b) with the true attitude, the given $\mathbf{r}_i$'s and $\sigma_1 = \sigma_2 = \sigma_3 = \sigma_4 = 0.001 \times \pi/180$ rad. Run 1,000 Monte Carlo runs using Davenport's method to compute the quaternion. Plot the small angle errors along with their respected 3σ bounds computed from the error-covariance given by the inverse of the Fisher information matrix in Eq. (5.139).

5.3. In this exercise you will compare Davenport's method, QUEST, the SVD method and FOAM for a number of test cases. Let the true attitude be given by

$$A^{\text{true}} = \begin{bmatrix} 0.352 & 0.864 & 0.360 \\ -0.864 & 0.152 & 0.460 \\ 0.360 & -0.480 & 0.800 \end{bmatrix}$$

Consider the following four cases:

a) $\mathbf{r}_1 = [1\ 0\ 0]^T$, $\mathbf{r}_2 = [0\ 1\ 0]^T$, and $\sigma_1 = \sigma_2 = 0.01$ rad.
b) $\mathbf{r}_1 = [1\ 0\ 0]^T$, $\mathbf{r}_2 = [1\ 0.01\ 0]^T/\|[1\ 0.01\ 0]^T\|$, and $\sigma_1 = \sigma_2 = 0.01$ rad.
c) $\mathbf{r}_1 = [1\ 0\ 0]^T$, $\mathbf{r}_2 = [1\ 0.01\ 0]^T/\|[1\ 0.01\ 0]^T\|$, and $\sigma_1 = \sigma_2 = 1 \times 10^{-6}$ rad.
d) $\mathbf{r}_1 = [1\ 0\ 0]^T$, $\mathbf{r}_2 = [0.96\ 0.28\ 0]^T/\|[0.96\ 0.28\ 0]^T\|$, $\sigma_1 = 0.001$ rad, and $\sigma_2 = 1 \times 10^{-6}$ rad.

To generate body measurements with errors use the QMM in Eq. (5.107b) with the true attitude, the given $\mathbf{r}_i$'s and aforementioned sigma values for each case. Run 1,000 Monte Carlo runs using Davenport's method, QUEST, the SVD method and FOAM to compute the attitude, A, for each case. Then compute the following error metric:

$$\vartheta_{\text{err}} = 2\sin^{-1}\left(\|A - A^{\text{true}}\|_F/\sqrt{8}\right)$$

Compute the numerical mean and standard deviation of ϑ_{err} for each algorithm and compare their relative accuracies.

5.4. Suppose that you wish to determine the average quaternion, denoted by $\mathbf{q}_{\text{ave}}$, from a set of weighted quaternions given by $\mathbf{q}_i$, $i = 1, 2 \ldots, N$, with weights w_i. Simply summing the weighted set of quaternions and dividing by N has two issues. First, it does not produce a unit vector in general. Second, changing the sign of any $\mathbf{q}_i$ should not change the average, but it is clear that the weighted sum approach does not have this property. To find an averaged quaternion that overcomes these issues the following loss function is chosen to be minimized [23]:

$$J(\mathbf{q}_{\text{ave}}) = \sum_{i=1}^{N} w_i \|A(\mathbf{q}_{\text{ave}}) - A(\mathbf{q}_i)\|_F^2, \quad \text{s.t. } \mathbf{q}_{\text{ave}}^T \mathbf{q}_{\text{ave}} = 1$$

where the Frobenius norm is described in Problem 5.1. Show that the solution for $\mathbf{q}_{\text{ave}}$ can be found by first forming the following matrix:

$$M \equiv \sum_{i=1}^{N} w_i \mathbf{q}_i \mathbf{q}_i^T$$

and then taking the eigenvector associated with the maximum eigenvalue of M to obtain $\mathbf{q}_{\text{ave}}$. For this problem you will use a different solution approach than the one shown in [23] though. Start with the following relation given from Problem 5.1: $\|A(\mathbf{q}_{\text{ave}}) - A(\mathbf{q}_i)\|_F^2 = 8\sin^2(\vartheta_i/2)$, where ϑ_i is the angle of rotation of the matrix $\delta A(\delta\mathbf{q}) \equiv A(\mathbf{q}_{\text{ave}})A^T(\mathbf{q}_i)$.

Using the eigenvector associated with the maximum eigenvalue of M is equivalent to Davenport's solution of Wahba's problem. But the QUEST solution cannot be used directly using the matrix M. How can this matrix be modified so that the QUEST solution can be used? Note that the matrix K in the QUEST algorithm is traceless. Show that your modification does not change the solution for $\mathbf{q}_{\text{ave}}$.

5.5. Use Eqs. (5.113) and (5.114) to demonstrate the star tracker accuracies for multiple stars claimed in Sect. 4.2.3. Assume there are five stars with the same error standard deviation $\sigma_i = \sigma$ at

$$\mathbf{b}_1 = \begin{bmatrix} 0 \\ 0 \\ 1 \end{bmatrix}, \quad \mathbf{b}_2 = \begin{bmatrix} s \\ 0 \\ c \end{bmatrix}, \quad \mathbf{b}_3 = \begin{bmatrix} -s \\ 0 \\ c \end{bmatrix}, \quad \mathbf{b}_4 = \begin{bmatrix} 0 \\ s \\ c \end{bmatrix}, \quad \mathbf{b}_5 = \begin{bmatrix} 0 \\ -s \\ c \end{bmatrix}$$

where $s \equiv \sin\theta$, $c \equiv \cos\theta$, and where we have omitted the superscript true. Compute the information matrix and invert it to find the covariance. The square roots of its diagonal elements are the standard deviations of the attitude errors about the three axes. You should find that

$$\sigma_x = \sigma_y = \frac{\sigma}{\sqrt{5 - 2s^2}} \approx \frac{\sigma}{\sqrt{N_{\text{stars}}}}$$

$$\sigma_z = \frac{\sigma}{2s} \approx \frac{\sigma}{\sqrt{N_{\text{stars}}}\sqrt{4/5}\,\theta}$$

The two approximations are both valid for small θ, and $\sqrt{4/5}\,\theta$ is the RMS angular distance of the stars from the boresight, given by $\mathbf{b}_1$.

5.6. Suppose that you are given unit Sun and magnetic field reference vectors, denoted by $\mathbf{r}_{\text{sun}}$ and $\mathbf{r}_{\text{mag}}$, respectively, and a body TAM unit-vector observation, denoted by $\mathbf{b}_{\text{mag}}$. The Sun sensor is a "slit" type design that tells only when the Sun is in a particular slit plane, but not orientation in that plane. So the body Sun unit-vector, denoted by $\mathbf{b}_{\text{sun}}$, is unknown but another unit vector, denoted by $\mathbf{v}_{\text{slit}}$, exists such that $\mathbf{b}_{\text{sun}} \cdot \mathbf{v}_{\text{slit}} = 0$. Assuming that $\mathbf{r}_{\text{sun}} \cdot \mathbf{r}_{\text{mag}} = \mathbf{b}_{\text{sun}} \cdot \mathbf{b}_{\text{mag}}$, find $\mathbf{b}_{\text{sun}}$ from the given unit vectors. Note this problem is equivalent to finding the intersection of a cone and a plane with the vertex of the cone constrained to lie in the plane.

5.7. Consider the following true and measured vectors for a star tracker model, given by Eq. (4.2):

$$\mathbf{b}^{\text{true}} = \frac{1}{\sqrt{1 + \|\boldsymbol{\gamma}^{\text{true}}\|^2}} \begin{bmatrix} \boldsymbol{\gamma}^{\text{true}} \\ 1 \end{bmatrix}, \quad \mathbf{b} = \frac{1}{\sqrt{1 + \|\boldsymbol{\gamma}\|^2}} \begin{bmatrix} \boldsymbol{\gamma} \\ 1 \end{bmatrix}$$

where $\boldsymbol{\gamma}^{\text{true}} \equiv [\tan^2\beta^{\text{true}}\ \ \tan^2\alpha^{\text{true}}]^T$ and $\boldsymbol{\gamma} \equiv [\tan^2\beta\ \ \tan^2\alpha]^T$. The measured and true vectors are related by $\boldsymbol{\gamma} = \boldsymbol{\gamma}^{\text{true}} + \Delta\boldsymbol{\gamma}$, where $\Delta\boldsymbol{\gamma}$ is a zero-mean Gaussian noise process with covariance R_γ. Using the analysis in Sect. 5.7 show that the following expression is valid to within first-order:

$$\mathbf{b} = \mathbf{b}^{\text{true}} - \frac{1}{\sqrt{1 + \|\boldsymbol{\gamma}^{\text{true}}\|^2}}[\mathbf{b}^{\text{true}}\times]^2 I_{3\times2}\,\Delta\boldsymbol{\gamma}$$

where

$$I_{3\times2} = \begin{bmatrix} 1\ 0 \\ 0\ 1 \\ 0\ 0 \end{bmatrix}$$

The covariance of $\mathbf{b}$ is then given by

$$R_b = \frac{1}{1 + \|\boldsymbol{\gamma}^{\text{true}}\|^2}[\mathbf{b}^{\text{true}}\times]^2 I_{3\times2}\, R_\gamma\, I_{3\times2}^T[\mathbf{b}^{\text{true}}\times]^2$$

Determine R_γ in terms of $\boldsymbol{\gamma}^{\text{true}} = [\gamma_1^{\text{true}}\ \ \gamma_2^{\text{true}}]^T$ so that R_b is exactly equal to the QMM given by Eq. (5.107b).

5.8. In this exercise noise will be added to both the body and reference vectors. This can occur when the errors in an assumed star catalog are significant compared to the sensor errors. Assume that the reference vectors are modeled by

$$\mathbf{r}_i^{\text{model}} = \frac{\mathbf{z}_i^{\text{true}} + \Delta\mathbf{z}_i}{\|\mathbf{z}_i^{\text{true}} + \Delta\mathbf{z}_i\|}$$

where $\mathbf{z}_i^{\text{true}}$ is some true position vector and $\Delta\mathbf{z}_i$ is a zero-mean Gaussian noise process with covariance R_{z_i}. Using the analysis in Sect. 5.7 show that the following expression is valid to within first-order:

$$\mathbf{r}_i^{\text{model}} = \mathbf{r}_i^{\text{true}} + \Delta\mathbf{r}_i$$

where $\mathbf{r}_i^{\text{true}} = \mathbf{z}_i^{\text{true}}/\|\mathbf{z}_i^{\text{true}}\|$ and

$$\Delta\mathbf{r}_i = -\frac{1}{\|\mathbf{z}_i^{\text{true}}\|}[\mathbf{r}_i^{\text{true}}\times]^2\Delta\mathbf{z}_i$$

The "measured" vector with errors is related to the true body vector through

$$\mathbf{r}_i \equiv (A^{\text{true}})^T\mathbf{b}_i^{\text{true}} + (A^{\text{true}})^T\Delta\mathbf{b}_i + \Delta\mathbf{r}_i$$

This vector will be used in the nonlinear least squares solution. Here it is assumed that $\Delta\mathbf{b}_i$ and $\Delta\mathbf{r}_i$ are uncorrelated. Using the QMM for $\Delta\mathbf{b}_i$ show that the covariance of $(A^{\text{true}})^T\Delta\mathbf{b}_i + \Delta\mathbf{r}_i$, denoted by R_i, can be given by

$$R_i = -\sigma_{b_i}^2 [\mathbf{r}_i^{\text{true}}\times]^2 + \frac{1}{\|\mathbf{z}_i^{\text{true}}\|^2}[\mathbf{r}_i^{\text{true}}\times]^2 R_{z_i} [\mathbf{r}_i^{\text{true}}\times]^2$$

Note that R_i is independent of the attitude matrix. Also, R_i is a singular matrix, but the rank-one approach of [3] can be employed to overcome this issue. Define the following matrix:

$$\bar{R}_i = R_i + \frac{1}{2}\text{tr}(R_i)(\mathbf{r}_i^{\text{true}})(\mathbf{r}_i^{\text{true}})^T$$

The loss function to be minimized is given by

$$J = \frac{1}{2}\sum_{i-1}^{N}(\mathbf{r}_i - A^T\mathbf{b}_i)^T\bar{R}_i^{-1}(\mathbf{r}_i - A^T\mathbf{b}_i)$$

Derive the attitude error-covariance.

Note that to determine an optimal solution for the attitude a total least squares [8] solution must be employed since errors exist both in the body and reference vectors. However, the nonlinear least squares solution in Example 12.4 can provide near optimal solutions. The true body vectors are given by

$$\mathbf{b}_1^{\text{true}} = \begin{bmatrix} 0 \\ 0 \\ 1 \end{bmatrix}, \quad \mathbf{b}_2^{\text{true}} = \begin{bmatrix} 0 \\ 0.1 \\ \sqrt{1-0.1^2} \end{bmatrix}$$

Note that the angle between these vectors is 5.74°. The MRP vector will be estimated for the attitude, as shown in Example 12.4. The true MRP and associated attitude matrix are given by

$$\mathbf{p}^{\text{true}} = \begin{bmatrix} \dfrac{1}{1+\sqrt{2}} \\ 0 \\ 0 \end{bmatrix}, \quad A(\mathbf{p}^{\text{true}}) = \begin{bmatrix} 1 & 0 & 0 \\ 0 & 0 & 1 \\ 0 & -1 & 0 \end{bmatrix}$$

Assume that $\|\mathbf{z}_1^{\text{true}}\| = 30$ and $\|\mathbf{z}_2^{\text{true}}\| = 50$. To generate body measurements use the QMM in Eq. (5.107b) with $\sigma_{b_i} = 0.001$ deg for $i = 1, 2$. The covariances for R_{z_1} and R_{z_1} are equal, denoted by R_z, which is given by

$$R_z = \begin{bmatrix} 7\times 10^{-6} & 7\times 10^{-7} & 1\times 10^{-5} \\ 7\times 10^{-7} & 3\times 10^{-5} & -5\times 10^{-6} \\ 1\times 10^{-5} & -5\times 10^{-6} & 2\times 10^{-5} \end{bmatrix}$$

Fig. 5.4 Ellipse with rotation

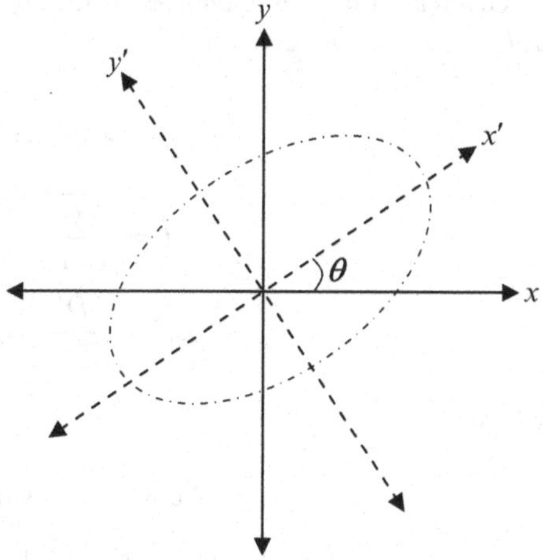

Using the approach outlined in Example 12.4 create synthetic measurements for both the body and reference vectors. Next, determine the estimated MRP using the estimate model $\hat{\mathbf{r}}_i = A^T(\hat{\mathbf{p}})\mathbf{b}_i$. Perform 1,000 Monte Carlo runs and show that the attitude errors are bounded from the 3σ bounds computed from the derived attitude error-covariance.

5.9. A problem that is closely related to the attitude determination problem involves determining ellipse parameters from measured data. Figure 5.4 depicts a general ellipse rotated by an angle θ. The basic equation of an ellipse is given by

$$\frac{(x'-x_0')^2}{a^2}+\frac{(y'-y_0')^2}{b^2}=1$$

where (x_0', y_0') denotes the origin of the ellipse and (a, b) are positive values. The coordinate transformation follows

$$x' = x\cos\theta + y\sin\theta$$

$$y' = -x\sin\theta + y\cos\theta$$

Show that the ellipse equation can be rewritten as

$$Ax^2 + Bxy + Cy^2 + Dx + Ey + F = 0$$

Many possibilities exist for this equation. Use the form that produces the following expression for F:

$$F = \frac{1}{2}\left[\frac{b}{a}x_0'^2 + \frac{a}{b}y_0'^2 - ab\right]$$

With your determined coefficients show that the following constraint is satisfied: $4AC - B^2 = 1$.

Given a set of coefficients A, B, C, D, E, and F, show that the formulas for θ, a, b, x_0', and y_0' are given by

$$\cot(2\theta) = \frac{A - C}{B}$$

$$a = \sqrt{\frac{Q'}{A'}}, \quad b = \sqrt{\frac{Q'}{C'}}$$

$$x_0' = -\frac{D'}{2A'}, \quad y_0' = -\frac{E'}{2C'}$$

where

$$\begin{aligned}
A' &= A\cos^2\theta + B\sin\theta\cos\theta + C\sin^2\theta \\
B' &= B(\cos^2\theta - \sin^2\theta) + 2(C - A)\sin\theta\cos\theta = 0 \\
C' &= A\sin^2\theta - B\sin\theta\cos\theta + C\cos^2\theta \\
D' &= D\cos\theta + E\sin\theta \\
E' &= -D\sin\theta + E\cos\theta \\
F' &= F \\
Q' &\equiv A'\left(\frac{D'}{2A'}\right)^2 + C'\left(\frac{E'}{2C'}\right)^2 - F'
\end{aligned}$$

(hint: show that the new variables follow the rotated ellipse equation: $A'x'^2 + B'x'y' + C'y'^2 + D'x' + E'y' + F' = 0$). Also, prove that $B' = 0$. Given the following quantities:

$$A = 3.0000, \; B = -6.9282, \; C = 7.0000$$

$$D = -5.5359, \; E = 17.5885, \; F = 4.0000$$

compute θ, x_0', y_0', a and b.

Suppose that a set of measurements for x and y exist, and we form the following vector of unknown parameters:

$$\mathbf{x}^{\text{true}} \equiv \begin{bmatrix} A^{\text{true}} & B^{\text{true}} & C^{\text{true}} & D^{\text{true}} & E^{\text{true}} & F^{\text{true}} \end{bmatrix}^T$$

Our goal is to determine an estimate of $\mathbf{x}^{\text{true}}$, denoted by $\mathbf{x}$, from this measured data set. Show that the minimum norm-squared loss function can be written as

$$J(\mathbf{x}) = \mathbf{x}^T H^T H \,\mathbf{x}$$

subject to

$$\mathbf{x}^T Z \,\mathbf{x} = 1$$

where the ith row of H is given by

$$H_i = \left[x_i^2 \; x_i \, y_i \; y_i^2 \; x_i \; y_i \; 1\right]$$

Determine the matrix Z that satisfies the constraint $4AC - B^2 = 1$. Find the form for the optimal solution for $\mathbf{x}$ using an eigenvalue/eigenvector approach. Note, a more robust approach involves using a reduced eigenvalue decomposition [16] or a singular value decomposition approach [14].

Write a computer program for your derived solution and perform a simulation to test your algorithm. Note that for a given set of x' values, y' can be determined using

$$y' = y_0' \pm b\sqrt{1 - \frac{(x' - x_0')^2}{a^2}}$$

where the $\pm$ term gives values for the upper and lower half of the ellipse. Use the computed θ, x_0', y_0', a and b from before and pick at least 1,200 samples of x' and y' values that span the entire ellipse. Then convert these to x and y using the inverse transformation

$$x = x' \cos\theta - y' \sin\theta$$
$$y = x' \sin\theta + y' \cos\theta$$

Using your formulated eigenvalue/eigenvector solution compute the optimal solution for $\mathbf{x}$. Next add noise to x and y using a zero-mean Gaussian white-noise process with standard deviation given by 0.01 and recompute your solution using measured values instead of the true ones.

5.10. Derive the expression given in Eq. (5.161). Also, derive the expression given in Eq. (5.167).

5.11. The loss function in Eq. (5.177) can be derived from a maximum likelihood approach by assuming that each a_{ij} is given by σ_{ij}^{-1}, which is the inverse of the variance of the assumed zero-mean Gaussian noise on z_{ij}. Using this loss function derive the attitude Fisher information matrix.

References

1. Andrews, S.F., Bilanow, S.: Recent flight results of the TRMM Kalman filter. In: AIAA Guidance, Navigation and Control Conference. Monterey, CA (2002). AIAA 2002-5047
2. Black, H.D.: A passive system for determining the attitude of a satellite. AIAA J. **2**(7), 1350–1351 (1964)
3. Cheng, Y., Crassidis, J.L., Markley, F.L.: Attitude estimation for large field-of-view sensors. J. Astronaut. Sci. **54**(3/4), 433–448 (2006)

4. Cheng, Y., Shuster, M.D.: The speed of attitude estimation. In: Akella, M.R., Gearhart, J.W., Bishop, R.H., Treder, A.J. (eds.) Proceedings of the AAS/AIAA Space Flight Mechanics Meeting 2007, Advances in the Astronautical Sciences, vol. 127, pp. 101–116. AAS/AIAA, Univelt, San Diego (2007)
5. Cheng, Y., Shuster, M.D.: Improvement to the implementation of the QUEST algorithm. J. Guid. Contr. Dynam. **37**(1), 301–305 (2014)
6. Cohen, C.E.: Attitude determination. In: Parkinson, B., Spilker, J. (eds.) Global Positioning System: Theory and Applications, Progress in Astronautics and Aeronautics, vol. 64, chap. 19. American Institute of Aeronautics and Astronautics, Washington, DC (1996)
7. Crassidis, J.L., Andrews, S.F., Markley, F.L., Ha, K.: Contingency designs for attitude determination of TRMM. In: Proceedings of the Flight Mechanics/Estimation Theory Symposium, pp. 419–433. NASA-Goddard Space Flight Center, Greenbelt (1995)
8. Crassidis, J.L., Junkins, J.L.: Optimal Estimation of Dynamic Systems, 2nd edn. CRC Press, Boca Raton (2012)
9. Crassidis, J.L., Markley, F.L.: New algorithm for attitude determination using Global Positioning System signals. J. Guid. Contr. Dynam. **20**(5), 891–896 (1997)
10. Crassidis, J.L., Markley, F.L., Lightsey, E.G.: Global Positioning System integer ambiguity resolution without attitude knowledge. J. Guid. Contr. Dynam. **22**(2), 212–218 (1999)
11. Farrell, J.L.: Attitude determination by Kalman filtering. Contractor Report NASA-CR-598, NASA Goddard Space Flight Center, Washington, DC (1964)
12. Farrell, J.L.: Attitude determination by Kalman filtering. Automatica **6**, 419–430 (1970)
13. Farrell, J.L., Stuelpnagel, J.C.: A least-squares estimate of satellite attitude. SIAM Rev. **7**(3), 384–386 (1966)
14. Gander, W., Golub, G.H., Strebel, R.: Least-squares fitting of circles and ellipses. In: editorial board Bulletin Belgian Mathematical Society (ed.) Numerical Analysis (in honour of Jean Meinguet), pp. 63–84 (1996)
15. Golub, G.H., Van Loan, C.F.: Matrix Computations, 3rd edn. The Johns Hopkins University Press, Baltimore (1996)
16. Halíř, R., Flusser, J.: Numerically stable direct least squares fitting of ellipses. In: 6th International Conference in Central Europe on Computer Graphics and Visualization, WSCG '98, pp. 125–132. University of West Bohemia, Campus Bory, Plzen - Bory (1998)
17. Horn, R.A., Johnson, C.R.: Matrix Analysis. Cambridge University Press, Cambridge (1985)
18. Lerner, G.M.: Three-axis attitude determination. In: Wertz, J.R. (ed.) Spacecraft Attitude Determination and Control, chap. 12. Kluwer Academic, Dordrecht (1978)
19. Markley, F.L.: Attitude determination using vector observations and the singular value decomposition. J. Astronaut. Sci. **36**(3), 245–258 (1988)
20. Markley, F.L.: Attitude determination using vector observations: A fast optimal matrix algorithm. J. Astronaut. Sci. **41**(2), 261–280 (1993)
21. Markley, F.L.: Fast quaternion attitude estimation from two vector measurements. J. Guid. Contr. Dynam. **25**(2), 411–414 (2002)
22. Markley, F.L.: Optimal attitude matrix from two vector measurements. J. Guid. Contr. Dynam. **31**(3), 765–768 (2008)
23. Markley, F.L., Cheng, Y., Crassidis, J.L., Oshman, Y.: Averaging quaternions. J. Guid. Contr. Dynam. **30**(4), 1193–1196 (2007)
24. Markley, F.L., Mortari, D.: Quaternion attitude estimation using vector observations. J. Astronaut. Sci. **48**(2/3), 359–380 (2000)
25. Mortari, D.: ESOQ: A closed-form solution of the Wahba problem. J. Astronaut. Sci. **45**(2), 195–204 (1997)
26. Mortari, D.: Second estimator of the optimal quaternion. J. Guid. Contr. Dynam. **23**(5), 885–888 (2000)
27. Potter, J.E., Vander Velde, W.E.: Optimum mixing of gyroscope and star tracker data. J. Spacecraft **5**(5), 536–540 (1968)
28. Psiaki, M.L., Mohiuddin, S.: Global Positioning System integer ambiguity resolution using factorized least-squares techniques. J. Guid. Contr. Dynam. **30**(2), 346–356 (2007)

29. Shuster, M.D.: Maximum likelihood estimation of spacecraft attitude. J. Astronaut. Sci. **37**(1), 79–88 (1989)
30. Shuster, M.D.: Deterministic three-axis attitude determination. J. Astronaut. Sci. **52**(3), 405–419 (2004)
31. Shuster, M.D.: The quest for better attitudes. J. Astronaut. Sci. **54**(3/4), 657–683 (2006)
32. Shuster, M.D.: The TRIAD algorithm as maximum likelihood estimation. J. Astronaut. Sci. **54**(1), 113–123 (2006)
33. Shuster, M.D.: The TASTE test. J. Astronaut. Sci. **57**(1/2), 61–71 (2009)
34. Shuster, M.D., Natanson, G.A.: Quaternion computation from a geometric point of view. J. Astronaut. Sci. **41**(4), 545–556 (1993)
35. Shuster, M.D., Oh, S.D.: Attitude determination from vector observations. J. Guid. Contr. **4**(1), 70–77 (1981)
36. Wahba, G.: A least-squares estimate of satellite attitude. SIAM Rev. **7**(3), 409 (1965)

Chapter 6
Filtering for Attitude Estimation and Calibration

Attitude estimation refers to the process of estimating the current attitude state of a system from a set of measured observations. The state estimation problem involves finding the best estimate of the true system state using a dynamic model and measurements that are both corrupted with random noise of known statistics. The variables to be estimated are usually collected into a state vector, which typically includes other variables in addition to the attitude. For example, star tracker measurements can be combined with a kinematics model, which is propagated using gyroscopic measurements. However, all gyros have inherent drift, or bias, which causes inaccuracies in the propagated model. A complementary filter is used to simultaneously estimate the attitude and gyro drift from the measurements. Filtering can generally provide a more accurate attitude estimate than static methods because it incorporates memory of past observations.

This chapter presents the basics of filtering for attitude estimation. It is assumed that the reader is familiar with the basic ideas of estimation/filtering theory as reviewed in Chap. 12. Reference [21] includes a survey of early attitude estimation approaches, and a survey of more modern attitude estimation approaches can be found in [12]. We begin with general discussions of attitude representations for Kalman filtering. This is followed by an overview of the equations for an extended Kalman filter (EKF) using the quaternion to represent the "global" attitude and a three-component representation of attitude errors. Alternative ways to use gyro data in attitude filters are then compared, followed by several applications. An EKF is developed for gyro calibration, which includes estimation of biases, scale factors, and misalignments, as well as the attitude. A "mission mode" EKF is then shown that estimates attitude and gyro biases, which is the estimator typically used for actual onboard applications. We then examine a very useful single-axis covariance analysis of the accuracy that attitude filters can be expected to provide. Finally, batch and realtime methods for three-axis magnetometer calibration are developed. The batch approaches involve a suboptimal linear solution and an optimal nonlinear least squares solution, and the realtime approach uses an EKF formulation.

F.L. Markley and J.L. Crassidis, *Fundamentals of Spacecraft Attitude Determination and Control*, Space Technology Library 33, DOI 10.1007/978-1-4939-0802-8_6,
© Springer Science+Business Media New York 2014

6.1 Attitude Representations for Kalman Filtering

Section 2.9 presented several different representations of spacecraft attitude, some of which are more suitable for filtering applications than others. The only representations that have seen widespread application are three-component representations and the four-component quaternion representation. We will briefly discuss the advantages and drawbacks of the different representations, after which we will concentrate on quaternion-based filters.

6.1.1 Three-Component Representations

Three-component representations are the most natural representations for filtering, because only three parameters are needed to represent rotations. As was pointed out in Sect. 2.9, though, all three-parameter representations of the rotation group have discontinuities or singularities. A filter using a three-dimensional attitude representation must provide some guarantee of avoiding these singular points.

The earliest Kalman filters for spacecraft attitude estimation used Euler angles, specifically a roll, pitch, yaw sequence of Tait-Bryan angles, which have a very intuitive meaning if they do not become too large [13, 14]. This is a very useful representation if the middle angle of the sequence, which is generally the pitch angle, stays well away from positive or negative rotations of 90°. Filters for pitch, roll, and yaw have been used mostly for Earth-pointing spacecraft, which typically have small pitch angles. One disadvantage of this representation is that it requires a fair number of trigonometric function evaluations, but this has become less of an issue with increasing computing power, especially in onboard computers.

The Gibbs vector or Rodrigues parameter representation has been used in a Kalman filter [18], but it is not well suited to filtering because of its inability to represent 180° rotations. It provides an excellent representation of small attitude errors, however, as we shall see.

The modified Rodrigues parameters (MRPs) are non-singular for rotations of less than 360°, and rotations greater than 180° can be represented by an MRP in the shadow set, as discussed in Sect. 2.9.5. This has advanced the use of MRPs in attitude estimation Kalman filters [10, 19]. The transformation to the shadow set is made at some angle greater than 180° to avoid "chattering" between the two MRP representations if the rotation angle dwells for an extended period in the vicinity of 180°. When transforming to the shadow set, it is necessary to transform the covariance matrix to the new parameters. This transformation is discussed in the literature.

The rotation vector is very similar to the MRPs in having a singularity for 360° rotations. When its magnitude $\|\boldsymbol{\vartheta}\|$ becomes greater than 180°, the representation can be switched to an equivalent rotation of magnitude $2\pi - \|\boldsymbol{\vartheta}\|$ in the opposite direction. The rotation vector has no clear advantage over the MRPs for filtering, though, and has the disadvantage of requiring the evaluation of trigonometric functions.

6.1.2 Additive Quaternion Representation

The quaternion has become the representation of choice for attitude estimation because it is the lowest-dimensional parameterization that is free from singularities. However, the quaternion must obey its normalization constraint, which can cause issues in the standard EKF.

The normalization problem is apparent in the Kalman filter update equation[1]

$$\hat{\mathbf{q}}^{+} = \hat{\mathbf{q}}^{-} + K[\mathbf{y} - \mathbf{h}(\hat{\mathbf{x}}^{-})] \tag{6.1}$$

where $\hat{\mathbf{x}}^{-}$ is the pre-update estimate of the state vector that includes the quaternion as four of its components. It is clear that $\hat{\mathbf{q}}^{+}$ and $\hat{\mathbf{q}}^{-}$ cannot both be normalized to unity unless there is some special relation between $\hat{\mathbf{q}}^{-}$ and $K[\mathbf{y} - \mathbf{h}(\hat{\mathbf{x}}^{-})]$. No such relation holds in the general case, so the filter update can spoil the normalization.

There is a deeper problem with quaternion normalization. An *unbiased estimator* has the property that the expectation of the estimated value is the true value, as discussed in Sect. 12.3.2, i.e.

$$E\{\hat{\mathbf{q}}\} = \mathbf{q}^{\text{true}} \tag{6.2}$$

We define the additive quaternion error as the algebraic difference between the true quaternion and its estimate

$$\Delta\mathbf{q} \equiv \mathbf{q}^{\text{true}} - \hat{\mathbf{q}} \tag{6.3}$$

so

$$\|\hat{\mathbf{q}}\|^2 = \|\mathbf{q}^{\text{true}} - \Delta\mathbf{q}\|^2 = \|\mathbf{q}^{\text{true}}\|^2 - 2\Delta\mathbf{q}^T\mathbf{q}^{\text{true}} + \|\Delta\mathbf{q}\|^2 \tag{6.4}$$

It follows trivially from Eq. (6.3) that the expectation of $\Delta\mathbf{q}$ is zero, as it is for every unbiased estimator, and the true quaternion is assumed to have unit norm, so the expectation of this equation is

$$E\{\|\hat{\mathbf{q}}\|^2\} = 1 + E\{\|\Delta\mathbf{q}\|^2\} = 1 + \text{tr}(P_{qq}) \tag{6.5}$$

where $P_{qq} \equiv E\{\Delta\mathbf{q}\Delta\mathbf{q}^T\}$ is the quaternion covariance. Thus we see that an unbiased estimate of the four-component quaternion must violate the unit norm constraint.

Suppose that we give up on the requirement that the estimator be unbiased and develop a biased estimator obeying the unit norm constraint, i.e. an estimator giving $\|\mathbf{q}^{\text{true}}\| = \|\hat{\mathbf{q}}\| = 1$. We will show that this leads to a different problem. For such an estimator Eq. (6.4) would give

[1] See Sect. 12.3.7.1 for a review of the Kalman filter.

$$2\Delta\mathbf{q}^T\mathbf{q}^{\text{true}} = \|\Delta\mathbf{q}\|^2 \tag{6.6}$$

Because the estimator is biased, the covariance is

$$P_{qq} = E\{\Delta\mathbf{q}\Delta\mathbf{q}^T\} - E\{\Delta\mathbf{q}\}E\{\Delta\mathbf{q}^T\} \tag{6.7}$$

Then

$$\begin{aligned}(\mathbf{q}^{\text{true}})^T P_{qq}\,\mathbf{q}^{\text{true}} &= E\{(\mathbf{q}^{\text{true}})^T\Delta\mathbf{q}\,\Delta\mathbf{q}^T\mathbf{q}^{\text{true}}\} - E\{(\mathbf{q}^{\text{true}})^T\Delta\mathbf{q}\}E\{\Delta\mathbf{q}^T\mathbf{q}^{\text{true}}\}\\ &= \frac{1}{4}\left[E\{\|\Delta\mathbf{q}\|^4\} - \left(E\{\|\Delta\mathbf{q}\|^2\}\right)^2\right] = \frac{1}{4}\left[E\{\|\Delta\mathbf{q}\|^4\} - \text{tr}(P_{qq})^2\right]\end{aligned} \tag{6.8}$$

This becomes very small as the attitude errors become small, indicating that the covariance matrix is ill-conditioned in that limit.[2] The ill-conditioning has been observed in toy models [24], numerical studies [5], and analysis [31]. It is worth noting that if the norm constraint had been linear, the quaternion covariance would have been truly singular.

At least four methods have been proposed to deal with these problems within the framework of additive quaternion filtering, but none is completely satisfactory.

The first method is to renormalize the estimate by brute force

$$\hat{\mathbf{q}} \equiv E\{\mathbf{q}^{\text{true}}\}/\|E\{\mathbf{q}^{\text{true}}\}\| \tag{6.9}$$

If the update $K[\mathbf{y} - \mathbf{h}(\hat{\mathbf{x}}^-)]$ is orthogonal to $\hat{\mathbf{q}}^-$, then Eq. (6.1) shows that the renormalization makes only a second-order correction in the quaternion errors. In this case brute force normalization is consistent with the EKF, which is a first-order estimator. If the update is not orthogonal to $\hat{\mathbf{q}}^-$, it is difficult to justify brute force normalization.

The second method is to modify the Kalman filter update equations to enforce the norm constraint by means of a Lagrange multiplier [42]. These two approaches both yield biased estimates of the quaternion.

The third method is to give up on enforcing the quaternion norm, and to define the attitude matrix as

$$A(\mathbf{q}) = \|\mathbf{q}\|^{-2}\left\{\left(q_4^2 - \|\mathbf{q}_{1:3}\|^2\right)I_3 + 2\mathbf{q}_{1:3}\,\mathbf{q}_{1:3}^T - 2q_4[\mathbf{q}_{1:3}\times]\right\} \tag{6.10}$$

This definition is guaranteed to provide an orthogonal attitude matrix. We call this the ray representation, because any quaternion along a ray through the origin corresponds to the same attitude matrix. This approach avoids the complications arising from enforcing the norm constraint, but introduces an unobservable degree of freedom, the quaternion norm. The result is that one eigenvalue of P_{qq} is unaffected

[2] But not singular, as was claimed in [21].

by measurements and thus retains its initial value, which can cause loss of numerical precision if the attitude estimates become very accurate. This method has been used successfully, however [32, 33].

The fourth method is to give up on enforcing the quaternion norm, but to continue to define the attitude matrix by Eq. (2.125), which is Eq. (6.10) without the factor of $\|\mathbf{q}\|^{-2}$. The attitude matrix is not orthogonal in this case. This is a significant drawback of the method, but the lack of orthogonality gives rise to measurement residuals resulting in Kalman filter updates that tend to drive the quaternion norm to unity. This method can lead to an ill-conditioned covariance matrix [24].

6.1.3 Multiplicative Quaternion Representation

The basic idea of the multiplicative EKF (MEKF) is to use the quaternion as the "global" attitude representation and use a three-component state vector $\boldsymbol{\delta\vartheta}$ for the "local" representation of attitude errors. Instead of writing the true quaternion as the sum of the estimated quaternion and an error quaternion, we write it as the product of an error quaternion and the estimate

$$\mathbf{q}^{\text{true}} = \boldsymbol{\delta}\mathbf{q}(\boldsymbol{\delta\vartheta}) \otimes \hat{\mathbf{q}} \tag{6.11}$$

Note that $\mathbf{q}^{\text{true}}$, $\boldsymbol{\delta}\mathbf{q}$, and $\hat{\mathbf{q}}$ are all properly normalized unit quaternions. This idea was first applied to the Space Precision Attitude Reference System (SPARS) in 1969 [29, 39], was later developed for NASA's Multimission Modular Spacecraft (MMS) [27], and has been widely applied to a great number of space missions. The appearance of $\boldsymbol{\delta}\mathbf{q}$ on the left of $\hat{\mathbf{q}}$ in Eq. (6.11) means that the attitude error vector is defined in the body reference frame. This is the most widely used form of the MEKF, but some authors write $\mathbf{q}^{\text{true}} = \hat{\mathbf{q}} \otimes \boldsymbol{\delta}\mathbf{q}'$ in place of Eq. (6.11), defining the attitude error vector to be in the reference frame [16].

We have written the error vector as $\boldsymbol{\delta\vartheta}$, but any of the error representations discussed in Sect. 2.10 can be used: the rotation vector, two times the vector part of the quaternion, two times the vector of Rodrigues parameters, four times the vector of MRPs, or the vector of roll, pitch, and yaw angles. The MEKF updates the error state

$$\Delta\mathbf{x} \equiv \begin{bmatrix} \boldsymbol{\delta\vartheta} \\ \Delta\boldsymbol{\xi} \end{bmatrix} \tag{6.12}$$

where $\boldsymbol{\xi}^{\text{true}} = \hat{\boldsymbol{\xi}} + \Delta\boldsymbol{\xi}$ is a vector of other variables to be estimated. Thus we have a conventional EKF that computes an unconstrained and unbiased estimate of $\Delta\mathbf{x}$. The correctly normalized four-component $\hat{\mathbf{q}}$ is not actually part of the EKF, but a reset operation moves the updates into this global variable to keep the error quaternion small and thus far away from any singularities [21, 23]. It is important to note that the estimate $\hat{\mathbf{q}}$ is not defined as an expectation, which is how the MEKF avoids the pitfalls of Sect. 6.1.2.[3]

If the vector $\boldsymbol{\xi}$ has n components, the MEKF covariance is a well-conditioned $(n+3)\times(n+3)$ matrix, while the covariance matrix of a filter estimating the full quaternion would be $(n+4)\times(n+4)$. The conceptual advantage of this dimensional reduction, as more truly representing the actual degrees of freedom of the system, has been debated at length [5, 23, 24, 31, 36, 37], but the computational advantages are indisputable. Another great advantage of the MEKF is that the covariance of the attitude error angles has a transparent physical interpretation. The covariance of estimators using other attitude representations has a less obvious interpretation unless the attitude matrix is close to the identity matrix.

Equation (6.11) is equivalent to

$$A^{\text{true}} = A(\mathbf{q}^{\text{true}}) = A(\boldsymbol{\delta\vartheta})A(\hat{\mathbf{q}}) = A(\boldsymbol{\delta\vartheta})\hat{A} \tag{6.13}$$

This shows that the attitude matrix could also be used as the global representation, but the quaternion has two advantages. The first is compactness; it only has four components rather than nine. The second is that if computational errors cause the quaternion norm to deviate from unity, it is easy to normalize the quaternion by dividing all its components by its norm. It is not so straightforward to restore the orthogonality constraint on the attitude matrix if computational errors cause the constraint to be violated.

6.2 Attitude Estimation

6.2.1 Kalman Filter Formulation

Our attitude filters all use some variant of the MEKF, which proceeds by iteration of three steps: measurement update, state vector reset, and propagation to the next measurement time. The measurement update step updates the error state vector. The reset moves the updated information from the error state to the global attitude representation and resets the components of the error state to zero. The propagation step propagates the global variables to the time of the next representation. The error state variables do not need to be propagated because they are identically zero over the propagation step. We will now discuss these three steps in more detail.

[3]Reference [23] writes $\mathbf{q}^{\text{true}} = \delta\mathbf{q}(\boldsymbol{\delta\vartheta}) \otimes \mathbf{q}_{\text{ref}}$, where $\mathbf{q}_{\text{ref}}$ is a "reference" quaternion, and derives the propagation and update equations to ensure that $E\{\boldsymbol{\delta\vartheta}\}$ is only nonzero between the measurement update and reset. It turns out that $\mathbf{q}_{\text{ref}}$ obeys the same equations as our $\hat{\mathbf{q}}$ and is the best attitude estimate, so the treatment here is mathematically equivalent, although less rigorous.

6.2.1.1 Measurement Update

The observation model is given by Eq. (12.103), which for attitude estimation is

$$\mathbf{y} = \mathbf{h}(\mathbf{q}^{\text{true}}, \boldsymbol{\xi}^{\text{true}}) + \mathbf{v} \tag{6.14}$$

where $\mathbf{v}$ is a vector of Gaussian errors with covariance matrix R. The MEKF updates the error state of Eq. (6.12), not the global state representation $\hat{\mathbf{q}}$ and $\hat{\boldsymbol{\xi}}$, so the measurement sensitivity matrix is

$$H(\mathbf{q}, \boldsymbol{\xi}) = \frac{\partial \mathbf{h}}{\partial(\Delta \mathbf{x})} = \begin{bmatrix} \dfrac{\partial \mathbf{h}}{\partial(\boldsymbol{\delta\vartheta})} & \dfrac{\partial \mathbf{h}}{\partial \boldsymbol{\xi}} \end{bmatrix} \equiv \begin{bmatrix} H_\vartheta & H_\xi \end{bmatrix} \tag{6.15}$$

The alternative representations for $\boldsymbol{\delta}\mathbf{q}(\boldsymbol{\delta\vartheta})$ are all equivalent to first order in $\boldsymbol{\delta\vartheta}$, as was shown in Sect. 2.10, so we can express the error quaternion to first order in the error vector as

$$\boldsymbol{\delta}\mathbf{q} \approx \begin{bmatrix} \boldsymbol{\delta\vartheta}/2 \\ 1 \end{bmatrix} = \mathbf{I}_q + \frac{1}{2}\begin{bmatrix} \boldsymbol{\delta\vartheta} \\ 0 \end{bmatrix} \tag{6.16}$$

Then Eqs. (6.11) and (2.98) give

$$\mathbf{q}^{\text{true}} \approx \left(\mathbf{I}_q + \frac{1}{2}\begin{bmatrix} \boldsymbol{\delta\vartheta} \\ 0 \end{bmatrix}\right) \otimes \hat{\mathbf{q}} = \hat{\mathbf{q}} + \frac{1}{2}\boldsymbol{\delta\vartheta} \otimes \hat{\mathbf{q}} = \hat{\mathbf{q}} + \frac{1}{2}\Xi(\hat{\mathbf{q}})\boldsymbol{\delta\vartheta} \tag{6.17}$$

It follows that the attitude part of the measurement sensitivity matrix can be evaluated by using the chain rule

$$H_\vartheta = \frac{\partial \mathbf{h}}{\partial \mathbf{q}} \frac{\partial \mathbf{q}}{\partial(\boldsymbol{\delta\vartheta})} = \frac{1}{2}\frac{\partial \mathbf{h}}{\partial \mathbf{q}} \Xi(\hat{\mathbf{q}}) \tag{6.18}$$

An even more convenient form of the measurement sensitivity matrix for the common case of vector measurements will be derived in Sect. 6.2.2.

Immediately after a reset or propagation, the quantities $\boldsymbol{\delta}\hat{\boldsymbol{\vartheta}}$ and $\Delta\hat{\boldsymbol{\xi}}$ are zero. If several measurements are processed at one time without an intervening reset, though, these quantities may have finite values. In order to avoid recalculating the nonlinear function $\mathbf{h}(\hat{\mathbf{q}}, \hat{\boldsymbol{\xi}})$, we use the first-order Taylor series to compute the expectation

$$E\{\mathbf{h}(\mathbf{q}^{\text{true}}, \boldsymbol{\xi}^{\text{true}})\} \approx \mathbf{h}(\hat{\mathbf{q}}, \hat{\boldsymbol{\xi}}) + H(\hat{\mathbf{q}}, \hat{\boldsymbol{\xi}}) \begin{bmatrix} \boldsymbol{\delta}\hat{\boldsymbol{\vartheta}} \\ \Delta\hat{\boldsymbol{\xi}} \end{bmatrix} \tag{6.19}$$

The state update for the kth measurement is then

$$\begin{bmatrix} \boldsymbol{\delta}\hat{\boldsymbol{\vartheta}}_k^+ \\ \Delta\hat{\boldsymbol{\xi}}_k^+ \end{bmatrix} = \begin{bmatrix} \boldsymbol{\delta}\hat{\boldsymbol{\vartheta}}_k^- \\ \Delta\hat{\boldsymbol{\xi}}_k^- \end{bmatrix} + K_k \left\{ \mathbf{y}_k - \mathbf{h}_k(\hat{\mathbf{q}}_k^-, \hat{\boldsymbol{\xi}}_k^-) - H_k(\hat{\mathbf{q}}_k^-, \hat{\boldsymbol{\xi}}_k^-) \begin{bmatrix} \boldsymbol{\delta}\hat{\boldsymbol{\vartheta}}_k^- \\ \Delta\hat{\boldsymbol{\xi}}_k^- \end{bmatrix} \right\} \tag{6.20}$$

The Kalman gain computation and covariance update have the standard Kalman filter forms.

Many modern-day devices, such as star trackers, provide quaternion-out capabilities along with associated error covariances. The MEKF measurement model for incorporating these quaternion "measurements" can be derived from

$$\mathbf{q}_k \otimes \hat{\mathbf{q}}_k^{-1} = \boldsymbol{\delta}\mathbf{q}(\boldsymbol{\delta\vartheta}_k) \tag{6.21}$$

where $\mathbf{q}_k$ is the quaternion measurement and we understand $\hat{\mathbf{q}}_k$ to be $\hat{\mathbf{q}}_k^-$. Because it is important that the measurement update should agree closely with the measurement residual computation, the measurement model should use the same parameterization for $\boldsymbol{\delta}\mathbf{q}(\boldsymbol{\delta\vartheta}_k)$ as is used in the reset, rather than a first-order approximation. The Rodrigues vector parameterization has the advantage that the observation model is insensitive to the sign ambiguity in the tracker output quaternion $\mathbf{q}_k$. Using this parameterization, the measurement used in the filter and its predicted value are given by Eqs. (2.136), (2.184), (6.19), and (6.21) as

$$\mathbf{y}_k = \boldsymbol{\delta\vartheta}_k = 2\,\frac{(\mathbf{q}_k \otimes \hat{\mathbf{q}}_k^{-1})_{1:3}}{(\mathbf{q}_k \otimes \hat{\mathbf{q}}_k^{-1})_4} \tag{6.22a}$$

$$\mathbf{h}_k(\hat{\mathbf{q}}_k^-, \hat{\boldsymbol{\xi}}_k^-) + H_k(\hat{\mathbf{q}}_k^-, \hat{\boldsymbol{\xi}}_k^-)\begin{bmatrix}\boldsymbol{\delta}\hat{\boldsymbol{\vartheta}}_k^- \\ \Delta\hat{\boldsymbol{\xi}}_k^-\end{bmatrix} = 2\,\frac{(\hat{\mathbf{q}}_k \otimes \hat{\mathbf{q}}_k^{-1})_{1:3}}{(\hat{\mathbf{q}}_k \otimes \hat{\mathbf{q}}_k^{-1})_4} + [I_3 \;\; 0_{3\times n}]\begin{bmatrix}\boldsymbol{\delta}\hat{\boldsymbol{\vartheta}}_k^- \\ \Delta\hat{\boldsymbol{\xi}}_k^-\end{bmatrix} = \boldsymbol{\delta}\hat{\boldsymbol{\vartheta}}_k^- \tag{6.22b}$$

The measurement sensitivity matrix is simply given by $H_k = [I_3 \;\; 0_{3\times n}]$, and the measurement covariance matrix, R_k, is a 3×3 matrix of attitude measurement error angles.

Using a quaternion-out approach simplifies the computations in the EKF, but issues such as estimator consistency need to be considered when using the EKF to fuse multiple measurement sets. See [7] for more details.

6.2.1.2 Reset

The discrete measurement update assigns finite post-update values to $\boldsymbol{\delta}\hat{\boldsymbol{\vartheta}}^+$ and $\Delta\hat{\boldsymbol{\xi}}^+$, but the components of the global state still retain the values $\hat{\mathbf{q}}^-$ and $\hat{\boldsymbol{\xi}}^-$. A reset procedure is used to move the update information to a post-update estimate global state vector $\hat{\mathbf{q}}^+$ and $\hat{\boldsymbol{\xi}}^+$, while simultaneously resetting $\boldsymbol{\delta}\hat{\boldsymbol{\vartheta}}$ and $\Delta\hat{\boldsymbol{\xi}}$ to zero. The reset does not change the overall estimate, so the reset must obey

$$\hat{\mathbf{q}}^+ = \boldsymbol{\delta}\mathbf{q}(\mathbf{0}_3) \otimes \hat{\mathbf{q}}^+ = \boldsymbol{\delta}\mathbf{q}(\boldsymbol{\delta}\hat{\boldsymbol{\vartheta}}^+) \otimes \hat{\mathbf{q}}^- \tag{6.23a}$$

$$\hat{\boldsymbol{\xi}}^+ = \hat{\boldsymbol{\xi}}^+ + \mathbf{0}_n = \hat{\boldsymbol{\xi}}^- + \Delta\hat{\boldsymbol{\xi}}^+ \tag{6.23b}$$

After the update, $\boldsymbol{\delta\vartheta}_k^-$ and $\Delta\boldsymbol{\xi}^-$ are reset to zero. Thus the reset moves information from one part of the estimate to another part. This reset rotates the reference frame for the attitude covariance, so that we might expect the covariance to be rotated, even though no new information is added. However, the covariance depends on the assumed statistics of the measurements, not on the actual measurements. Therefore, because the update is zero mean, the mean rotation caused by the reset is actually zero, and so the covariance is in fact not affected by the reset.[4]

If a reset is done after each measurement update, Eq. (6.20) simplifies to

$$\begin{bmatrix} \boldsymbol{\delta}\hat{\boldsymbol{\vartheta}}_k^+ \\ \Delta\hat{\boldsymbol{\xi}}_k^+ \end{bmatrix} = K_k \left[\mathbf{y}_k - \mathbf{h}_k(\hat{\mathbf{q}}_k^-, \hat{\boldsymbol{\xi}}_k^-) \right] \equiv \begin{bmatrix} K_{\vartheta k} \\ K_{\xi k} \end{bmatrix} \left[\mathbf{y}_k - \mathbf{h}_k(\hat{\mathbf{q}}_k^-, \hat{\boldsymbol{\xi}}_k^-) \right] \tag{6.24}$$

In this case, the reset of the non-attitude state can be done implicitly as part of the measurement update, giving

$$\hat{\boldsymbol{\xi}}_k^+ = \hat{\boldsymbol{\xi}}_k^- + K_{\xi k} \left[\mathbf{y}_k - \mathbf{h}_k(\hat{\mathbf{q}}_k^-, \hat{\boldsymbol{\xi}}_k^-) \right] \tag{6.25}$$

Then only the quaternion has to be reset explicitly. The reset is often delayed for computational efficiency until all the updates for a set of simultaneous measurements have been performed, though, in which case all the terms in Eq. (6.20) must be included. It is imperative to perform a reset either implicitly or explicitly before beginning the time propagation, however, to avoid the necessity of propagating $\boldsymbol{\delta\vartheta}_k^-$ and $\Delta\boldsymbol{\xi}^-$ between measurements.

The quaternion reset in Eq. (6.23a) is the special feature of the MEKF. This reset has to preserve the quaternion norm, so an exact unit-norm expression for the functional dependence of $\boldsymbol{\delta}\mathbf{q}$ on $\boldsymbol{\delta\vartheta}$ must be used, not the linear approximation of Eq. (6.16). Using the Rodrigues parameter vector has the practical advantage that the reset operation for this parameterization is

$$\hat{\mathbf{q}}^+ = \boldsymbol{\delta}\mathbf{q}(\boldsymbol{\delta}\hat{\boldsymbol{\vartheta}}^+) \otimes \hat{\mathbf{q}}^- = \frac{1}{\sqrt{1 + \|\boldsymbol{\delta}\hat{\boldsymbol{\vartheta}}^+/2\|^2}} \begin{bmatrix} \boldsymbol{\delta}\hat{\boldsymbol{\vartheta}}^+/2 \\ 1 \end{bmatrix} \otimes \hat{\mathbf{q}}^- \tag{6.26}$$

Using an argument similar to Eq. (6.17), this can be accomplished in two steps:

$$\mathbf{q}^* = \begin{bmatrix} \boldsymbol{\delta}\hat{\boldsymbol{\vartheta}}^+/2 \\ 1 \end{bmatrix} \otimes \hat{\mathbf{q}}^- = \hat{\mathbf{q}}^- + \frac{1}{2}\Xi(\hat{\mathbf{q}}^-)\boldsymbol{\delta}\hat{\boldsymbol{\vartheta}}^+ \tag{6.27}$$

followed by

$$\hat{\mathbf{q}}^+ = \frac{\mathbf{q}^*}{\|\mathbf{q}^*\|} \tag{6.28}$$

[4]Not everyone agrees with this statement; see [25] and [34].

The first step is just the usual linear Kalman update, and the second step is mathematically equivalent to a brute force normalization. Thus the MEKF using Rodrigues parameters for the error vector provides a theoretical justification for the brute force update. It has the additional advantage of completely avoiding the possibility of accumulated errors in the quaternion norm after many updates.

The Rodrigues parameters also have the conceptual advantage that they map the rotation group into three-dimensional Euclidean space, with the largest possible 180° attitude errors mapped to points at infinity. Thus probability distributions with infinitely long tails, such as Gaussian distributions, make sense in Rodrigues parameter space.

6.2.1.3 Propagation

An EKF must propagate the expectation and covariance of the state. The MEKF is unusual in propagating the expectations $\hat{\mathbf{q}}$ and $\hat{\boldsymbol{\xi}}$ and the covariance of the error-state vector in Eq. (6.12). We now derive the expressions needed for propagation, beginning by differentiating Eq. (6.11):

$$\dot{\mathbf{q}}^{\text{true}} = \delta\dot{\mathbf{q}} \otimes \hat{\mathbf{q}} + \delta\mathbf{q} \otimes \dot{\hat{\mathbf{q}}} \tag{6.29}$$

The true and estimated quaternions satisfy the kinematic equations

$$\dot{\mathbf{q}}^{\text{true}} = \frac{1}{2}\begin{bmatrix} \boldsymbol{\omega}^{\text{true}} \\ 0 \end{bmatrix} \otimes \mathbf{q}^{\text{true}} \tag{6.30a}$$

$$\dot{\hat{\mathbf{q}}} = \frac{1}{2}\begin{bmatrix} \hat{\boldsymbol{\omega}} \\ 0 \end{bmatrix} \otimes \hat{\mathbf{q}} \tag{6.30b}$$

where $\boldsymbol{\omega}^{\text{true}}$ and $\hat{\boldsymbol{\omega}}$ are the true and estimated angular rates, respectively. Substituting these equations and Eq. (6.11) into Eq. (6.29) gives

$$\frac{1}{2}\begin{bmatrix} \boldsymbol{\omega}^{\text{true}} \\ 0 \end{bmatrix} \otimes \delta\mathbf{q} \otimes \hat{\mathbf{q}} = \delta\dot{\mathbf{q}} \otimes \hat{\mathbf{q}} + \frac{1}{2}\delta\mathbf{q} \otimes \begin{bmatrix} \hat{\boldsymbol{\omega}} \\ 0 \end{bmatrix} \otimes \hat{\mathbf{q}} \tag{6.31}$$

Multiplying on the right by $\hat{\mathbf{q}}^{-1}$ and rearranging terms gives [21]

$$\delta\dot{\mathbf{q}} = \frac{1}{2}\left(\begin{bmatrix} \boldsymbol{\omega}^{\text{true}} \\ 0 \end{bmatrix} \otimes \delta\mathbf{q} - \delta\mathbf{q} \otimes \begin{bmatrix} \hat{\boldsymbol{\omega}} \\ 0 \end{bmatrix}\right) \tag{6.32}$$

Substituting $\boldsymbol{\omega}^{\text{true}} = \hat{\boldsymbol{\omega}} + \delta\boldsymbol{\omega}$ into Eq. (6.32), where $\delta\boldsymbol{\omega}$ is the error angular velocity, leads to

$$\delta\dot{\mathbf{q}} = \frac{1}{2}\left(\begin{bmatrix}\hat{\omega}\\0\end{bmatrix} \otimes \delta\mathbf{q} - \delta\mathbf{q} \otimes \begin{bmatrix}\hat{\omega}\\0\end{bmatrix}\right) + \frac{1}{2}\begin{bmatrix}\delta\omega\\0\end{bmatrix} \otimes \delta\mathbf{q}$$

$$= -\begin{bmatrix}\hat{\omega} \times \delta\mathbf{q}_{1:3}\\0\end{bmatrix} + \frac{1}{2}\begin{bmatrix}\delta\omega\\0\end{bmatrix} \otimes (\delta\mathbf{q} - \mathbf{I}_q + \mathbf{I}_q) \tag{6.33}$$

Note that this is an exact kinematic relationship since no linearizations have been performed yet. The only nonlinearity in the errors appears in the last term on the right hand side. Both $\boldsymbol{\delta\omega}$ and $\delta\mathbf{q} - \mathbf{I}_q$ are small, though, so we can ignore their product in the spirit of the linearized EKF, resulting in

$$\delta\dot{\mathbf{q}} = -\begin{bmatrix}\hat{\omega} \times \delta\mathbf{q}_{1:3}\\0\end{bmatrix} + \frac{1}{2}\begin{bmatrix}\delta\omega\\0\end{bmatrix} \tag{6.34}$$

The first three components of this, after substituting Eq. (6.16), are

$$\delta\dot{\boldsymbol{\vartheta}} = -\hat{\omega} \times \delta\boldsymbol{\vartheta} + \delta\boldsymbol{\omega} \tag{6.35}$$

and the fourth component is $\delta\dot{q}_4 = 0$. Equation (6.35), which is just Eq. (3.48) in modified notation, is the equation needed to propagate the covariance of the attitude error-angle covariance. Detailed models will be provided during the presentations of specific estimators.

The expectation of Eq. (6.35) is

$$\delta\dot{\hat{\boldsymbol{\vartheta}}} = -\hat{\omega} \times \delta\hat{\boldsymbol{\vartheta}} \tag{6.36}$$

because $\boldsymbol{\delta\omega}$ has zero expectation. This says that if $\delta\hat{\boldsymbol{\vartheta}}$ is zero at the beginning of a propagation it will remain zero through the propagation, which is equivalent to saying that $\delta\hat{\mathbf{q}}$ will be equal to the identity quaternion throughout the propagation.

6.2.1.4 Gyros for Dynamic Model Replacement

Precise angular rate information is required for precise pointing and attitude maneuvers, and also for filtering noisy attitude sensor data. In principle, this could be provided by a dynamic model based on the equations in Chap. 3. Many spacecraft, including virtually all spacecraft with stringent pointing and/or maneuvering requirements, are provided with accurate gyros, which are the most crucial of all the attitude sensors. The usual Kalman filter update equations can be employed to include gyro data in an EKF as measurements.

A great number of attitude filters incorporate gyro information as part of the dynamic model rather than using than using the gyro information as a Kalman measurement update. This alternative is often referred to as using gyros in the dynamic model replacement mode. The reasons for favoring this method are twofold. Firstly, gyro information may well be much more accurate than the available models of rotational dynamics and torques, and inaccurate dynamic models could actually corrupt the gyro data. The second reason for using gyros in dynamic replacement mode, which is particularly important for onboard filtering, is that it requires much less computation. A filter including actual rotational dynamic models and treating the gyro data as measurements should be more accurate in theory, but it can be inferior in practice.

The basic idea of dynamic model replacement is to use Eq. (4.31a) or (4.34) to get the true angular velocity $\boldsymbol{\omega}^{\text{true}}$ in terms of the gyro-sensed rate $\boldsymbol{\omega}$. This true angular rate or its expectation is then substituted directly into Eqs. (6.30) and (6.36). The details of this method will be presented in the following discussions of specific filters.

6.2.2 Gyro Calibration Kalman Smoother

Gyros must be calibrated to provide accurate rate measurements. Both sequential filtering [30] and batch [28] approaches can be used for this calibration. The traditional attitude on-orbit calibration approach uses the 6-state EKF discussed in the following section to estimate the current attitude and gyro biases simultaneously [21]. This filter assumes unvarying alignment of the sensors involved in attitude estimation. However, sensor misalignment is inevitable and would contribute to unreliable attitude estimates [38]. More stringent attitude pointing accuracy requires misalignments to be estimated and incorporated into the attitude estimator. The importance of proper calibration for use in fault detection or rate derivation has been noted in several papers [3, 4, 41].

Although several methods exist for gyro calibration, we present a batch process based on the smoother shown in Sect. 12.3.7.3. A sequential EKF is first executed forwards in time and then smoothed estimates are provided by employing Eq. (12.139) backwards in time with final conditions given by Eq. (12.138). A calibration model for a 3-axis gyro often comprises a set of 3 biases and a 3×3 scale factor/misalignment matrix containing 3 scale factors and 6 misalignments, for a total of 12 calibration parameters [30]. The term "bias" implies a constant offset in the measurements, but in reality this offset actually drifts and is typically modeled using a random walk process. However, in most cases, this drift occurs slowly over many orbits. The other calibration parameters can vary as well but typically not as much as the gyro drift. Here a 15-state model is derived to estimate the attitude as well as all gyro calibration parameters. Unlike magnetometers, all gyro calibration procedures require an external attitude or independent rate sensor. Here it is assumed that star tracker measurements are available. Once the calibration parameters are determined using the smoother, then a realtime 6-state filter can be employed for the spacecraft mission mode. Periodic calibration of the gyro is typically done throughout the spacecraft lifetime, at time intervals dictated by the required accuracy in the attitude estimates or rate estimates for other purposes, such as jitter control.

The 16-component global state vector is made up of the true quaternion, $\mathbf{q}^{\text{true}}$, gyro drift biases, $\boldsymbol{\beta}^{\text{true}}$, scale factors, $\mathbf{s}^{\text{true}}$, and misalignments, $\mathbf{k}_U^{\text{true}}$ and $\mathbf{k}_L^{\text{true}}$. The 15-component state error vector is

$$\Delta\mathbf{x}(t) \equiv \left[\delta\boldsymbol{\vartheta}^T(t)\ \Delta\boldsymbol{\beta}^T(t)\ \Delta\mathbf{s}^T(t)\ \Delta\mathbf{k}_U^T(t)\ \Delta\mathbf{k}_L^T(t)\right]^T \tag{6.37}$$

where $\delta\boldsymbol{\vartheta}$ is the local vector of small attitude errors and the other components are the errors in the gyro biases, scale factors, and misalignments defined as the difference between their true and estimated values by $\Delta\boldsymbol{\beta} \equiv \boldsymbol{\beta}^{\text{true}} - \hat{\boldsymbol{\beta}}$, $\Delta\mathbf{s} \equiv \mathbf{s}^{\text{true}} - \hat{\mathbf{s}}$, $\Delta\mathbf{k}_U \equiv \mathbf{k}_U^{\text{true}} - \hat{\mathbf{k}}_U$, and $\Delta\mathbf{k}_L \equiv \mathbf{k}_L^{\text{true}} - \hat{\mathbf{k}}_L$.

The gyro model is given by Eqs. (4.34) and (4.31b) from Sect. 4.7.1:

$$\boldsymbol{\omega} = (I_3 + S^{\text{true}})\boldsymbol{\omega}^{\text{true}} + \boldsymbol{\beta}^{\text{true}} + \boldsymbol{\eta}_v \tag{6.38a}$$

$$\dot{\boldsymbol{\beta}}^{\text{true}} = \boldsymbol{\eta}_u \tag{6.38b}$$

where the spectral densities of $\boldsymbol{\eta}_v$ and $\boldsymbol{\eta}_u$ are $\sigma_v^2 I_3$ and $\sigma_v^2 I_3$, respectively. The matrix S^{true} is written as

$$S^{\text{true}} \equiv \begin{bmatrix} s_1^{\text{true}} & k_{U1}^{\text{true}} & k_{U2}^{\text{true}} \\ k_{L1}^{\text{true}} & s_2^{\text{true}} & k_{U3}^{\text{true}} \\ k_{L2}^{\text{true}} & k_{L3}^{\text{true}} & s_3^{\text{true}} \end{bmatrix} \tag{6.39}$$

with true vectors given by $\mathbf{s}^{\text{true}} \equiv [s_1^{\text{true}}\ s_2^{\text{true}}\ s_3^{\text{true}}]^T$, $\mathbf{k}_U^{\text{true}} \equiv [k_{U1}^{\text{true}}\ k_{U2}^{\text{true}}\ k_{U3}^{\text{true}}]^T$, and $\mathbf{k}_L^{\text{true}} \equiv [k_{L1}^{\text{true}}\ k_{L2}^{\text{true}}\ k_{L3}^{\text{true}}]^T$. The dynamics of $\mathbf{s}^{\text{true}}$, $\mathbf{k}_U^{\text{true}}$, and $\mathbf{k}_L^{\text{true}}$ are given by

$$\dot{\mathbf{s}}^{\text{true}} = \boldsymbol{\eta}_s \tag{6.40a}$$

$$\dot{\mathbf{k}}_U^{\text{true}} = \boldsymbol{\eta}_U \tag{6.40b}$$

$$\dot{\mathbf{k}}_L^{\text{true}} = \boldsymbol{\eta}_L \tag{6.40c}$$

where the spectral densities of $\boldsymbol{\eta}_s$, $\boldsymbol{\eta}_U$, and $\boldsymbol{\eta}_L$ are $\sigma_s^2 I_3$, $\sigma_U^2 I_3$, and $\sigma_L^2 I_3$, respectively. The error-state vector obeys the linearized dynamic equation

$$\Delta\dot{\mathbf{x}}(t) = F(t)\,\Delta\mathbf{x}(t) + G(t)\,\mathbf{w}(t) \tag{6.41}$$

where $F(t)$ is the Jacobian of $\mathbf{f}(\mathbf{x}, t)$ and

$$\mathbf{w}(t) \equiv \left[\boldsymbol{\eta}_v^T(t)\ \boldsymbol{\eta}_u^T(t)\ \boldsymbol{\eta}_s^T(t)\ \boldsymbol{\eta}_U^T(t)\ \boldsymbol{\eta}_L^T(t)\right]^T \tag{6.42}$$

The matrices $F(t)$, $G(t)$, and the spectral density $Q(t)$ of $\mathbf{w}(t)$ are now derived for the 15-state filter.

If the error terms in S^{true} are small, which is almost always a valid assumption, then $(I_3 + S^{\text{true}})^{-1} \approx (I_3 - S^{\text{true}})$, leading to

$$\boldsymbol{\omega}^{\text{true}} = (I_3 - S^{\text{true}})(\boldsymbol{\omega} - \boldsymbol{\beta}^{\text{true}} - \boldsymbol{\eta}_v) \tag{6.43a}$$

$$\hat{\boldsymbol{\omega}} = (I_3 - \hat{S})(\boldsymbol{\omega} - \hat{\boldsymbol{\beta}}) \tag{6.43b}$$

Then $\delta\boldsymbol{\omega}$ is given by

$$\delta\boldsymbol{\omega} = -\Delta S(\boldsymbol{\omega} - \hat{\boldsymbol{\beta}} - \Delta\boldsymbol{\beta} - \boldsymbol{\eta}_v) - (I_3 - \hat{S})(\Delta\boldsymbol{\beta} + \boldsymbol{\eta}_v) \tag{6.44}$$

where $\Delta S \equiv S^{\text{true}} - \hat{S}$. Ignoring second-order terms leads to

$$\delta\boldsymbol{\omega} = -\text{diag}(\boldsymbol{\omega} - \hat{\boldsymbol{\beta}})\Delta\mathbf{s} - \hat{U}\Delta\mathbf{k}_U - \hat{L}\Delta\mathbf{k}_L - (I_3 - \hat{S})(\Delta\boldsymbol{\beta} + \boldsymbol{\eta}_v) \tag{6.45}$$

with

$$\hat{U} = \begin{bmatrix} \omega_2 - \hat{\beta}_2 & \omega_3 - \hat{\beta}_3 & 0 \\ 0 & 0 & \omega_3 - \hat{\beta}_3 \\ 0 & 0 & 0 \end{bmatrix} \tag{6.46a}$$

$$\hat{L} = \begin{bmatrix} 0 & 0 & 0 \\ \omega_1 - \hat{\beta}_1 & 0 & 0 \\ 0 & \omega_1 - \hat{\beta}_1 & \omega_2 - \hat{\beta}_2 \end{bmatrix} \tag{6.46b}$$

where diag is defined in Eq. (2.8) and $\Delta\mathbf{s}$, $\Delta\mathbf{k}_U$, and $\Delta\mathbf{k}_L$ are implicitly defined by Eq. (6.39). Hence, the matrices $F(t)$, $G(t)$, and $Q(t)$ are given by

$$F(t) = \begin{bmatrix} -[\hat{\boldsymbol{\omega}}(t)\times] & -(I_3 - \hat{S}) & -\text{diag}(\boldsymbol{\omega} - \hat{\boldsymbol{\beta}}) & -\hat{U} & -\hat{L} \\ 0_{3\times3} & 0_{3\times3} & 0_{3\times3} & 0_{3\times3} & 0_{3\times3} \\ 0_{3\times3} & 0_{3\times3} & 0_{3\times3} & 0_{3\times3} & 0_{3\times3} \\ 0_{3\times3} & 0_{3\times3} & 0_{3\times3} & 0_{3\times3} & 0_{3\times3} \\ 0_{3\times3} & 0_{3\times3} & 0_{3\times3} & 0_{3\times3} & 0_{3\times3} \end{bmatrix} \tag{6.47a}$$

$$G(t) = \begin{bmatrix} -(I_3 - \hat{S}) & 0_{3\times3} & 0_{3\times3} & 0_{3\times3} & 0_{3\times3} \\ 0_{3\times3} & I_3 & 0_{3\times3} & 0_{3\times3} & 0_{3\times3} \\ 0_{3\times3} & 0_{3\times3} & I_3 & 0_{3\times3} & 0_{3\times3} \\ 0_{3\times3} & 0_{3\times3} & 0_{3\times3} & I_3 & 0_{3\times3} \\ 0_{3\times3} & 0_{3\times3} & 0_{3\times3} & 0_{3\times3} & I_3 \end{bmatrix} \tag{6.47b}$$

$$Q(t) = \text{blkdiag}\left(\left[\sigma_v^2 I_3 \;\; \sigma_u^2 I_3 \;\; \sigma_s^2 I_3 \;\; \sigma_U^2 I_3 \;\; \sigma_L^2 I_3\right]\right) \tag{6.47c}$$

where blkdiag denotes a block diagonal matrix of appropriate dimension.

Assume that N vector observations are available at time t_k. We concatenate them to form the $3N$-dimensional measurement vector

$$\mathbf{y}_k = \left[\begin{bmatrix} A(\mathbf{q}^{\text{true}})\mathbf{r}_1 \\ A(\mathbf{q}^{\text{true}})\mathbf{r}_2 \\ \vdots \\ A(\mathbf{q}^{\text{true}})\mathbf{r}_N \end{bmatrix}\right|_{t_k} + \left.\begin{bmatrix} \boldsymbol{\nu}_1 \\ \boldsymbol{\nu}_2 \\ \vdots \\ \boldsymbol{\nu}_N \end{bmatrix}\right|_{t_k} \equiv \mathbf{h}_k(\mathbf{x}_k^{\text{true}}) + \mathbf{v}_k \tag{6.48a}$$

$$R = \text{blkdiag}\left[R_1\ R_2 \cdots R_3\right] \tag{6.48b}$$

where R_i is the covariance of $\boldsymbol{\nu}_i$. We often make the simplifying assumption that the measurement errors are isotropic so that $R_i = \sigma_i^2 I_3$. The errors of unit vector measurements are not isotropic, though, and the QUEST measurement model (QMM) from Eq. (5.107) should be used instead. But this produces a singular matrix to be inverted in the EKF. It turns out that the assumption of isotropic errors can be justified, as will be shown in the following section.

We now derive the observation sensitivity matrix $H_k(\hat{\mathbf{x}}_k^-)$ for this observation. The true attitude matrix, $A(\mathbf{q}^{\text{true}})$, is related to the a priori attitude, $A(\hat{\mathbf{q}}^-)$, through

$$A(\mathbf{q}^{\text{true}}) = A(\delta\mathbf{q})A(\hat{\mathbf{q}}^-) \tag{6.49}$$

The first-order approximation of the error-attitude matrix, $A(\delta\mathbf{q})$, is given by

$$A(\delta\mathbf{q}) \approx I_3 - [\delta\boldsymbol{\vartheta}\times] \tag{6.50}$$

For a single sensor the true and estimated body vectors are given by

$$\mathbf{b}^{\text{true}} = A(\mathbf{q}^{\text{true}})\mathbf{r} \tag{6.51a}$$

$$\hat{\mathbf{b}}^- = A(\hat{\mathbf{q}}^-)\mathbf{r} \tag{6.51b}$$

Substituting Eqs. (6.49) and (6.50) into Eq. (6.51) yields

$$\Delta\mathbf{b} \equiv \mathbf{b}^{\text{true}} - \hat{\mathbf{b}}^- = -[\delta\boldsymbol{\vartheta}\times]A(\hat{\mathbf{q}}^-)\mathbf{r} = [\hat{\mathbf{b}}^-\times]\delta\boldsymbol{\vartheta} \tag{6.52}$$

The sensitivity matrix for all measurement sets is therefore given by

$$H_k(\hat{\mathbf{x}}_k^-) = \left.\begin{bmatrix} [\hat{\mathbf{b}}_1^-\times] & 0_{3\times12} \\ [\hat{\mathbf{b}}_2^-\times] & 0_{3\times12} \\ \vdots & \vdots \\ [\hat{\mathbf{b}}_N^-\times] & 0_{3\times12} \end{bmatrix}\right|_{t_k} \tag{6.53}$$

Table 6.1 Extended Kalman filter for gyro calibration

Initialize	$\hat{\mathbf{q}}(t_0) = \mathbf{q}_0, \quad \hat{\boldsymbol{\beta}}(t_0) = \boldsymbol{\beta}_0, \quad \hat{\mathbf{s}}(t_0) = \mathbf{s}_0$ $\hat{\mathbf{k}}_U(t_0) = \mathbf{k}_{U_0}, \quad \hat{\mathbf{k}}_L(t_0) = \mathbf{k}_{L_0}, \quad P(t_0) = P_0$
Gain	$K_k = P_k^- H_k^T(\hat{\mathbf{x}}_k^-)[H_k(\hat{\mathbf{x}}_k^-)P_k^- H_k^T(\hat{\mathbf{x}}_k^-) + R_k]^{-1}$ $H_k(\hat{\mathbf{x}}_k^-) = \left.\begin{bmatrix} [A(\hat{\mathbf{q}}^-)\mathbf{r}_1\times] & 0_{3\times 12} \\ \vdots & \vdots \\ [A(\hat{\mathbf{q}}^-)\mathbf{r}_N\times] & 0_{3\times 12} \end{bmatrix}\right\|_{t_k}$
Update	$P_k^+ = [I - K_k H_k(\hat{\mathbf{x}}_k^-)]P_k^-$ $\delta\hat{\boldsymbol{\vartheta}}_k^- = \mathbf{0}_3$ $\hat{\mathbf{x}}_k^+ = \hat{\mathbf{x}}_k^- + K_k[\mathbf{y}_k - \mathbf{h}_k(\hat{\mathbf{x}}_k^-)]$ $\hat{\mathbf{x}}_k \equiv \left[\delta\hat{\boldsymbol{\vartheta}}_k^T \ \hat{\boldsymbol{\beta}}_k^T \ \hat{\mathbf{s}}_k^T \ \hat{\mathbf{k}}_{U_k}^T \ \hat{\mathbf{k}}_{L_k}^T\right]^T$ $\mathbf{h}_k(\hat{\mathbf{x}}_k^-) = \left.\begin{bmatrix} A(\hat{\mathbf{q}}^-)\mathbf{r}_1 \\ A(\hat{\mathbf{q}}^-)\mathbf{r}_2 \\ \vdots \\ A(\hat{\mathbf{q}}^-)\mathbf{r}_N \end{bmatrix}\right\|_{t_k}$ $\hat{\mathbf{q}}^* = \hat{\mathbf{q}}_k^- + \frac{1}{2}\Xi(\hat{\mathbf{q}}_k^-)\delta\hat{\boldsymbol{\vartheta}}_k^+$ $\hat{\mathbf{q}}_k^+ = \mathbf{q}^*/\|\mathbf{q}^*\|$
Propagation	$\hat{\boldsymbol{\omega}}(t) = [I_3 - \hat{S}(t)][\boldsymbol{\omega}(t) - \hat{\boldsymbol{\beta}}(t)]$ $\dot{\hat{\mathbf{q}}}(t) = \frac{1}{2}\Xi(\hat{\mathbf{q}}(t))\hat{\boldsymbol{\omega}}(t)$ $\dot{P}(t) = F(t)P(t) + P(t)F^T(t) + G(t)Q(t)G^T(t)$

The final part in the EKF involves the quaternion and bias updates. The state update follows Eq. (6.25), which assumes that the state is reset before every measurement update and that the non-attitude part of the state is reset implicitly. Thus

$$\hat{\mathbf{x}}_k^+ = \hat{\mathbf{x}}_k^- + K_k[\mathbf{y}_k - \mathbf{h}_k(\hat{\mathbf{x}}_k^-)] \tag{6.54}$$

where $\hat{\mathbf{x}}_k \equiv \left[\delta\hat{\boldsymbol{\vartheta}}_k^T \ \hat{\boldsymbol{\beta}}_k^T \ \hat{\mathbf{s}}_k^T \ \hat{\mathbf{k}}_{U_k}^T \ \hat{\mathbf{k}}_{L_k}^T\right]^T$ and $\mathbf{h}_k(\hat{\mathbf{x}}_k^-)$ is the estimated observation, given by

$$\mathbf{h}_k(\hat{\mathbf{x}}_k^-) = \left.\begin{bmatrix} \hat{\mathbf{b}}_1^- \\ \hat{\mathbf{b}}_2^- \\ \vdots \\ \hat{\mathbf{b}}_N^- \end{bmatrix}\right\|_{t_k} \tag{6.55}$$

The attitude estimation algorithm is summarized in Table 6.1. The filter is first initialized with a known state (the bias initial condition is usually assumed zero) and error covariance matrix. The first three diagonal elements of the error covariance matrix correspond to attitude errors. Then, the Kalman gain is computed using the measurement-error covariance R_k and sensitivity matrix in Eq. (6.53). The state error covariance follows the standard EKF update, while the state update is computed using Eq. (6.54). The quaternion is reset following Eqs. (6.27) and (6.28). Finally, the estimated angular velocity is used to propagate the quaternion kinematic model and error covariance in the EKF.

A discrete-time propagation, given by Eq. (12.131), can be used for the covariance matrix in order to reduce the computational load. The first step is to set up the following $2n \times 2n$ matrix, where $n = 15$ in the present case [40]:

$$\mathscr{A} = \begin{bmatrix} -F & GQG^T \\ 0_{n\times n} & F^T \end{bmatrix} \Delta t \tag{6.56}$$

The matrix exponential is then calculated:

$$\mathscr{B} = e^{\mathscr{A}} = \begin{bmatrix} \mathscr{B}_{11} & \mathscr{B}_{12} \\ 0_{n\times n} & \mathscr{B}_{22} \end{bmatrix} = \begin{bmatrix} \mathscr{B}_{11} & \Phi_k^{-1} Q_k \\ 0_{n\times n} & \Phi_k^T \end{bmatrix} \tag{6.57}$$

The state transition and covariance matrices are then given as

$$\Phi_k = \mathscr{B}_{22}^T \tag{6.58a}$$

$$Q_k = \Phi_k \mathscr{B}_{12} \tag{6.58b}$$

Note that Eq. (6.56) is only valid for constant system and covariance matrices, but this is a good approximation for small Δt. Often adequate results are provided by an even simpler first-order approximation:

$$\Phi_k \approx I_n + \Delta t\, F(t) \tag{6.59a}$$

$$Q_k \approx \Delta t\, GQG^T \tag{6.59b}$$

Then Eq. (3.17) gives the discrete-time quaternion propagation as

$$\hat{\mathbf{q}}_{k+1}^- = \exp[(\Delta\boldsymbol{\theta}/2)\otimes]\hat{\mathbf{q}}_k^+ \approx \bar{\Theta}(\hat{\omega}_k^+)\hat{\mathbf{q}}_k^+ \tag{6.60}$$

with

$$\bar{\Theta}(\hat{\omega}_k^+) \equiv \begin{bmatrix} \cos\left(\frac{1}{2}\|\hat{\omega}_k^+\|\,\Delta t\right) I_3 - \left[\hat{\boldsymbol{\psi}}_k^+ \times\right] & \hat{\boldsymbol{\psi}}_k^+ \\ -\hat{\boldsymbol{\psi}}_k^{+T} & \cos\left(\frac{1}{2}\|\hat{\omega}_k^+\|\,\Delta t\right) \end{bmatrix} \tag{6.61}$$

where

$$\hat{\boldsymbol{\psi}}_k^+ \equiv \frac{\sin\left(\frac{1}{2}\|\hat{\boldsymbol{\omega}}_k^+\|\,\Delta t\right)\hat{\boldsymbol{\omega}}_k^+}{\|\hat{\boldsymbol{\omega}}_k^+\|} \tag{6.62}$$

Example 6.1. In this example, simulation results are presented for the aforementioned gyro calibration algorithm. A 90-min simulation run is shown. Quaternion measurements are assumed to exist, sampled every second, and the covariance of the measurements is given by $R_k = 36 I_3$ arcsec^2. This assumes that errors in each axis are equivalent. Using one star tracker typically produces errors along the boresight axis that are an order of magnitude larger than the other axes. Using two star trackers that are orthogonal produces errors that are more nearly isotropic. Also, two star trackers are typically employed for redundancy purposes. The spacecraft angular velocity is given by $\boldsymbol{\omega}^{\text{true}} = 0.1 \times [\sin(0.01t) \;\; \sin(0.0085t) \;\; \cos(0.0085t)]^T$ deg/s. This angular velocity is used to increase the observability of the calibration parameters. Missions typically perform calibration maneuvers to ensure proper observability. The other parameters used in the simulation are given by

- Initial gyro bias: 0.1 deg/h for each axis,
- Gyro scale factors: $s_1 = 1{,}500$ ppm, $s_2 = 1{,}000$ ppm, $s_3 = 1{,}500$ ppm,
- Gyro misalignments: $k_{U1} = 1{,}000$ ppm, $k_{U2} = 1{,}500$ ppm, $k_{U3} = 2{,}000$ ppm, $k_{L1} = 500$ ppm, $k_{L2} = 1{,}000$ ppm, $k_{L3} = 1{,}500$ ppm,
- Gyro noise parameters: $\sigma_u = \sqrt{10} \times 10^{-10}$ rad/s$^{3/2}$, $\sigma_v = \sqrt{10} \times 10^{-7}$ rad/s$^{1/2}$, $\sigma_s = \sigma_U = \sigma_L = 0$,
- Initial quaternion: $\mathbf{q}_0^{\text{true}} = \sqrt{2}/2 \times [1 \;\; 0 \;\; 0 \;\; 1]^T$.

where ppm denotes parts per million. The gyros are also assumed to be sampled every second. The EKF attitude is initialized using the measurement quaternion and the initial covariance is given by $(6/3{,}600 \times \pi/180)^2 I_3$ rad^2. All gyro calibration parameters are assumed to be initialized to zero. The initial covariances for these parameters are: initial gyro bias covariance of $(0.2/3{,}600 \times \pi/180)^2 I_3$ (rad/s)^2, initial scale factor covariance of $(0.002/3)^2 I_3$ rad^2, initial upper misalignment covariance of $(0.002/3)^2 I_3$ rad^2, and initial lower misalignment covariance of $(0.002/3)^2 I_3$ rad^2.

The EKF errors with their respective 3σ bounds for the attitude, gyro biases, scale factors and upper misalignments are shown in Fig. 6.1. Roll, pitch, and yaw denote the components of $\boldsymbol{\delta\vartheta}$. The attitude errors converge in about 15 min and good filtering performance is obtained. The gyro biases are also well estimated with 3σ bounds of about 0.01 deg/h in each axis. The scale factors and misalignments take slightly longer to converge than the attitude and gyro biases, but they are also well estimated with 3σ bounds of about 25 μrad in each axis. The EKF estimates are then employed in a smoother to further refine the estimates. The smoothed estimates and covariances of each parameter are nearly constant over the entire simulation run. Estimates with their respective 3σ bounds are shown in Table 6.2. Clearly, as expected, the smoother is able to provide more accurate results than the EKF.

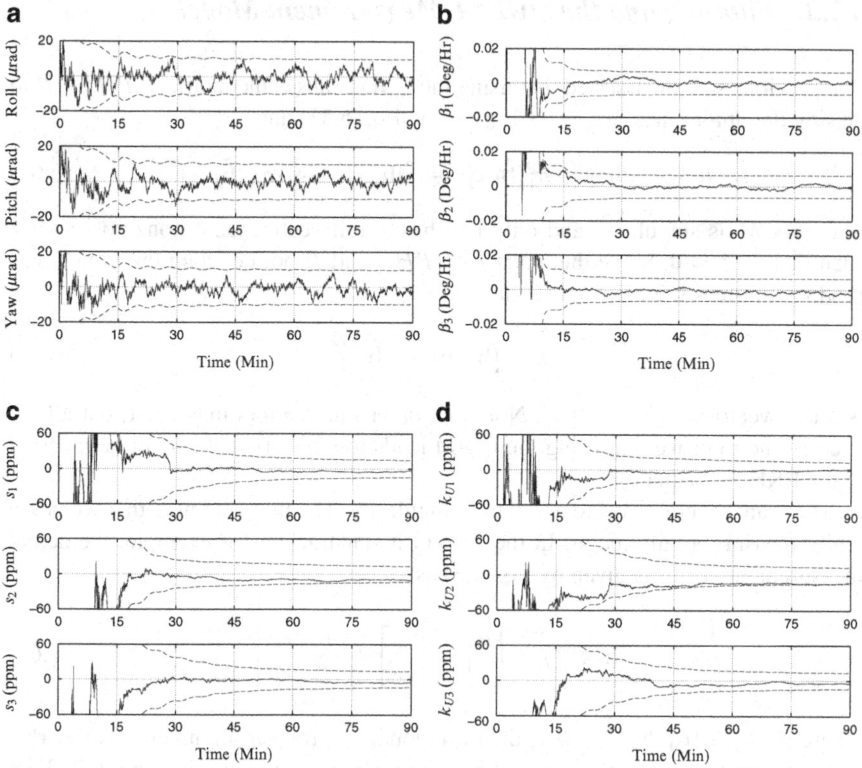

Fig. 6.1 Extended Kalman filter results. (**a**) Attitude errors and 3σ bounds. (**b**) Bias errors and 3σ bounds. (**c**) Scale errors and 3σ bounds. (**d**) Upper misalignment errors and 3σ bounds

Table 6.2 Smoother results for gyro calibration

Parameter	Truth	Estimate
β_1 (deg/h)	0.1	0.0981 ± 0.0065
β_2 (deg/h)	0.1	0.1029 ± 0.0065
β_3 (deg/h)	0.1	0.0999 ± 0.0065
s_1 (ppm)	1,500	$1{,}497 \pm 14.5800$
s_2 (ppm)	1,000	997 ± 13.5651
s_3 (ppm)	1,500	$1{,}488 \pm 13.6671$
k_{U1} (ppm)	1,000	991 ± 13.5936
k_{U2} (ppm)	1,500	$1{,}501 \pm 13.6984$
k_{U3} (ppm)	2,000	$1{,}999 \pm 13.7233$
k_{L1} (ppm)	500	499 ± 14.6252
k_{L2} (ppm)	1,000	994 ± 14.6197
k_{L3} (ppm)	1,500	$1{,}497 \pm 13.6126$

6.2.3 Filtering and the QUEST Measurement Model

The previous section asserted that using the QMM causes the matrix $HPH^T + R$ to be singular at all times, where H is given by Eq. (6.53) and

$$R = \text{blkdiag}\left[-\sigma_1^2[\hat{\mathbf{b}}_1\times]^2 \; -\sigma_2^2[\hat{\mathbf{b}}_2\times]^2 \cdots -\sigma_N^2[\hat{\mathbf{b}}_N\times]^2\right] \tag{6.63}$$

A matrix M is singular if and only if it has a null vector, i.e. a nonzero vector $\mathbf{z}$ such that $M\,\mathbf{z} = \mathbf{0}$. Since the matrices HPH^T and R both contain the same cross product matrices it is easy to see that

$$\mathbf{z} = \left[\hat{\mathbf{b}}_1^T \; \hat{\mathbf{b}}_2^T \cdots \hat{\mathbf{b}}_N^T\right]^T \tag{6.64}$$

is a null vector of $HPH^T + R$. Note that other null vectors may exist, but all we need is one to prove that using the QMM leads to a matrix that cannot be inverted in the EKF gain equation.

There are several ways to avoid this singularity. The first is to note that we never really measure a unit vector. In the case of a star tracker, for example, the actual measurement vector is given by Eq. (4.3) as

$$\mathbf{h}(\mathbf{x}^{\text{true}}) = \begin{bmatrix} u \\ v \end{bmatrix} = \begin{bmatrix} u_0 \\ v_0 \end{bmatrix} + \frac{f}{r_3^{\text{st}}}\begin{bmatrix} r_1^{\text{st}} \\ r_2^{\text{st}} \end{bmatrix} \tag{6.65}$$

where $\mathbf{r}^{\text{st}} = BA(\mathbf{q}^{\text{true}})\mathbf{r}^{\text{true}}$, B is the orthogonal star tracker alignment matrix, $\mathbf{r}^{\text{true}}$ is the star unit vector in the celestial coordinate frame, and the subscripts $1, 2, 3$ are Cartesian indices. The attitude part of the sensitivity matrix for this measurement model is found by the chain rule to be

$$H_\vartheta = \frac{\partial \mathbf{h}}{\partial \mathbf{r}^{\text{st}}}\frac{\partial \mathbf{r}^{\text{st}}}{\partial(\delta\boldsymbol{\vartheta})} = \frac{f}{(\hat{r}_3^{\text{st}})^2}\begin{bmatrix} \hat{r}_3^{\text{st}} & 0 & -\hat{r}_1^{\text{st}} \\ 0 & \hat{r}_3^{\text{st}} & -\hat{r}_2^{\text{st}} \end{bmatrix}[\hat{\mathbf{r}}^{\text{st}}\times]B \tag{6.66}$$

Then R and $HPH^T + R$ are well-conditioned 2×2 matrices.

It is often desirable to compute a unit vector from the sensor measurements for input to the filter, in order to provide a universal filter interface from a variety of sensors. The nonsingular unit vector measurement covariance matrix of Eq. (5.126) could be used in place of the QMM, but this would produce an ill-conditioned matrix $HPH^T + R$, which is not much of an improvement. Fortunately, Malcolm Shuster found a way out of this dilemma.

Shuster's analysis of this singularity issue leads to a remarkably simple solution [35]. Let $\boldsymbol{\chi}$ be a unit vector that is very close to $A^{\text{true}}\mathbf{r}$, and let $\{\boldsymbol{\chi},\, \boldsymbol{\alpha}(\boldsymbol{\chi}),\, \boldsymbol{\beta}(\boldsymbol{\chi})\}$ be a right-handed orthogonal triad. We can determine $\mathbf{b}$ by measuring only two components nearly perpendicular to $A^{\text{true}}\mathbf{r}$:

$$\zeta_1 = \boldsymbol{\alpha}(\boldsymbol{\chi}) \cdot \mathbf{b} \tag{6.67a}$$

$$\zeta_2 = \boldsymbol{\beta}(\boldsymbol{\chi}) \cdot \mathbf{b} \tag{6.67b}$$

The measurement vector is given in terms of the projection matrix

$$U(\boldsymbol{\chi}) \equiv \left[\boldsymbol{\alpha}(\boldsymbol{\chi})\ \boldsymbol{\beta}(\boldsymbol{\chi})\right]^T \tag{6.68}$$

by

$$\boldsymbol{\zeta} \equiv [\zeta_1\ \zeta_2]^T = U(\boldsymbol{\chi})\mathbf{b} \tag{6.69}$$

The projection matrix satisfies the identities

$$U(\boldsymbol{\chi})U^T(\boldsymbol{\chi}) = I_2 \tag{6.70a}$$

$$U^T(\boldsymbol{\chi})\,U(\boldsymbol{\chi}) = I_3 - \boldsymbol{\chi}\,\boldsymbol{\chi}^T \tag{6.70b}$$

Substituting the measurement model $\mathbf{b} = A^{\text{true}}\mathbf{r} + \boldsymbol{\nu}$ into Eq. (6.69) leads to

$$\boldsymbol{\zeta} = U(\boldsymbol{\chi})A^{\text{true}}\mathbf{r} + \mathbf{e} \tag{6.71}$$

where $\mathbf{e} = U(\boldsymbol{\chi})\boldsymbol{\nu}$. If $\boldsymbol{\nu}$ obeys the QMM, $E\{\boldsymbol{\nu}\boldsymbol{\nu}^T\} = \sigma^2[I_3 - (A^{\text{true}}\mathbf{r})(A^{\text{true}}\mathbf{r})^T]$, then the projected error $\mathbf{e}$ satisfies

$$E\{\mathbf{e}\} = \mathbf{0} \tag{6.72a}$$

$$E\left\{\mathbf{e}\,\mathbf{e}^T\right\} = \sigma^2\left\{I_2 - [U(\boldsymbol{\chi})A^{\text{true}}\mathbf{r}][U(\boldsymbol{\chi})A^{\text{true}}\mathbf{r}]^T\right\} \tag{6.72b}$$

Since $\boldsymbol{\chi}$ is very close to $A^{\text{true}}\mathbf{r}$ then $\|U(\boldsymbol{\chi})A^{\text{true}}\mathbf{r}\| \ll 1$ and we can neglect the second term in Eq. (6.72b), yielding

$$R_\zeta \equiv E\left\{\mathbf{e}\,\mathbf{e}^T\right\} = \sigma^2 I_2 \tag{6.73}$$

which is a nonsingular matrix.

The vector $\boldsymbol{\chi}$ has been undefined, except for being close to $A^{\text{true}}\mathbf{r}$. This will now be specifically defined in the Kalman filter setting. Substituting Eqs. (6.49) and (6.50) into Eq. (6.71) leads to

$$\boldsymbol{\zeta} = U(\boldsymbol{\chi})(I_3 - [\delta\boldsymbol{\vartheta}\times])\hat{\mathbf{b}}^- + \mathbf{e} \tag{6.74a}$$

$$= U(\boldsymbol{\chi})(\hat{\mathbf{b}}^- + \hat{\mathbf{b}}^- \times \delta\boldsymbol{\vartheta}) + \mathbf{e} \tag{6.74b}$$

If we choose $\boldsymbol{\chi} = \hat{\mathbf{b}}^-$, the first term in Eq. (6.74b) vanishes because $U(\hat{\mathbf{b}}^-)\hat{\mathbf{b}}^- = \mathbf{0}$. Defining $U^- \equiv U(\hat{\mathbf{b}}^-)$ gives

$$\boldsymbol{\zeta} \equiv U^-\mathbf{b} = U^-[\hat{\mathbf{b}}^-\times]\delta\boldsymbol{\vartheta} + \mathbf{e} \tag{6.75}$$

Now define the following error vector: $\Delta\mathbf{x} \equiv [\delta\boldsymbol{\vartheta}^T \;\; \Delta\boldsymbol{\xi}^T]^T$, where $\Delta\boldsymbol{\xi}$ is an $n \times 1$ vector of non-attitude error states, such as the error gyro biases. Then the linearized measurement equation becomes

$$\boldsymbol{\zeta} = H\,\Delta\mathbf{x} + \mathbf{e} \tag{6.76}$$

where H is given by

$$H = \begin{bmatrix} W & 0_{2\times n} \end{bmatrix} \tag{6.77}$$

with $W \equiv U^-[\hat{\mathbf{b}}^-\times]$.

The predicted innovations covariance is given by

$$\begin{aligned} E^- &= H\,P^- H^T + R_\zeta = W\,P^-_{\vartheta\vartheta} W^T + \sigma^2 I_2 \\ &= U^-\left([\hat{\mathbf{b}}^-\times]P^-_{\vartheta\vartheta}[\hat{\mathbf{b}}^-\times]^T + \sigma^2 I_3\right)(U^-)^T \end{aligned} \tag{6.78}$$

where $P^-_{\vartheta\vartheta}$ is the upper-left 3×3 corner of P^-. Because $\boldsymbol{\chi}$ is very close to $A^{\text{true}}\mathbf{r}$, the inverse of this is, to a very good approximation,

$$(E^-)^{-1} = U^-\left([\hat{\mathbf{b}}^-\times]P^-_{\vartheta\vartheta}[\hat{\mathbf{b}}^-\times]^T + \sigma^2 I_3\right)^{-1}(U^-)^T \tag{6.79}$$

Note that this would be trivial if U^- were orthogonal, but it is not.

The Kalman gain is given by $K = P^- H^T(H\,P^- H^T + R_\zeta)^{-1}$. If we define

$$\mathscr{H} \equiv \begin{bmatrix} [\hat{\mathbf{b}}^-\times] & 0_{3\times n} \end{bmatrix} \tag{6.80a}$$

$$\mathscr{K} \equiv P^-\mathscr{H}^T(\mathscr{H}\,P^-\mathscr{H}^T + \sigma^2 I_3)^{-1} \tag{6.80b}$$

then we have

$$\mathscr{H}\,P^-\mathscr{H}^T + \sigma^2 I_3 = [\hat{\mathbf{b}}^-\times]P^-_{\vartheta\vartheta}[\hat{\mathbf{b}}^-\times]^T + \sigma^2 I_3 \tag{6.81a}$$

$$K\,H = \mathscr{K}\,\mathscr{H} \tag{6.81b}$$

Equation (6.81) indicates that the QMM can be effectively replaced by $\sigma^2 I_3$, in the sense that it leads to the same filter that uses the projected measurements in Eq. (6.67)! Note that the projection matrix U^- does not appear in $\mathscr{H}$ or $\mathscr{K}$. We have recovered the usual Kalman filter equations for a unit vector measurement, but with a nonsingular isotropic measurement covariance matrix in place of the singular QMM expression. Shuster shows that all other equations in the original EKF remain unchanged. This analysis can easily be extended to multiple observations.

Table 6.3 Extended Kalman filter for attitude estimation

Initialize	$\hat{\mathbf{q}}(t_0) = \hat{\mathbf{q}}_0, \quad \hat{\boldsymbol{\beta}}(t_0) = \hat{\boldsymbol{\beta}}_0$ $P(t_0) = P_0$
Gain	$K_k = P_k^- H_k^T(\hat{\mathbf{x}}_k^-)[H_k(\hat{\mathbf{x}}_k^-)P_k^- H_k^T(\hat{\mathbf{x}}_k^-) + R_k]^{-1}$ $H_k(\hat{\mathbf{x}}_k^-) = \left.\begin{bmatrix} [A(\hat{\mathbf{q}}^-)\mathbf{r}_1\times] & 0_{3\times3} \\ \vdots & \vdots \\ [A(\hat{\mathbf{q}}^-)\mathbf{r}_N\times] & 0_{3\times3} \end{bmatrix}\right\|_{t_k}$
Update	$P_k^+ = [I - K_k H_k(\hat{\mathbf{x}}_k^-)]P_k^-$ $\Delta\hat{\mathbf{x}}_k^+ = K_k[\mathbf{y}_k - \mathbf{h}_k(\hat{\mathbf{x}}_k^-)]$ $\Delta\hat{\mathbf{x}}_k^+ \equiv \left[\delta\hat{\boldsymbol{\vartheta}}_k^{+T} \;\; \Delta\hat{\boldsymbol{\beta}}_k^{+T}\right]^T$ $\mathbf{h}_k(\hat{\mathbf{x}}_k^-) = \left.\begin{bmatrix} A(\hat{\mathbf{q}}^-)\mathbf{r}_1 \\ A(\hat{\mathbf{q}}^-)\mathbf{r}_2 \\ \vdots \\ A(\hat{\mathbf{q}}^-)\mathbf{r}_N \end{bmatrix}\right\|_{t_k}$ $\hat{\mathbf{q}}^* = \hat{\mathbf{q}}_k^- + \frac{1}{2}\Xi(\hat{\mathbf{q}}_k^-)\delta\hat{\boldsymbol{\vartheta}}_k^+$ $\hat{\mathbf{q}}_k^+ = \mathbf{q}^*/\lVert\mathbf{q}^*\rVert$ $\hat{\boldsymbol{\beta}}_k^+ = \hat{\boldsymbol{\beta}}_k^- + \Delta\hat{\boldsymbol{\beta}}_k^+$
Propagation	$\hat{\boldsymbol{\omega}}(t) = \boldsymbol{\omega}(t) - \hat{\boldsymbol{\beta}}(t)$ $\dot{\hat{\mathbf{q}}}(t) = \frac{1}{2}\Xi(\hat{\mathbf{q}}(t))\,\hat{\boldsymbol{\omega}}(t)$ $\dot{P}(t) = F(t)\,P(t) + P(t)\,F^T(t) + G(t)\,Q(t)\,G^T(t)$

6.2.4 Mission Mode Kalman Filter

The EKF described in the Sect. 6.2.2 can be used for full gyro calibration, and improved estimates can be obtained using a smoother. As mentioned previously, the mission mode EKF estimates only attitude and gyro biases, assuming that the scale factors and misalignments have already been determined. The 6-state EKF attitude estimator is shown in Table 6.3. Note that in this formulation $\Delta\hat{\mathbf{x}}$ is used and only one update is performed at each time step, so that the measurement update follows Eq. (6.24) and an explicit update of the bias vector is performed. This formulation and the one in the next section are equivalent and both have been seen in the open literature. Thus we feel it is important to see how each approach is implemented.

The matrices $F(t)$, $G(t)$, and $Q(t)$ are given by

$$F(t) = \begin{bmatrix} -[\hat{\omega}(t)\times] & -I_3 \\ 0_{3\times3} & 0_{3\times3} \end{bmatrix} \tag{6.82a}$$

$$G(t) = \begin{bmatrix} -I_3 & 0_{3\times3} \\ 0_{3\times3} & I_3 \end{bmatrix} \tag{6.82b}$$

$$Q(t) = \begin{bmatrix} \sigma_v^2 I_3 & 0_{3\times3} \\ 0_{3\times3} & \sigma_u^2 I_3 \end{bmatrix} \tag{6.82c}$$

Closed-form expressions for the discrete-time matrices are possible in this case. The discrete error-state transition matrix is given by

$$\Phi = \begin{bmatrix} \Phi_{11} & \Phi_{12} \\ \Phi_{21} & \Phi_{22} \end{bmatrix} \tag{6.83a}$$

$$\Phi_{11} = I_3 - [\hat{\omega}\times]\frac{\sin(\|\hat{\omega}\|\,\Delta t)}{\|\hat{\omega}\|} + [\hat{\omega}\times]^2\frac{\{1-\cos(\|\hat{\omega}\|\,\Delta t)\}}{\|\hat{\omega}\|^2} \tag{6.83b}$$

$$\Phi_{12} = [\hat{\omega}\times]\frac{\{1-\cos(\|\hat{\omega}\|\,\Delta t)\}}{\|\hat{\omega}\|^2} - I_3\Delta t - [\hat{\omega}\times]^2\frac{\{\|\hat{\omega}\|\,\Delta t - \sin(\|\hat{\omega}\|\,\Delta t)\}}{\|\hat{\omega}\|^3} \tag{6.83c}$$

$$\Phi_{21} = 0_{3\times3} \tag{6.83d}$$

$$\Phi_{22} = I_3 \tag{6.83e}$$

It is left as an exercise to derive these equations and to show that Φ_{11} is an orthogonal matrix.

The conversion from the spectral density $Q(t)$ to discrete-time covariance Q_k is given by [8]

$$Q_k = \int_0^{\Delta t} \Phi(t)\,G(t)\,Q(t)G^T(t)\Phi^T(t)\,dt \tag{6.84}$$

It is assumed in the ensuing derivations for Q_k that $\hat{\omega}$ is constant throughout the sampling interval. Partition the matrix Q_k into 3×3 sub-matrices:

$$Q_k = \begin{bmatrix} Q_{11_k} & Q_{12_k} \\ Q_{12_k}^T & Q_{22_k} \end{bmatrix} \tag{6.85}$$

Performing the multiplication $B(t) \equiv \Phi(t)\,G(t)\,Q(t)G^T(t)\Phi^T(t)$ gives

$$B(t) = \begin{bmatrix} \sigma_v^2\Phi_{11}\Phi_{11}^T + \sigma_u^2\Phi_{12}\Phi_{12}^T & \sigma_v^2\Phi_{11}\Phi_{21}^T + \sigma_u^2\Phi_{12}\Phi_{22}^T \\ \sigma_v^2\Phi_{21}\Phi_{11}^T + \sigma_u^2\Phi_{22}\Phi_{12}^T & \sigma_v^2\Phi_{21}\Phi_{21}^T + \sigma_u^2\Phi_{22}\Phi_{22}^T \end{bmatrix} \tag{6.86}$$

Using Eqs. (6.83d) and (6.83e) and the orthogonality of Φ_{11} simplifies $B(t)$ to

$$B(t) = \begin{bmatrix} \sigma_v^2 I_3 + \sigma_u^2 \Phi_{12}\Phi_{12}^T & \sigma_u^2 \Phi_{12} \\ \sigma_u^2 \Phi_{12}^T & \sigma_u^2 I_3 \end{bmatrix} \tag{6.87}$$

Therefore, Q_{22_k} is simply given by

$$Q_{22_k} = (\sigma_u^2 \Delta t) I_3 \tag{6.88}$$

Integrating Φ_{12} gives

$$\begin{aligned} Q_{12_k} = \sigma_u^2 \int_0^{\Delta t} \Phi_{12}\, dt = \sigma_u^2 \Bigg\{ & [\hat{\omega}_k \times] \frac{\|\hat{\omega}_k\| \Delta t - \sin(\|\hat{\omega}_k\|\, \Delta t)}{\|\hat{\omega}_k\|^3} - \frac{1}{2}\Delta t^2 I_3 \\ & - [\hat{\omega}_k \times]^2 \frac{\frac{1}{2}\|\hat{\omega}_k\|^2 \Delta t^2 + \cos(\|\hat{\omega}_k\|\, \Delta t) - 1}{\|\hat{\omega}_k\|^4} \Bigg\} \end{aligned} \tag{6.89}$$

If the sampling rate is below Nyquist's limit (for example, with a safety factor of 10 we require $\|\hat{\omega}\| \Delta t < \pi/10$ for all time), then $Q_{12_k} \approx -\frac{1}{2}\sigma_u^2 \Delta t^2 I_3$.

Performing the multiplication $\Phi_{12}\Phi_{12}^T$ yields

$$\begin{aligned} \Phi_{12}\Phi_{12}^T &= -[\hat{\omega}\times]^2 \frac{1 - 2\cos(\|\hat{\omega}\|\, t) + \cos^2(\|\hat{\omega}\|\, t)}{\|\hat{\omega}\|^4} \\ &\quad - [\hat{\omega}\times]^2 \frac{\|\hat{\omega}\|^2 t^2 - 2\|\hat{\omega}\|\, t \sin(\|\hat{\omega}\|\, t) + \sin^2(\|\hat{\omega}\|\, t)}{\|\hat{\omega}\|^4} \\ &\quad + 2[\hat{\omega}\times]^2 \frac{\|\hat{\omega}\|\, t^2 - t\sin(\|\hat{\omega}\|\, t)}{\|\hat{\omega}\|^3} + t^2 I_3 \\ &= -[\hat{\omega}\times]^2 \frac{2 - 2\cos(\|\hat{\omega}\|\, t) - \|\hat{\omega}\|^2 t^2}{\|\hat{\omega}\|^4} + t^2 I_3 \end{aligned} \tag{6.90}$$

The integral of $\Phi_{12}\Phi_{12}^T\, dt$ is

$$\int_0^{\Delta t} \Phi_{12}\Phi_{12}^T\, dt = -[\hat{\omega}\times]^2 \frac{2\|\hat{\omega}\|\, \Delta t - 2\sin(\|\hat{\omega}\|\, \Delta t) - \frac{1}{3}\|\hat{\omega}\|^3 \Delta t^3}{\|\hat{\omega}\|^5} + \frac{1}{3}\Delta t^3 I_3 \tag{6.91}$$

Therefore, Q_{11_k} is given by

$$\begin{aligned} Q_{11_k} &= (\sigma_v^2 \Delta t) I_3 \\ &\quad + \sigma_u^2 \left\{ \frac{1}{3}\Delta t^3 I_3 - [\hat{\omega}_k \times]^2 \frac{2\|\hat{\omega}_k\|\, \Delta t - 2\sin(\|\hat{\omega}_k\|\, \Delta t) - \frac{1}{3}\|\hat{\omega}_k\|^3 \Delta t^3}{\|\hat{\omega}_k\|^5} \right\} \end{aligned} \tag{6.92}$$

If the sampling rate is below Nyquist's limit, then $Q_{11_k} \approx \left(\sigma_v^2 \Delta t + \frac{1}{3}\sigma_u^2 \Delta t^3\right) I_3$. Then, the discrete process noise covariance is approximated by

$$Q_k \approx \begin{bmatrix} \left(\sigma_v^2 \Delta t + \dfrac{1}{3}\sigma_u^2 \Delta t^3\right) I_3 & -\left(\dfrac{1}{2}\sigma_u^2 \Delta t^2\right) I_3 \\ -\left(\dfrac{1}{2}\sigma_u^2 \Delta t^2\right) I_3 & \left(\sigma_u^2 \Delta t\right) I_3 \end{bmatrix} \tag{6.93}$$

It should be noted that Eq. (6.93) is exact when $F(t)$ is given by

$$F(t) = \begin{bmatrix} 0_{3\times 3} & -I_3 \\ 0_{3\times 3} & 0_{3\times 3} \end{bmatrix} \tag{6.94}$$

Equation (6.93) is often used because the sampling rate is often below Nyquist's limit. These discrete-time forms make the EKF especially suitable for onboard implementation.

6.2.5 *Murrell's Version*

The gain calculation in the filter shown in Table 6.1 requires inverting a $3N \times 3N$ matrix. Murrell's variation of the filter can be used to avoid this expensive computation [27]. Even though the EKF involves nonlinear models, a linear update is still performed. Therefore, linear tools such as the principle of superposition are applicable. Murrell's filter uses this principle to process one 3×1 vector observation at a time, delaying the reset until all N observations have been processed. In this case, all the terms of Eq. (6.20) must be included. A flow diagram of Murrell's approach is given in Fig. 6.2. The first step involves propagating the quaternion, gyro bias, and error covariance to the current observation time. Then the attitude matrix is computed and the error state vector is initialized to zero. Next, the error covariance and state quantities are updated using a single vector observation. This update is continued (replacing the propagated error covariance and state vector with the updated values) until all vector observations are processed. Finally, a reset moves the updated values into the global state representation, and the global state and error covariance are propagated to the next observation time. Therefore, this approach reduces taking an inverse of a $3N \times 3N$ matrix to taking an inverse of a 3×3 matrix N times, significantly decreasing the computational load.

Example 6.2. This example is a simulation of an EKF using a typical star tracker to determine the attitude of a spacecraft in a 90-min low-Earth orbit. The spacecraft rotates at one revolution per orbit about its $-x$ body axis, which is the $\mathbf{i}_3$ vector of the inertial reference frame shown in Fig. 2.3. The star tracker's boresight is assumed to be along the z (yaw) body axis pointed in the anti-nadir direction, and is initially

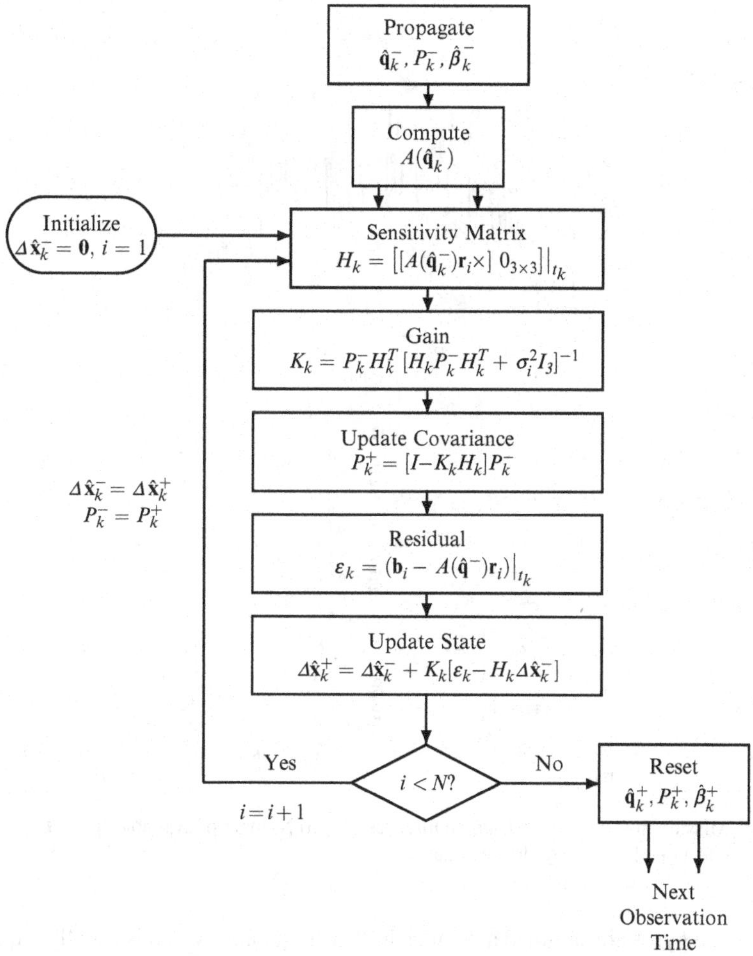

Fig. 6.2 Computationally efficient attitude estimation algorithm

aligned with the $\mathbf{i}_1$ vector of the inertial reference frame. The star tracker can sense up to 10 stars in a $6^\circ \times 6^\circ$ field of view. The catalog contains stars that can be sensed up to a magnitude of 6.0 (larger magnitudes indicate dimmer stars).

The star tracker body observations are obtained by using the measurement model presented in Sect. 4.2 with $\sigma = (0.005/3)^\circ$. We also assume that no Sun intrusions are present (although this is not truly realistic). Star images are taken at 1-s intervals. The goal of the EKF is to estimate the attitude and gyro biases by filtering the star tracker measurements. The noise parameters for the gyro measurements are given by $\sigma_u = \sqrt{10} \times 10^{-10}$ rad/s$^{3/2}$ and $\sigma_v = \sqrt{10} \times 10^{-7}$ rad/s$^{1/2}$. The initial bias for each axis is given by 0.1 deg/h. Also, the gyro measurements are sampled at the same rate as the star tracker measurements (i.e. at 1 Hz). We should note that in

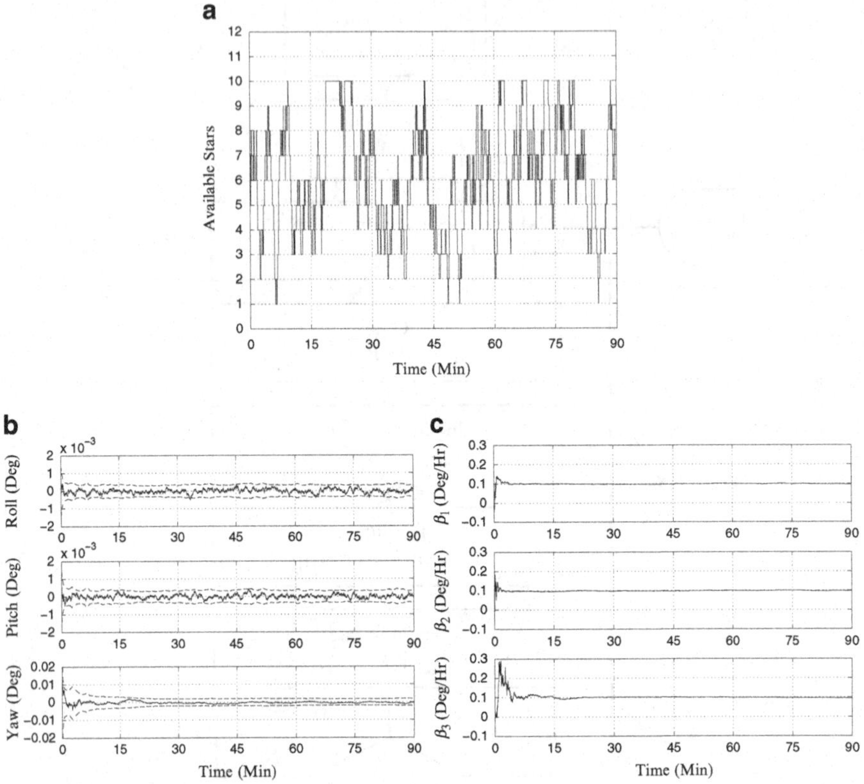

Fig. 6.3 Mission mode extended Kalman filter results. (**a**) Number of available stars. (**b**) Attitude errors and 3σ bounds. (**c**) Gyro drift estimates

practice the gyros are sampled at a much higher frequency, which is usually required for jitter control. The initial covariance for the attitude error is set to 0.1^2 deg^2, and the initial covariance for the gyro drift is set to 0.2^2 (deg/h)^2. Converting these quantities to radians and seconds gives the initial attitude and gyro drift covariances for each axis: $P_0^a = 3.0462 \times 10^{-6}$ rad^2 and $P_0^b = 9.4018 \times 10^{-13}$ (rad/s)^2, so that the initial covariance matrix is given by $P_0 = \text{diag}[P_0^a \; P_0^a \; P_0^a \; P_0^b \; P_0^b \; P_0^b]$. The initial attitude estimate for the EKF is given by the true quaternion, which is given by $\mathbf{q}_0^{\text{true}} = \sqrt{2}/2 \times [0 \;\; 1 \;\; 0 \;\; 1]^T$. The initial gyro drift biases in the EKF are set to zero.

Figure 6.3a shows a plot of the number of available stars. Note that at a few times only one star is available, so the static attitude determination methods of Chap. 5 cannot be employed at these times. However, this is not an issue with the EKF since an update is still possible even with one star. Note that an EKF-based solution is possible unless there is never more than one star available and it is always the same star. Figure 6.3b shows a plot of the attitude errors and associated 3σ bounds.

Clearly, the computed 3σ bounds do indeed bound the attitude errors. Note that the yaw errors are much larger than the roll and pitch errors because the boresight of the star tracker is along the yaw axis. Figure 6.3c shows that the EKF is able to accurately estimate the gyro drift bias. The "drift" in this plot looks very steady, due to the fact that a high-grade three-axis gyro has been used in the simulation. This example clearly shows the power of the EKF, which has been successfully applied for attitude estimation of many spacecraft. The unscented Kalman filter provides a more robust approach to initial condition errors [11].

6.3 Farrenkopf's Steady-State Analysis

In this section we consider the case of single-axis attitude estimation with an angle measurement and gyro outputs used in place of a dynamic model. The single angle case is useful because Farrenkopf found a closed-form solution for its Kalman filter error-covariance [15]. We will use an alternative derivation that includes an additional source of gyro errors not included in Farrenkopf's original analysis [26].

The true single-axis attitude angle θ^{true} obeys the kinematic equation

$$\dot{\theta}^{\text{true}} = \omega^{\text{true}} \tag{6.95}$$

where ω^{true} is the true angular velocity. A rate-integrating gyro measures an angle θ^{RIG}, which is the integral of

$$\dot{\theta}^{\text{RIG}} = \omega = \omega^{\text{true}} + \beta^{\text{true}} + \eta_v \tag{6.96}$$

where ω is the rate sensed by the gyro, which is corrupted by the gyro drift β^{true} and a zero-mean Gaussian white-noise process η_v with spectral density σ_v^2. The drift rate is modeled by a random walk process, given by

$$\dot{\beta}^{\text{true}} = \eta_u \tag{6.97}$$

where η_u is a zero-mean Gaussian white-noise process with spectral density given by σ_u^2. Thus the three-component state $\mathbf{x} = [\theta^{\text{true}}\ \beta^{\text{true}}\ \theta^{\text{RIG}}]^T$ obeys the discrete-time propagation equation

$$\mathbf{x}(t+\delta t) = \begin{bmatrix} 1 & 0 & 0 \\ 0 & 1 & 0 \\ 0 & \delta t & 1 \end{bmatrix} \mathbf{x}(t) + \begin{bmatrix} 1 \\ 0 \\ 1 \end{bmatrix} \delta\theta^{\text{true}} + \begin{bmatrix} 0 \\ N_u \\ N_v \end{bmatrix} \tag{6.98}$$

In this equation δt is the interval between gyro measurements and is *not* assumed to be infinitesimal. The angular motion $\delta\theta^{\text{true}}$, the integral of ω^{true} over δt, is also not infinitesimal. The quantities N_u and N_v are defined by

$$N_u = \int_t^{t+\delta t} \eta_u(t')dt' \tag{6.99a}$$

$$N_v = \int_t^{t+\delta t} \left[\eta_v(t') + (t + \delta t - t')\eta_u(t')\right] dt' \tag{6.99b}$$

Now we will use the gyro measurement in place of the unknown angular motion $\delta\theta^{\text{true}}$. This is what we mean by saying that we use the gyros as a replacement for a dynamic model. The gyro measurement at the end of the propagation interval is

$$\theta^{\text{out}}(t + \delta t) = \theta^{\text{RIG}}(t + \delta t) + v_e = \begin{bmatrix} 0 & \delta t & 1 \end{bmatrix} \mathbf{x}(t) + \delta\theta^{\text{true}} + N_v + v_e \tag{6.100}$$

where v_e is a zero-mean Gaussian measurement noise with variance σ_e^2. We assume that η_u, η_v, and v_e are uncorrelated. Solving this equation for $\delta\theta^{\text{true}}$ in terms of $\theta^{\text{out}}(t + \delta t)$ and substituting back into Eq. (6.98) gives

$$\begin{aligned}\mathbf{x}(t + \delta t) &= \begin{bmatrix} 1 & 0 & 0 \\ 0 & 1 & 0 \\ 0 & \delta t & 1 \end{bmatrix} \mathbf{x}(t) + \begin{bmatrix} 1 \\ 0 \\ 1 \end{bmatrix} \left\{\theta^{\text{out}}(t+\delta t) - \begin{bmatrix} 0 & \delta t & 1 \end{bmatrix} \mathbf{x}(t) - N_v - v_e\right\} + \begin{bmatrix} 0 \\ N_u \\ N_v \end{bmatrix} \\ &= \begin{bmatrix} 1 & -\delta t & -1 \\ 0 & 1 & 0 \\ 0 & 0 & 0 \end{bmatrix} \mathbf{x}(t) + \begin{bmatrix} 1 \\ 0 \\ 1 \end{bmatrix} \theta^{\text{out}}(t + \delta t) + \begin{bmatrix} -N_v - v_e \\ N_u \\ -v_e \end{bmatrix}\end{aligned} \tag{6.101}$$

This is the discrete-time state propagation equation with the gyros used in dynamic model replacement mode. Because η_u, η_v, and v_e have zero mean, the expectation $\hat{\mathbf{x}} \equiv E\{\mathbf{x}\}$ and the state error vector $\Delta\mathbf{x} \equiv \mathbf{x} - \hat{\mathbf{x}}$ obey

$$\hat{\mathbf{x}}(t + \delta t) = \Phi(\delta t)\,\hat{\mathbf{x}}(t) + \begin{bmatrix} 1 \\ 0 \\ 1 \end{bmatrix} \theta^{\text{out}}(t + \delta t) \tag{6.102a}$$

$$\Delta\mathbf{x}(t + \delta t) = \Phi(\delta t)\,\Delta\mathbf{x}(t) + \begin{bmatrix} -N_v - v_e \\ N_u \\ -v_e \end{bmatrix} \tag{6.102b}$$

where

$$\Phi(\delta t) = \begin{bmatrix} 1 & -\delta t & -1 \\ 0 & 1 & 0 \\ 0 & 0 & 0 \end{bmatrix} \tag{6.103}$$

The covariance $P \equiv E\{\Delta\mathbf{x}\Delta\mathbf{x}^T\}$ propagates according to

$$P(t+\delta t) = \Phi(\delta t)P(t)\Phi^T(\delta t) + Q(\delta t) \tag{6.104}$$

where

$$Q(\delta t) = E\left\{\begin{bmatrix} -N_v - v_e \\ N_u \\ -v_e \end{bmatrix}\begin{bmatrix} -N_v - v_e \\ N_u \\ -v_e \end{bmatrix}^T\right\}$$
$$= \begin{bmatrix} \sigma_v^2\delta t + \frac{1}{3}\sigma_u^2\delta t^3 + \sigma_e^2 & -\frac{1}{2}\sigma_u^2\delta t^2 & \sigma_e^2 \\ -\frac{1}{2}\sigma_u^2\delta t^2 & \sigma_u^2\delta t & 0 \\ \sigma_e^2 & 0 & \sigma_e^2 \end{bmatrix} \tag{6.105}$$

The expectation calculations involved in $Q(\delta t)$ are very similar to the computations in Sect. 4.7.1. It is easy to show by mathematical induction that propagation by n steps gives

$$P(t+n\delta t) = \Phi(n\delta t)P(t)\Phi^T(n\delta t) + Q(n\delta t) \tag{6.106}$$

This equation has two remarkable properties. The first is that it depends only on the total propagation time $n\delta t$ and not on n and δt separately. The second is that the noise term σ_e does not accumulate, so that the covariance only depends on the output noise of the last readout.

Now assume that we have measurements from an angle sensor at an interval Δt that is some multiple of the gyro measurement time. It doesn't matter what the angle sensor is, but we assume that its measurement variance is the scalar σ_n^2. We want to find a steady-state solution for the covariance, which means that we have the same covariance P^- before every angle sensor measurement and the same covariance P^+ after every measurement update. These obey the Kalman filter propagation and update equations (see Table 12.2)

$$P^- = \Phi(\Delta t)P^+\Phi^T(\Delta t) + Q(\Delta t) \tag{6.107a}$$

$$P^+ = P^- - P^-H^T(HP^-H^T + \sigma_n^2)^{-1}HP^- \tag{6.107b}$$

The measurement sensitivity matrix is $H = [1\ 0\ 0]$, so HP^-H^T is a scalar. Equation (6.107) are equivalent to 12 coupled scalar equations for the 12 independent components of the symmetric covariance matrices P^- and P^+. Solving them is straightforward but difficult, so we will just state the solution:

$$P^- = \sigma_n^2 \begin{bmatrix} \zeta^2 - 1 & -\zeta\, S_u/\Delta t & S_e^2 \\ -\zeta\, S_u/\Delta t & \left[\zeta - (1+S_e^2)\zeta^{-1} + \frac{1}{2}S_u\right] S_u/\Delta t^2 & 0 \\ S_e^2 & 0 & S_e^2 \end{bmatrix} \tag{6.108a}$$

$$P^+ = \sigma_n^2 \begin{bmatrix} 1 - \zeta^{-2} & -\zeta^{-1} S_u/\Delta t & \zeta^{-2} S_e^2 \\ -\zeta^{-1} S_u/\Delta t & \left[\zeta - (1+S_e^2)\zeta^{-1} - \frac{1}{2}S_u\right] S_u/\Delta t^2 & \zeta^{-1} S_e^2 S_u/\Delta t \\ \zeta^{-2} S_e^2 & \zeta^{-1} S_e^2 S_u (\Delta t)^{-1} & S_e^2(1 - \zeta^{-2} S_e^2) \end{bmatrix} \tag{6.108b}$$

where

$$S_u \equiv \sigma_u \Delta t^{3/2}/\sigma_n \tag{6.109a}$$

$$S_v \equiv \sigma_v \Delta t^{1/2}/\sigma_n \tag{6.109b}$$

$$S_e \equiv \sigma_e/\sigma_n \tag{6.109c}$$

and where ζ obeys the quartic equation

$$\begin{aligned} 0 &= \zeta^4 - S_u\zeta^3 - \left[2(1+S_e^2) + S_v^2 - S_u^2/6\right]\zeta^2 - (1+S_e^2)S_u\zeta + (1+S_e^2)^2 \\ &= \left[\zeta^2 - 2(\gamma + S_u/4)\zeta + 1 + S_e^2\right]\left[\zeta^2 + 2(\gamma - S_u/4)\zeta + 1 + S_e^2\right] \end{aligned} \tag{6.110}$$

with

$$\gamma \equiv \sqrt{1 + S_e^2 + \frac{1}{4}S_v^2 + \frac{1}{48}S_u^2} \tag{6.111}$$

Equation (6.110) has four roots, but the only root giving a positive definite P^+ is the largest one, namely

$$\zeta = \gamma + \frac{1}{4}S_u + \frac{1}{2}\sqrt{2\gamma S_u + S_v^2 + \frac{1}{3}S_u^2} \tag{6.112}$$

We are not interested in the entire covariance matrix. The quantities of greatest interest are the standard deviation of the angle estimate, which is the square root of the $1-1$ component of the covariance, and the standard deviation of the drift estimate, which is the square root of the $2-2$ component. The standard deviations prior to the sensor update are

$$\sigma_\theta^- = \sigma_n\sqrt{\zeta^2 - 1} \tag{6.113a}$$

$$\sigma_\beta^- = \sqrt{2\sigma_n\sigma_u}\,\Delta t^{-1/4}\sqrt{\gamma - (1+S_e^2)\zeta^{-1} + S_u/2} \tag{6.113b}$$

and after the update

$$\sigma_\theta^+ = \sigma_n\sqrt{1 - \zeta^{-2}} \tag{6.114a}$$

$$\sigma_\beta^+ = \sqrt{2\sigma_n\sigma_u}\,\Delta t^{-1/4}\sqrt{\gamma - (1+S_e^2)\zeta^{-1}} \tag{6.114b}$$

If $\sigma_e = 0$, the third columns and third rows of Q, P^-, and P^+ are all zero, reflecting the fact that the gyro measurements give perfect knowledge of θ^{RIG}. In this case, the covariance propagation and update equations are equivalent to

$$\tilde{P}^- = \tilde{\Phi}(\Delta t)\tilde{P}^+\tilde{\Phi}^T(\Delta t) + \tilde{Q}(\Delta t) \tag{6.115a}$$

$$\tilde{P}^+ = \tilde{P}^- - \tilde{P}^-\tilde{H}^T(\tilde{H}\tilde{P}^-\tilde{H}^T + \sigma_n^2)^{-1}\tilde{H}\tilde{P}^- \tag{6.115b}$$

where $\tilde{P}^-$, $\tilde{P}^-$, $\tilde{P}^+$, $\tilde{Q}$, and $\tilde{\Phi}$ are the upper-left 2×2 blocks of P^-, P^+, Q, and Φ with $\sigma_e = 0$, respectively, and $\tilde{H} = [\,1\ 0\,]$. This is the model originally analyzed by Farrenkopf, and it is the basis of most of the Kalman filters used on spacecraft and presented in this text. It corresponds to the dynamical model of Eqs. (6.95) and (6.38) with $S^{\text{true}} = 0$, namely

$$\dot{\theta}^{\text{true}} = \omega - \beta^{\text{true}} - \eta_v \tag{6.116a}$$

$$\dot{\beta}^{\text{true}} = \eta_u \tag{6.116b}$$

This is the dynamical model that was used in Farrenkopf's original paper, and it has been used by guidance and control engineers for many years [15]. It was the basis for the analysis in Sect. 4.7.1 and will be employed extensively in this chapter. It is useful to retain the dependence on gyro electronic noise σ_e for covariance estimates, though.

Figure 6.4 shows the pre- and post-update angle standard deviations for angle sensor update times between 0.01 and 100 s for parameters characteristic of a ring laser gyro with very low drift: $\sigma_v = 0.025\ \text{deg/h}^{1/2} = 7.27\ \mu\text{rad/s}^{1/2}$ and $\sigma_u = 3.7 \times 10^{-3}\ \text{deg/h}^{3/2} = 3 \times 10^{-4}\ \mu\text{rad/s}^{3/2}$. The angle sensor is assumed to be a star tracker having measurement noise with standard deviation $\sigma_n = 15\ \mu\text{rad}$. The solid curves are for $\sigma_e = 5\ \mu\text{rad}$, and the dashed curves are for $\sigma_e = 0$. It is clear from the figure that gyro output white noise is only important for frequent star tracker updates, and that it causes the estimation accuracy to fall more slowly in that limit.

In the limiting case of very frequent updates, the pre-update and post-update attitude error standard deviations both approach constant values if $\sigma_e \neq 0$. In this case $\zeta \approx \gamma \approx \sqrt{1 + S_e^2}$ and

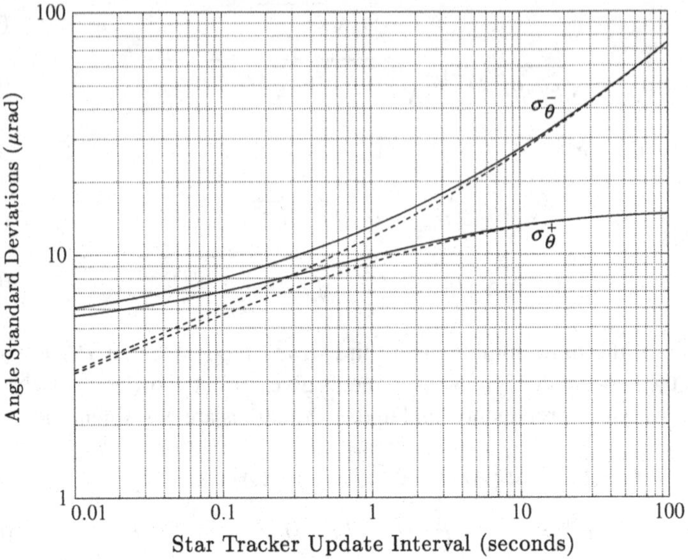

Fig. 6.4 Steady-state attitude estimation standard deviations, *solid curves* for $\sigma_e = 5\,\mu$rad, *dashed curves* for $\sigma_e = 0$

$$\sigma_\theta^- \to \sigma_e, \quad \text{and} \quad \sigma_\theta^+ \to \frac{\sigma_e \sigma_n}{\sqrt{\sigma_e^2 + \sigma_n^2}} \tag{6.117}$$

If $\sigma_e = 0$, the frequent-update case approaches the continuous-update limit, given by

$$\sigma_\theta^\pm \to \sigma_n (S_v^2 + 2S_u)^{1/4} = \Delta t^{1/4} \sigma_n^{1/2} \left(\sigma_v^2 + 2\sigma_n \sigma_u \Delta t^{1/2}\right)^{1/4} \tag{6.118}$$

The even simpler limiting form when the contribution of σ_u to the attitude error is negligible is given by

$$\sigma_\theta^\pm \to \Delta t^{1/4} (\sigma_n \sigma_v)^{1/2} \tag{6.119}$$

which indicates a one-half power dependence on both σ_n and σ_v, and a one-fourth power dependence on the update time Δt. This shows why it is extremely difficult to improve the attitude performance by simply increasing the update frequency. To lowest order in σ_u, the standard deviation of the gyro drift bias approaches

$$\sigma_\beta^\pm \to (\sigma_u \sigma_v)^{1/2} \tag{6.120}$$

in the frequent update limit, regardless of whether or not σ_e is zero. The fact that the attitude covariance has a "floor" determined by σ_e reveals a shortcoming of using gyros in model replacement mode. Rotational inertia prevents the spacecraft from executing motions on the order of σ_e on very short time scales, and a filter incorporating actual spacecraft dynamics would reflect this fact. The extra computational effort cannot be justified if Farrenkopf's covariance estimates indicate that mission requirements can be met without dynamic modeling.

Farrenkopf's equations bound the expected attitude and bias errors to provide an initial estimate of attitude performance. Using the noise parameters of Example 6.2 in Eq. (6.119) gives an approximate 3σ bound of 6.96 µrad for the attitude error, which is very to close the actual solution of 7.18 µrad. Even though the actual model is not a single-axis model, Farrenkopf's analysis can provide good estimates for various gyro parameters and sampling intervals. Converting 6.96 µrad to degrees gives 4×10^{-4} deg, which closely matches the roll and pitch errors of the results shown in Fig. 6.3b.

6.4 Magnetometer Calibration

A paramount issue to the attitude accuracy obtained using magnetometer measurements is the precision of the onboard calibration. The accuracy obtained using a three-axis magnetometer (TAM) depends on a number of factors, including: biases, scale factors, and non-orthogonality corrections. Scale factors and non-orthogonality corrections occur because the individual magnetometer axes are not orthonormal, typically due to thermal gradients within the magnetometer or to mechanical stress from the spacecraft [1]. Magnetometer calibration is often accomplished using batch methods, where an entire set of data must be stored to determine the unknown parameters. This process is often repeated many times during the lifetime of a spacecraft in order to ensure that the best possible precision is obtained from magnetometer measurements.

The magnetometer calibration problem is easy to solve if an accurate attitude estimate is known a priori. However, this is generally not the case. Fortunately, the norms of the body-measurement and geomagnetic-reference vectors provide an attitude-independent scalar observation. For the noise-free case, these norms are identical because the attitude matrix preserves the length of a vector. This process is also known as "scalar checking" [22]. Unfortunately, even for the simpler magnetometer-bias determination problem, the loss function to be minimized is quartic in nature. Gambhir proposed the most common technique to overcome this difficulty, a "centering" approximation to yield a quadratic loss function that can be minimized using simple linear least squares [17]. Alonso and Shuster expand upon Gambhir's approach by using a second step that employs the centered estimate as an initial value to an iterative nonlinear least squares approach. Their algorithm, called "TWOSTEP" [2], has been shown to perform well when other algorithms fail due to divergence problems. The algorithm presented here is Alonso and Shuster's extension of this approach to perform a complete calibration involving biases as well as scale factors and non-orthogonality corrections [1]. All of these approaches are based on batch processing. A realtime approach is shown in [9].

6.4.1 Measurement Model

In this section the TAM measurement model and attitude-independent observation are summarized. More details on these concepts can be found in [1]. The magnetometer measurements can be modeled as

$$\mathbf{B}_k = (I_3 + D^{\text{true}})^{-1}(\mathcal{O}^T A_k^{\text{true}} \mathbf{R}_k + \mathbf{b}^{\text{true}} + \boldsymbol{\epsilon}_k), \quad k = 1, 2, \ldots, N \tag{6.121}$$

where $\mathbf{B}_k$ is the measurement of the magnetic field by the magnetometer at time t_k, $\mathbf{R}_k$ is the corresponding value of the geomagnetic field with respect to the Earth-Centered/Earth-Fixed (ECEF) system, A_k^{true} is the unknown attitude matrix of the magnetometer with respect to the Earth-fixed coordinates, D^{true} is an unknown fully-populated matrix of scale factors (the diagonal elements) and non-orthogonality corrections (the off-diagonal elements), $\mathcal{O}$ is an orthogonal matrix (see [1] for a discussion on the physical connotations of this matrix), $\mathbf{b}^{\text{true}}$ is the bias vector, $\boldsymbol{\epsilon}_k$ is the measurement noise vector that is assumed to be a zero-mean Gaussian process with covariance Σ_k, and N is the total number of available measurements. The matrix D^{true} can be assumed to be symmetric without loss of generality, because any skew-symmetric contribution is equivalent to a rotation that can be absorbed into $\mathcal{O}$:

$$D^{\text{true}} = \begin{bmatrix} D_{11}^{\text{true}} & D_{12}^{\text{true}} & D_{13}^{\text{true}} \\ D_{12}^{\text{true}} & D_{22}^{\text{true}} & D_{23}^{\text{true}} \\ D_{13}^{\text{true}} & D_{23}^{\text{true}} & D_{33}^{\text{true}} \end{bmatrix} \tag{6.122}$$

The goal of the full attitude-independent calibration problem is to estimate D^{true} and $\mathbf{b}^{\text{true}}$, because an attitude-independent method cannot observe $\mathcal{O}$.

An attitude-independent approach is possible by computing

$$\begin{aligned} y_k &\equiv \|\mathbf{B}_k\|^2 - \|\mathbf{R}_k\|^2 \\ &= -\mathbf{B}_k^T[2D^{\text{true}} + (D^{\text{true}})^2]\mathbf{B}_k + 2\mathbf{B}_k^T(I_3 + D^{\text{true}})\mathbf{b}^{\text{true}} - \|\mathbf{b}^{\text{true}}\|^2 + v_k \end{aligned} \tag{6.123}$$

where the effective noise is given by

$$v_k \equiv 2[(I_3 + D^{\text{true}})\mathbf{B}_k - \mathbf{b}^{\text{true}}]^T \boldsymbol{\epsilon}_k - \|\boldsymbol{\epsilon}_k\|^2 \tag{6.124}$$

The effective noise is approximately Gaussian with mean $\mu_k \equiv E\{v_k\}$ and variance $\sigma_k^2 \equiv E\{v_k^2\} - \mu_k^2$ given by

$$\mu_k = -\text{tr}(\Sigma_k) \tag{6.125a}$$

$$\sigma_k^2 = 4[(I_3 + D^{\text{true}})\mathbf{B}_k - \mathbf{b}^{\text{true}}]^T \Sigma_k [(I_3 + D^{\text{true}})\mathbf{B}_k - \mathbf{b}^{\text{true}}] + 2\left(\text{tr}\, \Sigma_k^2\right) \tag{6.125b}$$

$$\Sigma_k = E\left\{\boldsymbol{\epsilon}_k \boldsymbol{\epsilon}_k^T\right\} \tag{6.125c}$$

Note that the measurement variance is a function of the unknown parameters.

The attitude-independent measurement in Eq. (6.123) is highly nonlinear in the calibration parameters. We can define a modified state vector so that the nonlinear dependence appears in only one term. To accomplish this task the following quantities are first defined:

$$E^{\text{true}} \equiv 2D^{\text{true}} + (D^{\text{true}})^2 = \begin{bmatrix} E_{11}^{\text{true}} & E_{12}^{\text{true}} & E_{13}^{\text{true}} \\ E_{12}^{\text{true}} & E_{22}^{\text{true}} & E_{23}^{\text{true}} \\ E_{13}^{\text{true}} & E_{23}^{\text{true}} & E_{33}^{\text{true}} \end{bmatrix} \tag{6.126a}$$

$$\mathbf{c}^{\text{true}} \equiv (I_3 + D^{\text{true}})\mathbf{b}^{\text{true}} \tag{6.126b}$$

$$\mathbf{S}_k \equiv \left[B_{1_k}^2 \;\; B_{2_k}^2 \;\; B_{3_k}^2 \;\; 2B_{1_k}B_{2_k} \;\; 2B_{1_k}B_{3_k} \;\; 2B_{2_k}B_{3_k} \right]^T \tag{6.126c}$$

$$\mathbf{E}^{\text{true}} \equiv \left[E_{11}^{\text{true}} \;\; E_{22}^{\text{true}} \;\; E_{33}^{\text{true}} \;\; E_{12}^{\text{true}} \;\; E_{13}^{\text{true}} \;\; E_{23}^{\text{true}} \right]^T \tag{6.126d}$$

Note that E^{true} is also symmetric. The attitude-independent measurement can now be written as

$$y_k = L_k \mathbf{x}'^{\text{true}} - \|\mathbf{b}(\mathbf{x}'^{\text{true}})\|^2 + v_k \tag{6.127}$$

where the row vector L_k and the modified state vector $\mathbf{x}'^{\text{true}}$ are defined by

$$L_k \equiv \left[2\mathbf{B}_k^T \;\; -\mathbf{S}_k^T \right] \tag{6.128a}$$

$$\mathbf{x}'^{\text{true}} \equiv \left[(\mathbf{c}^{\text{true}})^T \;\; (\mathbf{E}^{\text{true}})^T \right]^T \tag{6.128b}$$

Now the only nonlinear dependence on $\mathbf{x}'^{\text{true}}$ is in the term $\|\mathbf{b}(\mathbf{x}'^{\text{true}})\|^2$.

Equation (6.127) can be solved by iteration. We start with an initial estimate $\hat{\mathbf{x}}'_0$, which can be zero, and compute

$$\hat{\mathbf{x}}'_i = P' \sum_{k=1}^{N} \frac{1}{\sigma_k^2} \left(y_k + \|\mathbf{b}(\hat{\mathbf{x}}'_{i-1})\|^2 - \mu_k \right) L_k^T \quad \text{for } i = 1, 2, \ldots \tag{6.129}$$

where N is the number of observations and

$$P' = \left(\sum_{k=1}^{N} \sigma_k^{-2} L_k^T L_k \right)^{-1} \tag{6.130}$$

The iteration is terminated when $\hat{\mathbf{x}}'_i - \hat{\mathbf{x}}'_{i-1}$ is less than some chosen tolerance.

This computation requires the conversion from $\hat{\mathbf{x}}'$ back to the variables $\mathbf{b}$ and D [1]. The conversion of the true and estimated quantities is identical. Since E is symmetric its singular value decomposition is given by

$$E = U\,S\,U^T \tag{6.131}$$

where U is orthogonal and S is diagonal with elements s_1, s_2, and s_3. To determine D first compute the following elements of a diagonal matrix W:

$$w_j = -1 + \sqrt{1 + s_j}, \quad j = 1, 2, 3 \tag{6.132}$$

Then the matrix D is given by

$$D = U\,W\,U^T \tag{6.133}$$

The vector $\mathbf{b}$ is simply given by

$$\mathbf{b} = (I_3 + D)^{-1}\mathbf{c} \tag{6.134}$$

6.4.2 Centered Solution

The centering approximation effectively removes the nonlinear dependence of $\|\mathbf{b}\|^2$ in Eq. (6.127), making it possible to develop a solution for the calibration parameters that does not require iteration [17]. We define *center variables*, indicated by an overbar, and *centered variables*, indicated by a breve, as follows:

$$\bar{\sigma}^2 \equiv \left(\sum_{k=1}^{N} \sigma_k^{-2}\right)^{-1} \tag{6.135a}$$

$$\bar{L} \equiv \bar{\sigma}^2 \sum_{k=1}^{N} \sigma_k^{-2} L_k, \quad \breve{L}_k \equiv L_k - \bar{L} \tag{6.135b}$$

$$\bar{y} \equiv \bar{\sigma}^2 \sum_{k=1}^{N} \sigma_k^{-2} y_k, \quad \breve{y}_k \equiv y_k - \bar{y} \tag{6.135c}$$

$$\bar{v} \equiv \bar{\sigma}^2 \sum_{k=1}^{N} \sigma_k^{-2} v_k, \quad \breve{v}_k \equiv v_k - \bar{v} \tag{6.135d}$$

$$\bar{\mu} \equiv \bar{\sigma}^2 \sum_{k=1}^{N} \sigma_k^{-2} \mu_k, \quad \breve{\mu}_k \equiv \mu_k - \bar{\mu} \tag{6.135e}$$

The center variables and centered variables have the measurement models

$$\bar{y} = \bar{L}\,\mathbf{x}'^{\text{true}} - \|\mathbf{b}(\mathbf{x}'^{\text{true}})\|^2 + \bar{v} \tag{6.136a}$$

$$\breve{y}_k = \breve{L}_k \mathbf{x}'^{\text{true}} + \breve{v}_k \tag{6.136b}$$

Note that Eq. (6.136b) is now linear in $\mathbf{x}'^{\text{true}}$, so linear least squares can be employed. The centering process has introduced correlations among the noise terms $\breve{v}_k$,[5] but these are generally ignored in practice, giving the centered estimate, $\hat{\breve{\mathbf{x}}}'$ and its approximate covariance, $\breve{P}'$, as [2]

$$\breve{P}' = \left(\sum_{k=1}^{N} \sigma_k^{-2} \breve{L}_k^T \breve{L}_k\right)^{-1} \tag{6.137a}$$

$$\hat{\breve{\mathbf{x}}}' = \breve{P}' \sum_{k=1}^{N} \sigma_k^{-2} (\breve{y}_k - \breve{\mu}_k) \breve{L}_k^T \tag{6.137b}$$

Since the centered solution is linear, a sequential formulation can be derived to provide realtime estimates. First, the sequential formulas for the averaged quantities are given by

$$\bar{L}_{k+1} = \frac{1}{\sigma_{k+1}^2 + \bar{\sigma}_k^2} \left(\sigma_{k+1}^2 \bar{L}_k + \bar{\sigma}_k^2 L_{k+1}\right) \tag{6.138a}$$

$$\bar{y}_{k+1} = \frac{1}{\sigma_{k+1}^2 + \bar{\sigma}_k^2} \left(\sigma_{k+1}^2 \bar{y}_k + \bar{\sigma}_k^2 y_{k+1}\right) \tag{6.138b}$$

$$\bar{\mu}_{k+1} = \frac{1}{\sigma_{k+1}^2 + \bar{\sigma}_k^2} \left(\sigma_{k+1}^2 \bar{\mu}_k + \bar{\sigma}_k^2 \mu_{k+1}\right) \tag{6.138c}$$

where

$$\bar{\sigma}_{k+1}^2 = \left(\bar{\sigma}_k^{-2} + \sigma_{k+1}^{-2}\right)^{-1} \tag{6.139}$$

Next, the following centered variables are defined:

$$\breve{L}_{k+1} \equiv L_{k+1} - \bar{L}_{k+1} \tag{6.140a}$$

$$\breve{y}_{k+1} \equiv y_{k+1} - \bar{y}_{k+1} \tag{6.140b}$$

$$\breve{\mu}_{k+1} \equiv \mu_{k+1} - \bar{\mu}_{k+1} \tag{6.140c}$$

Finally, the sequential formulas for the centered estimate of $\hat{\breve{\mathbf{x}}}'$ and covariance $\breve{P}'$ are given by

$$K_{k+1} \equiv \breve{P}_k' \breve{L}_{k+1}^T \left(\breve{L}_{k+1} \breve{P}_k' \breve{L}_{k+1}^T + \sigma_{k+1}^2\right)^{-1} \tag{6.141a}$$

$$\hat{\breve{\mathbf{x}}}_{k+1}' = \hat{\breve{\mathbf{x}}}_k' + K_{k+1}\left(\breve{y}_{k+1} - \breve{\mu}_{k+1} - \breve{L}_{k+1} \hat{\breve{\mathbf{x}}}_k'\right) \tag{6.141b}$$

[5] Note, in particular, that $\sum_{k=1}^{N} \sigma_k^{-2} \breve{v}_k = 0$.

$$\breve{P}'_{k+1} = (I_9 - K_{k+1}\breve{L}_{k+1})\breve{P}'_k \tag{6.141c}$$

Note that only an inverse of a scalar quantity is required in this process, which can be initialized using a small batch of data. Also, there is an approach for determining σ^2_{k+1} that uses the previous estimate in Eq. (6.125b).

6.4.3 The TWOSTEP Algorithm

The second step of the TWOSTEP algorithm uses a nonlinear maximum likelihood minimization to find the optimal estimate of $\mathbf{x}'^{\text{true}}$. The first step computes the centered solution to initialize the nonlinear least squares iteration process. The data-dependent part of the negative log-likelihood function to be minimized in the second step is:

$$J(\mathbf{x}'^{\text{true}}) = \frac{1}{2}\sum_{k=1}^{N}\left[\frac{1}{\sigma_k^2}(y_k - L_k\mathbf{x}'^{\text{true}} + \|\mathbf{b}(\mathbf{x}'^{\text{true}})\|^2 - \mu_k)^2 + \log\sigma_k^2\right] \tag{6.142}$$

The partial derivative of $h(\mathbf{x}') \equiv L_k\mathbf{x}' - \|\mathbf{b}(\mathbf{x}')\|^2$ with respect to $\mathbf{x}'$ is required for the nonlinear least squares iterations. The partial of $L_k\mathbf{x}'$ is simply given by L_k^T. The partial derivatives of $\|\mathbf{b}(\mathbf{x}')\|^2$ are given by

$$\frac{\partial\|\mathbf{b}(\mathbf{x}')\|^2}{\partial c_m} = 2[(I_3 + E)^{-1}\mathbf{c}]_m \tag{6.143a}$$

$$\frac{\partial\|\mathbf{b}(\mathbf{x}')\|^2}{\partial E_{mn}} = -(2 - \delta_{mn})[(I_3 + E)^{-1}\mathbf{c}]_m[(I_3 + E)^{-1}\mathbf{c}]_n \tag{6.143b}$$

where $[(I_3 + E)^{-1}\mathbf{c}]_m$ is the mth element of $(I_3 + E)^{-1}\mathbf{c}$. As mentioned previously, the least squares solution process typically ignores the nonlinear dependence of σ_k^2 on the unknown parameters.

Once estimates $\hat{\mathbf{c}}$ and $\hat{E}$ for $\mathbf{c}^{\text{true}}$ and E^{true} are found, the estimates $\hat{\mathbf{b}}$ and $\hat{D}$ for $\mathbf{b}^{\text{true}}$ and D^{true} can be determined using Eqs. (6.133) and (6.134). Define

$$\hat{\mathbf{x}} \equiv \begin{bmatrix}\hat{\mathbf{b}}^T & \hat{\mathbf{D}}^T\end{bmatrix}^T \tag{6.144a}$$

$$\hat{\mathbf{D}} \equiv \begin{bmatrix}\hat{D}_{11} & \hat{D}_{22} & \hat{D}_{33} & \hat{D}_{12} & \hat{D}_{13} & \hat{D}_{23}\end{bmatrix}^T \tag{6.144b}$$

The covariance associated with $\hat{\mathbf{x}}$, denoted P, must be computed from P', the covariance of $\hat{\mathbf{x}}'$, which is given by Eqs. (12.98) and (12.99) as the inverse of the Fisher information matrix. The conversion to P is computed by

$$P = \frac{\partial(\hat{\mathbf{b}}, \hat{\mathbf{D}})}{\partial(\hat{\mathbf{c}}, \hat{\mathbf{E}})} P' \left[\frac{\partial(\hat{\mathbf{b}}, \hat{\mathbf{D}})}{\partial(\hat{\mathbf{c}}, \hat{\mathbf{E}})}\right]^T \tag{6.145}$$

Note that in actuality the true variables should be used instead of the estimated ones in Eq. (6.145). However, since the true variables are unknown the estimated ones are used instead. The partials can be computed using

$$\frac{\partial(\hat{\mathbf{b}},\hat{\mathbf{D}})}{\partial(\hat{\mathbf{c}},\hat{\mathbf{E}})} = \left[\frac{\partial(\hat{\mathbf{c}},\hat{\mathbf{E}})}{\partial(\hat{\mathbf{b}},\hat{\mathbf{D}})}\right]^{-1} = \begin{bmatrix} I_3+\hat{D} & M_{cD}(\hat{\mathbf{b}}) \\ 0_{6\times 3} & M_{ED}(\hat{D}) \end{bmatrix}^{-1} \tag{6.146a}$$

$$M_{cD}(\hat{\mathbf{b}}) = \begin{bmatrix} \hat{b}_1 & 0 & 0 & \hat{b}_2 & \hat{b}_3 & 0 \\ 0 & \hat{b}_2 & 0 & \hat{b}_1 & 0 & \hat{b}_3 \\ 0 & 0 & \hat{b}_3 & 0 & \hat{b}_1 & \hat{b}_2 \end{bmatrix} \tag{6.146b}$$

$$M_{ED}(\hat{D}) = 2\,I_6 + \begin{bmatrix} 2\hat{D}_{11} & 0 & 0 & 2\hat{D}_{12} & 2\hat{D}_{13} & 0 \\ 0 & 2\hat{D}_{22} & 0 & 2\hat{D}_{12} & 0 & 2\hat{D}_{23} \\ 0 & 0 & 2\hat{D}_{33} & 0 & 2\hat{D}_{13} & 2\hat{D}_{23} \\ \hat{D}_{12} & \hat{D}_{12} & 0 & \hat{D}_{11}+\hat{D}_{22} & \hat{D}_{23} & \hat{D}_{13} \\ \hat{D}_{13} & 0 & \hat{D}_{13} & \hat{D}_{23} & \hat{D}_{11}+\hat{D}_{33} & \hat{D}_{12} \\ 0 & \hat{D}_{23} & \hat{D}_{23} & \hat{D}_{13} & \hat{D}_{12} & \hat{D}_{22}+\hat{D}_{33} \end{bmatrix} \tag{6.146c}$$

A strategy to handle the dependence of σ_k^2 on the parameters is to first assume $\mathbf{b}^{\text{true}}$ and D^{true} are zero to compute σ_k^2 and employ the nonlinear least squares iteration process. Then use the estimates of $\mathbf{b}^{\text{true}}$ and D^{true} to determine a new σ_k^2 and employ another nonlinear least squares iteration process. This refinement strategy continues until the estimates no longer change. In many cases this refinement strategy needs only one iteration.

Example 6.3. In this example, results of the TWOSTEP and centered formulations are shown using simulated data. The simulated spacecraft is modeled after the TRMM spacecraft. This is an Earth-pointing spacecraft (rotating about its y-axis) in low-Earth orbit with an inclination of 35° [3]. The geomagnetic field is simulated using a 10th-order International Geomagnetic Reference Field (IGRF) model [20]. The magnetometer-body and geomagnetic-reference vectors for the simulated runs each have a magnitude of about 50 micro-Tesla (μT). The measurement noise is assumed to be white and Gaussian, and the covariance is taken to be isotropic with a standard deviation of 0.05 μT. The measurements are sampled every 10 s over an 8-h span. The true values for the bias **b** and elements of the D matrix are shown in Table 6.4. Large values for the biases are used to test the robustness of the centered and TWOSTEP algorithms.

One thousand runs have been executed, which provide a Monte Carlo simulation. Shown in Table 6.4 are the averaged batch solutions given by the TWOSTEP and centered algorithms, each with their computed 3σ bounds. Both the centered and TWOSTEP algorithms do a good job at estimating all parameters in the mean sense, but the centered algorithm has larger variations than the TWOSTEP algorithm.

Table 6.4 Results using simulated magnetic field data

Parameter	Truth	Centered	TWOSTEP
b_1	5 μT	4.9972 ± 0.3373	4.9983 ± 0.0445
b_2	3 μT	3.0157 ± 0.2549	2.9841 ± 0.0525
b_3	6 μT	5.9955 ± 0.5447	5.9972 ± 0.0464
D_{11}	0.05	0.0500 ± 0.0035	0.0500 ± 0.0002
D_{22}	0.10	0.1001 ± 0.0023	0.0994 ± 0.0020
D_{33}	0.05	0.0500 ± 0.0094	0.0500 ± 0.0003
D_{12}	0.05	0.0499 ± 0.0027	0.0499 ± 0.0010
D_{13}	0.05	0.0499 ± 0.0054	0.0499 ± 0.0002
D_{23}	0.05	0.0499 ± 0.0045	0.0499 ± 0.0010

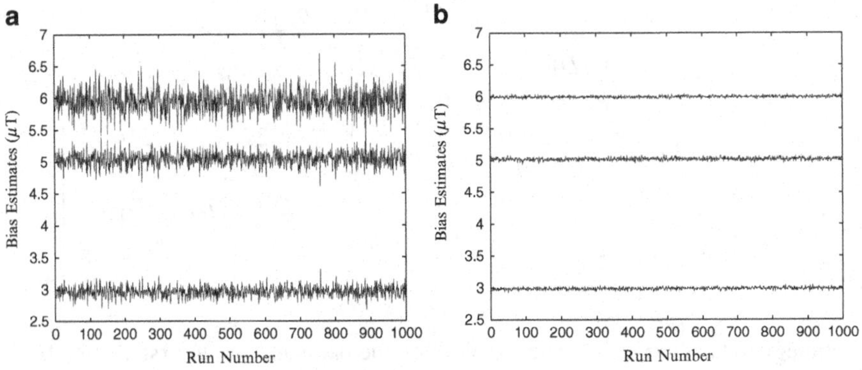

Fig. 6.5 (**a**) Centered and (**b**) TWOSTEP bias estimates

This can also be seen in Fig. 6.5, which plots the bias estimates for each Monte Carlo run. Results using colored noise, which is more realistic, can be found in [1].

6.4.4 Extended Kalman Filter Approach

This section presents an EKF to determine the calibration parameters in real time [9]. An advantage of this formulation over the sequential centered approach is that it computes $\hat{\mathbf{b}}$ and $\hat{\mathbf{D}}$ directly without a conversion from $\hat{\mathbf{c}}$ and $\hat{\mathbf{E}}$. Since the parameters to be estimated are constant, the state dynamic model is $\dot{\hat{\mathbf{x}}}(t) = \mathbf{0}$, where $\hat{\mathbf{x}}$ is the state vector defined by Eq. (6.144). The measurement model of Eq. (6.123) is written as $y_k = h_k(\mathbf{x}^{\text{true}}) + v_k$, where using Eq. (6.126) gives

$$h_k(\mathbf{x}^{\text{true}}) = -\mathbf{S}_k^T \mathbf{E}^{\text{true}}(\mathbf{D}^{\text{true}}) + 2\mathbf{B}_k^T (I_3 + D^{\text{true}})\mathbf{b}^{\text{true}} - \|\mathbf{b}^{\text{true}}\|^2 \tag{6.147}$$

Since no process noise appears in the state model, the updated quantities (state and covariance) are given by their respective propagated quantities. The EKF equations, which can be found in Chap. 12, then reduce to

$$\hat{\mathbf{x}}_{k+1} = \hat{\mathbf{x}}_k + K_k[y_{k+1} - h_{k+1}(\hat{\mathbf{x}}_k)] \tag{6.148a}$$

$$P_{k+1} = [I_9 - K_k H_{k+1}(\hat{\mathbf{x}}_k)] P_k \tag{6.148b}$$

$$K_k = P_k H_{k+1}^T(\hat{\mathbf{x}}_k) \left[H_{k+1}(\hat{\mathbf{x}}_k) P_k H_{k+1}^T(\hat{\mathbf{x}}_k) + \sigma_{k+1}^2(\hat{\mathbf{x}}_k) \right]^{-1} \tag{6.148c}$$

where P is the covariance of the estimated parameters for $\mathbf{b}^{\text{true}}$ and $\mathbf{D}^{\text{true}}$. The state dependence of the measurement variance is shown through Eq. (6.125b). The 1×9 measurement sensitivity matrix $H(\mathbf{x})$ is given by

$$H(\mathbf{x}) \equiv \frac{\partial h(\mathbf{x})}{\partial \mathbf{x}} = \left[2\mathbf{B}^T (I_3 + D) - 2\mathbf{b}^T \quad -\mathbf{S}^T M_{ED}(\hat{D}) + 2J \right] \tag{6.149}$$

where $\mathbf{S}$ is defined by Eq. (6.126c), $M_{ED}(\hat{D})$ by Eq. (6.146c), and

$$J \equiv \left[B_1 b_1 \quad B_2 b_2 \quad B_3 b_3 \quad B_1 b_2 + B_2 b_1 \quad B_1 b_3 + B_3 b_1 \quad B_2 b_3 + B_3 b_2 \right] \tag{6.150}$$

The sensitivity matrix $H(\hat{\mathbf{x}})$ in the EKF evaluates $H(\mathbf{x})$ at its current estimate, and the notations $h_{k+1}(\hat{\mathbf{x}}_k)$, $H_{k+1}(\hat{\mathbf{x}}_k)$, and $\sigma_{k+1}^2(\hat{\mathbf{x}}_k)$ denote an evaluation at the $k+1$ time-step measurement using $\mathbf{B}_{k+1}$ and at the k time-step estimate using $\hat{\mathbf{x}}_k$.

6.4.5 TRACE Spacecraft Results

In this section, EKF results using real data from the Transition Region And Coronal Explorer (TRACE) spacecraft are shown. TRACE is in a Sun-synchronous low-Earth orbit. The data collected for the spacecraft is given during an inertial pointing mode. The errors associated with the geomagnetic field model are typically spatially correlated and may be non-Gaussian in nature [6]. This violates the assumptions for all the estimators shown in this text. We still assume that the measurement noise is white and Gaussian, but the standard deviation is now increased to a value of 0.3 μT, which bounds the errors in a practical sense. The measurements are sampled every 3 s over a 6-h span.

The EKF initializes every component of the state vector to zero at time $t = 0$. The initial covariance matrix is diagonal, given by

$$P_0 = \text{blkdiag}\left([I_3 \ \ 0.0001 I_6]\right) \ \ (\mu\text{T})^2 \tag{6.151}$$

where blkdiag denotes a block diagonal matrix of appropriate dimension. Figure 6.6a shows EKF estimates for the bias vector. Figure 6.6b shows the 3σ bounds for the bias estimates. Note that the bias estimates with the larger 3σ bounds have greater variability, which is due to the relative observability between parameters (the second bias is the least observable parameter in this case). Similar results are obtained for the D matrix parameters.

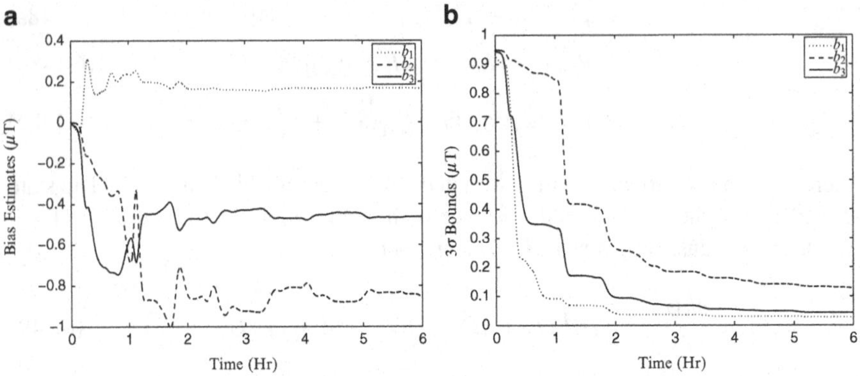

Fig. 6.6 TRACE bias estimates and 3σ bounds. (**a**) EKF bias estimates. (**b**) EKF 3σ bounds

Fig. 6.7 Norm residual using real TRACE data

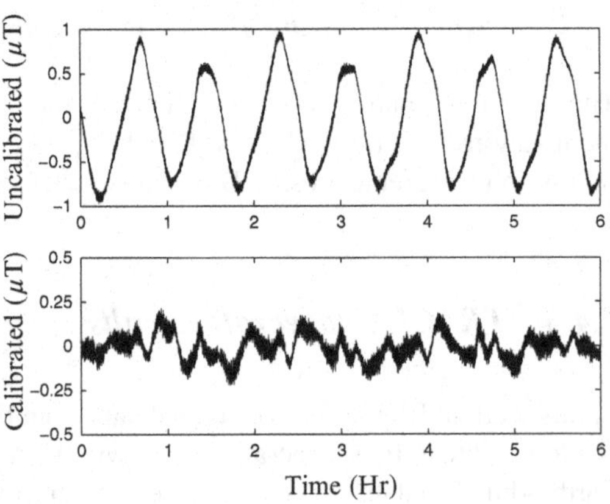

An investigation of the residuals between the norm of the estimated vector, using the uncalibrated and calibrated body vectors, and the geomagnetic-reference vector is useful to check the consistency of the results. A plot of these residuals is shown in Fig. 6.7. A spectrum analysis shows the presence of sinusoidal variations with periods equivalent to the orbital period ($\approx$ 90 min) and higher-order harmonics (see [6] for a model of this process). Clearly, the uncalibrated results have higher residuals than the calibrated results. This shows the power of using an onboard calibration algorithm that can be executed in real time.

Problems

6.1. Derive the relations in Eq. (6.33).

6.2. Derive the measurement sensitivity matrix in Eq. (6.66).

6.3. Prove that Eq. (6.79) is a good approximation to Eq. (6.78) if χ is very close to $A^{\text{true}}\mathbf{r}$.

Table 6.5 Orbital elements for GPS satellites

GPS	Orbital elements
1	$a = 26{,}560$ km, $e = 2.6221 \times 10^{-3}$, $M_0 = -82.277$ deg, $i = 55.037$ deg $\Omega_0 = 46.492$ deg, $\dot{\Omega} = -4.4332 \times 10^{-7}$ deg/s, $\omega = 24.008$ deg
2	$a = 26{,}561$ km, $e = 1.2955 \times 10^{-2}$, $M_0 = -38.066$ deg, $i = 53.808$ deg $\Omega_0 = 44.973$ deg, $\dot{\Omega} = -4.5511 \times 10^{-7}$ deg/s, $\omega = -144.31$ deg
3	$a = 26{,}561$ km, $e = 1.6708 \times 10^{-2}$, $M_0 = -27.429$ deg, $i = 53.692$ deg $\Omega_0 = -22.198$ deg, $\dot{\Omega} = -4.8392 \times 10^{-7}$ deg/s, $\omega = 77.836$ deg
4	$a = 26{,}560$ km, $e = 1.0645 \times 10^{-2}$, $M_0 = 154.02$ deg, $i = 53.751$ deg $\Omega_0 = 45.899$ deg, $\dot{\Omega} = -4.5708 \times 10^{-7}$ deg/s, $\omega = 61.387$ deg
5	$a = 26{,}561$ km, $e = 3.5214 \times 10^{-3}$, $M_0 = 89.435$ deg, $i = 54.277$ deg $\Omega_0 = 106.11$ deg, $\dot{\Omega} = -4.6624 \times 10^{-7}$ deg/s, $\omega = 21.164$ deg
6	$a = 26{,}560$ km, $e = 8.3880 \times 10^{-3}$, $M_0 = 78.699$ deg, $i = 54.067$ deg $\Omega_0 = -17.551$ deg, $\dot{\Omega} = -4.7738 \times 10^{-7}$ deg/s, $\omega = -10.890$ deg

6.4. The state transition matrix obeys the following relationships:

$$\Phi(t, t) = 0$$

$$\frac{d}{dt}\Phi(t, \tau) = F(t)\Phi(t, \tau)$$

Prove that the discrete error-state transition matrix in Eq. (6.83) obeys these relationships.

6.5. Prove that Φ_{11} defined by Eq. (6.83b) is an orthogonal matrix.

6.6. Reproduce the simulation results of Example 6.2 using Murrell's approach.

6.7. In this problem you will develop an EKF program for attitude estimation using GPS phase difference measurements discussed in Sect. 5.9. First, propagate six GPS satellites using the algorithm in Table 4.1 with orbital elements given in Table 6.5 and a time of applicability given by $t_a = 61{,}440$ s. Next, create the GPS receiver spacecraft orbit in GCI coordinates using the algorithm in Table 10.1 with the following orbital elements: $a = 6777.3$ km, $e = 1.584 \times 10^{-4}$, $M_0 = 93.4312°$, $i = 34.9654°$, $\Omega = 346.4609°$, and $\omega = 266.6234°$. Convert this GCI-based orbit into ECEF coordinates using Eq. (2.67) with an epoch of January 5, 2014 at 17:03:47. Use an interval of 0.1 s with a total run time of 600 s for both the GPS satellites and the receiver spacecraft. Then, create sightline vectors using Eq. (4.23).

Assuming an initial quaternion of $\mathbf{q}_0^{\text{true}} = \sqrt{2}/2 \times [0\ 1\ 0\ 1]^T$, propagate the attitude using an angular velocity of $\boldsymbol{\omega}(t) = \pi/180 \times [0\ 0\ 0.1]^T$ rad/s for all time. Create synthetic phase difference measurements using the following equation:

$$z_{ij} = \mathbf{b}_i^T A(\mathbf{q}^{\text{true}})\mathbf{s}_j + v_{ij}$$

where v_{ij} is a zero-mean Gaussian noise process with standard deviation of 0.026 cm. The three baseline vectors are given by

$$\mathbf{b}_1 = \begin{bmatrix} 100 \; 0 \; 0 \end{bmatrix}^T \text{ cm}$$

$$\mathbf{b}_2 = \begin{bmatrix} 0 \; 100 \; 0 \end{bmatrix}^T \text{ cm}$$

$$\mathbf{b}_3 = \begin{bmatrix} 100 \; 100 \; 0 \end{bmatrix}^T \text{ cm}$$

Create synthetic gyro measurements using the model shown in Eq. (4.54) with $S = 0$, $\sigma_u = \sqrt{10} \times 10^{-10}$ rad/s$^{3/2}$, and $\sigma_v = \sqrt{10} \times 10^{-7}$ rad/s$^{1/2}$. Sample the gyro measurements using an interval of 0.1 s. The initial bias for each axis is given by 0.1 deg/h. The initial covariance for the attitude error is set to $(1/3)^2$ deg^2, and the initial covariance for the gyro drift is set to 1^2 (deg/h)^2. Converting these quantities to (rad/s)^2 gives the following initial attitude and gyro drift covariances for each axis: $P_0^a = 0.3385 \times 10^{-4}$ rad^2 and $P_0^b = 0.2350 \times 10^{-10}$ (rad/s)^2, so that the initial covariance is given by

$$P_0 = \text{diag}\begin{bmatrix} P_0^a \; P_0^a \; P_0^a \; P_0^b \; P_0^b \; P_0^b \end{bmatrix}$$

Use the discrete-time covariance propagation in Eq. (12.132) with $\Upsilon_k = I_6$, and Φ_k and Q_k given by Eqs. (6.83) and (6.93), respectively. The initial attitude condition in the EKF is given by the true quaternion. The initial gyro bias conditions in the EKF are set to zero. Plot the attitude estimate errors along with their respective 3σ bounds and gyro bias estimates obtained from the EKF simulation results.

6.8. Another linearization approach for the MEKF involves defining the attitude errors in inertial space by using [16]

$$\mathbf{q}^{\text{true}} = \hat{\mathbf{q}} \otimes \delta\mathbf{q}(\delta\boldsymbol{\vartheta})$$

Show that the linearized $F(t)$ and $G(t)$ matrices for the mission mode Kalman filter are given by

$$F(t) = \begin{bmatrix} 0_{3\times3} & -A^T(\hat{\mathbf{q}}) \\ 0_{3\times3} & 0_{3\times3} \end{bmatrix}, \quad G(t) = \begin{bmatrix} -A^T(\hat{\mathbf{q}}) & 0_{3\times3} \\ 0_{3\times3} & I_3 \end{bmatrix}$$

Derive the new $H_k(\hat{\mathbf{x}}_k^-)$ matrix and quaternion update equation for this approach. Redo Example 6.2 using this new filter and compare the results to the standard MEKF given in Table. Note that Eq. (6.11) is used to compute the attitude errors for both filters. So the attitude portion of the error-covariance for the inertial-space MEKF must be rotated from inertial space to body space for this comparison.

6.9. Derive the discrete error-state transition matrix from the matrix $F(t)$ given in Problem 6.8. Also, derive the discrete-time covariance using Eq. (6.84) from your derived discrete error-state transition matrix and the matrix $G(t)$ in Problem 6.8.

6.10. In this exercise you will use the MEKF to estimate the attitude from TAM and Sun sensors as well as gyro measurements. The first step involves determining the position of the spacecraft. Assume the following orbital elements: $a = 6777.2745$ km, $e = 0.0001584$, $M_0 = 93.4312°$, $i = 34.9654°$, $\Omega = 346.4609°$, and $\omega = 266.6234°$. The epoch is March 31, 2011 at 02:32:41. Determine the position and velocity of the spacecraft using Table 10.1 with a time interval of 10 s and a total run time of 8 h. Next, compute the reference magnetic field using the dipole model in Eq. (11.5) and the reference Sun vector and Sun availability using the methods shown in Sect. 11.3. The true initial attitude is given by Eq. (2.79). The constant angular velocity vector is $\boldsymbol{\omega} = [0 \ \ -1.13156 \times 10^{-3} \ \ 0]^T$ rad/s.

Simulate TAM sensor noise using a Gaussian white-noise process with a mean of zero and a standard deviation of 50 nT per axis. Here is assumed that one Sun sensor is used and its axes are aligned with the body axes of the spacecraft. The Sun sensor noise is also modeled by a Gaussian white-noise process with a mean of zero and a standard deviation of 0.05° per axis. Simulate gyro measurements using Eqs. (4.31) and (4.32) of Sect. 4.7.1, with $\sigma_u = \sqrt{10} \times 10^{-10}$ rad/s$^{3/2}$, and $\sigma_v = \sqrt{10} \times 10^{-7}$ rad/s$^{1/2}$. The initial bias for each axis is given by 0.1 deg/h. The initial covariance for the attitude error is set to $(1/3)^2$ deg^2 per axis, and the initial covariance for the gyro drift is set to 1^2 (deg/h)^2 per axis. Converting these quantities to radians and seconds gives the following initial attitude and gyro drift covariances for each axis: $P_0^a = 0.3385 \times 10^{-4}$ rad^2 and $P_0^b = 0.2350 \times 10^{-10}$ (rad/s)^2, so that the initial covariance is given by

$$P_0 = \text{diag}\left[P_0^a \ P_0^a \ P_0^a \ P_0^b \ P_0^b \ P_0^b\right]$$

Use the discrete-time covariance propagation in Eq. (12.132) with $\Upsilon_k = I_6$, and Φ_k and Q_k given by Eqs. (6.83) and (6.93), respectively. All sensors are sampled at 10-s intervals. Run the MEKF using the TAM, available Sun and gyro measurements. Set the initial attitude estimate to the true quaternion and set all initial bias estimates to zero. Plot the attitude errors along with their respective 3σ bounds, as well as the estimated gyro biases.

6.11. Perform the following quaternion multiplication in Eq. (6.33):

$$\frac{1}{2}\begin{bmatrix} \delta\boldsymbol{\omega} \\ 0 \end{bmatrix} \otimes \begin{bmatrix} \delta\mathbf{q}_{1:3} \\ \delta q_4 \end{bmatrix}$$

and show the exact nonlinear differential equations for $\delta\dot{\mathbf{q}}_{1:3}$ and $\delta\dot{q}_4$. Next, consider the following error-MRP:

$$\delta\mathbf{p} = \frac{\delta\mathbf{q}_{1:3}}{1 + \delta q_4}$$

Take the time derivative of $\delta\mathbf{p}$, substitute in the quantities for $\delta\mathbf{q}_{1:3}$, δq_4, and their respective derivatives, in terms of $\delta\mathbf{p}$, to develop the exact nonlinear differential equation for $\delta\dot{\mathbf{p}}$. Next using the approximation $\delta\mathbf{p} \approx \delta\boldsymbol{\vartheta}/4$, show that the first-order equation for $\delta\dot{\boldsymbol{\vartheta}}$ is given by Eq. (6.35).

6.12. In this exercise you will develop an MEKF based on the MRPs. Here it is assumed that MRP measurements are given using the following discrete-time measurement model:

$$\mathbf{p} = \mathbf{v}_p \otimes \mathbf{p}^{\text{true}} = \frac{\left(1 - \|\mathbf{p}^{\text{true}}\|^2\right)\mathbf{v}_p + \left(1 - \|\mathbf{v}_p\|^2\right)\mathbf{p}^{\text{true}} - 2\,\mathbf{v}_p \times \mathbf{p}^{\text{true}}}{1 + \|\mathbf{p}^{\text{true}}\|^2\|\mathbf{v}_p\|^2 - 2\,\mathbf{v}_p \cdot \mathbf{p}^{\text{true}}}$$

where $\mathbf{v}_p$ is a zero-mean Gaussian white-noise process with covariance R_p. Assuming that the signal-to-noise ratio is large, show that this equation can be approximated by

$$\mathbf{p} \approx \mathbf{p}^{\text{true}} + \bar{R}\,\mathbf{v}_p$$

where

$$\bar{R} \equiv \left(1 - \|\mathbf{p}^{\text{true}}\|^2\right) I_3 + 2\,[\mathbf{p}^{\text{true}}\times] + 2\,(\mathbf{p}^{\text{true}})\,(\mathbf{p}^{\text{true}})^T$$

Therefore, the covariance of $\mathbf{p}$, denoted by R, is given by $R = \bar{R}\,R_p\,\bar{R}^T$. The multiplicative error-MRP is given by $\delta\mathbf{p} = \mathbf{p}^{\text{true}} \otimes \hat{\mathbf{p}}^{-1}$. The small angle approximation is given by $\delta\mathbf{p} \approx \delta\boldsymbol{\vartheta}/4$, where the kinematic equation for $\delta\boldsymbol{\vartheta}$ is given by Eq. (6.35). The mission mode multiplicative MRP propagation equations are given in Table 6.3, replacing the quaternion kinematics with the MRP kinematics, given by Eq. (3.24). Derive the multiplicative MRP update equation using a state vector given by $\Delta\hat{\mathbf{x}}_k^+ \equiv [\delta\hat{\boldsymbol{\vartheta}}_k^{+T} \;\; \Delta\hat{\boldsymbol{\beta}}_k^{+T}]^T$ and the MRP measurement model shown above.

Next, program the MEKF MRP equations to estimate the attitude and gyro biases from MRP measurements. The initial condition for the true MRP is given by $\mathbf{p}_0^{\text{true}} = [0.3\;\; 0.1\;\; -0.5]^T$ and the true angular rate for all time is given by $\boldsymbol{\omega}^{\text{true}}(t) = [-0.2\;\; 0.2\;\; -0.192]^T$ deg/s. Generate synthetic MRP measurements using a covariance of $R_p = (20/3{,}600) \times (\pi/180) I_3$ and take MRP samples at 5-s intervals. Simulate gyro measurements using Eqs. (4.31) and (4.32) of Sect. 4.7.1, with $\sigma_u = \sqrt{10} \times 10^{-10}$ rad/s$^{3/2}$, and $\sigma_v = \sqrt{10} \times 10^{-7}$ rad/s$^{1/2}$. The initial bias is given by $\boldsymbol{\beta}_0^{\text{true}} = [-1\;\; 2\;\; -3]^T$ deg/h. Sample the gyro measurements at 0.5-s intervals. Set the initial state estimate to zeros for both the MRPs and gyro biases. Set the initial covariance matrix to

$$P_0 = \text{diag}[P_0^a \;\; P_0^a \;\; P_0^a \;\; P_0^b \;\; P_0^b \;\; P_0^b]$$

with $P_0^a = 16 \times 0.175$ rad^2 and $P_0^b = 0.005$ (rad/s)^2. Use the discrete-time covariance propagation in Eq. (12.132) with $\Upsilon_k = I_6$, and Φ_k and Q_k given by

Eqs. (6.83) and (6.93), respectively. Use a total simulation run time of 200 min. Plot the attitude errors along with their respective 3σ bounds, as well as the estimated gyro biases.

6.13. This problem has two parts:

a) Derive Eq. (6.99b).
b) Derive the Q matrix in Eq. (6.105).

6.14. Prove Eq. (6.106) by mathematical induction.

6.15. Show that Eqs. (6.108) and (6.110) satisfy Eqs. (6.107).

6.16. Derive the relations in Eqs. (6.113) and (6.114).

6.17. Show that the dynamic model of Eq. (6.116) leads to the covariance propagation and update given by (6.115).

6.18. Derive the asymptotic limits in Eqs. (6.117)–(6.120).

6.19. Redo the simulation shown in Example 6.3. Pick various truth values for the biases and D_{ij} calibration parameters as well as various noise levels to assess the robustness of the centered and TWOSTEP algorithms.

References

1. Alonso, R., Shuster, M.D.: Complete linear attitude-independent magnetometer calibration. J. Astronaut. Sci. **50**(4), 477–490 (2002)
2. Alonso, R., Shuster, M.D.: TWOSTEP: A fast robust algorithm for attitude-independent magnetometer-bias determination. J. Astronaut. Sci. **50**(4), 433–451 (2002)
3. Andrews, S.F., Bilanow, S.: Recent flight results of the TRMM Kalman filter. In: AIAA Guidance, Navigation and Control Conference. Monterey (2002). AIAA 2002-5047
4. Andrews, S.F., Morgenstern, W.M.: Design, implementation, testing, and flight results of the TRMM Kalman filter. In: AIAA Guidance, Navigation and Control Conference. Boston (1998). AIAA 1998-4509
5. Carmi, A., Oshman, Y.: On the covariance singularity of quaternion estimators. In: AIAA Guidance, Navigation and Control Conference. Hilton Head (2007). AIAA 2007-6814
6. Crassidis, J.L., Andrews, S.F., Markley, F.L., Ha, K.: Contingency designs for attitude determination of TRMM. In: Proceedings of the Flight Mechanics/Estimation Theory Symposium, pp. 419–433. NASA-Goddard Space Flight Center, Greenbelt (1995)
7. Crassidis, J.L., Cheng, Y., Nebelecky, C.K., Fosbury, A.M.: Decentralized attitude estimation using a quaternion covariance intersection approach. J. Astronaut. Sci. **57**(1/2), 113–128 (2009)
8. Crassidis, J.L., Junkins, J.L.: Optimal Estimation of Dynamic Systems, 2nd edn. CRC Press, Boca Raton (2012)
9. Crassidis, J.L., Lai, K.L., Harman, R.R.: Real-time attitude-independent three-axis magnetometer calibration. J. Guid. Contr. Dynam. **28**(1), 115–120 (2005)
10. Crassidis, J.L., Markley, F.L.: Attitude estimation using modified Rodrigues parameters. In: Proceedings of the Flight Mechanics/Estimation Theory Symposium, pp. 71–83. NASA-Goddard Space Flight Center, Greenbelt (1996)
11. Crassidis, J.L., Markley, F.L.: Unscented filtering for spacecraft attitude estimation. J. Guid. Contr. Dynam. **26**(4), 536–542 (2003)

12. Crassidis, J.L., Markley, F.L., Cheng, Y.: Survey of nonlinear attitude estimation methods. J. Guid. Contr. Dynam. **30**(1), 12–28 (2007)
13. Farrell, J.L.: Attitude determination by Kalman filtering. Contractor Report NASA-CR-598, NASA Goddard Space Flight Center, Washington, DC (1964)
14. Farrell, J.L.: Attitude determination by Kalman filtering. Automatica **6**, 419–430 (1970)
15. Farrenkopf, R.L.: Analytic steady-state accuracy solutions for two common spacecraft attitude estimators. J. Guid. Contr. **1**(4), 282–284 (1978)
16. Gai, E., Daly, K., Harrison, J., Lemos, L.: Star-sensor-based attiude/attitude rate estimator. J. Guid. Contr. Dynam. **8**(5), 560–565 (1985)
17. Gambhir, B.: Determination of magnetometer biases using Module RESIDG. Tech. Rep. 3000-32700-01TN, Computer Sciences Corporation (1975)
18. Idan, M.: Estimation of Rodrigues parameters from vector observations. IEEE Trans. Aero. Electron. Syst. **32**(2), 578–586 (1996)
19. Karlgaard, C.D., Schaub, H.: Nonsingular attitude filtering using modified Rodrigues parameters. J. Astronaut. Sci. **57**(4), 777–791 (2010)
20. Langel, R.A.: The main field. In: Jacobs, J.A. (ed.) Geomagnetism, pp. 249–512. Academic Press, Orlando (1987)
21. Lefferts, E.J., Markley, F.L., Shuster, M.D.: Kalman filtering for spacecraft attitude estimation. J. Guid. Contr. Dynam. **5**(5), 417–429 (1982)
22. Lerner, G.M.: Three-axis attitude determination. In: Wertz, J.R. (ed.) Spacecraft Attitude Determination and Control, chap. 12. Kluwer Academic, Dordrecht (1978)
23. Markley, F.L.: Attitude error representations for Kalman filtering. J. Guid. Contr. Dynam. **26**(2), 311–317 (2003)
24. Markley, F.L.: Attitude estimation or quaternion estimation? J. Astronaut. Sci. **52**(1 & 2), 221–238 (2004)
25. Markley, F.L.: Lessons learned. J. Astronaut. Sci. **57**(1 & 2), 3–29 (2009)
26. Markley, F.L., Reynolds, R.G.: Analytic steady-state accuracy of a spacecraft attitude estimator. J. Guid. Contr. Dynam. **23**(6), 1065–1067 (2000)
27. Murrell, J.W.: Precision attitude determination for multimission spacecraft. In: Proceedings of the AIAA Guidance, Navigation, and Control Conference, pp. 70–87. Palo Alto (1978)
28. Pandiyan, R., Solaiappan, A., Malik, N.: A one step batch filter for estimating gyroscope calibration parameters using star vectors. In: AIAA/AAS Astrodynamics Specialist Conference and Exhibit. Providence (2004). AIAA 04-4858
29. Pauling, D.C., Jackson, D.B., Brown, C.D.: SPARS algorithms and simulation results. In: Proceedings of the Symposium on Spacecraft Attitude Estimation, vol. 1, pp. 293–317. Aerospace Corporation, El Segundo (1969)
30. Pittelkau, M.E.: Kalman filtering for spacecraft system alignment calibration. J. Guid. Contr. Dynam. **24**(6) (2001)
31. Pittelkau, M.E.: An analysis of the quaternion attitude determination filter. J. Astronaut. Sci. **51**(1) (2003)
32. Psiaki, M.L., Klatt, E.M., Kintner Jr., P.M., Powell, S.P.: Attitude estimation for a flexible spacecraft in unstable spin. J. Guid. Contr. Dynam. **25**(1), 88–95 (2002)
33. Psiaki, M.L., Theiler, J., Bloch, J., Ryan, S., Dill, R.W., Warner, R.E.: ALEXIS spacecraft attitude reconstruction with thermal/flexible motions due to launch damage. J. Guid. Contr. Dynam. **20**(5), 1033–1041 (1997)
34. Reynolds, R.G.: Asymptotically optimal attitude filtering with guaranteed covergence. J. Guid. Contr. Dynam. **31**(1), 114–122 (2008)
35. Shuster, M.D.: Kalman filtering of spacecraft attitude and the QUEST model. J. Astronaut. Sci. **38**(3), 377–393 (1990)
36. Shuster, M.D.: Constraint in attitude estimation part I: Constrained estimation. J. Astronaut. Sci. **51**(1), 51–74 (2003)
37. Shuster, M.D.: Constraint in attitude estimation part II: Unconstrained estimation. J. Astronaut. Sci. **51**(1), 75–101 (2003)

38. Shuster, M.D., Pitone, D.S.: Batch estimation of spacecraft sensor alignments, II. Absolute alignment estimation. J. Astronaut. Sci. **39**(4), 547–571 (1991)
39. Toda, N.F., Heiss, J.L., Schlee, F.H.: SPARS: The system, algorithms, and test results. In: Proceedings of the Symposium on Spacecraft Attitude Estimation, vol. 1, pp. 361–370. Aerospace Corporation, El Segundo (1969)
40. van Loan, C.F.: Computing integrals involving the matrix exponential. IEEE Trans. Automat. Contr. **AC-23**(3), 396–404 (1978)
41. Ward, D.K., Davis, G.T., O'Donnell Jr., J.R.: The Microwave Anisotropy Probe guidance, navigation and control hardware suite. In: AIAA Guidance, Navigation and Control Conference. Monterey (2002). AIAA 2002-4579
42. Zanetti, R., Majji, M., Bishop, R.H., Mortari, D.: Norm-constrained Kalman filtering. J. Guid. Contr. Dynam. **32**(5), 1458–1465 (2009)

Chapter 7
Attitude Control

7.1 Introduction

Spacecraft attitude control is essential to meet mission pointing requirements, such as required science modes and thruster pointing requirements for orbital maneuvers. Early spacecraft mission designs used passive spin stabilization to hold one axis relatively fixed by spinning the spacecraft around that axis, usually the axis of maximum moment of inertia. Spin stabilization was mostly used due to the limited control actuation and lack of sophisticated computer technology to implement complex control laws. Spin-stabilized spacecraft are very stable, but they have to be sensitively balanced; every component has to be designed and located with spacecraft balance in mind. This can be extremely difficult to accomplish to the required accuracy. In most cases the last few weights are added and adjusted only after actual flight hardware is delivered and installed, and the spacecraft is experimentally spin tested. Allowances must also be made for everything onboard that can move during flight.

In the modern era advancements in sensors, actuators, and computer processors allow for three-axis stabilized spacecraft designs, although spinners are still used to this day for many missions. Attitude control law theory also has been extensively studied and advanced, allowing for guaranteed control stability even with nonlinear attitude dynamics. The control of spacecraft for large angle slewing maneuvers poses a difficult problem, however. These difficulties include the highly nonlinear characteristics of the governing equations, control rate, and saturation constraints and limits, and incomplete state knowledge due to sensor failure or omission. The control of spacecraft with large angle slews can be accomplished by either open-loop or closed-loop schemes. Open-loop schemes usually require a pre-determined pointing maneuver and are typically determined using optimal control techniques, which involve the solution of a two-point boundary value problem (e.g. the time optimal maneuver problem [37]). Open-loop schemes are sensitive to spacecraft

F.L. Markley and J.L. Crassidis, *Fundamentals of Spacecraft Attitude Determination and Control*, Space Technology Library 33, DOI 10.1007/978-1-4939-0802-8_7,
© Springer Science+Business Media New York 2014

parameter uncertainties and unexpected disturbances [47]. Closed-loop systems can account for parameter uncertainties and disturbances, and thus provide a more robust design methodology.

For many years now, much effort has been devoted to the closed-loop design of large angle slews. In [50] a number of simple control schemes are derived using quaternion and angular velocity (rate) feedback. Asymptotic stability is shown by using a Lyapunov function analysis for all cases. Reference [45] expands upon these formulations by deriving simple control laws based on both a Gibbs vector parameterization and a modified Rodrigues parameterization, each with rate feedback. Lyapunov functions are shown for all the controllers developed in [45] as well. Other full state feedback techniques have been developed that are based on variable-structure (sliding-mode) control, which uses a feedback linearizing technique and an additional term aimed at dealing with model uncertainty [41]. This type of control has been successfully applied for large angle maneuvers using a Gibbs vector parameterization [18], a quaternion parameterization [15, 46], and a modified Rodrigues parameterization [13]. Another robust control scheme using a nonlinear H_∞ control methodology has been developed in [22]. This scheme involves the solution of Hamilton-Jacobi-Isaacs inequalities, which essentially determines feedback gains for the full state feedback control problem so that the spacecraft is stabilized in the presence of uncertainties and disturbances. Another class of controllers involves adaptive techniques, which update the model during operation based on measured performances (e.g. see [41]). An adaptive scheme which estimates external torques by tracking a Lyapunov function has been developed by [35]. This method has been shown to be very robust in the presence of spacecraft modeling errors and disturbances.

The aforementioned techniques all utilize full state knowledge (i.e. attitude and rate feedback). The problem of controlling a spacecraft without full state feedback is more complex. The basic approaches used to solve this problem can be divided into methods which estimate the unmeasured states using a filter algorithm and methods which develop control laws directly from output feedback. Filtering methods of Chap. 6, such as the extended Kalman filter, have been successfully applied on numerous spacecraft systems without the use of rate-integrating gyro measurements (e.g. see [9, 11, 14]). An advantage of these methods is that the attitude may be estimated by using only one set of vector attitude observations (such as magnetometer observations). However, these methods are usually much less accurate than methods that use gyro measurements. A more direct technique has been developed in [25], which solves the attitude problem without rate knowledge. This method is based on a passivity approach, which replaces the rate feedback by a nonlinear filter of the quaternion. A model-based filter reconstructing the angular velocity is not needed in this case.

This chapter presents the fundamental concepts for modern spacecraft control designs. First, attitude control laws are developed using both external torques and reaction wheels for both regulation and tracking cases. An example from a real mission, the Wilkinson Microwave Anisotropy Probe (WMAP), is shown for attitude tracking. Next, attitude thruster control is discussed using pulse-width pulse-frequency modulation. This is followed by attitude control using magnetic torquers for both detumbling and momentum dumping. Then, the effects of noise on the control systems are discussed and mitigation approaches are shown. Finally, an actual attitude control system design, based on the SAMPEX spacecraft, is shown. This spacecraft design is useful to show the combined effects of using filters shown in Chap. 6 with linear controllers to meet mission designs.

7.2 Attitude Control: Regulation Case

Regulation control is defined as bringing the attitude to some fixed location (usually the identity quaternion) and the angular velocity to zero. The quaternion attitude kinematics and Euler's rotational equation of motion are given in Chap. 3 (removing superscripts here for brevity):

$$\dot{\mathbf{q}} = \frac{1}{2}\Xi(\mathbf{q})\boldsymbol{\omega} = \frac{1}{2}\Omega(\boldsymbol{\omega})\mathbf{q} \tag{7.1a}$$

$$J\dot{\boldsymbol{\omega}} = -[\boldsymbol{\omega}\times]J\boldsymbol{\omega} + \mathbf{L} \tag{7.1b}$$

The goal is to drive the actual quaternion to some commanded and constant quaternion denoted by $\mathbf{q}_c$. This requires that the angular velocity approach zero. The error quaternion is given by

$$\delta\mathbf{q} \equiv \begin{bmatrix} \delta\mathbf{q}_{1:3} \\ \delta q_4 \end{bmatrix} = \mathbf{q} \otimes \mathbf{q}_c^{-1} \tag{7.2}$$

where

$$\delta\mathbf{q}_{1:3} = \Xi^T(\mathbf{q}_c)\mathbf{q} \tag{7.3a}$$

$$\delta q_4 = \mathbf{q}^T\mathbf{q}_c \tag{7.3b}$$

Taking the time derivative of Eq. (7.2) gives

$$\delta\dot{\mathbf{q}} = \dot{\mathbf{q}} \otimes \mathbf{q}_c^{-1} \tag{7.4}$$

Substituting Eq. (7.1a) into Eq. (7.4) and using Eq. (7.2) leads to

$$\delta\dot{\mathbf{q}} = \frac{1}{2}\Omega(\boldsymbol{\omega})\delta\mathbf{q} \tag{7.5}$$

Using the definition of $\Omega(\boldsymbol{\omega})$, the differential equations for $\delta\mathbf{q}_{1:3}$ and δq_4 are specifically given by

$$\delta\dot{\mathbf{q}}_{1:3} = \frac{1}{2}[\delta\mathbf{q}_{1:3}\times]\boldsymbol{\omega} + \frac{1}{2}\delta q_4\,\boldsymbol{\omega} \tag{7.6a}$$

$$\delta\dot{q}_4 = -\frac{1}{2}\delta\mathbf{q}_{1:3}^T\boldsymbol{\omega} \tag{7.6b}$$

The goal of the controller is to drive $\boldsymbol{\omega}$ to zero and $\delta\mathbf{q}$ to the identity quaternion $\mathbf{I}_q = [0\,0\,0\,1]^T$.

Several feedback controllers are presented in [50]. The first is given by

$$\mathbf{L} = -k_p\,\delta\mathbf{q}_{1:3} - k_d\,\boldsymbol{\omega} \tag{7.7}$$

where k_p and k_d are positive scalar gains. Substituting Eq. (7.7) into Eq. (7.1b) gives the closed-loop system governed by Eq. (7.5) and

$$\dot{\boldsymbol{\omega}} = -J^{-1}\left([\boldsymbol{\omega}\times]J\boldsymbol{\omega} + k_p\delta\mathbf{q}_{1:3} + k_d\boldsymbol{\omega}\right) \tag{7.8}$$

The only equilibrium point is $[\delta\mathbf{q}_{1:3}^T\;\;\boldsymbol{\omega}^T]^T = \mathbf{0}$.

Stability is proven using Lyapunov's direct method. Reference [50] uses a difference between the actual and command quaternions in the definition of the candidate Lyapunov function. Here, a multiplicative approach is employed, which leads to the same result as in [50] but is more intuitive in terms of the quaternion error kinematics. Define the following candidate Lyapunov function:

$$V = \frac{1}{4}\boldsymbol{\omega}^T J\boldsymbol{\omega} + \frac{1}{2}k_p\,\delta\mathbf{q}_{1:3}^T\delta\mathbf{q}_{1:3} + \frac{1}{2}k_p\,(1-\delta q_4)^2 \geq 0 \tag{7.9}$$

Note that $V = 0$ when $\boldsymbol{\omega} = \mathbf{0}$ and $\delta\mathbf{q} = \mathbf{I}_q$, which is the equilibrium point. The time derivative of V is given by

$$\dot{V} = \frac{1}{2}\boldsymbol{\omega}^T J\dot{\boldsymbol{\omega}} + k_p\,\delta\mathbf{q}_{1:3}^T\delta\dot{\mathbf{q}}_{1:3} - k_p\,(1-\delta q_4)\delta\dot{q}_4 \tag{7.10}$$

Substituting Eqs. (7.6) and (7.8) into Eq. (7.10) gives

$$\begin{aligned}\dot{V} &= -\frac{1}{2}\left(\boldsymbol{\omega}^T\delta\mathbf{q}_{1:3}\right)\left[k_p + k_p\delta q_4 - k_p(1+\delta q_4)\right] - \frac{1}{2}k_d\,\boldsymbol{\omega}^T\boldsymbol{\omega}\\ &= -\frac{1}{2}k_d\,\boldsymbol{\omega}^T\boldsymbol{\omega} \leq 0\end{aligned} \tag{7.11}$$

Thus, the closed-loop system is stable since $\dot{V} \leq 0$.

Asymptotic stability can be proven using LaSalle's theorem (see Sect. 12.2.2). The equality in Eq. (7.11) is given when $\boldsymbol{\omega} = \mathbf{0}$, where $\delta\mathbf{q}_{1:3}$ can be anything. We must check that the system cannot remain in a state where $\dot{V} = 0$ while $\delta\mathbf{q}_{1:3} \neq \mathbf{0}$. Equation (7.11) guarantees that $\lim_{t\to\infty}\boldsymbol{\omega} = \mathbf{0}$. The closed-loop dynamics in Eq. (7.8) shows that this asymptotic condition can only be achieved if $\lim_{t\to\infty} \delta\mathbf{q}_{1:3} = \mathbf{0}$ also. Thus, this control law asymptotically reorients the spacecraft to the desired attitude from an arbitrary initial orientation. Note that δq_4 can be ± 1, but this does not matter since both signs produce the same attitude.

However, the control law in Eq. (7.7) does not guarantee that the shortest path is provided to the final orientation. This becomes an issue when the fourth component of the initial error quaternion is negative. Reference [50] shows two other control laws that overcome this issue. One of them is a slight modification of Eq. (7.7), given by

$$\mathbf{L} = -k_p \operatorname{sign}(\delta q_4)\delta\mathbf{q}_{1:3} - k_d\,\boldsymbol{\omega} \tag{7.12}$$

Note that if $\delta q_4 < 0$ then a positive feedback term is introduced, which provides the shorter path to reach the desired equilibrium point. Thus Eq. (7.12) is always preferred over Eq. (7.7) unless $\delta q_4 = 0$ exactly, which is not a concern for practical applications.

The control law in Eq. (7.7) is linear in the state. Nonlinear control laws can also be used. Consider the following control law:

$$\mathbf{L} = -k_p\,\delta\mathbf{q}_{1:3} - k_d\,(1 \pm \delta\mathbf{q}_{1:3}^T\delta\mathbf{q}_{1:3})\boldsymbol{\omega} \tag{7.13}$$

This law also produces a globally asymptotic stable response, which can be proven using Lyapunov's direct method (which is left as an exercise for the reader). When the minus sign is used then by the quaternion unity constraint we have $1 - \delta\mathbf{q}_{1:3}^T\delta\mathbf{q}_{1:3} = \delta q_4^2$. The shortest distance control law can be achieved by using

$$\mathbf{L} = -k_p \operatorname{sign}(\delta q_4)\delta\mathbf{q}_{1:3} - k_d\,(1 \pm \delta\mathbf{q}_{1:3}^T\delta\mathbf{q}_{1:3})\boldsymbol{\omega} \tag{7.14}$$

The aforementioned control laws can also be used with reaction wheels. For the reaction-wheel-only case Eq. (3.147) expresses Euler's rotational equation as

$$J\dot{\boldsymbol{\omega}} = -[\boldsymbol{\omega}\times](J\boldsymbol{\omega} + \mathbf{h}) - \dot{\mathbf{h}} \tag{7.15}$$

where J now includes the transverse inertia of the wheels, $\mathbf{h} \equiv \mathbf{H}_B^w$ is the wheel angular momentum, and $\dot{\mathbf{h}}$ is the wheel torque. Note that the total angular momentum, $J\boldsymbol{\omega} + \mathbf{h}$, is conserved as discussed in Sect. 3.3.5.1. An equivalent but more useful form of Eq. (7.15) for control purposes is given by

$$J\dot{\boldsymbol{\omega}} = -[\boldsymbol{\omega}\times]J\boldsymbol{\omega} + \bar{\mathbf{L}} \tag{7.16a}$$

$$\dot{\mathbf{h}} = -[\boldsymbol{\omega}\times]\mathbf{h} - \bar{\mathbf{L}} \tag{7.16b}$$

where $\bar{\mathbf{L}}$ is an effective wheel torque input. Adding Eqs. (7.16a) and (7.16b) shows that these two equations are consistent with Eq. (7.15). Now define the following control laws for the wheel torques:

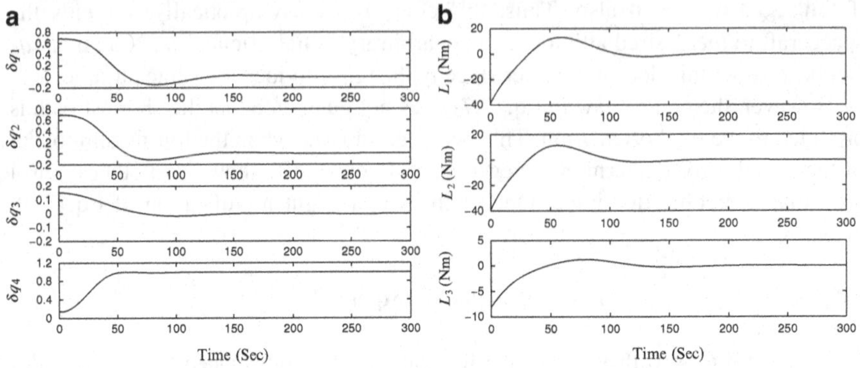

Fig. 7.1 (**a**) Quaternion errors and (**b**) control torques

$$\bar{\mathbf{L}} = -k_p\,\text{sign}(\delta q_4)\boldsymbol{\delta}\mathbf{q}_{1:3} - k_d\,\boldsymbol{\omega} \tag{7.17a}$$

$$\bar{\mathbf{L}} = -k_p\,\text{sign}(\delta q_4)\boldsymbol{\delta}\mathbf{q}_{1:3} - k_d\,(1 \pm \boldsymbol{\delta}\mathbf{q}_{1:3}^T\boldsymbol{\delta}\mathbf{q}_{1:3})\boldsymbol{\omega} \tag{7.17b}$$

Using the candidate Lyapunov function in Eq. (7.9) proves that these control laws provide asymptotic stability with reaction wheels.

Example 7.1. In this example the control law in Eq. (7.12) is used to perform a reorientation maneuver with large initial errors. The inertia matrix of the spacecraft is given by

$$J = \begin{bmatrix} 10000 & 0 & 0 \\ 0 & 9000 & 0 \\ 0 & 0 & 12000 \end{bmatrix} \text{kg-m}^2$$

The initial quaternion is given by $\mathbf{q}(t_0) = [0.6853\ 0.6953\ 0.1531\ 0.1531]^T$ and the initial angular velocity is given by $\boldsymbol{\omega}(t_0) = [0.5300\ 0.5300\ 0.0530]^T$ deg/s. The desired quaternion is the identity quaternion. The control gains are set to $k_p = 50$ and $k_d = 500$. A plot of the quaternion errors is shown in Fig. 7.1a. The fourth error-quaternion component approaches 1 while the other three components approach zero. The control torques are shown in Fig. 7.1b. Note the large control torques at the beginning of the maneuver, which are due to the large initial errors. From these plots it is clear that the control law provides a reorientation maneuver to the desired attitude.

Example 7.2. In this example the control law in Eq. (7.17a) is used to perform a reorientation maneuver using reaction wheels. The inertia matrix of the spacecraft is given by

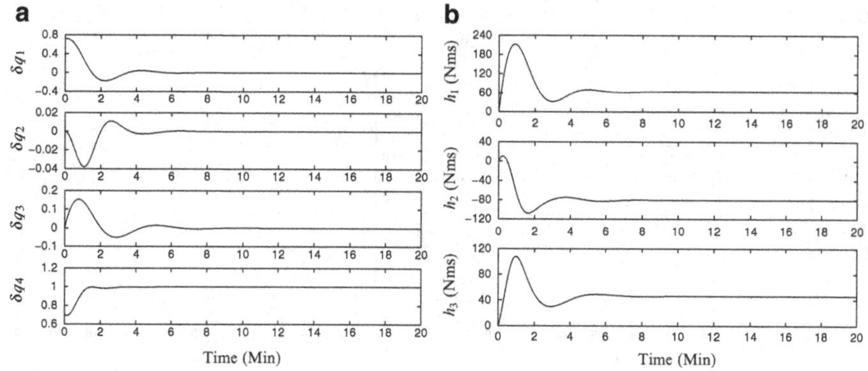

Fig. 7.2 (**a**) Quaternion errors and (**b**) wheel momenta

$$J = \begin{bmatrix} 6400 & -76.4 & -25.6 \\ -76.4 & 4730 & -40 \\ -25.6 & -40 & 8160 \end{bmatrix} \text{kg-m}^2$$

The initial quaternion is given by $\mathbf{q}(t_0) = \sqrt{2}/2\,[1\ \ 0\ \ 0\ \ 1]^T$ and the initial angular velocity is given by $\boldsymbol{\omega}(t_0) = [0.01\ 0.01\ 0.01]^T$ rad/s. The desired quaternion is the identity quaternion. The initial wheel momentum is set to $\mathbf{h}(t_0) = \mathbf{0}$. The gains are set to $k_p = 10$ and $k_d = 150$. A plot of the quaternion errors is shown in Fig. 7.2a. The fourth error-quaternion component approaches 1 while the other three components approach zero. The wheel momenta are shown in Fig. 7.2b. Note that wheel momenta do not decrease to zero once the desired attitude is achieved. This is due to the conservation of momentum in the spacecraft. For this case, since $\mathbf{h}(t_0) = \mathbf{0}$ then the norm of the momentum is $\|J\boldsymbol{\omega}(t_0)\| = 112.4586$ Nms. The angular velocity of the spacecraft still achieves its desired zero value at the end of the maneuver and the momentum of the spacecraft is reoriented to perform this maneuver. As before, from these plots it is clear that the control law provides a reorientation maneuver to the desired attitude.

We next convert the momenta from this example to two specific implementations of the four-wheel configurations discussed in Sect. 4.8.3, using the pseudoinverses of the distribution matrices:

$$\mathscr{W}_4 = \frac{1}{\sqrt{2}}\begin{bmatrix} 1 & -1 & 0 & 0 \\ 1 & 1 & 1 & 1 \\ 0 & 0 & 1 & -1 \end{bmatrix}, \quad \mathscr{W}_{\text{N}} = \begin{bmatrix} 1 & 0 & 0 & 1/\sqrt{3} \\ 0 & 1 & 0 & 1/\sqrt{3} \\ 0 & 0 & 1 & 1/\sqrt{3} \end{bmatrix}$$

Plots of the converted momenta are shown in Fig. 7.3. Clearly, the different configurations produce different momenta for each wheel. The Euclidean norm of the wheel momenta for the pyramid configuration at the final time is 96.8044 Nms,

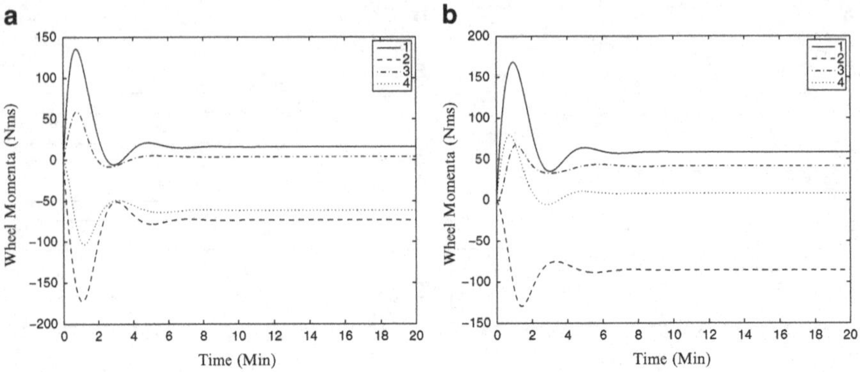

Fig. 7.3 Comparison between wheel configurations. (**a**) Pyramid configuration. (**b**) NASA standard configuration

while the norm of the wheel momenta for the NASA standard configuration at the final time is 111.8695 Nms.

7.3 Attitude Control: Tracking Case

The previous section described approaches to maneuver a spacecraft to a fixed attitude while driving the angular velocity to zero. This has applications to inertially pointing spacecraft like the Hubble Space Telescope. This section discusses various forms of attitude tracking that are used to control the spacecraft to follow a desired time-varying trajectory. Note that tracking also encompasses regulation by simply setting the desired attitude to a constant and the desired angular velocity to zero. Optimal control methods that minimize a user-defined cost function are based on solving two-point boundary value problems [21]. They provide the control trajectory over the desired time interval and cannot be executed in real time. They are very useful for analysis purposes but less useful for real spacecraft mission designs, e.g. time-optimal control to provide rapid re-pointing of the spacecraft [37].

Feedback methods are more suitable for actual attitude tracking control applications because they can be executed in real time. Feedback-based attitude tracking control is a widely studied topic. Most controllers are model based, e.g. they typically require knowledge of the inertia matrix. No one controller can work for every spacecraft mission objective in terms of pointing and jitter requirements. Some are more sensitive to noise effects and/or modeling errors, while others are sensitive to external torque disturbances, such as torques induced by solar radiation pressure. So called "robust" controllers have been developed to mitigate these sensitivities. Here, the use of variable-structure (sliding-mode) control is shown as an introduction to spacecraft tracking. Sliding-mode control is robust to arbitrary model parameter inaccuracies, but comes at the price of possibly high control activity [41]. Section 12.2.3 presents a brief overview of sliding-mode control.

The sliding-mode controller developed in [15] is shown here for both the case of external torques and internal reaction wheels. The dynamic equation is given by Eq. (7.1b) for the external torque case and by Eq. (7.16) for the reaction wheel case. First we must select the sliding surface. An obvious choice is to use a form that is similar to the right side of Eq. (7.7). Here the difference between the actual angular velocity and the commanded angular velocity, denoted by $\boldsymbol{\omega}_c$, is used so that the sliding surface vector is given by

$$\mathbf{s} = (\boldsymbol{\omega} - \boldsymbol{\omega}_c) + k\, \delta\mathbf{q}_{1:3} \tag{7.18}$$

where k is some positive scalar constant. Note that here the commanded quaternion $\mathbf{q}_c$ may be time varying, but it must be consistent with the commanded angular velocity, i.e. $\mathbf{q}_c$ must be derived from the quaternion kinematics driven by $\boldsymbol{\omega}_c$. If $\mathbf{s} = \mathbf{0}$ then the actual attitude and angular velocity will track the commanded inputs. Reference [15] shows that Eq. (7.18) can actually be derived from optimal control theory.

Taking the time derivative of Eq. (7.18) gives

$$\dot{\mathbf{s}} = (\dot{\boldsymbol{\omega}} - \dot{\boldsymbol{\omega}}_c) + k\, \delta\dot{\mathbf{q}}_{1:3} \tag{7.19}$$

Evaluating Eq. (6.32) with $\boldsymbol{\omega} \equiv \boldsymbol{\omega}^{\text{true}}$ and $\boldsymbol{\omega}_c \equiv \hat{\boldsymbol{\omega}}$ gives an expression for $\delta\dot{\mathbf{q}}_{1:3}$:

$$\delta\dot{\mathbf{q}}_{1:3} = \frac{1}{2}\delta q_4(\boldsymbol{\omega} - \boldsymbol{\omega}_c) + \frac{1}{2}\delta\mathbf{q}_{1:3}\times(\boldsymbol{\omega} + \boldsymbol{\omega}_c) \tag{7.20}$$

Substituting Eqs. (7.1b) and (7.20) into Eq. (7.19) gives

$$\dot{\mathbf{s}} = -J^{-1}[\boldsymbol{\omega}\times]J\boldsymbol{\omega} + J^{-1}\mathbf{L}_e - \dot{\boldsymbol{\omega}}_c + \frac{k}{2}\left[\delta q_4(\boldsymbol{\omega} - \boldsymbol{\omega}_c) + \delta\mathbf{q}_{1:3}\times(\boldsymbol{\omega} + \boldsymbol{\omega}_c)\right] \tag{7.21}$$

Note that $\mathbf{L}$ is replaced with $\mathbf{L}_e$ because the equivalent control law must first be derived, as explained in Sect. 12.2.3. As also explained in that section the time derivative of the Lyapunov function in Eq. (12.66) requires that $\dot{\mathbf{s}} = \mathbf{0}$. Using this condition gives the following equivalent control law:

$$\mathbf{L}_e = J\left\{\frac{k}{2}\left[\delta q_4(\boldsymbol{\omega}_c - \boldsymbol{\omega}) - \delta\mathbf{q}_{1:3}\times(\boldsymbol{\omega} + \boldsymbol{\omega}_c)\right] + \dot{\boldsymbol{\omega}}_c\right\} + [\boldsymbol{\omega}\times]J\boldsymbol{\omega} \tag{7.22}$$

As explained in Sect. 12.2.3, a discontinuous term is added across the sliding surface to account for model uncertainties. In order to reduce the resulting chattering phenomenon, a saturation function is used instead of the signum function. The resulting control law will provide the sliding-mode attitude tracking law, however it is not guaranteed to reorient the spacecraft in the shortest distance. Reference [15] proves that the following sliding-mode control law reorients the spacecraft in the shortest distance:

$$\mathbf{s} = (\boldsymbol{\omega} - \boldsymbol{\omega}_c) + k\,\text{sign}(\delta q_4)\boldsymbol{\delta}\mathbf{q}_{1:3} \tag{7.23a}$$

$$\mathbf{L} = J\left\{\frac{k}{2}\left[|\delta q_4|(\boldsymbol{\omega}_c - \boldsymbol{\omega}) - \text{sign}(\delta q_4)\,\boldsymbol{\delta}\mathbf{q}_{1:3}\times(\boldsymbol{\omega} + \boldsymbol{\omega}_c)\right] + \dot{\boldsymbol{\omega}}_c - G\,\bar{\mathbf{s}}\right\} + [\boldsymbol{\omega}\times]J\boldsymbol{\omega} \tag{7.23b}$$

where G is a positive definite matrix and the ith component of $\bar{\mathbf{s}}$ is given by

$$\bar{s}_i = \text{sat}(s_i, \epsilon_i), \quad i = 1, 2, 3 \tag{7.24}$$

where ϵ_i is a positive quantity and s_i is the ith component of the sliding vector given by Eq. (7.23a). The saturation function is defined by

$$\text{sat}(s_i, \epsilon_i) \equiv \begin{cases} 1 & \text{for} \quad s_i > \epsilon_i \\ s_i/\epsilon_i & \text{for} \quad |s_i| \le \epsilon_i \\ -1 & \text{for} \quad s_i < -\epsilon_i \end{cases} \tag{7.25}$$

Note that $\bar{\mathbf{s}}$ drives the system to the sliding surface. If reaction wheels are used to control the spacecraft, then as in Sect. 7.2 the same control law is used with $\bar{\mathbf{L}}$ replacing $\mathbf{L}$.

The robustness of the sliding-mode controller is now shown. We first define the following bounded modeling errors for the inertia matrix:

$$J^{-1} = \hat{J}^{-1} + \delta J^{-1} \tag{7.26}$$

where $\hat{J}$ is the nominal inertia matrix. Next, neglecting the gyroscopic term in Euler's rotational equation and adding an external disturbance input yields

$$\dot{\boldsymbol{\omega}} = J^{-1}\mathbf{L} + J^{-1}\mathbf{d} \tag{7.27}$$

where $\mathbf{d}$ denotes a bounded disturbance input. Under these conditions the time-derivative of the sliding manifold can be approximated by

$$\dot{\mathbf{s}} = \delta J^{-1}\hat{J}\left\{\frac{k}{2}\left[|\delta q_4|(\boldsymbol{\omega}_c - \boldsymbol{\omega}) - \text{sign}(\delta q_4)\,\boldsymbol{\delta}\mathbf{q}_{1:3}\times(\boldsymbol{\omega} + \boldsymbol{\omega}_c)\right] + \dot{\boldsymbol{\omega}}_c - G\,\bar{\mathbf{s}}\right\} - J^{-1}\hat{J}G\,\bar{\mathbf{s}} + J^{-1}\mathbf{d} \tag{7.28}$$

where it is assumed that the higher-order perturbations in the inertia matrix are small. We also assume that the thickness of the boundary layer ϵ and the gain G are sufficiently large to keep the time derivative of the associated Lyapunov function negative definite in the presence of modeling errors and external disturbances. Then the dynamics of the sliding manifold can be approximated by

$$\dot{\mathbf{s}} \approx -\frac{1}{\epsilon} J^{-1} \hat{J} G \, \mathbf{s} + J^{-1} \mathbf{d} \tag{7.29}$$

Therefore, $\mathbf{s}$ will satisfy the inequality

$$\|\mathbf{s}\| < \|\epsilon \, (\hat{J} G)^{-1}\| \times \|\mathbf{d}\|_{\max} \tag{7.30}$$

if the time derivative of the sliding manifold is small after all transients have decayed. Equation (7.30) is valid using either external torques or reaction wheels in the control system.

Example 7.3. In this example the sliding-mode controller is used to control the attitude of the WMAP spacecraft using quaternion and angular velocity observations. The WMAP mission created a full-sky map of the cosmic microwave background and measured its anisotropy with 0.3° angular resolution in order to answer fundamental cosmological questions such as the age of the universe, the value of the Hubble constant, and the existence and nature of dark matter.

WMAP is in a Lissajous orbit [23] about the Earth-Sun L_2 Lagrange point with an approximately 180-day period (see Sect. 10.5). Because of its distance, 1.5 million km from Earth, this orbit affords great protection from the Earth's microwave emission, magnetic fields, and other disturbances, with the dominant disturbance torque being solar radiation pressure. It also provides for a very stable thermal environment and near 100 % observing efficiency, since the Sun, Earth, and Moon are always behind the instrument's field of view. In this orbit WMAP sees a Sun/Earth angle between 2° and 10°. The instrument scans an annulus in the hemisphere away from the Sun, so the universe is scanned twice as the Earth revolves once around the Sun. Reference [27] presents an overview of the WMAP attitude control system.

The spacecraft instruments are shown in Fig. 7.4. The spacecraft orbit and attitude specifications are shown in Fig. 7.5. To provide the scan pattern, the spacecraft spins about the z-axis at 0.464 rpm, and the z-axis cones about the Sun-line at 1 rev/h. A 22.5° ± 0.25° angle between the z-axis and the Sun direction must be maintained to provide a constant power input, and to provide constant temperatures for alignment stability and science quality. The spacecraft's attitude is defined by a $3-1-3$ Euler angle rotation relative to a rotating, Sun-referenced frame. The three commanded Euler angles are ϕ_c, θ_c, and ψ_c, and the desired states for the observing mode are

$$\dot{\phi}_c = 1 \, \frac{\text{rev}}{\text{h}} = 0.001745 \, \frac{\text{rad}}{\text{s}}$$

$$\theta_c = 22.5° = 0.3927 \, \text{rad}$$

$$\dot{\psi}_c = 0.464 \, \text{rpm} = 0.04859 \, \frac{\text{rad}}{\text{s}}$$

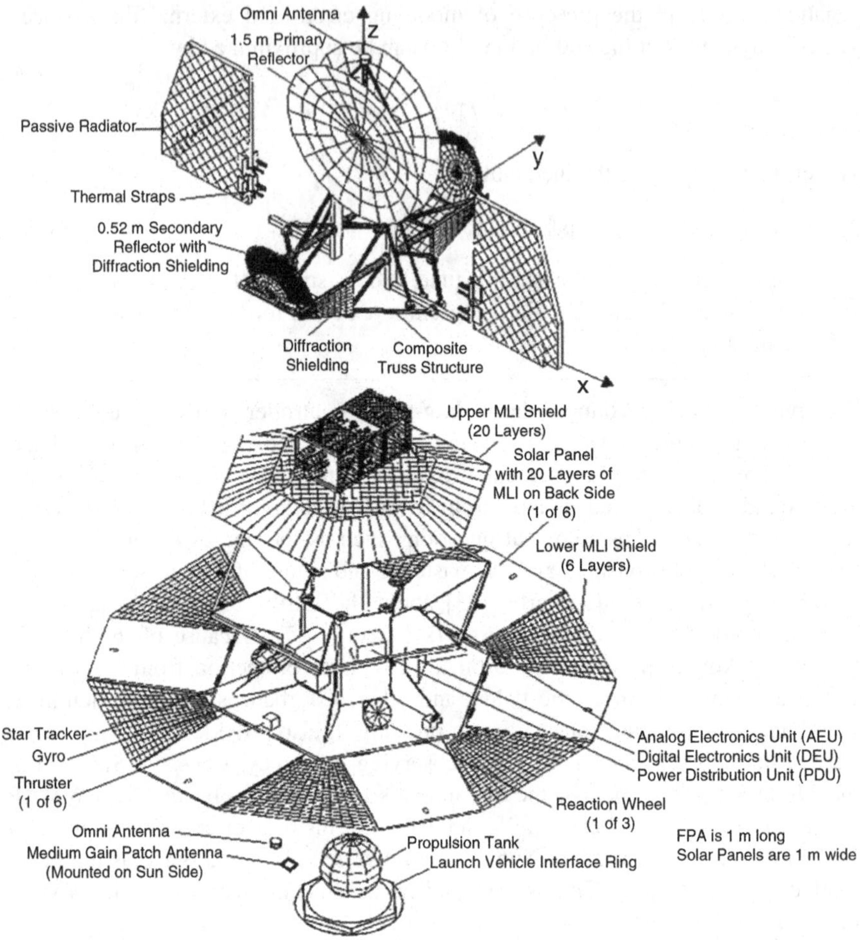

Fig. 7.4 WMAP spacecraft

The desired Euler angles for ϕ_c and ψ_c are determined by integrating the Euler rates. The scan pattern can be simulated by first multiplying the transpose of the $3-1-3$ attitude matrix, shown in Table 9.1, by the vector $[\cos\theta_c \;\; 0 \;\; \sin\theta_c]^T$, which gives the following components:

$$a_1 = \cos\phi_c \cos\theta_c \cos\psi_c - \sin\phi_c \cos^2\theta_c \sin\psi_c + \sin\phi_c \sin^2\theta_c$$

$$a_2 = \sin\phi_c \cos\theta_c \cos\psi_c + \cos\phi_c \cos^2\theta_c \sin\psi_c - \cos\phi_c \sin^2\theta_c$$

$$a_3 = \cos\theta_c \sin\theta_c (\sin\psi_c + 1)$$

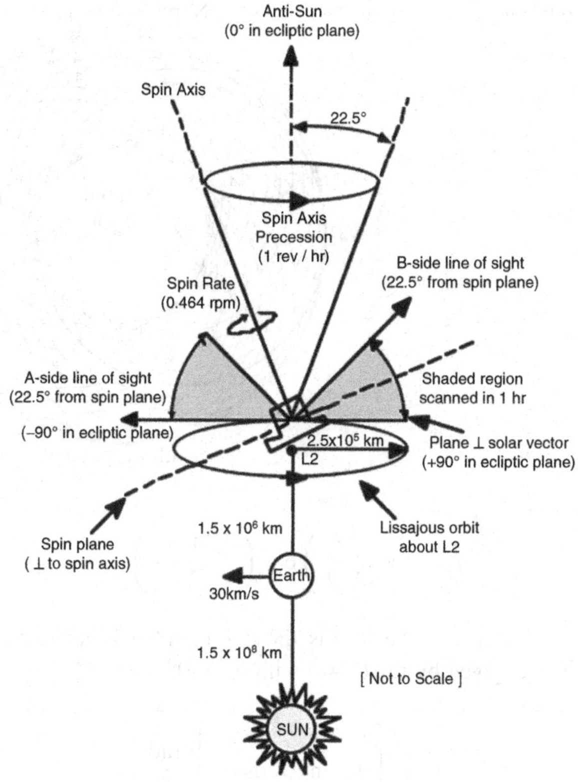

Fig. 7.5 WMAP mission specifications

The x and y coordinates are given by

$$x = \frac{a_1}{1 + a_3}, \quad y = \frac{a_2}{1 + a_3}$$

Figure 7.6 shows the scan pattern for one complete precession (1 h), displayed in ecliptic coordinates in which the ecliptic equator runs horizontally across the map. The bold circle shows the path for a single spin (2.2 min).

The commanded quaternion is determined using

$$q_{c_1} = \sin\left(\frac{\theta_c}{2}\right)\cos\left(\frac{\phi_c - \psi_c}{2}\right)$$

$$q_{c_2} = \sin\left(\frac{\theta_c}{2}\right)\sin\left(\frac{\phi_c - \psi_c}{2}\right)$$

$$q_{c_3} = \cos\left(\frac{\theta_c}{2}\right)\sin\left(\frac{\phi_c + \psi_c}{2}\right)$$

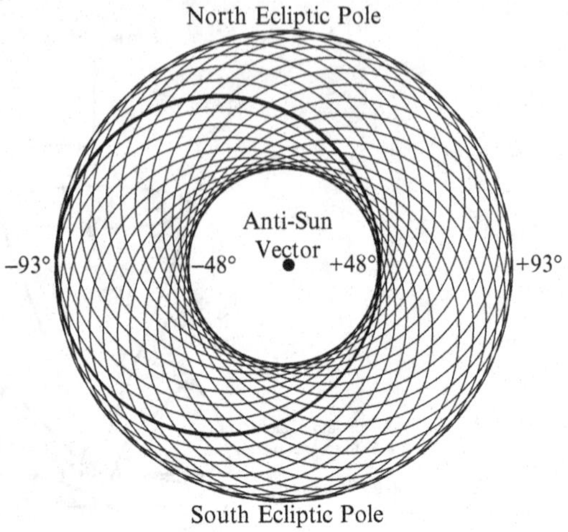

Fig. 7.6 WMAP mission scan pattern

$$q_{c_4} = \cos\left(\frac{\theta_c}{2}\right)\cos\left(\frac{\phi_c + \psi_c}{2}\right)$$

The kinematic equation that transforms the commanded Euler rates to the commanded body rates is given by Eq. (3.41) with $\dot{\theta}_c = 0$:

$$\boldsymbol{\omega}_c = \begin{bmatrix} \dot{\phi}_c \sin\theta_c \sin\psi_c \\ \dot{\phi}_c \sin\theta_c \cos\psi_c \\ \dot{\psi}_c \end{bmatrix} \frac{\text{rad}}{\text{s}}$$

Its derivative, which is required for the sliding-mode controller, is given by

$$\dot{\boldsymbol{\omega}}_c = \dot{\phi}_c \dot{\psi}_c \begin{bmatrix} \sin\theta_c \cos\psi_c \\ -\sin\theta_c \sin\psi_c \\ 0 \end{bmatrix} \frac{\text{rad}}{\text{s}^2}$$

The initial quaternion for the simulation test is given by

$$\mathbf{q}(t_0) = [0 \;\; 0 \;\; \sin(\Phi/2) \;\; \cos(\Phi/2)]^T \otimes \mathbf{q}_c(t_0)$$

The initial angular velocity is set to $\boldsymbol{\omega}(t_0) = \mathbf{0}$. Three reaction wheels are used to control the spacecraft. The initial wheel momentum is set to $\mathbf{h}(t_0) = \mathbf{0}$. The true inertia matrix of the spacecraft is given by

$$J = \begin{bmatrix} 399 & -2.81 & -1.31 \\ -2.81 & 377 & 2.54 \\ -1.31 & 2.54 & 377 \end{bmatrix} \text{kg-m}^2$$

The nominal (assumed) inertia matrix used in the controller is given by

$$\hat{J} = \begin{bmatrix} 380 & -2.81 & -1.31 \\ -2.81 & 360 & 2.54 \\ -1.31 & 2.54 & 340 \end{bmatrix} \text{ kg-m}^2$$

The following external disturbance is also added:

$$\mathbf{d} = \begin{bmatrix} 0.005\sin(0.05\,t) \\ 0.003 \\ 0.005\cos(0.05\,t) \end{bmatrix} \text{ Nm}$$

The simulation uses the following parameters: $\Phi = 60°$, $k = 0.015$, $G = 0.15 I_3$, and $\epsilon = 0.01$. A plot of the angle errors is given in Fig. 7.7a. Convergence occurs in about 20 min. The slight oscillations are due to the disturbance. A plot of the angular velocity errors is shown in Fig. 7.7b, which clearly shows that the desired angular velocity profile is achieved. The wheel momenta are shown in Fig. 7.7c. Note that the total angular momentum for the spacecraft is zero because $\boldsymbol{\omega}(t_0) = \mathbf{0}$ and $\mathbf{h}(t_0) = \mathbf{0}$. Therefore any increase in $J\boldsymbol{\omega}$ causes an equal negative value for $\mathbf{h}$. Figure 7.7d presents plots of the upper bound given by Eq. (7.30) and the actual norm of the sliding vector. This shows the effectiveness of using Eq. (7.30) to accurately bound the sliding manifold errors.

7.3.1 Alternative Formulation

We now describe another tracking control law that we will use when we consider the effects of noise [30]. First, define the following angular velocity difference:

$$\delta\boldsymbol{\omega} = \boldsymbol{\omega} - \delta A\,\boldsymbol{\omega}_c \tag{7.31}$$

where $\delta A = A\,A_c^T$. The time derivative of δA is given by

$$\begin{aligned} \delta\dot{A} &= \dot{A}\,A_c^T + A\,\dot{A}_c^T = -[\boldsymbol{\omega}\times]\delta A + \delta A[\boldsymbol{\omega}_c\times] \\ &= -[\boldsymbol{\omega}\times]\delta A + \delta A[\delta A^T(\boldsymbol{\omega} - \delta\boldsymbol{\omega})\times] = -[\delta\boldsymbol{\omega}\times]\delta A \end{aligned} \tag{7.32}$$

where Eq. (2.63) has been used. The corresponding error quaternion kinematics equation is given by Eq. (7.5).

Taking the time derivative of Eq. (7.31) and left multiplying by the inertia matrix gives

$$J\,\delta\dot{\boldsymbol{\omega}} = J\dot{\boldsymbol{\omega}} - J\,\delta\dot{A}\,\boldsymbol{\omega}_c - J\,\delta A\,\dot{\boldsymbol{\omega}}_c \tag{7.33}$$

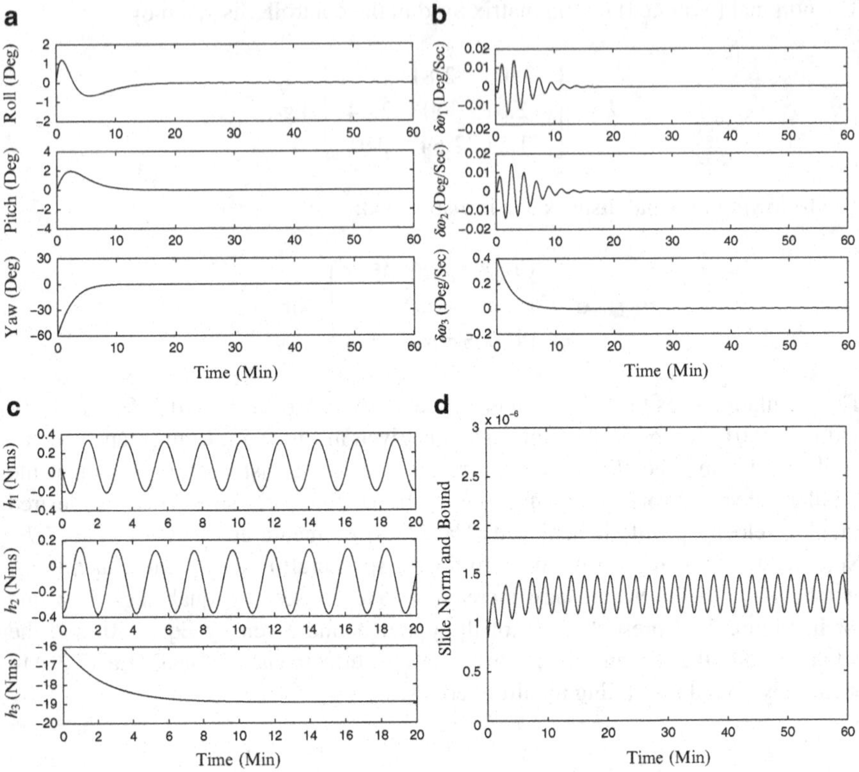

Fig. 7.7 WMAP simulation results. (**a**) Angle errors. (**b**) Angular velocity errors. (**c**) Wheel momenta. (**d**) Slide norm and upper bound

Substituting Eqs. (7.1b) and (7.3.1) into Eq. (7.33), and after some simple algebraic manipulations leads to the following dynamic equation:

$$J\,\delta\dot{\boldsymbol{\omega}} = \Sigma(\delta\boldsymbol{\omega},\, \boldsymbol{\omega}_c,\, \delta A)\delta\boldsymbol{\omega} - [(\delta A\,\boldsymbol{\omega}_c)\times]J\,\delta A\,\boldsymbol{\omega}_c - J\,\delta A\,\dot{\boldsymbol{\omega}}_c + \mathbf{L} \tag{7.34}$$

where $\Sigma(\delta\boldsymbol{\omega},\, \boldsymbol{\omega}_c,\, \delta A)$ is the skew-symmetric matrix

$$\Sigma(\delta\boldsymbol{\omega},\, \boldsymbol{\omega}_c,\, \delta A) \equiv [(J\,\delta\boldsymbol{\omega})\times] + [(J\delta A\,\boldsymbol{\omega}_c)\times] - [(\delta A\,\boldsymbol{\omega}_c)\times]J - J[(\delta A\,\boldsymbol{\omega}_c)\times] \tag{7.35}$$

As with the sliding-mode control approach, $\boldsymbol{\omega}_c$ can be any possible commanded trajectory. A stabilizing control law is given by

$$\mathbf{L} = [(\delta A\,\boldsymbol{\omega}_c)\times]J\,\delta A\,\boldsymbol{\omega}_c + J\,\delta A\,\dot{\boldsymbol{\omega}}_c - k_p\,\delta\mathbf{q}_{1:3} - k_d\,\delta\boldsymbol{\omega} \tag{7.36}$$

The stability proof for this control law is left as an exercise for the reader. Note that when $\boldsymbol{\omega}_c = \mathbf{0}$ then Eq. (7.36) is identical to Eq. (7.7).

7.4 Attitude Thruster Control

Previous sections have presented a theoretical framework for the use of external torques to control the attitude. External-torque control is rarely used for fine-pointing but is needed to dump momentum in order to prevent saturation of control wheels, and is often used for initial attitude acquisition after launch and for safehold modes to cope with reaction wheel failures. The only external means of applying full three-axis external torques is by the use of thrusters. Magnetic torquers can also generate torques, but not in all three axes simultaneously. Magnetic torquers are often used for low-Earth orbits while thrusters are used for high-Earth orbits or interplanetary missions where magnetic fields are weak and unpredictable. Use of thrusters in autonomous safehold modes is dangerous, however, because thrusters can impart a large amount of angular momentum to the system, which magnetic control torques can remove only with great difficulty (if at all) after the thruster propellant has been expended.

The most common attitude control method by thrusters uses pulse-width pulse-frequency (PWPF) modulation. The PWPF modulator translates the continuous commanded control torque signal to an on/off signal. The pulse width is achieved by modulating the width of the activated reaction pulse proportionally to the level of the torque command input, while the pulse frequency modulates the distance between the pulses. Usually the width is very short and the frequency is set to a level that is much faster than the spacecraft rigid-body dynamics.

Figure 7.8 shows the main components of a PWPF modulator, which includes a Schmidt trigger, a lag filter, and a feedback loop. The Schmidt trigger is an on/off relay with a deadband and hysteresis. When a positive input to the Schmidt trigger is greater than U_{on}, the trigger output is U_m. When the input falls below U_{off}, the trigger output is 0. This response is also reflected for negative inputs, i.e. when a negative input to the Schmidt trigger is less than $-U_{\text{on}}$, the trigger output is $-U_m$, and when the input is above $-U_{\text{off}}$ the trigger output is 0.

The variable K_p is a gain used to amplify or reduce the commanded input. The filter is a simple first-order filter with time constant τ_m and filter gain K_m. The hysteresis effect for positive inputs is as follows. The output of the filter increases until it reaches the U_{on} value and the signal is set to the prescribe U_m

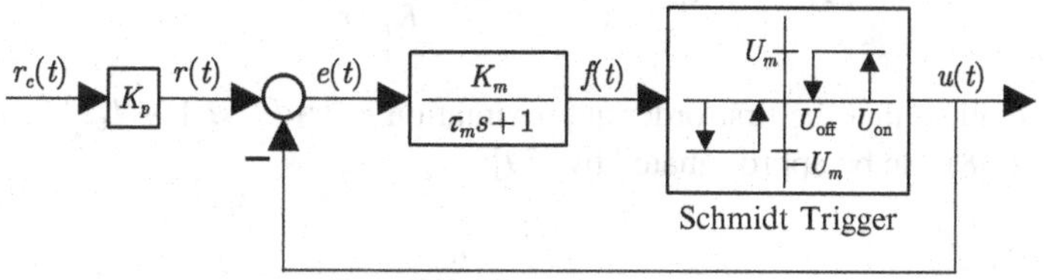

Fig. 7.8 Pulse-width pulse-frequency modulated system

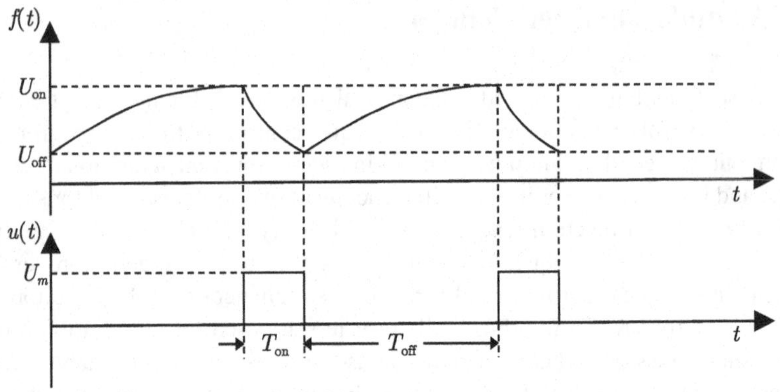

Fig. 7.9 Behavior of a pulse-width pulse-frequency modulator with filter

value. Because of the negative feedback loop, the signal then begins to decrease until it reaches the U_{off} value and the signal is set to 0. The width of the hysteresis is given by $h \equiv U_{\text{on}} - U_{\text{off}}$. The behavior of this process is shown in Fig. 7.9.

The PWPF modulator is a popular choice for attitude control thrusters because of its good fuel efficiency characteristics. If the spacecraft can be considered to be a rigid body then static behavior can be used to analyze the characteristics of the PWPF modulator. Reference [1] uses a describing function approach to analyze the case when flexible dynamics are present. In the static case the input, $e(t)$, is considered to be a constant here set to simply e, and the PWPF has nearly a linear duty cycle. Then $f(t)$ is given by

$$f(t) = f(0) + [K_m\, e - f(0)](1 - e^{-t/\tau_m}) \tag{7.37}$$

Note that as time approaches infinity, $f(t)$ approaches $K_m\, e$. Let $e = r - u$, where $r = K_p\, r_c$ is a constant input into the summer in Fig. 7.8 and u is the constant output. The on-time, otherwise known as the pulse width, is denoted by T_{on}. This can be found by setting $f(0) = U_{\text{on}}$ and $f(T_{\text{on}}) = U_{\text{off}}$ in Eq. (7.37) and solving for T_{on}, which yields

$$T_{\text{on}} = -\tau_m \ln\left(1 - \frac{h}{U_{\text{on}} - K_m(r-u)}\right) \tag{7.38}$$

If T_{on} is small so that the first-order approximation $e^{-T_{\text{on}}/\tau_m} \approx 1 - T_{\text{on}}/\tau_m$ is good, then Eq. (7.38) can be approximated by [39]

$$T_{\text{on}} \approx \tau_m \frac{h}{U_{\text{on}} - K_m(r-u)} \tag{7.39}$$

The output is zero when the thruster is off, so the off-time can be found by setting $e = r$, $f(0) = U_{\text{off}}$, and $f(T_{\text{off}}) = U_{\text{on}}$ in Eq. (7.37) and solving for T_{off}, yielding

$$T_{\text{off}} = -\tau_m \ln\left(1 - \frac{h}{K_m\, r - U_{\text{off}}}\right) \tag{7.40}$$

The first-order approximation for the exponential gives

$$T_{\text{off}} = \tau_m \frac{h}{K_m\, r - U_{\text{off}}} \tag{7.41}$$

The output frequency, denoted by f, is calculated using

$$f = \frac{1}{T_{\text{on}} + T_{\text{off}}} \tag{7.42}$$

The duty cycle, denoted by DC, can be used to determine how well the modulator output follows its input. This quantity is given by

$$DC = \frac{T_{\text{on}}}{T_{\text{on}} + T_{\text{off}}} \tag{7.43}$$

The internal deadband is defined as the magnitude of the signal required to activate the Schmidt trigger. Assuming zero dynamics in the filter and $K_p = 1$, from Fig. 7.8, this occurs when $K_m\, r \geq U_{\text{on}}$. Thus, the internal deadband, denoted by r_{DB}, is given by

$$r_{\text{DB}} = \frac{U_{\text{on}}}{K_m} \tag{7.44}$$

Equation (7.44) indicates that increasing K_m can reduce the size of deadband. In order to ensure that U_{on} is the upper bound of the deadband, $K_m > 1$ should be chosen. The minimum pulse width, denoted by Δ, is found by substituting Eq. (7.44) for r in Eq. (7.38) and setting $u = U_m$, which yields

$$\Delta = -\tau_m \ln\left(1 - \frac{h}{K_m U_m}\right) \tag{7.45}$$

The saturation level, denoted by r_{sat}, is obtained when the thrusters are on all the time. This is determined by equating the maximum value of the filter output, $K_m(r_{\text{sat}} - U_m)$, to the Schmidt trigger off condition, U_{off}. Solving for r_{sat} gives

$$r_{\text{sat}} = U_m + \frac{U_{\text{off}}}{K_m} \tag{7.46}$$

Table 7.1 Recommended ranges for PWPF parameters

Parameter	Static analysis	Dynamic analysis	Recommended
K_m	$2 < K_m < 7$	N/A	$2 < K_m < 7$
τ_m	$0.1 < \tau_m < 1$	$0.1 < \tau_m < 0.5$	$0.1 < \tau_m < 0.5$
U_{on}	$U_{\text{on}} > 0.3$	N/A	$U_{\text{on}} > 0.3$
U_{off}	$U_{\text{off}} < 0.8\,U_{\text{on}}$	N/A	$U_{\text{off}} < 0.8\,U_{\text{on}}$
K_p	N/A	$K_p \geq 20$	$K_p \geq 20$

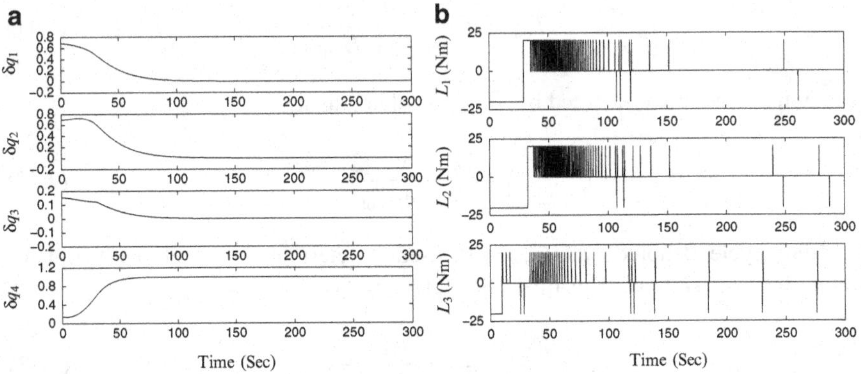

Fig. 7.10 (**a**) Quaternion errors and (**b**) control torques using PWPF thrusters

Equation (7.46) gives the maximum value for the pseudo-linear region. Reference [4] provides recommended parameter values based on static and dynamic tests, which are shown in Table 7.1. Although these values may not be suitable for every application, they provide a good starting point for the design of a PWPF thruster.

Example 7.4. In this example a PWPF controller and the control law in Eq. (7.12) are used for the spacecraft with parameters described in Example 7.1. Three PWPF thrusters are used with each $r_c(t)$ given by L_i, $i = 1, 2, 3$, in Eq. (7.12). The output of each PWPF thruster is the corresponding torque input to the spacecraft. The initial quaternion is again given by $\mathbf{q} = [0.6853\ 0.6953\ 0.1531\ 0.1531]^T$ and the initial angular velocity is again given by $\boldsymbol{\omega} = [0.5300\ 0.5300\ 0.0530]^T$ deg/s. The desired quaternion is the identity quaternion. The gains are $k_p = 50$ and $k_d = 500$. The PWPF parameters are $K_m = 5$, $\tau_m = 0.5$, $U_{\text{on}} = 10$, $U_{\text{off}} = 6$, $K_p = 275$, and $U_m = 20$. A plot of the quaternion errors is shown in Fig. 7.10a. The fourth error-quaternion component approaches 1 while the other three components approach zero. The control torques, shown in Fig. 7.10b, are large at the beginning of the maneuver and have signs equivalent to the ones shown in Fig. 7.1b. From these plots it is clear that the PWPF thrusters also provide a reorientation maneuver to the desired attitude. Fewer thruster firings are required at the end of the maneuver. Gains and the PWPF parameters are often scheduled in practice at various phases of the maneuver to provide the desired steady-state performance. Reference [24] provides a study on optimal tuning of the PWPF parameters.

7.5 Magnetic Torque Attitude Control

Attitude control using magnetic torquers was first proposed in the early 1960s [48]. One of the main uses of magnetic torquers is to dump excess momentum induced by external disturbances. This is typically required to insure that wheels do not saturate. Recall that wheels only redistribute a spacecraft's angular momentum since they are internal body mechanisms. When disturbances are present excess momentum can build up in the spacecraft. The secular (orbit-averaged) component of external disturbance torques would lead to saturation of the momentum capacity of the reaction wheels, so either mass expulsion or magnetic control torques are needed to dump excess wheel angular momentum. Magnetic control torques have several advantages for near-Earth missions, including smoothness of application, essentially unlimited mission life (due to the absence of expendables), and absence of catastrophic failure modes [43]. It is important to note that magnetic torquers need not compensate for the entire disturbance torque in such an application, but only for its secular component [8]. Other uses include detumbling, initial acquisition, precession control, nutation damping, and momentum control.

Magnetic torquers use the Earth's magnetic field to produce a torque. As described in Sect. 11.1 in Chap. 11, the Earth's magnetic field magnitude decreases as the inverse cube of the distance from the center of the Earth. Thus the magnetic torque will be several orders of magnitude smaller at high-Earth orbits, such as geosynchronous orbits, than at low-Earth orbits. Magnetic control torques are typically on the order of 10^{-5} to 10^{-4} Nm for low-Earth orbits, depending on a number of factors such as orbit inclination. Another issue is that the torques are constrained to lie in a two-dimensional plane orthogonal to the magnetic field, so only two out of three axes can be controlled at a given time instant. However, full three-axis control is available provided that the spacecraft's orbital plane does not coincide with the geomagnetic equatorial plane and does not contain the magnetic poles [7].

Reference [38] provides a good example, which is summarized here. Suppose that one magnetic torquer is aligned with the spin axis of the spacecraft. Activating this torquer will cause the spin axis to precess about the direction of the magnetic field when the magnetic field has a component that is perpendicular to the spin axis. If the spacecraft orbit is in the magnetic equatorial plane then the spin axis can only be precessed about the direction of the North magnetic pole. In this case three-axis attitude control is not possible. However, if the orbit plane is offset from the magnetic equatorial plane, then the direction of the magnetic field will change

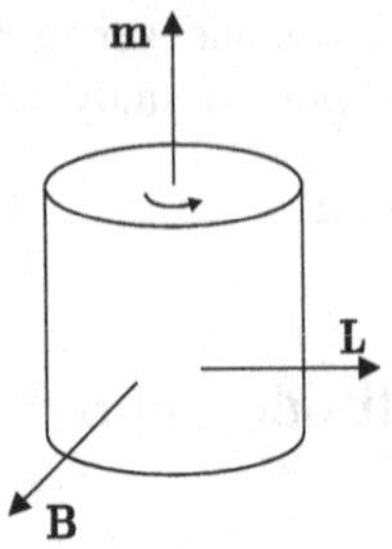

Fig. 7.11 Magnetic torquer control

during the orbit and magnetic attitude control is possible. The magnetic equatorial plane is only 11° from the Earth's equator.[1] Thus for equatorial orbits the axis pointing in the direction of motion has the most control authority. For polar orbits the axis orthogonal to the direction of motion and to the Earth-pointing direction has the most control authority.

A survey of magnetic spacecraft attitude control can be found in [40]. Figure 7.11 depicts the magnetic control scheme for the aforementioned spin-axis control. The torque generated by the magnetic torquers is given by

$$\mathbf{L} = \mathbf{m} \times \mathbf{B} \tag{7.47}$$

where $\mathbf{m}$ is the commanded magnetic dipole moment generated by the torquers and $\mathbf{B}$ is the local geomagnetic field expressed in body-frame coordinates. This is related to the magnetic field vector, $\mathbf{R}$, expressed in reference-frame coordinates through the attitude matrix: $\mathbf{B} = A\mathbf{R}$. The vector $\mathbf{R}$ depends on the spacecraft's orbital position. The magnetic moment is given in units of Am^2 and the magnetic flux density of the geomagnetic field is given in units of $1\,\text{Wb/m}^2 = 1$ Tesla (T) $= 10^4$ Gauss (G).

7.5.1 Detumbling

In this section a magnetic control law is developed that can be used to detumble a spacecraft (i.e. null its angular velocity vector). Detailed theory behind this control law can be found in [6]. The control is effected by commanding a magnetic dipole moment:

$$\mathbf{m} = \frac{k}{\|\mathbf{B}\|}\,\omega \times \mathbf{b} \tag{7.48}$$

[1]See Sect. 11.1.

where $\mathbf{b} = \mathbf{B}/\|\mathbf{B}\|$, $\boldsymbol{\omega}$ is the angular velocity and k is a positive scalar gain. This gives a control torque

$$\mathbf{L} = \frac{k}{\|\mathbf{B}\|}(\boldsymbol{\omega} \times \mathbf{b}) \times \mathbf{B} = k\,(\boldsymbol{\omega} \times \mathbf{b}) \times \mathbf{b} = -k\,(I_3 - \mathbf{b}\,\mathbf{b}^T)\,\boldsymbol{\omega} \tag{7.49}$$

The control torque is clearly perpendicular to $\mathbf{b}$. To prove the stability of this control law consider the following candidate Lyapunov function:

$$V = \frac{1}{2}\boldsymbol{\omega}^T J \boldsymbol{\omega} \tag{7.50}$$

Using Eq. (7.1b) with Eq. (7.49), $\dot{V}$ can be shown to be given by

$$\dot{V} = -k\,\boldsymbol{\omega}^T (I_3 - \mathbf{b}\,\mathbf{b}^T)\,\boldsymbol{\omega} \tag{7.51}$$

Since the eigenvalues of $(I_3 - \mathbf{b}\,\mathbf{b}^T)$ are always 0, 1, and 1, then $\dot{V}$ is only negative semi-definite. Stated another way, when $\boldsymbol{\omega}$ is parallel to $\mathbf{b}$ then $\dot{V} = 0$. This is not a concern in practice, though [6].

If no angular velocity information is available, we use Eq. (3.14) for the magnetic field vector, which in the notation of this section is

$$\dot{\mathbf{B}} = A\dot{\mathbf{R}} - \boldsymbol{\omega} \times \mathbf{B} \tag{7.52}$$

We assume, as is the case for the initial stages of detumbling, that $\dot{\mathbf{R}} \ll \dot{\mathbf{B}}$, so a good approximation to Eq. (7.48) is given by

$$\mathbf{m} = -\frac{k}{\|\mathbf{B}\|}\dot{\mathbf{B}} \tag{7.53}$$

where $\mathbf{B}$ is the field sensed by onboard magnetometers. This is an alternative version of the well-known B-dot control shown in [43]. As stated in [6] global asymptotic stability cannot be proven using Eq. (7.53). But, the absolute angular velocity can be reduced down to a value of the same order of magnitude as the orbit rate (around 10^{-3} rad/s) as in the case when the standard B-dot command law is used. Note that in practical application, $\dot{\mathbf{B}}$ would be computed by a finite difference approach, which introduces more noise in the control signal. A filter could be employed to reduce the noise levels, but B-dot control is often implemented as a bang-bang control law. Assume that we have n torquers, and that the ith torquer can produce a maximum dipole of $\pm m_i^{\max}$ in a direction specified by the unit vector $\mathbf{u}_i$. Then the bang-bang B-dot detumbling control commands are

$$m_i = -m_i^{\max}\,\text{sign}(\mathbf{u}_i \cdot \dot{\mathbf{B}}) \quad \text{for } i = 1, \ldots, n \tag{7.54}$$

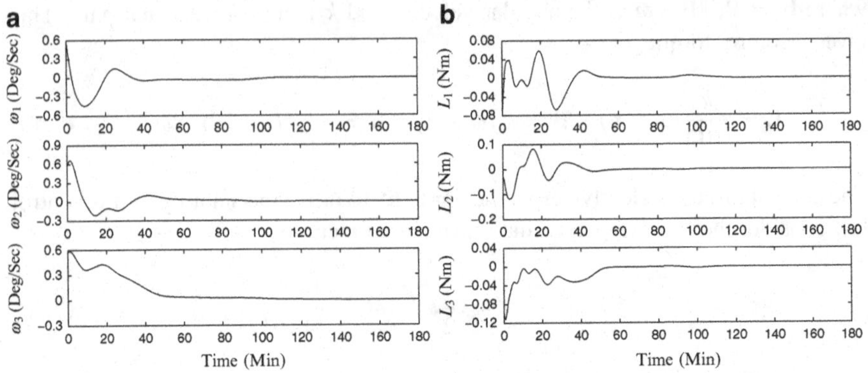

Fig. 7.12 Detumbling simulation results. (**a**) Angular velocities. (**b**) Control torques

To avoid feedback from the torquers to the magnetometers, the computation of $\dot{\mathbf{B}}$ is inhibited for some period after any m_i changes sign.

Reference [6] provides a gain expression based on analyzing the closed-loop dynamics of the component of $\boldsymbol{\omega}$ perpendicular to the Earth's magnetic field:

$$k = \frac{4\pi}{T_{\text{orb}}}(1 + \sin \xi_m)\, J_{\text{min}} \tag{7.55}$$

where T_{orb} is the orbital period in seconds, ξ_m is the inclination of the spacecraft orbit relative to the geomagnetic equatorial plane and J_{min} is the minimum principal moment of inertia. Note that k is always positive. Constant positive values can also be used to achieve design specifications if desired.

Example 7.5. In this example the control law in Eq. (7.49) is used to null the angular velocity of a rotating spacecraft. The gain is given by Eq. (7.55). The inertia of the spacecraft is given in Example 7.2. Note that the minimum principal moment of inertia is given by 4726.01952 kg-m^2. The GCI inertial position and velocity of the spacecraft are given by $\mathbf{r}_0 = [1029.7743\ \ 6699.3469\ \ 3.7896]^T$ km and $\mathbf{v}_0 = [-6.2119\ \ 0.9524\ \ 4.3946]^T$ km/s. The epoch time is May 10, 2011 at 4:56:36.9191 a.m. This information is required to generate the reference magnetic field. The initial quaternion is given by $\mathbf{q}(t_0) = \sqrt{2}/2\,[1\ \ 0\ \ 0\ \ 1]^T$ and the initial angular velocity is given by $\boldsymbol{\omega}(t_0) = [0.01\ 0.01\ 0.01]^T$ rad/s. A plot of the angular velocity trajectories is shown in Fig. 7.12a and the associated magnetic control torques are shown in Fig. 7.12b. Clearly, the control law is able to detumble the spacecraft.

7.5.2 Momentum Dumping

As stated previously the most common use of magnetic torquers for low-Earth orbiting spacecraft is momentum dumping. Excess momentum is usually built up in the spacecraft through external disturbances, which are non-conservative. A periodic disturbance torque along one spacecraft axis results in a cyclic variation in the angular velocity along that axis, while a constant (secular) disturbance results in a linear increase in angular velocity, where the wheel is accelerated at a constant rate in order to transfer the excess momentum from the external disturbance to the wheel [26]. Eventually saturation of the wheels will occur due to the excess momentum, which can only be dumped through external torques.

A common approach to design a magnetic torquer control law for momentum dumping is to command a magnetic dipole moment [8]:

$$\mathbf{m} = \frac{k}{\|\mathbf{B}\|}\,\mathbf{h} \times \mathbf{b} \tag{7.56}$$

which is just like the detumbling control of Eq. (7.48) with the angular velocity vector replaced by the wheel angular momentum. The resulting torque is given by

$$\mathbf{L} = -k\,(I_3 - \mathbf{b}\,\mathbf{b}^T)\,\mathbf{h} \tag{7.57}$$

As with the detumbling case a torque cannot be exerted when **h** is parallel to **b**, but as before, this is not a concern for practical applications.

Example 7.6. In this example the control law in Eq. (7.57) is used to reduce the wheel momentum for the simulation shown in Example 7.2. All the parameters, such as the inertia matrix, the initial quaternion, the initial angular velocity, the initial wheel momentum, and control gains are identical to the ones shown in Example 7.2. The initial position, velocity and epoch are the same as the ones shown in Example 7.5. The goal is to reduce the overall momentum by about half in less than 3 h. To accomplish this goal the gain k in Eq. (7.57) is set to 0.0001.

The momentum dumping control law is not turned on until the 30-min mark. This allows the spacecraft to first complete its reorientation maneuver. A plot of the quaternion errors is shown in Fig. 7.13. As before, the fourth error-quaternion component approaches 1 while the other three components approach zero. The wheel momenta are shown in Fig. 7.13. Note that the wheel momenta begin to approach zero after the momentum dumping control law is executed. The commanded dipoles are shown in Fig. 7.13. The spacecraft momentum is shown in Fig. 7.13. At the end of 3 h we clearly see that the momentum is reduced by the desired goal. The magnitude of control torques, not shown here, is around 10^{-3} Nm. This is about an order of magnitude higher than a typical magnetic control torque for momentum dumping. In practice the gain k would be set to a much lower level because reducing momentum by half over the short time period shown in this

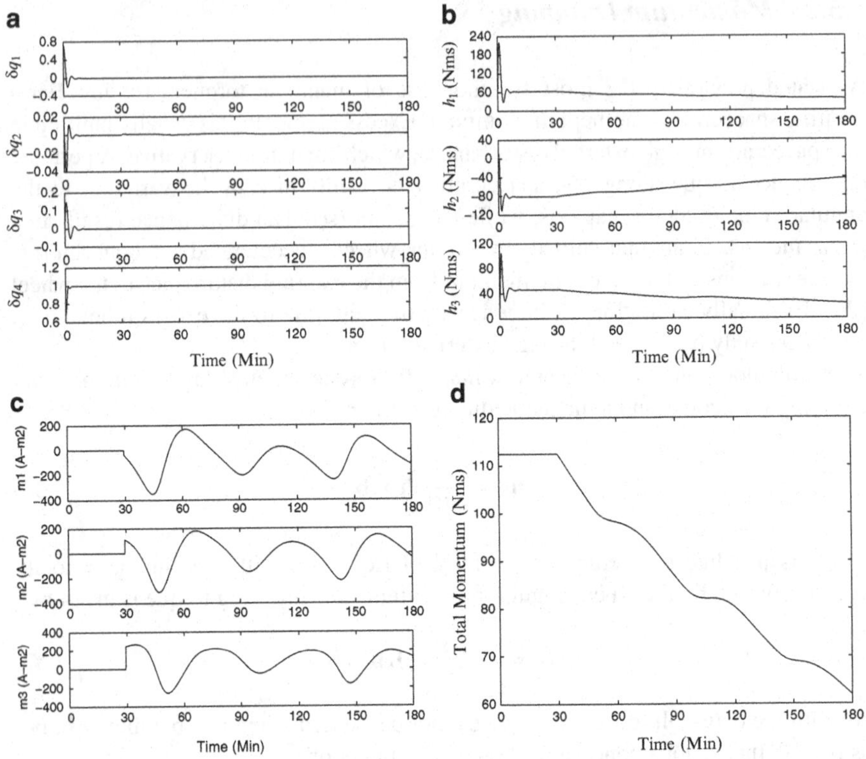

Fig. 7.13 Momentum dumping simulation results. (**a**) Quaternion errors. (**b**) Wheel momenta. (**c**) Commanded dipoles. (**d**) Spacecraft momentum

example is typically not required. The magnetic control law is usually continuously active to dump excess momentum due to external disturbances during the mission mode.

7.6 Effects of Noise

None of the control simulations presented in the previous sections of this chapter incorporates measurement noise, which is always present in practice for any attitude sensor. In most cases the filtering methods in Chap. 6 are employed to filter noisy measurements and the estimates are used in the feedback controllers. For linear systems replacing the "true" values with the Kalman filter estimates in the control law turns out to actually be the optimal approach, as is proven by the *Separation Theorem* [12], also known as the *Certainty Equivalence Principle* [5, 16, 42] (see Sect. 12.3.9). This theorem states that the solution of the overall optimal control problem with incomplete state knowledge is given by the solution of two separate sub-problems: (1) the estimation problem solved using the Kalman filter to provide optimal state estimates, and (2) the control problem using the optimal state estimates, which is derived from the standard optimal control results. Another way to show this separation of the overall control design involves the eigenvalue separation property [2], which states that the eigenvalues of the overall closed-loop system are given by the eigenvalues of the control system together with those of the state estimator system. Unfortunately, no such theorem is available for general nonlinear systems. Still, the combined Kalman filter and feedback control approach works well for most attitude control systems.

The quaternion is the parameterization of choice for both estimation and control because it avoids singularity issues. As explained previously, the mapping of quaternions to rotations is globally two-to-one because $\mathbf{q}$ and $-\mathbf{q}$ represent the same attitude. Neglecting this property in the control law can induce a phenomenon called *unwinding*, which may produce a control law that does not guarantee that the shortest path is provided to the final orientation, as described in Sect. 7.2. To overcome this path issue a discontinuous set of quaternions is used through the signum function, as also shown in previous sections of this chapter. The problem with the signum function is that it is not robust when noise is present, meaning that noise can destroy any global attractivity property [33].

Reference [30] provides an excellent example of the noise and unwinding issues. We consider an attitude regulation problem whose goal is to drive the attitude error to zero, which means to drive the error quaternion to $\pm\mathbf{I}_q$, using the angular velocity as the input. The error quaternion kinematics are given by

$$\boldsymbol{\delta}\dot{\mathbf{q}}_{1:3} = \frac{1}{2}[\boldsymbol{\delta}\mathbf{q}_{1:3}\times]\boldsymbol{\omega} + \frac{1}{2}\delta q_4\,\boldsymbol{\omega} \tag{7.58a}$$

$$\delta\dot{q}_4 = -\frac{1}{2}\boldsymbol{\delta}\mathbf{q}_{1:3}^T\boldsymbol{\omega} \tag{7.58b}$$

Consider the control law given by $\boldsymbol{\omega} = -\boldsymbol{\delta}\mathbf{q}_{1:3}$ and the following candidate Lyapunov function:

$$V_1(\boldsymbol{\delta}\mathbf{q}) = 2\,(1-\delta q_4) = (1-\delta q_4)^2 + \|\boldsymbol{\delta}\mathbf{q}_{1:3}\|^2 \tag{7.59}$$

Clearly $V_1(\boldsymbol{\delta}\mathbf{q}) \geq 0$ with equality if and only if $\boldsymbol{\delta}\mathbf{q} = \mathbf{I}_q$, and $V_1(\boldsymbol{\delta}\mathbf{q})$ achieves its maximum when $\boldsymbol{\delta}\mathbf{q} = -\mathbf{I}_q$. The time derivative of $V_1(\boldsymbol{\delta}\mathbf{q})$ with $\boldsymbol{\omega} = -\boldsymbol{\delta}\mathbf{q}_{1:3}$ is

$$\dot{V}_1(\boldsymbol{\delta}\mathbf{q}) = -\|\boldsymbol{\delta}\mathbf{q}_{1:3}\|^2 \tag{7.60}$$

which is almost always negative. However, there are two equilibrium points: a stable one at $\boldsymbol{\delta}\mathbf{q} = \mathbf{I}_q$ and an unstable one at $\boldsymbol{\delta}\mathbf{q} = -\mathbf{I}_q$. This leads to the classic unwinding issue, which is depicted in Fig. 7.14a. If δq_4 is initially negative, the controller increases the attitude error before driving it to zero.

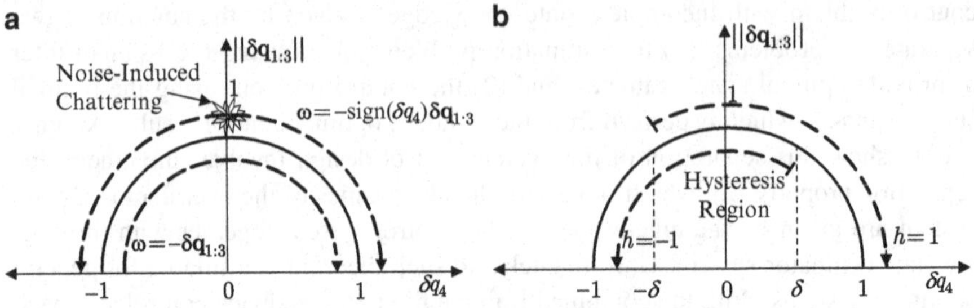

Fig. 7.14 Noise chattering and hysteretic regulation. (**a**) Quaternion attitude control. (**b**) Hysteretic regulation

Now consider the control law $\boldsymbol{\omega} = -\delta q_4 \boldsymbol{\delta} \mathbf{q}_{1:3}$ and the following candidate Lyapunov function:

$$V_2(\boldsymbol{\delta}\mathbf{q}) = 1 - \delta q_4^2 = \|\boldsymbol{\delta}\mathbf{q}_{1:3}\|^2 \tag{7.61}$$

In this case $V_2(\boldsymbol{\delta}\mathbf{q}) = 0$ if and only if $\boldsymbol{\delta}\mathbf{q} = \pm\mathbf{I}_q$, and its maximum value is achieved when $\delta q_4 = 0$, which corresponds to a 180° rotation. Taking the time derivative of $V_2(\boldsymbol{\delta}\mathbf{q})$ with $\boldsymbol{\omega} = -\delta q_4 \boldsymbol{\delta}\mathbf{q}_{1:3}$ leads to

$$\dot{V}_2(\boldsymbol{\delta}\mathbf{q}) = -\delta q_4^2 \|\boldsymbol{\delta}\mathbf{q}_{1:3}\|^2 \tag{7.62}$$

which is always negative except at the two stable equilibrium points $\boldsymbol{\delta}\mathbf{q} = \pm\mathbf{I}_q$. Also note that $V_2(\boldsymbol{\delta}\mathbf{q}) = V_2(-\boldsymbol{\delta}\mathbf{q})$ and $\boldsymbol{\omega}(\boldsymbol{\delta}\mathbf{q}) = \boldsymbol{\omega}(-\boldsymbol{\delta}\mathbf{q})$. This clearly solves the unwinding problem, but $\boldsymbol{\omega} = \mathbf{0}$ when $\delta q_4 = 0$ so convergence takes longer to achieve as initial conditions are closer and closer to 180° rotations.

In an attempt to overcome the convergence and unwinding issues, the control law $\boldsymbol{\omega} = -\text{sign}(\delta q_4)\boldsymbol{\delta}\mathbf{q}_{1:3}$ is chosen with the following candidate Lyapunov function:

$$V_3(\boldsymbol{\delta}\mathbf{q}) = \begin{cases} 2\,(1-\delta q_4) = (1-\delta q_4)^2 + \|\boldsymbol{\delta}\mathbf{q}_{1:3}\|^2 & \delta q_4 \geq 0 \\ 2\,(1+\delta q_4) = (1+\delta q_4)^2 + \|\boldsymbol{\delta}\mathbf{q}_{1:3}\|^2 & \delta q_4 < 0 \end{cases} \tag{7.63}$$

Taking the time derivative of $V_3(\boldsymbol{\delta}\mathbf{q})$ with $\boldsymbol{\omega} = -\text{sign}(\delta q_4)\boldsymbol{\delta}\mathbf{q}_{1:3}$ leads to

$$\dot{V}_3(\boldsymbol{\delta}\mathbf{q}) = -\|\boldsymbol{\delta}\mathbf{q}_{1:3}\|^2 \tag{7.64}$$

which is always negative except at the two stable equilibrium points $\boldsymbol{\delta}\mathbf{q} = \pm\mathbf{I}_q$. Thus this control law achieves global asymptotic stability, but it is not robust to measurement noise. Reference [29] shows that for an arbitrarily small noise signal, with initial conditions close to 180° rotations, this control law keeps the state near the discontinuity for all time. This is the noise-induced chattering problem depicted in Fig. 7.14a.

To solve the unwinding and noise-induced chattering problems [30] develops a hybrid controller, which incorporates hysteresis-based switching using a single binary logic variable for each quaternion error state. The strategy is depicted in Fig. 7.14b, where $\delta \in (0,\ 1)$ denotes the hysteresis half-width. Define the following function:

$$\overline{\text{sign}}(s) = \begin{cases} \text{sign}(s) & |s| > 0 \\ \{-1,\ 1\} & s = 0 \end{cases} \tag{7.65}$$

When $s = 0$ the value of $\overline{\text{sign}}(s)$ is $+1$ if s approaches zero from the positive side and -1 if s approaches from the negative side. Now consider the following control law: $\boldsymbol{\omega} = -h\,\boldsymbol{\delta}\mathbf{q}_{1:3}$, where $h \in \{-1,\ 1\}$ and the dynamics of h are given by

$$\dot{h} = 0 \quad \text{when } (\boldsymbol{\delta}\mathbf{q},\ h) \in \{h\,\delta q_4 \geq -\delta\} \tag{7.66a}$$

$$h^+ = \overline{\text{sign}}(\delta q_4) \quad \text{when } (\boldsymbol{\delta}\mathbf{q},\ h) \in \{h\,\delta q_4 \leq -\delta\} \tag{7.66b}$$

where h^+ denotes the value of the logic variable after being updated. This function represents a hysteretic regulation in the control law. It is a hybrid approach generalizing the control laws $\boldsymbol{\omega} = -\boldsymbol{\delta}\mathbf{q}_{1:3}$ and $\boldsymbol{\omega} = -\text{sign}(\delta q_4)\boldsymbol{\delta}\mathbf{q}_{1:3}$, where δ in Eq. (7.66) manages a tradeoff between robustness to measurement noise and hysteresis-induced inefficiency. The control law $\boldsymbol{\omega} = -\text{sign}(\delta q_4)\boldsymbol{\delta}\mathbf{q}_{1:3}$ is recovered when $\delta = 0$. Setting δ to value greater than or equal to 1 gives a simple control law with the full unwinding effect.

Reference [30] applies the aforementioned hybrid approach to the control law given in Eq. (7.36). The hybrid control law is given by

$$\mathbf{L} = [(\delta A\,\boldsymbol{\omega}_c)\times]J\delta A\,\boldsymbol{\omega}_c + J\delta A\,\dot{\boldsymbol{\omega}}_c - k_p h\,\boldsymbol{\delta}\mathbf{q}_{1:3} - k_d\,\boldsymbol{\delta\omega} \tag{7.67}$$

where $\delta \in (0,\ 1)$ and the dynamics of h are given by Eq. (7.66).

Example 7.7. This example reproduces the example results of [30]. It involves a regulation case that takes any initial quaternion and angular velocity to the identity quaternion and zero angular velocity, respectively. Let $\mathbf{p} = [1\ \ 2\ \ 3]^T/\|[1\ \ 2\ \ 3]\|$. The inertia matrix is given by $J = 10 \times \text{diag}([p_1\ p_2\ p_3])$ and the control gains are given by $k_p = k_d = 1$. The hysteresis half-width is chosen to be $\delta = 0.4$. Synthetic quaternion measurements are generated using

$$\mathbf{q} = \frac{\mathbf{q}^{\text{true}} + m\,\mathbf{v}}{\|\mathbf{q}^{\text{true}} + m\,\mathbf{v}\|}$$

where $\mathbf{v}$ is a zero-mean Gaussian white-noise process with covariance given by the identity matrix and m is drawn from a uniform distribution on the interval $[0,\ 0.2]$. These noise parameters produce a 1σ attitude measurement error of about $6.5°$ in each axis, which is not realistic for any spacecraft attitude sensor; however,

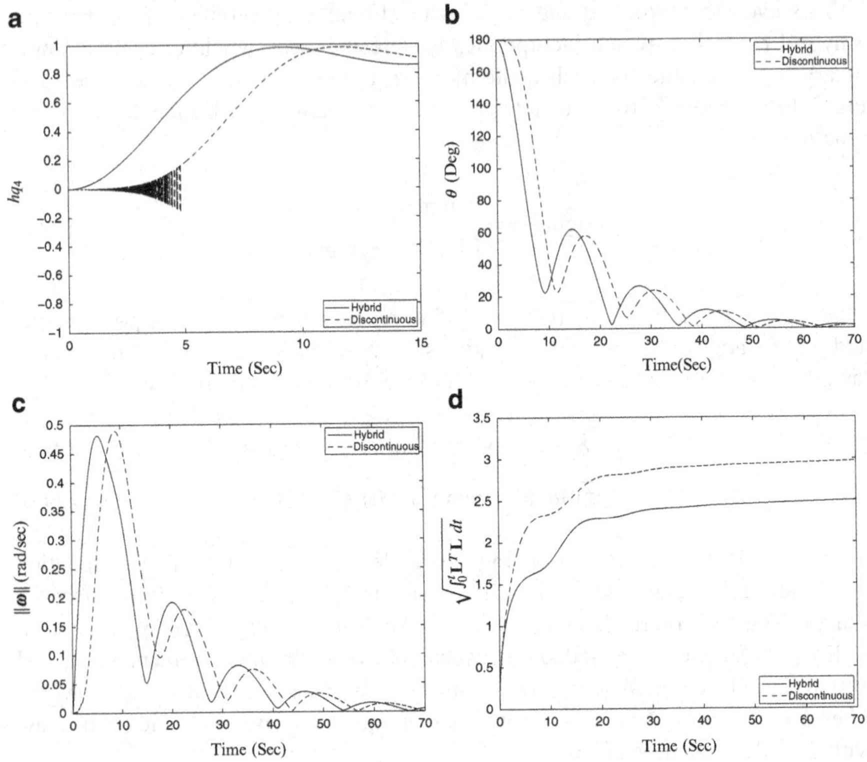

Fig. 7.15 Noise sensitivity results. (**a**) Scalar component quaternion errors. (**b**) Rotation angle errors. (**c**) Norm of angular velocity errors. (**d**) Control effort

this noise level more distinctly shows the effectiveness of the hybrid control law. No noise is added to the true velocity vector. Finally, the initial condition for h is given by $h(0) = 1$.

In the first simulation the initial quaternion and angular velocity are given by $\mathbf{q}^{\text{true}}(0) = [\mathbf{p}^T \;\; 0]^T$ and $\boldsymbol{\omega}^{\text{true}}(0) = \mathbf{0}$, respectively. Plots of $h\, q_4$, the rotation angle error $\theta = 2\cos^{-1}|q_4|$, norm of the angular velocity, and control effort are shown in Fig. 7.15. Setting $\delta = 0$ gives the discontinuous controller case. The unwinding controller case is not shown because its results are identical to the hybrid controller results because h does not change in the hybrid case. Figure 7.15a shows the sensitivity to noise for the discontinuous controller case where the chattering behavior is clearly visible. This behavior causes a lag in the response, which in turn requires more control effort as shown by Fig. 7.15d. The hybrid controller is clearly more robust to measurement noise and requires less control effort than the discontinuous controller.

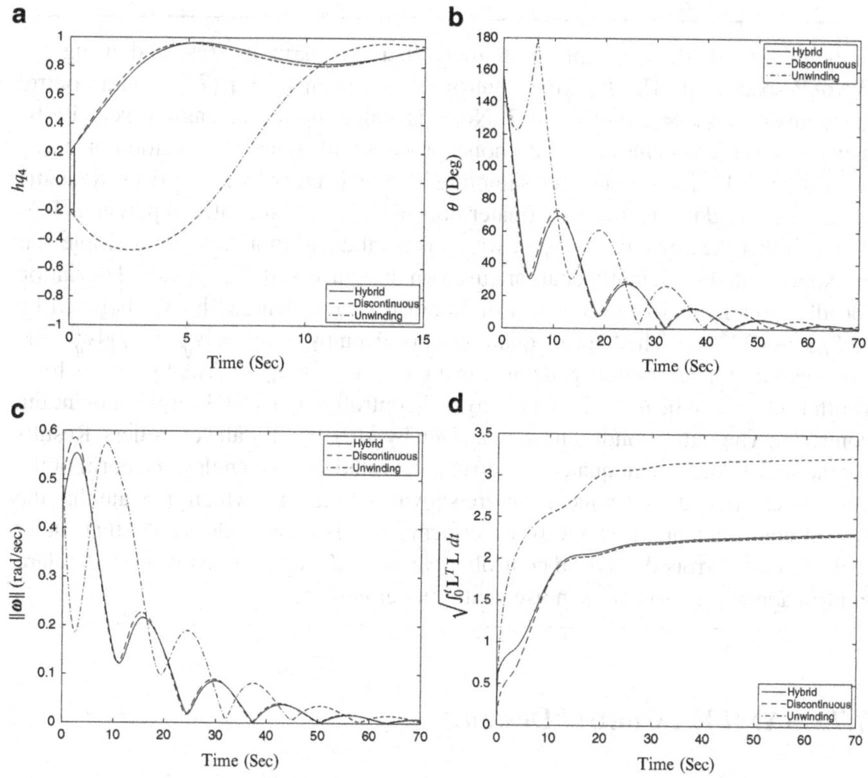

Fig. 7.16 Effects of unwinding. (**a**) Scalar component quaternion errors. (**b**) Rotation angle errors. (**c**) Norm of angular velocity errors. (**d**) Control effort

The second simulation shows how control laws that exhibit unwinding can resist a "beneficial" angular velocity. All parameters are identical to the first simulation except that the initial quaternion is given by $\mathbf{q}^{\text{true}}(0) = [\sqrt{1-0.2^2}\mathbf{p}^T \quad -0.2]^T$ and the initial angular velocity is $\boldsymbol{\omega}^{\text{true}}(0) = 0.5\mathbf{p}$. Figure 7.16 shows the results for the hybrid, discontinuous, and unwinding controller cases. The unwinding controller case is given by setting $\delta > 1$. The initial velocity is specifically chosen so that it is in the direction that decreases the angle between the initial quaternion and commanded quaternion. The discontinuous controller immediately pulls the quaternion toward $-\mathbf{I}_q$ (note that h immediately jumps to -1 and stays there). The hybrid controller initially pulls the quaternion toward $+\mathbf{I}_q$, but after the initial angular velocity pushes the attitude past the hysteresis width, its value of h switches and then pulls the quaternion toward $-\mathbf{I}_q$. The unwinding-inducing controller always pulls the quaternion toward $\mathbf{I}_q$. Figure 7.16d shows that the unwinding-inducing controller requires the most control effort, as expected, while the discontinuous and hybrid controller give comparable control efforts.

Example 7.8. In this example a more realistic scenario is presented using the WMAP spacecraft. The tracking control law is given by Eq. (7.36) with control gains given by $k_p = 5$ and $k_d = 3$. Noise is added to the true angular velocity by using a zero-mean Gaussian white-noise process with standard deviation given by $\sqrt{\Delta t}\sqrt{10}\times 10^{-7}$ rad/s, where the sampling interval is given by $\Delta t = 0.1$ s. Realistic noise is also added to the true quaternion using a multiplicative approach. It is assumed that the noise is isotropic, which is a valid assumption when multiple star trackers pointed sufficiently apart are used on the spacecraft. The standard deviation for all components of the 3×1 attitude noise vector, denoted by $\mathbf{v}_q$, is given by $(0.5/3) \times 10^{-3}$ deg. The "noise" quaternion is given by $\mathbf{q}_{\text{noise}} = [\mathbf{v}_q^T \ \ 1]^T / \|[\mathbf{v}_q^T \ \ 1]\|$. The measured quaternion is generated using $\mathbf{q} = \mathbf{q}^{\text{true}} \otimes \mathbf{q}_{\text{noise}}$. The hysteresis half-width is chosen to be $\delta = 0.4$ for the hybrid controller and $\delta = 3$ for the unwinding controller. The initial condition for h is given by $h(0) = 1$ for all controllers. Results for the scalar component quaternion errors, the rotation error angles, the norm of the velocity errors and the torque norm are shown in Fig. 7.17, which indicate that the hybrid controller provides the best performance. This example shows that under realistic noise errors the hybrid controller can provide superior results over standard control approaches to handle noise chattering effects.

7.7 SAMPEX Control Design

This section presents the attitude determination and control design for a real spacecraft called the Solar, Anomalous and Magnetospheric Particle Explorer (SAMPEX) which was the first of the NASA Small Explorer (SMEX) missions. It was launched on July 3, 1992 into an 82° inclination orbit with an apogee of 670 km and a perigee of 520 km, with a planned mission life of 3 years. Its scientific instruments included the Heavy Ion Large Telescope (HILT), the Low Energy Ion Composition Analyzer (LEICA), the Mass Spectrometer Telescope (MAST), and the Proton/Electron Telescope (PET). Over its lifetime, SAMPEX provided the first continuous record of high sensitivity measurements of energetic ions and relativistic electrons over almost two solar cycles. New insights were given into the acceleration, transport and loss processes in the Earth's magnetosphere, driven by high speed streams or coronal mass ejections in the solar wind. SAMPEX also discovered doubly charged anomalous cosmic rays of interstellar origin, limiting the time scale for acceleration of these ions in the outer heliosphere to a few years. One of the significant achievements of SAMPEX was to confirm the existence of the trapped component of the anomalous cosmic rays. During its first year it confirmed that these rays are singly charged and located in a narrow belt of trapped cosmic rays within the inner of the two Van Allen radiation belts. This discovery provided proof of a third radiation belt. Furthermore, SAMPEX also successfully addressed important scientific objectives concerning solar energetic particles because it was launched

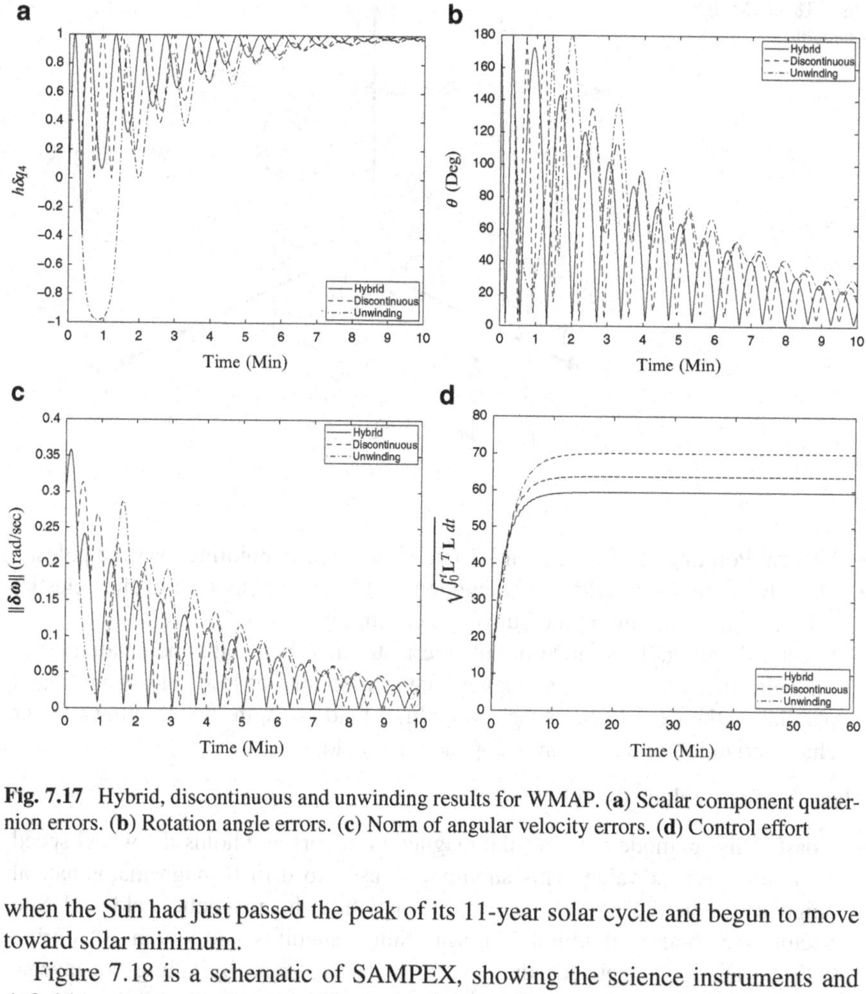

Fig. 7.17 Hybrid, discontinuous and unwinding results for WMAP. (**a**) Scalar component quaternion errors. (**b**) Rotation angle errors. (**c**) Norm of angular velocity errors. (**d**) Control effort

when the Sun had just passed the peak of its 11-year solar cycle and begun to move toward solar minimum.

Figure 7.18 is a schematic of SAMPEX, showing the science instruments and definitions of the body axes. The attitude control system (ACS) consists of one reaction wheel, three magnetic torquer bars, one two-axis fine Sun sensor, five coarse Sun sensors, and one three-axis magnetometer (TAM). The spin axis of the reaction wheel and the boresight of the fine Sun sensor are along the body y-axis, or pitch axis, $\mathbf{j} \equiv [0\ 1\ 0]^T$.[2] Attitude determination is performed using the TRIAD algorithm of Sect. 5.1 with data from the Sun sensor and TAM. The three-axis attitude requirement is 2°. During its lifetime SAMPEX had three primary science modes and included a number of submodes [44]. All science modes have a Sun angle constraint to maintain the body y-axis to within ± 5° of the Sun line in order to keep the solar arrays pointed towards the Sun. The primary science modes include the following:

[2]This has no relation to the orbit plane because SAMPEX is not Earth-pointing.

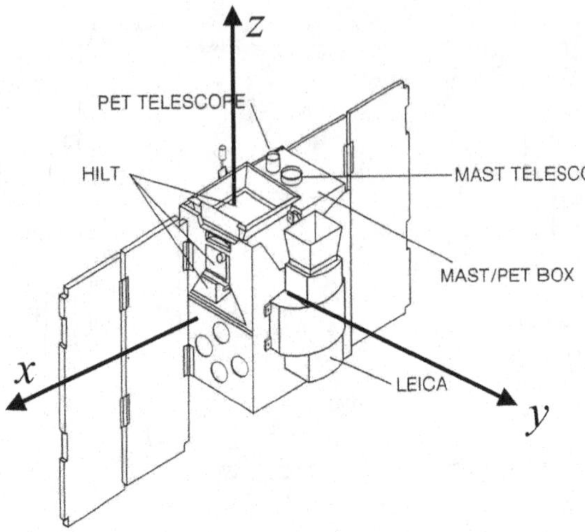

Fig. 7.18 SAMPEX schematic

- Vertical Pointing. This mode tends to maximize zenith pointing over the poles.
- Orbit Rate Rotation (ORR). This mode provides a smooth scan of the celestial sphere while maintaining the Sun angle constraint.
- Special Pointing. This mode orients the instrument boresights perpendicular to the field lines of the Earth's magnetic field in regions of low field strength and parallel to the field lines in regions of high field strength, which allows better characterization of heavy ions trapped by the field.

The submodes include the following:

- Coast. This submode turns off the magnetic torquers and holds the wheel speed at its most recent value. This submode is used to drift through mathematical singularities in the attitude determination when the magnetic field and Sun vectors are co-aligned within 5° during Sun availability and within 40° during eclipse. The angle between these vectors can be found by taking the inverse cosine of the dot product of the inertial magnetic and Sun vectors.
- Eclipse. This submode is enabled when the Sun presence flag returns a value of false. It turns off the magnetic torquers and determines the Sun vector from the assumption that the reaction wheel angular momentum keeps the spacecraft y-axis inertially fixed. This assumption is reasonable for short periods of time (<1 h). The body Sun vector is assumed to be given by $\mathbf{b}_{\text{sun}} = \mathbf{j}$ and attitude determination is performed normally using the TAM vector and assumed Sun body-vector. Note that the eclipse duration can be up to 45 min per orbit in shadow seasons.

Two other modes are also used. The first is a magnetometer calibration mode that senses the unwanted component of the signal on the TAM produced by the torquer bars and then estimates the coupling matrix values to compensate for the contamination. The other is a Sunpoint-only mode that is a digital implementation of the spacecraft's analog safehold mode.

SAMPEX's science mission was terminated on June 30, 2004, after it had exceeded its 3 year mission life goal by a factor of almost 4, but the spacecraft continued to be operated as a testbed. On August 18, 2007 the reaction wheel began to fail after more than 15 years of continuous operation. The spacecraft was then successfully placed into a spin stabilized mode described in [44] until it reentered the Earth's atmosphere on November 13, 2012.

7.7.1 Attitude Determination

As previously mentioned SAMPEX uses a TAM and a Sun sensor in the ACS hardware. The Sun sensor has a field of view of ±64° and outputs 8 bits of Gray code data for each axis with a resolution of 0.5°, which dominates the sensor noise. This is converted to binary and the two binary counts, N_a and N_b, are converted to the Sun's image plane coordinates through [19]

$$x = s_x N_a - b_x \tag{7.68a}$$

$$z = s_z N_b - b_z \tag{7.68b}$$

where (s_x, s_z) are scale factors in units of cm/count and (b_x, b_z) are biases in units of cm. For the SAMPEX Sun sensor these parameters are given by $s_x = s_z = 0.002754$ and $b_x = b_z = -0.350625$. The body Sun vector is computed using [10]

$$\mathbf{b}_{\text{sun}} = \frac{1}{\sqrt{x^2 + z^2 + h^2}} \begin{bmatrix} -n\,x \\ \sqrt{h^2 - (n^2 - 1)(x^2 + z^2)} \\ -n\,z \end{bmatrix} \tag{7.69}$$

where n is the refraction index of the glass and h is the glass thickness in cm. For SAMPEX $n = 1.4553$ and $h = 0.448$. To simulate the Sun sensor from the inertial Sun vector $\mathbf{r}_{\text{sun}}$, this vector is first converted to body coordinates using the attitude matrix: $\mathbf{b}_{\text{sun}} \equiv [b_{\text{sun}_1} \; b_{\text{sun}_2} \; b_{\text{sun}_3}]^T = A\,\mathbf{r}_{\text{sun}}$. Then the components x and z are computed using

$$z = \pm \frac{h\, b_{\text{sun}_3}}{\sqrt{n^2 - b_{\text{sun}_1}^2 - b_{\text{sun}_3}^2}} \tag{7.70a}$$

$$x = \frac{z\, b_{\text{sun}_1}}{b_{\text{sun}_3}} \tag{7.70b}$$

where the sign of z is easily determined by the known body motion. A Sun sensor measures an azimuth ϕ, and a coelevation θ, which are computed by

$$\phi = \text{atan2}(x, z) \tag{7.71a}$$

$$\theta = \tan^{-1}\left[\frac{n\sqrt{x^2+z^2}}{\sqrt{h^2-(n^2-1)(x^2+z^2)}}\right] \tag{7.71b}$$

Measurements are obtained by adding zero-mean Gaussian white-noise to ϕ and θ with standard deviations σ_ϕ and σ_θ, respectively. For SAMPEX the standard deviation of both of the variables is given by 0.01°. Next, the 0.5° resolution is simulated using the MATLAB "round" command, i.e. round$(meas/res) * res$, where $meas$ is the measurement including noise and res is the resolution. Note that σ_ϕ, σ_θ, and res must all be converted to radians. To produce the simulated body vector measurements, first the tangent of the angles for the rotations around the $-x$ and z axes, denoted by α and β, are computed using

$$\tan\alpha = \tan\theta\sin\phi \tag{7.72a}$$

$$\tan\beta = \tan\theta\cos\phi \tag{7.72b}$$

where ϕ and θ here are the measured values, including noise and the 0.5° resolution. The measured unit vector is then given by

$$\mathbf{b}_{\text{sun}} = \pm\frac{1}{\sqrt{1+\tan^2\alpha+\tan^2\beta}}\begin{bmatrix}\tan\alpha\\ 1\\ \tan\beta\end{bmatrix} \tag{7.73}$$

where the sign is easily determined from the true body vector.

A triaxial search coil magnetometer is used on SAMPEX, which outputs 12 bit words for each axis. Its resolution is 31.25 nT and its range is ±64,000 nT. The TAM body vector is denoted by $\mathbf{b}_{\text{mag}}$ and the respective inertial vector is denoted by $\mathbf{r}_{\text{mag}}$, which is computed using a 10th-order geomagnetic field model. TAM measurements are computed by first converting the inertial vector into body coordinates and then adding zero-mean Gaussian white-noise. For the SAMPEX TAM the covariance matrix of this noise is assumed to be isotropic with standard deviation given by 30 nT per axis. The SAMPEX algorithm contains a procedure to calibrate the torque-rod/magnetometer coupling matrix, denoted by C. It turns off all torquer bars and obtains an uncontaminated measurement, denoted by $\mathbf{b}^u_{\text{mag}}$. Then it sequentially turns on each torquer bar with a 10 Am^2 excitation and obtains a contaminated measurement, denoted by $\mathbf{b}^c_{\text{mag}}$. The relationship of these vectors is given by

$$\mathbf{b}^c_{\text{mag}} = \mathbf{b}^u_{\text{mag}} + C\,\mathbf{m} \tag{7.74}$$

where $\mathbf{m}$ is the dipole moment of the torquer rods computed from the torque magnetic assembly current feedback. The matrix components of C can be computed using a simple least-squares procedure with multiple observations of $\mathbf{b}^u_{\text{mag}}$ and $\mathbf{b}^c_{\text{mag}}$.

Once the TRIAD solution produces an attitude matrix, the spacecraft angular velocity is computed by first using a first-order finite difference to estimate the derivative of the attitude matrix:

$$\dot{A} \approx \frac{A_{k+1} - A_k}{\Delta t} \tag{7.75}$$

where $\Delta t = 0.5$ s is the sampling interval. Then the cross product matrix of the angular velocity is determined using the average values from the two possible solutions of the off-diagonal elements:

$$-[\boldsymbol{\omega}_{k+1}\times] = \frac{1}{2}\left[\dot{A}\,A^T - (\dot{A}\,A^T)^T\right] = \frac{1}{2\,\Delta t}(A_{k+1}A_k^T - A_k A_{k+1}^T) \tag{7.76}$$

The total spacecraft angular momentum is given by

$$\mathbf{H} = J\boldsymbol{\omega} + H^w\mathbf{j} \tag{7.77}$$

where J is the inertia matrix and H^w is the momentum of the wheel. For SAMPEX the inertia matrix is given by

$$J = \begin{bmatrix} 14.1005 & -0.1898 & -0.9897 \\ -0.1898 & 19.2526 & -0.4881 \\ -0.9897 & -0.4881 & 12.0668 \end{bmatrix} \text{ kg-m}^2 \tag{7.78}$$

The wheel momentum is computed using the measured angular velocity of the wheel, denoted by ω^w, with $H^w = J^w\omega^w$, where the inertia of the wheel is given by $J^w = 4.1488 \times 10^{-3}$ kg-m^2.

The computed values of $\mathbf{H}$ are very noisy due to the 0.5° resolution in the Sun sensor. A simple Kalman filter is used to provide filtered estimates of these computed values. The angular momentum dynamics is given by

$$\dot{\mathbf{H}} = -\boldsymbol{\omega} \times \mathbf{H} + \mathbf{L}_{\text{mag}} \tag{7.79}$$

where $\mathbf{L}_{\text{mag}}$ is the magnetic control torque in body coordinates. Using a first-order finite difference to approximate $\dot{\mathbf{H}}$ and a first-order finite difference to approximate $\dot{A}$ in $-[\boldsymbol{\omega}\times] = \dot{A}\,A^T \approx (A_{k+1}A_k^T - I_3)/\Delta t$ leads to the following discrete-time prediction equation [31]:

$$\mathbf{H}^-_{k+1} = A_{k+1}A_k^T\mathbf{H}^+_k + \Delta t\,\mathbf{L}_{\text{mag}_k} \tag{7.80}$$

The update equation for the Kalman filter is given by

$$\mathbf{H}_k^+ = (1 - K)\mathbf{H}_k^- + K\,\mathbf{H}_{\text{derived}_k} \tag{7.81}$$

where $\mathbf{H}_{\text{derived}_k}$ is given by Eq. (7.77) using computed values for the spacecraft and wheel angular velocities, and respective known inertia quantities. A Kalman gain of $K = 0.01$ is chosen, which provides good filtered estimates without excessive lag in the control signal.

7.7.2 *Magnetic Torque Control Law*

The magnetic torquers are used to control the magnitude and direction of the system angular momentum when SAMPEX is in sunlight. The commanded angular momentum magnitude, denoted by H_c, is 0.81349 Nms.[3] The desired direction is both along the body y-axis to damp spacecraft nutation and along the Sun line to keep the solar arrays Sun-pointing. An undesired component of the spacecraft angular momentum, denoted by $\Delta\mathbf{H}$, can be computed as

$$\Delta\mathbf{H} = (\mathbf{H} - H_c\,\mathbf{j}) + (\mathbf{H} - H_c\,\mathbf{b}_{\text{sun}}) = 2\mathbf{H} - H_c\,(\mathbf{j} + \mathbf{b}_{\text{sun}}) \tag{7.82}$$

The magnetic torquers are commanded to have dipole moment given by

$$\mathbf{m} = k_{\text{mag}}\Delta\mathbf{H} \times \mathbf{b}_{\text{mag}} \tag{7.83}$$

where k_{mag} is a constant gain, given by 7.376×10^{-13} for SAMPEX. The control torque is computed using Eq. (7.47):

$$\mathbf{L}_{\text{mag}} = \mathbf{m} \times \mathbf{b}_{\text{mag}} \tag{7.84}$$

The measured values of $\mathbf{b}_{\text{mag}}$ and $\mathbf{b}_{\text{sun}}$ are used to compute the control torque. As previously mentioned, magnetic control is turned off during eclipse because controlling the angular momentum is undesirable when Sun sensor data are unavailable. Angular momentum conservation keeps the solar arrays pointing toward the Sun when it is behind the Earth.

7.7.3 *Science Modes*

Reaction wheel control is used to align the instrument boresights, which are along the spacecraft z-axis, with a target vector denoted by $\mathbf{u}$. This section shows details of the three science modes of SAMPEX, each of which has a different target

[3] This is equal to 0.6 ft-lb-s and has been erroneously given as 0.6 Nms in the literature.

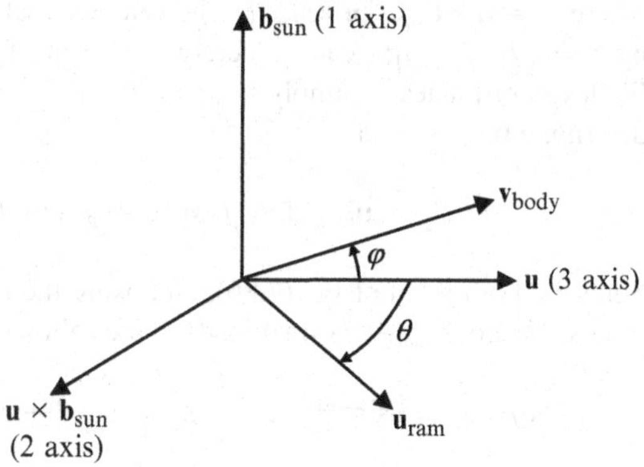

Fig. 7.19 Flatley coordinate system

vector. All science modes have a rotation about the body y-axis, so the target vector is always in the body x-z plane. All three science modes also employ a velocity avoidance constraint. This constraint was added due to orbital debris and micrometeoroid fluxes that were 50–100 times higher than the flux tables used in the original SAMPEX proposal. The SAMPEX HILT sensor includes a flow-through isobutane proportional counter that is susceptible to penetration by debris and other fluxes [20]. Keeping the HILT sensor pointed away from the velocity vector, which the direction of maximum flux, compensates for the higher than expected fluxes.

Since all three science modes use the velocity avoidance algorithm, it is discussed first. The RAM angle is defined as the angle between the HILT boresight and the velocity vector. An optimal RAM angle is 90° since it does not degrade any of the science modes while providing an estimated 89 % chance of survival for the HILT sensor over a 3 year mission period. This RAM angle was actually later reduced to 80° and the HILT sensor still functioned properly for nearly 20 years!

The velocity avoidance algorithm can be easily developed by using a new coordinate system, dubbed the "Flatley Coordinate System," and shown in Fig. 7.19. The body Sun vector is its first axis, the target vector is its third axis, and their cross product is its second axis. The angle φ is the angle between the target vector and the unit velocity vector expressed in body coordinates, denoted by $\mathbf{v}_{body}$; and θ is the angle between the target vector and the RAM vector, $\mathbf{u}_{ram}$, which is the target vector corrected for velocity avoidance. Note that both $\mathbf{u}$ and $\mathbf{u}_{ram}$ are perpendicular to the Sun vector. If $\mathbf{u} \cdot \mathbf{v}_{body} \leq \cos \varphi_{min}$, where φ_{min} is the minimum RAM angle, then the velocity avoidance algorithm is not necessary.

The attitude matrix that rotates vectors from the body frame to the Flatley frame is given by

$$A_B^F = \begin{bmatrix} \mathbf{b}_{sun}^T \\ \mathbf{w}^T \\ \mathbf{u}^T \end{bmatrix} \tag{7.85}$$

where $\mathbf{w} \equiv \mathbf{u} \times \mathbf{b}_{\text{sun}}/\|\mathbf{u} \times \mathbf{b}_{\text{sun}}\|$. The velocity vector in Flatley coordinates, denoted by $\mathbf{f} = [f_1 \ \ f_2 \ \ f_3]^T$, can be determined using $\mathbf{f} = A_B^F \mathbf{v}_{\text{body}}$. The vector $\mathbf{u}_{\text{ram}}$ in Flatley coordinates is simply $\mathbf{u}_{\text{ram}} = [0 \ \ \sin\theta \ \ \cos\theta]^T$. The desired constraint to determine $\mathbf{u}_{\text{ram}}$ is given by

$$\mathbf{u}_{\text{ram}} \cdot \mathbf{f} = f_2 \sin\theta + f_3 \cos\theta = \cos\varphi_{\text{min}} \tag{7.86}$$

Squaring both sides of Eq. (7.86), and using the relations $\cos^2\theta = 1 - \sin^2\theta$ and $f_3 \cos\theta = \cos\varphi_{\text{min}} - f_2 \sin\theta$ leads to the following quadratic equation for $\sin\theta$:

$$(f_2^2 + f_3^2)\sin^2\theta - 2f_2 \cos\varphi_{\text{min}} \sin\theta + \cos^2\varphi_{\text{min}} - f_3^2 = 0 \tag{7.87}$$

The solution for $\sin\theta$ is given by

$$\sin\theta = \frac{f_2 \cos\varphi_{\text{min}} \pm |f_3|\sqrt{f_2^2 + f_3^2 - \cos^2\varphi_{\text{min}}}}{f_2^2 + f_3^2} \tag{7.88}$$

If $f_2 \geq 0$ then $\sin\theta < 0$ and the negative sign is chosen in the radical, otherwise the positive sign is chosen since if $f_2 < 0$ then $\sin\theta > 0$ is true. The vector $\mathbf{u}_{\text{ram}}$ can now be computed using simple sine and cosine relationships from Fig. 7.19:

$$\mathbf{u}_{\text{ram}} = (\mathbf{u} \times \mathbf{b}_{\text{sun}}) \sin\theta + \mathbf{u} \cos\theta \tag{7.89}$$

Note that if $\theta = 0$ then $\mathbf{u}_{\text{ram}} = \mathbf{u}$ as expected and that if $\theta = 0$ then Eq. (7.86) shows that $f_3 = \cos\varphi_{\text{min}}$. In practice the measured values for $\mathbf{b}_{\text{sun}}$ are used. The attitude matrix is required to compute the velocity vector in body coordinates. Finally, the velocity vector in inertial coordinates is computed using an onboard orbit propagator, which is updated from ground observations.

For all three modes the pitch error angle, which is used in the control system, is given by

$$e = \operatorname{atan2}(-u_{\text{ram}_1}, u_{\text{ram}_3}) \tag{7.90}$$

where u_{ram_1} and u_{ram_3} are the first and third components of $\mathbf{u}_{\text{ram}}$, respectively. Note that in some cases the velocity avoidance algorithm is turned off. For these cases we simply set $\sin\theta = 0$ and $\cos\theta = 1$ in Eq. (7.89), so $\mathbf{u}_{\text{ram}} = \mathbf{u}$. Equation (7.90) can still be used in the control law.

7.7.3.1 Vertical Pointing Mode

The vertical pointing mode minimizes the angle between the spacecraft z-axis and the zenith vector within the Sun pointing constraint [19]. This mode has the undesirable property of pointing directly into the velocity vector twice per orbit when the Sun is in the orbit plane, but the velocity avoidance algorithm overcomes this problem. The target vector in inertial coordinates is given by

$$\mathbf{u}_{\text{ref}} = \frac{\mathbf{r}_{\text{sun}} \times (\mathbf{r} \times \mathbf{r}_{\text{sun}})}{\|\mathbf{r}_{\text{sun}} \times (\mathbf{r} \times \mathbf{r}_{\text{sun}})\|} \tag{7.91}$$

where $\mathbf{r}_{\text{sun}}$ is the Sun unit vector in inertial coordinates and $\mathbf{r}$ is the spacecraft position vector in inertial coordinates. The body target vector is found by using $\mathbf{u} = A\,\mathbf{u}_{\text{ref}}$ where A is the computed attitude matrix from the TRIAD algorithm. This mode clearly has the spacecraft z-axis as close to zenith as possible while remaining perpendicular to the Sun.

7.7.3.2 Orbit Rate Rotation Mode

In the ORR mode the spacecraft z-axis rotates at one revolution per orbit in a plane perpendicular to the Sun vector, which provides a smooth scan of the celestial sphere, while maintaining the y-axis Sun pointing requirement. At the same time, the z-axis is desired to point as close to North as possible at the northernmost point in the orbit, South as possible at the southernmost point, and parallel to the equator at the equatorial crossings [19]. To develop the target vector we first define the North pole vector in inertial coordinates: $\mathbf{p} = [0 \quad 0 \quad 1]^T$ and also compute the unit orbit normal vector:

$$\mathbf{n} = \frac{\mathbf{r} \times \mathbf{v}}{\|\mathbf{r} \times \mathbf{v}\|} \tag{7.92}$$

where $\mathbf{r}$ is the inertial position vector and $\mathbf{v}$ is the inertial velocity vector. We now compute the orbit angle as seen from the northernmost point in the orbit using the following vectors:

$$\mathbf{a} = \frac{\mathbf{p} \times \mathbf{n}}{\|\mathbf{p} \times \mathbf{n}\|} \tag{7.93a}$$

$$\mathbf{c} = \mathbf{n} \times \mathbf{a} \tag{7.93b}$$

Note that $\mathbf{a}$ is the vector in the direction of the ascending node and the vector $\mathbf{c}$ is in the direction of the northernmost point in the orbit. The sine and cosine of the orbit angle, defined by α, can be computed from

$$\sin\alpha = -\frac{\mathbf{r} \cdot \mathbf{a}}{\|\mathbf{r}\|} \tag{7.94a}$$

$$\cos\alpha = \frac{\mathbf{r} \cdot \mathbf{c}}{\|\mathbf{r}\|} \tag{7.94b}$$

The two vectors $\mathbf{w} = \mathbf{c}\times\mathbf{r}_{\text{sun}}/\|\mathbf{c}\times\mathbf{r}_{\text{sun}}\|$ and $\mathbf{r}_{\text{sun}}\times\mathbf{w}$ provide an orthonormal basis for the plane perpendicular to the Sun. Note that $\mathbf{w}$ is a vector that points perpendicular to the Sun and lies in the equatorial plane. Thus when the spacecraft is near the equator, it is desired to have the target vector point along $\mathbf{w}$, which corresponds to orbit angles of $\alpha = \pi/2$ and $\alpha = 3\pi/2$. The vector $\mathbf{r}_{\text{sun}} \times \mathbf{w}$ is also perpendicular to the Sun and points as close as possible to the northernmost point given the Sun constraint. Thus when the spacecraft is near the poles it is desired to have the target vector point along $\mathbf{r}_{\text{sun}} \times \mathbf{w}$, which corresponds to orbit angles of $\alpha = 0$ and $\alpha = \pi$. Since it is desired to rotate the body y-axis about the positive Sun line, the orientation of the orbit normal relative to the Sun line must be considered in the target vector. If the Sun passes through the orbit plane when the spacecraft is near the equator, this will cause a 180° flip, which we would like to prevent. The following target vector, expressed in inertial coordinates, has all these properties:

$$\mathbf{u}_{\text{ref}} = (\mathbf{r}_{\text{sun}} \times \mathbf{w}) \cos\alpha + T_s\, \mathbf{w} \sin\alpha \tag{7.95}$$

where T_s is the target sign variable, which has initial value of $\text{sign}(\mathbf{r}_{\text{sun}} \cdot \mathbf{n})$. If the Sun passes through the orbit plane, the next time the spacecraft comes within 0.5° of the northernmost or southernmost point of the orbit, whichever comes first, then the variable T_s will change sign, which will keep the spacecraft rotating about the Sun line without causing a flip. The body target vector is again found by using $\mathbf{u} = A\,\mathbf{u}_{\text{ref}}$ where A is the computed attitude matrix from the TRIAD algorithm.

Reference [20] shows the following limiting cases for the ORR mode. When the Sun is perpendicular to the orbit plane, the ORR mode reduces to a zenith pointing mode, as shown in Fig. 7.20a. Since the target vector is given by Eq. (7.95) the rotation about the z-axis along the orbital path is zenith pointing, as shown in Fig. 7.20b. For the case when the Sun is parallel to the orbit plane, the ORR mode becomes a zenith pointing mode over the poles and points in the $\mathbf{r} \times \mathbf{p}$ direction at the equator. Figure 7.21a shows that $\mathbf{w} = \mathbf{c} \times \mathbf{r}_{\text{sun}}$, and since the target vector is given by Eq. (7.95) the orientation of the z-axis can be determined throughout the orbital path, as shown in Fig. 7.21.

7.7.3.3 Special Pointing Mode

After the discovery of a third radiation belt it was desired to change the pointing algorithm to orient the instrument boresights perpendicular to the magnetic field line while passing through a region in the South Atlantic containing a high concentration of trapped particles. Figure 7.22 served as the "requirements document" from Dan Baker, a project scientist, and Glenn Mason, the SAMPEX Principal Investigator, which led to the SAMPEX special pointing mode. Details of this mode, documented in [28], are repeated here. In order to satisfy these revised science requirements, the pointing mode was modified using the magnetic field strength $\|\mathbf{r}_{\text{mag}}\|$ as a delimiter to point the spacecraft perpendicular to the magnetic field vector whenever the field strength is determined to be less than some specified value, which was set to 3 ×

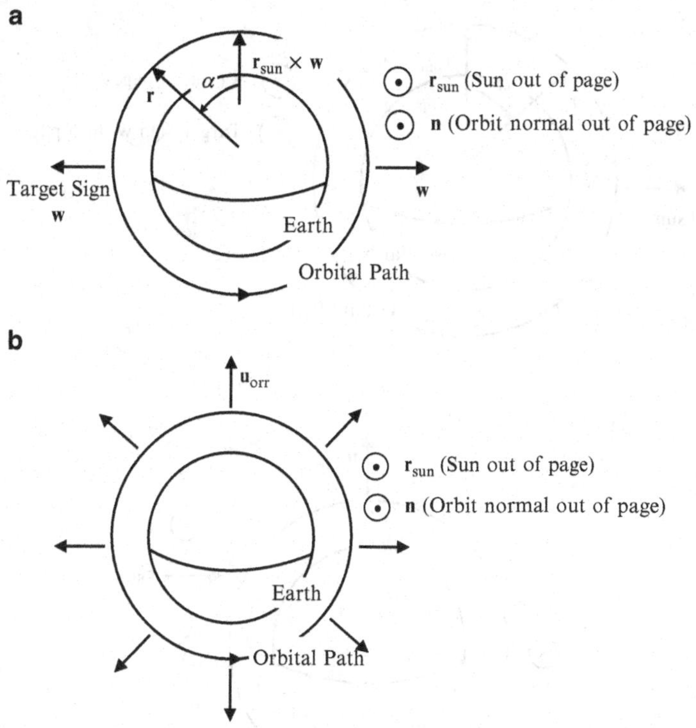

Fig. 7.20 Sun is perpendicular to the orbit plane. (**a**) Orbit rate rotation mode geometry. (**b**) Orbit rate rotation mode target pointing vector along the orbital path

10^4 nT. This pointing uses magnetometer data to determine the field direction in the spacecraft reference frame. Since the magnetometer data was judged to provide a good pointing reference, the specification was changed in the high-field regions to point as close to the magnetic field vector as possible, consistent with the constraint that the y-axis be pointed at the Sun. At northern latitudes, the desired orientation is anti-parallel to the field, and in the South the orientation is parallel to the field. In both cases, then, the spacecraft points away from the Earth in the polar regions.

In the low-field region the target vector is to be perpendicular to both the Sun vector and the magnetic field vector. This requirement is obviously satisfied by choosing the target vector in the direction of the cross product $\mathbf{b}_{sun} \times \mathbf{b}_{mag}$. The negative of this vector clearly satisfies the same requirement. One and only one of these two vectors is more than 90° from the velocity vector, and we choose this one to satisfy the velocity avoidance requirement. Since the magnetic control keeps the Sun vector within a few degrees of the y-axis vector $\mathbf{j}$, the cross product $\mathbf{b}_{sun} \times \mathbf{b}_{mag}$ can be well approximated by $\mathbf{j} \times \mathbf{b}_{mag}$, so the target vector is computed from the components of $\mathbf{b}_{mag}$ as

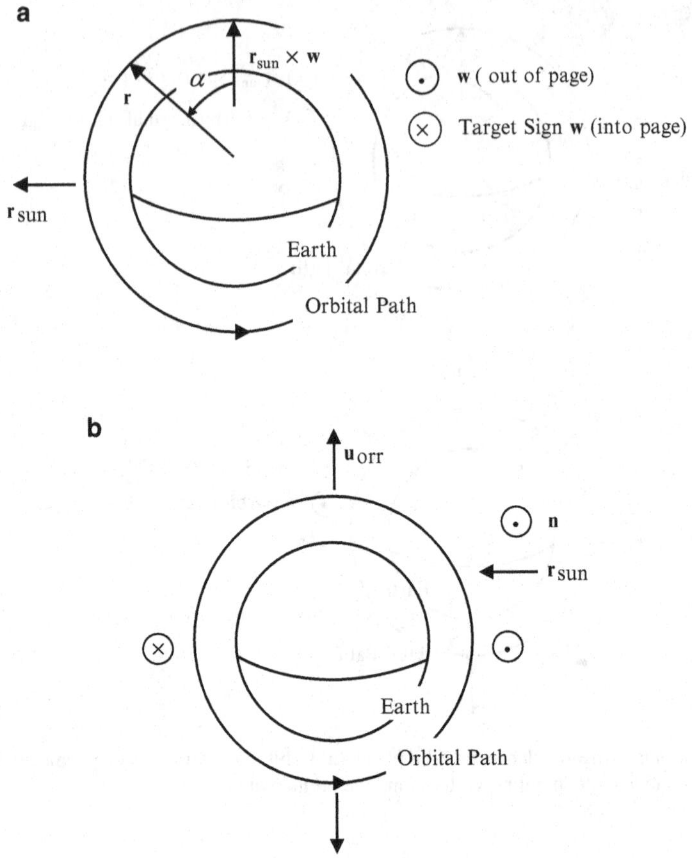

Fig. 7.21 Sun is parallel to the orbit plane. (**a**) Orbit rate rotation mode geometry. (**b**) Orbit rate rotation mode target pointing vector along the orbital path

$$\mathbf{u} = \frac{\pm 1}{\sqrt{b_{\text{mag}_1}^2 + b_{\text{mag}_3}^2}} \begin{bmatrix} b_{\text{mag}_3} \\ 0 \\ -b_{\text{mag}_1} \end{bmatrix} \tag{7.96}$$

The upper sign is chosen if $\mathbf{v} \cdot (\mathbf{r}_{\text{sun}} \times \mathbf{r}_{\text{mag}}) \leq 0$ and the lower sign is chosen if $\mathbf{v} \cdot (\mathbf{r}_{\text{sun}} \times \mathbf{r}_{\text{mag}}) > 0$, where $\mathbf{v}$ is the spacecraft velocity vector in inertial coordinates computed from the onboard ephemeris. Note that binary decisions are based on velocity and magnetic field vectors computed in the inertial frame from the ephemeris, in order to prevent toggling arising from noisy measurements; but that the TAM-sensed magnetic field vector in the body frame is used to compute the actual pointing vector.

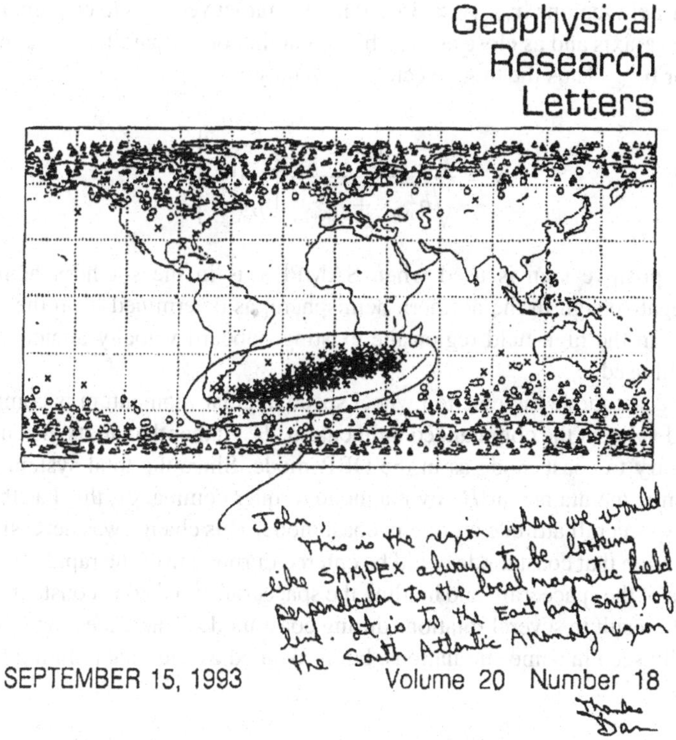

Fig. 7.22 Requirements specification for SAMPEX special pointing mode

The angle error in Eq. (7.90) is generally small, but it is on the order of 90° during transitions between parallel pointing and perpendicular pointing. In these cases, the wheel is commanded in the direction that requires the smallest rotation to null the angle error. Due to changing geometry however, some time during a passage through a low field region, a 180° pitch maneuver is generally required to satisfy the avoidance requirement. The special pointing algorithm assures that any 180° turns will be executed in a direction away from the velocity vector. Thus a computed pitch error angle magnitude greater than 2.5 rad is taken to signify a large reorientation maneuver. The sign of the x-axis component of the spacecraft velocity vector $\mathbf{v}$ in the body frame is then used to determine the direction of this maneuver, such that the instrument boresights are rotated away from the velocity vector rather than toward it.

In the high-field region, $\|\mathbf{r}_{\text{mag1}}\| > 3 \times 10^4$ nT, the target vector is to be perpendicular to the body y-axis and as close as possible to parallel or antiparallel to the magnetic field vector $\mathbf{b}_{\text{mag}}$. Thus the target vector is given by

$$\mathbf{u} = \frac{\pm 1}{\sqrt{b_{\text{mag}_1}^2 + b_{\text{mag}_3}^2}} \begin{bmatrix} b_{\text{mag}_1} \\ 0 \\ b_{\text{mag}_3} \end{bmatrix} \tag{7.97}$$

where the positive sign is used when SAMPEX is in the southern hemisphere and the negative sign in the northern hemisphere, as determined from the onboard ephemeris. In the high-field region the existing onboard velocity avoidance algorithm is still used.

During coast mode the reaction wheel speed angular momentum is commanded to the fixed value of the commanded momentum, H_c, rather than to its instantaneous value at entry to coast mode as in the ORR mode. Since the total system angular momentum is maintained at H_c by magnetic torquer commands, this has the effect of halting spacecraft attitude motion in coast mode. This change was necessitated by the observation that coast mode could be entered during one of the rapid 90° or 180° maneuvers of the spacecraft, and holding the spacecraft y-axis rate constant at a high value could result in several rotations during coast mode. This undesirable behavior was actually seen in some simulations, but is avoided by the final pointing law.

7.7.4 Reaction Wheel Control Law

The single reaction wheel is used to drive down the angle error shown in Eq. (7.90). A proportional-integral-derivative (PID) control law [17] is used to command the wheel torque, denoted by L^w. The derivative signal is computed using a simple finite-difference approach:

$$\dot{e} \approx \frac{e_k - e_{k-1}}{\Delta t} \equiv e_{\text{der}_k} \tag{7.98}$$

The integral portion, denoted by e_{int}, is computed by simply using

$$e_{\text{int}_k} = e_{\text{int}_{k-1}} + e_k\, \Delta t \tag{7.99}$$

The torque control law is then given by

$$L_k^w = k_p\, e_k + k_d\, e_{\text{der}_k} + k_i\, e_{\text{int}_k} \tag{7.100}$$

where k_p, k_d and k_i are the proportional, derivative and integral gains, respectively, given as follows for the various modes:

- All Modes: $k_p = \omega_n^2 J_{22}$, $k_d = 2\,\zeta\,\omega_n J_{22}$, and $k_i = (k_p\,k_d)/(10\,J_{22})$.
- Vertical and ORR Modes: $\zeta = \sqrt{2}/2$ and $\omega_n = 0.02$.
- Special Pointing Mode: $\zeta = \sqrt{2}/2$ and $\omega_n = 0.01$.

An integral limit is also imposed for all modes, which is set to $3 \times 10^{-3}/k_i$. Note that unlike the magnetic control system no filter is used to filter noisy measurements in the wheel control law. This is because the measurement errors do not significantly affect the overall wheel control performance to meet the desired specifications. This is an important point to make about the design phase of any spacecraft control system. Start with the simplest design for analysis purposes and then add one component, such as a Kalman filter, at a time to see how the performance improves. Once the desired specifications are met within some desired confidence and safety factor then it is usually best to not add any more components. A system that is over-designed may lead to unknown errors that may subsequently cause catastrophic failures. The SAMPEX design shows how simplicity often can lead to a successful control design that far exceeds initial lifetime performance specifications.

7.7.5 Simulations

This section provides simulation results for all three modes. The epoch is September 16, 2011 at midnight. The simulation runtime is 5 h with a sampling interval of 5 s for all sensors and actuators. The position and velocity vectors from the onboard ephemeris are given by

$$\mathbf{r}(t_0) = \begin{bmatrix} 3335.973299 \\ 2571.763319 \\ -5370.931739 \end{bmatrix} \text{ km}, \quad \mathbf{v}(t_0) = \begin{bmatrix} 3.530941 \\ 4.977268 \\ 4.566940 \end{bmatrix} \text{ km/s} \tag{7.101}$$

For these conditions the orbital altitude varies from an apogee of 450 km to a perigee of 390 km, which is far lower than the initial orbit insertion because of atmospheric drag that decayed the orbit over the lifetime of the mission. But this orbit is still sufficient to perform the required science modes. The onboard orbit propagator includes the effects of drag and higher-order gravity terms. Here only the effects of J_2 are simulated because drag has an insignificant effect on the orbit over 5 h. The dynamic model is given by [36]

$$\ddot{\mathbf{r}} = -\frac{\mu}{\|\mathbf{r}\|^3} + \mathbf{a}_{J_2} \tag{7.102}$$

where $\mathbf{a}_{J_2}$ is given by Eq. (10.103a) with J_2 from Table 10.2, $\mu = 3.98601 \times 10^5\,\text{km}^3/\text{s}^2$, and $R_\oplus = 6378.1363$ km. A plot of SAMPEX longitude and latitude for this simulation is given in Fig. 7.23. The inclination of 82° is clearly seen in this plot. A 10th-order IGRF model is used to generate the reference magnetic field using

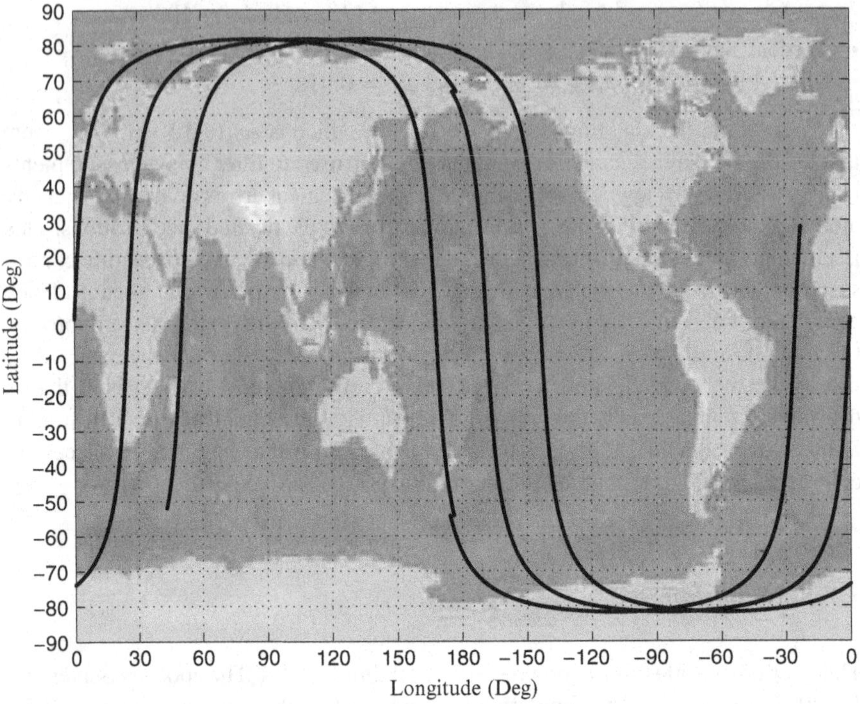

Fig. 7.23 SAMPEX ground track

the inertial position of the satellite. The inertial Sun vector is calculated simply from the epoch. All measurements are produced using the noise and resolution parameters discussed previously.

The rotation dynamic models are given by

$$\dot{\mathbf{q}} = \frac{1}{2}\Xi(\mathbf{q})\boldsymbol{\omega} \tag{7.103a}$$

$$\dot{\mathbf{H}} = -\boldsymbol{\omega} \times \mathbf{H} + \mathbf{L}_{\text{mag}} \tag{7.103b}$$

$$\dot{H}^w = L^w \tag{7.103c}$$

where $\boldsymbol{\omega} = J^{-1}(\mathbf{H} - H^w\mathbf{j})$. For all simulations the initial quaternion is given by the identity quaternion, and the system and wheel momenta are set to zero. For the first 2 h the wheel is turned off in order to reorient the spacecraft so that the y-axis is pointed towards the Sun using magnetic torquers only.

As previously mentioned the onboard attitude determination algorithm used the TRIAD algorithm, which ignores some part of the measurement. In order to provide more accurate attitude estimates Davenport's q method is applied here (see Sect. 5.3). Using this approach allows us to increase the Kalman gain to 0.04 instead

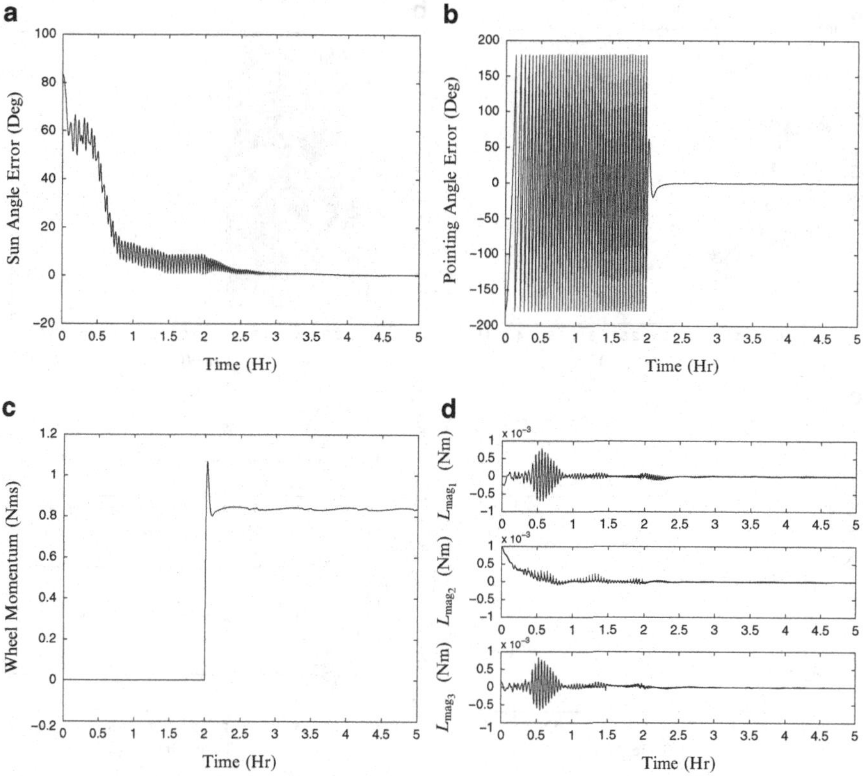

Fig. 7.24 SAMPEX vertical pointing mode. (**a**) Sun angle error. (**b**) Pointing angle error. (**c**) Wheel momentum. (**d**) Magnetic control torques

of 0.01, which actually gives better filtered estimates than the TRIAD method using a lower Kalman gain. Simulation results for the vertical pointing mode are shown in Fig. 7.24. The Sun error angle is shown in Fig. 7.24a. After 2 h the Sun error angle is about 10°. Figure 7.24b shows the pointing error. Note that since the wheel is turned off for the first 2 h there are large fluctuations in the error angle. The wheel momentum is shown in Fig. 7.24c. After 2 h the wheel is turned on and drives the error angle to near zero. The magnetic control torques are shown in Fig. 7.24d. Note the large control torques at the beginning of the simulation run, which are required to reorient the y-axis along the Sun line. After that period only minimal control torques are required. Simulation results for the orbit rate rotation mode are shown in Fig. 7.25. Results are very similar to the vertical pointing mode results. Simulation results for the special pointing mode are shown in Fig. 7.26. Note that unlike the other two modes, the error angle is not always maintained at zero. The fluctuations are due to achieving the "best possible" angle given the constraints of the geometry from the orbit position and spacecraft orientation requirements, particularly the Sun

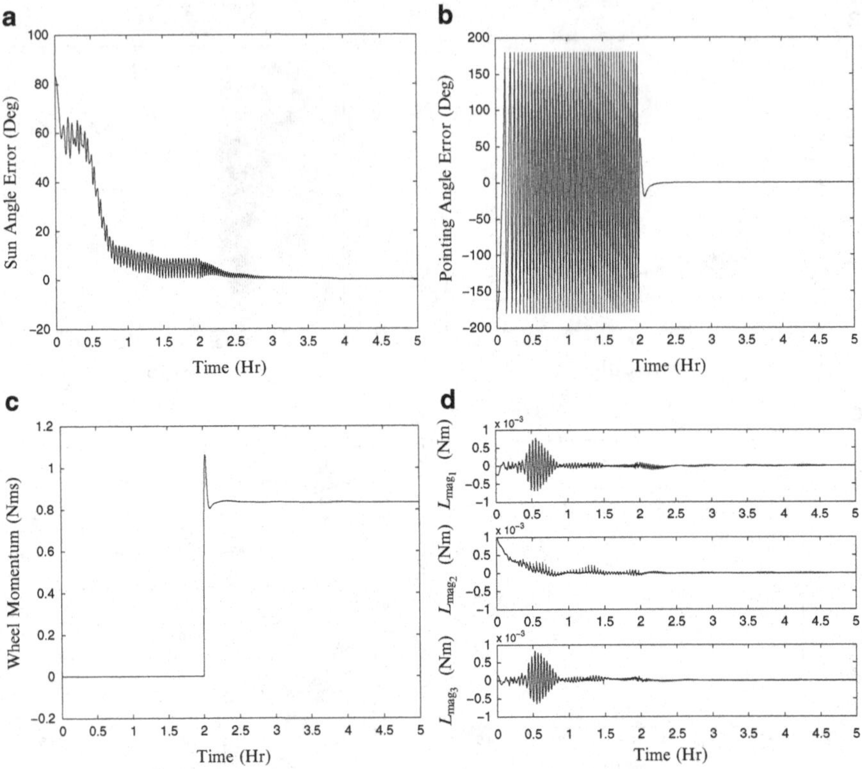

Fig. 7.25 SAMPEX orbit rate rotation mode. (**a**) Sun angle error. (**b**) Pointing angle error. (**c**) Wheel momentum. (**d**) Magnetic control torques

line pointing requirement. Still, the desired pointing is achieved enough to satisfy the special pointing mode requirements.

It is important to note that although the control laws and filter developed for SAMPEX were very simple, they were proven to be extremely effective. Extended Kalman filters are widely used for onboard attitude estimators, even though filter stability cannot be guaranteed. Many, if not most, modern-day attitude controllers are very simple PID controllers. For example, the actual WMAP spacecraft uses a simple PID controller that is not asymptotically stable from a theoretical point of view [3]. This leads to a "bias" in the control error signal, which was overcome by using a simple feed-forward term. Still the attitude control design meets the desired mission objectives because attitude knowledge accuracy is more important than pointing accuracy for the WMAP spacecraft. The SAMPEX and WMAP attitude control designs epitomize the notion of "keep it simple" if sufficient confidence can be obtained that the controllers will meet mission objectives. Future spacecraft control designs will most likely gravitate more towards Lyapunov-based controllers as confidence grows in their ability to achieve desired mission objectives. Although

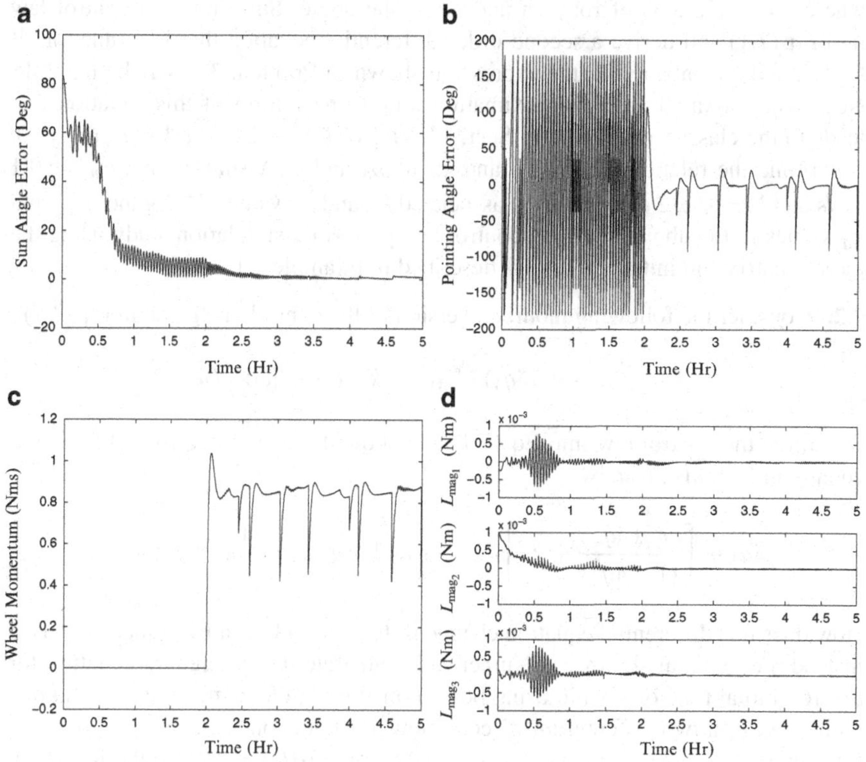

Fig. 7.26 SAMPEX special pointing mode. (**a**) Sun angle error. (**b**) Pointing angle error. (**c**) Wheel momentum. (**d**) Magnetic control torques

more complicated than simple PID controllers, these more advanced control laws may require fewer contingency mode analyses than simpler controllers, which may in fact reduce the time to design the actual spacecraft control system.

Problems

7.1. Consider the following modified version of the control law given in Eq. (7.7) [49]:

$$\mathbf{L} = -k_p\, J\,\boldsymbol{\delta}\mathbf{q}_{1:3} - k_d\, J\,\boldsymbol{\omega} + [\boldsymbol{\omega}\times]J\,\boldsymbol{\omega}$$

The error-quaternion given in Eq. (7.2) is represented by

$$\boldsymbol{\delta}\mathbf{q} \equiv \begin{bmatrix} \boldsymbol{\delta}\mathbf{q}_{1:3} \\ \delta q_4 \end{bmatrix} = \begin{bmatrix} \Delta\mathbf{e}\sin(\delta\vartheta/2) \\ \cos(\delta\vartheta/2) \end{bmatrix}$$

where $\Delta\mathbf{e}$ is the axis of rotation and $\delta\vartheta$ is the angle. Substitute the control law into Eq. (7.1) and derive a second-order differential equation that is a function of $\delta\vartheta$, k_p and k_d only. (Hint: the expression shown in Problem 3.3 may be helpful). Next, using a small angle approximation find a linear form of this equation that leads to the classic form given in Sect. 12.2.1.1 of $\delta\ddot{\vartheta}^2 + 2\,\zeta\,\omega_n\,\delta\dot{\vartheta} + \omega_n^2\delta\vartheta = 0$. Determine the relationships of k_p and k_d to ω_n and ζ. Assuming that $\omega_n = 0.1$ rad/s and $\zeta = \sqrt{2}/2$ determine the associated k_p and k_d values. Using these k_p and k_d values in the above modified control law perform a simulation study using the inertia matrix and initial conditions described in Example 7.1.

7.2. Consider the following modified version of the control law given in Eq. (7.7):

$$\mathbf{L} = -k_p\,\mathrm{sign}(\delta q_4)J\,\delta\mathbf{q}_{1:3} - k_d\,J\boldsymbol{\omega} + [\boldsymbol{\omega}\times]J\boldsymbol{\omega}$$

Substitute the control law into Eq. (7.1) and show that the closed-loop differential equation for δq_4 is given by

$$\delta\ddot{q}_4 + \left[\frac{\delta q_4\,\delta\dot{q}_4}{(1-\delta q_4^2)} + k_d\right]\delta\dot{q}_4 + \frac{1}{2}k_p\,|\delta q_4|\,\delta q_4 = \frac{1}{2}k_p\,\mathrm{sign}(\delta q_4)$$

How does this differential equation change if the control law in Example 7.1 is used instead, i.e. without $\mathrm{sign}(\delta q_4)$? Numerically integrate the differential equation for δq_4 for a total time of 30 min using both control laws; one with $\mathrm{sign}(\delta q_4)$ and one without it. Run two different initial conditions for both control laws: (1) $\delta q_4(t_0) = 0.5$ and $\delta\dot{q}_4(t_0) = 0$, and (2) $\delta q_4(t_0) = -0.5$ and $\delta\dot{q}_4(t_0) = 0$. Use the following control gains for both control laws: $k_p = 0.01$ and $k_d = 2$. Compare the trajectories of δq_4 over time in all cases and discuss how using $\mathrm{sign}(\delta q_4)$ produces the shortest distance. Also discuss the case using the following initial conditions: $\delta q_4(t_0) = 0.5$ and $\delta\dot{q}_4(t_0) = -1$.

7.3. Redo the simulation shown in Example 7.1. Next, using the same gains from this example, implement the controller shown in Eq. (7.14). Try both the plus and minus signs in the control law. Compute the function in Eq. (7.9) for each of the three controllers and plot them over time. Also plot the rotation angle of error, $\delta\vartheta$, for each of the three controllers. Discuss the differences seen in the responses.

7.4. Prove that the control law in Eq. (7.13) produces a globally asymptotic stable response for the closed-loop system.

7.5. Another attitude regulation control law involves using the attitude matrix directly [34]:

$$\mathbf{L} = -k_p\,\mathbf{s} - K_d\,\boldsymbol{\omega}$$

where k_p is a positive scalar, K_d is a diagonal matrix with positive elements and

$$\mathbf{s} = \sum_{i=1}^{3} r_i \, [(\delta A \, \mathbf{e}_i)\times]\mathbf{e}_i$$

where $\mathbf{e}_1 = [1 \;\; 0 \;\; 0]^T$, $\mathbf{e}_2 = [0 \;\; 1 \;\; 0]^T$, $\mathbf{e}_3 = [0 \;\; 0 \;\; 1]^T$, and r_1, r_2, and r_3 are positive scalars. The attitude error, δA, is given by $\delta A = A\,A_c^T$, where A_c is a constant attitude matrix. Prove that this control law produces an asymptotically stable closed-loop system by using the following candidate Lyapunov function:

$$V = \frac{1}{2}\boldsymbol{\omega}^T J \boldsymbol{\omega} + k_p \, \mathrm{tr}(R - R\,\delta A)$$

where R is a diagonal matrix with elements given by r_1, r_2, and r_3. First show that $\mathrm{tr}(R - R\,\delta A) \geq 0$. Note that the equality is given only when $\delta A = I_3$.

Next, consider the following gains:

$$k_p \equiv \frac{\alpha}{\mathrm{tr}R}$$

$$K_d \equiv \beta \begin{bmatrix} \dfrac{1}{1+|\omega_1|} & 0 & 0 \\ 0 & \dfrac{1}{1+|\omega_2|} & 0 \\ 0 & 0 & \dfrac{1}{1+|\omega_3|} \end{bmatrix}$$

where α and β are positive scalars. Show that $\|\mathbf{L}(t)\|_\infty \leq \alpha + \beta$. Begin by using the following relation:

$$\begin{aligned} \|\mathbf{L}(t)\|_\infty &= \|k_p \, \mathbf{s} + K_d \, \boldsymbol{\omega}\|_\infty \\ &\leq k_p \, \|\mathbf{s}\|_\infty + \|K_d \, \boldsymbol{\omega}\|_\infty \end{aligned}$$

Perform a simulation using the inertia matrix and initial conditions given in Example 7.1 with this control law. Choose the following gains: $\alpha = 50$, $\beta = 500$, and $R = I_3$.

7.6. Prove the relation given in Eq. (7.20). Next, show that the time derivative of δq_4 is $\delta \dot{q}_4 = \frac{1}{2}(\boldsymbol{\omega}_c - \boldsymbol{\omega})^T \delta \mathbf{q}_{1:3}$.

7.7. Prove that the control law in Eq. (7.36) produces a globally asymptotic stable response for the closed-loop tracking system.

7.8. In this exercise you will design a controller that tracks an LVLH attitude. Consider the following orbital elements: $a = 26{,}559$ km, $e = 0.704482$, $M_0 = 12.9979°$, $i = 63.1706°$, $\Omega = 206.346°$, and $\omega = 281.646°$. Note that this is a highly eccentric orbit. Compute the initial position and velocity using Table 10.1, and then propagate the orbit for an 18-h period in 10-s intervals. Next, compute the desired LVLH attitude using Eq. (2.79) and angular rate using Eq. (3.175). Also find an analytical expression for the derivative of Eq. (3.175). Assume that the actual initial attitude is given by the identity matrix and the actual initial angular rate is given by zero. The inertia matrix is given by

$$J = \begin{bmatrix} 10000 & 0 & 0 \\ 0 & 9000 & 0 \\ 0 & 0 & 12000 \end{bmatrix} \text{kg-m}^2$$

Finally, using the control law given by Eq. (7.36) pick values of k_p and k_d so that the actual attitude and angular rate converges to the desired attitude and angular rate in about 10 min. Show plots of the orbit, the first three components of the error quaternion and the angular velocity errors using Eq. (7.31).

7.9. Consider the case of tracking a desired quaternion with kinematics given by

$$\dot{\mathbf{q}}_c = \frac{1}{2}\Xi(\mathbf{q}_c)\boldsymbol{\omega}_c$$

where $\boldsymbol{\omega}_c$ is the desired angular velocity vector. The error quaternion is given by $\delta\mathbf{q} = \mathbf{q} \otimes \mathbf{q}_c^{-1}$. Let us assume that the closed-loop kinematic equation is desired to have the following prescribed *linear* form [32]:

$$\delta\ddot{\mathbf{q}}_{1:3} + L_2\delta\dot{\mathbf{q}}_{1:3} + L_1\delta\mathbf{q}_{1:3} = \mathbf{0}$$

where L_1 and L_2 are 3×3 gain matrices. Determine the control law that will yield this closed-loop linear form. Next, assuming that both L_1 and L_2 are each given by a scalar times identity matrix, with $L_1 = \ell_1 I_3$, $L_2 = \ell_2 I_3$, show that your derived control law reduces down to

$$\mathbf{L} = [\boldsymbol{\omega}\times]J\boldsymbol{\omega} + J\left\{\delta A\,\dot{\boldsymbol{\omega}}_c - [\boldsymbol{\omega}\times]\,\delta A\,\boldsymbol{\omega}_c - \ell_2\delta\boldsymbol{\omega} - 2\left[\frac{4\ell_1 - (\delta\boldsymbol{\omega}^T\delta\boldsymbol{\omega})}{4\,\delta q_4}\right]\delta\mathbf{q}_{1:3}\right\}$$

with

$$\delta A = A(\mathbf{q})A^T(\mathbf{q}_c)$$
$$\delta\boldsymbol{\omega} = \boldsymbol{\omega} - \delta A\,\boldsymbol{\omega}_c$$

This control law is clearly singular when $\delta q_4 = 0$. Do the same issues arise in the control law that uses general 3×3 L_1 and L_2 gain matrices?

7.10. Redo the simulation shown in Example 7.4. Pick different values for the PWPF parameters to see how they affect the overall performance.

7.11. Redo the simulation shown in Example 7.6. Pick different values for k to see how they affect the overall performance.

7.12. Redo the simulation shown in Example 7.7. Pick different values for the various parameters in the control law as well as different noise levels to see how they affect the overall performance.

References

1. Agrawal, B.N., McClelland, R.S., Song, G.: Attitude control of flexible spacecraft using pulse-width pulse-frequency modulated thrusters. Space Tech. **17**(1), 15–34 (1997)
2. Anderson, B.D.O., Moore, J.B.: Optimal Control: Linear Quadratic Methods. Prentice Hall, Englewood Cliffs (1990)
3. Andrews, S.F., Campbell, C.E., Ericsson-Jackson, A.J., Markley, F.L., O'Donnell Jr., J.R.: MAP attitude control system design and analysis. In: Proceedings of the Flight Mechanics/Estimation Theory Symposium, pp. 445–456. NASA-Goddard Space Flight Center, Greenbelt (1997)
4. Arantes, G., Martins-Filho, L.S., Santana, A.C.: Optimal on-off attitude control for the Brazilian Multimission Platform satellite. Math. Probl. Eng. **2009**(1) (2009)
5. Åström, K.J.: Introduction to Stochastic Control Theory. Academic Press, New York (1970)
6. Avanzini, G., Giulietti, F.: Magnetic detumbling of a rigid spacecraft. J. Guid. Contr. Dynam. **35**(4), 1326–1334 (2012)
7. Bhat, S.P., Dham, A.S.: Controllability of spacecraft attitude under magnetic actuation. In: Proceedings of the 42nd IEEE Conference on Decision and Control, pp. 2383–2388. Maui (2003)
8. Camillo, P.J., Markley, F.L.: Orbit-averaged behavior of magnetic control laws for momentum unloading. J. Guid. Contr. **3**(6), 563–568 (1980)
9. Challa, M.S., Natanson, G.A., Baker, D.E., Deutschmann, J.K.: Advantages of estimating rate corrections during dynamic propagation of spacecraft rates-applications to real-time attitude determination of SAMPEX. In: Proceedings of the Flight Mechanics/Estimation Theory Symposium, pp. 481–495. NASA-Goddard Space Flight Center, Greenbelt (1994)
10. Chen, L.C., Lerner, G.M.: Three-axis attitude determination. In: Wertz, J.R. (ed.) Sun Sensor Models, chap. 7. Kluwer Academic, Dordrecht (1978)
11. Chu, D., Harvie, E.: Accuracy of the ERBS definitive attitude determination system in the presence of propagation noise. In: Proceedings of the Flight Mechanics/Estimation Theory Symposium, pp. 97–114. NASA-Goddard Space Flight Center, Greenbelt (1990)
12. Crassidis, J.L., Junkins, J.L.: Optimal Estimation of Dynamic Systems, 2nd edn. CRC Press, Boca Raton (2012)
13. Crassidis, J.L., Markley, F.L.: Sliding mode control using modified Rodrigues parameters. J. Guid. Contr. Dynam. **19**(6), 1381–1383 (1996)
14. Crassidis, J.L., Markley, F.L.: Predictive filtering for attitude estimation without rate sensors. J. Guid. Contr. Dynam. **20**(3), 522–527 (1997)
15. Crassidis, J.L., Vadali, S.R., Markley, F.L.: Optimal variable-structure control tracking of spacecraft maneuvers. J. Guid. Contr. Dynam. **23**(3), 564–566 (2000)
16. Davis, M.: Linear Estimation and Stochastic Control. Chapman and Hall, London (1977)
17. Dorf, R.C., Bishop, R.H.: Modern Control Systems. Addison Wesley Longman, Menlo Park (1998)
18. Dwyer, T.A.W., Sira-Ramirez, H.: Variable structure control of spacecraft reorientation maneuvers. J. Guid. Contr. Dynam. **11**(3), 262–270 (1988)
19. Flatley, T.W., Forden, J.K., Henretty, D.A., Lightsey, E.G., Markley, F.L.: On-board attitude determination and control algorithms for SAMPEX. In: Proceedings of the Flight Mechanics/Estimation Theory Symposium, pp. 379–398. NASA-Goddard Space Flight Center, Greenbelt (1990)

20. Frakes, J.P., Henretty, D.A., Flatley, T.W., Markley, F.L., Forden, J.K., Lightsey, E.G.: SAMPEX science pointing modes with velocity avoidance. In: Proceedings of the 2nd AAS/AIAA Spaceflight Mechanics Meeting, pp. 949–966. Colorado Springs (1992)
21. Junkins, J.L., Turner, J.D.: Optimal Spacecraft Rotational Maneuvers. Elsevier, New York (1986)
22. Kang, W.: Nonlinear H_∞ control and its application to rigid spacecraft. IEEE Trans. Automat. Contr. **40**(7), 1281–1285 (1995)
23. Koon, W.S., Lo, M.W., Marsden, J.E., Ross, S.D.: Dynamical Systems, The Three-Body Problem and Space Mission Design. Marsden Books, Pasadena (2011)
24. Krøvel, T.D.: Optimal tuning of PWPF modulator for attitude control. Master's thesis, Norwegian University of Science and Technology, Department of Engineering Cybernetics, Trondheim (2005)
25. Lizarralde, F., Wen, J.T.Y.: Attitude control without angular velocity measurement: A passivity approach. IEEE Trans. Automat. Contr. **41**(3), 468–472 (1996)
26. Lovera, M.: Optimal magnetic momentum control for inertially pointing spacecraft. Eur. J. Contr. **7**(1), 30–39 (2001)
27. Markley, F.L., Andrews, S.F., O'Donnell Jr., J.R., Ward, D.K.: Attitude control system of the Wilkinson Microwave Anisotropy Probe. J. Guid. Contr. Dynam. **28**(3), 385–397 (2005)
28. Markley, F.L., Flatley, T.W., Leoutsakos, T.: SAMPEX special pointing mode. In: Proceedings of the Flight Mechanics/Estimation Theory Symposium, pp. 201–215. NASA-Goddard Space Flight Center, Greenbelt (1995)
29. Mayhew, C.G., Sanfelice, R.G., Teel, A.R.: Robust global asymptotic attitude stabilization of a rigid body by quaternion-based hybrid feedback. In: Joint 48th IEEE Conference on Decision and Control and 28th Chinese Control Conference, pp. 2522–2527. Shanghai (2009)
30. Mayhew, C.G., Sanfelice, R.G., Teel, A.R.: Quaternion-based hybrid control for robust global attitude tracking. IEEE Trans. Automat. Contr. **AC-56**(11), 2555–2565 (2011)
31. McCullough, J.D., Flatley, T.W., Henretty, D.A., Markley, F.L., San, J.K.: Testing of the on-board attitude determination and control algorithms for SAMPEX. In: Proceedings of the Flight Mechanics/Estimation Theory Symposium, pp. 55–68. NASA-Goddard Space Flight Center, Greenbelt (1992)
32. Paielli, R.A., Bach, R.E.: Attitude control with realization of linear error dynamics. J. Guid. Contr. Dynam. **16**(1), 182–189 (1993)
33. Sanfelice, R.G., Messina, M.J., Tuna, S.E., Teel, A.R.: Robust hybrid controllers for continuous-time systems with applications to obstacle avoidance and regulation to disconnected set of points. In: American Control Conference, pp. 3352–3357. Minneapolis (2006)
34. Sanyal, A., Fosbury, A., Chaturvedi, N., Bernstein, D.S.: Inertia-free spacecraft attitude tracking with disturbance rejection and almost global stabilization. J. Guid. Contr. Dynam. **32**(4), 1167–1178 (2009)
35. Schaub, H., Akella, M.R., Junkins, J.L.: Adaptive control of nonlinear attitude motions realizing linear closed loop dynamics. J. Guid. Contr. Dynam. **24**(1), 95–100 (2001)
36. Schaub, H., Junkins, J.L.: Analytical Mechanics of Aerospace Systems, 2nd edn. American Institute of Aeronautics and Astronautics, New York (2009)
37. Scrivener, S.L., Thompson, R.C.: Survey of time-optimal attitude maneuvers. J. Guid. Contr. Dynam. **17**(2), 225–233 (1994)
38. Shuster, M.D., Dellinger, W.F.: Spacecraft attitude determination and control. In: V.L. Pisacane (ed.) Fundamentals of Space Systems, 2nd edn., chap. 5. Oxford University Press, New York (2005)
39. Sidi, M.J.: Spacecraft Dynamics and Control: A Practical Engineering Approach. Cambridge University Press, New York (2006)
40. Silani, E., Lovera, M.: Magnetic spacecraft attitude control: A survey and some new results. Contr. Eng. Pract. **13**(3), 357–371 (2005)
41. Slotine, J.J.E., Li, W.: Applied Nonlinear Control. Prentice Hall, Englewood Cliffs (1991)
42. Stengel, R.F.: Optimal Control and Estimation. Dover Publications, New York (1994)

43. Stickler, A.C., Alfriend, K.T.: Elementary magnetic attitude control system. J. Spacecraft Rockets **13**(5), 282–287 (1976)
44. Tsai, D.C., Markley, F.L., Watson, T.P.: SAMPEX spin stabilized mode. In: SpaceOps Conference. Heidelberg, Germany (2008). AIAA 2008–3435
45. Tsiotras, P.: Stabilization and optimality results for the attitude control problem. J. Guid. Contr. Dynam. **19**(4), 772–779 (1996)
46. Vadali, S.R.: Variable structure control of spacecraft large angle maneuvers. J. Guid. Contr. Dynam. **9**(2), 235–239 (1986)
47. Vadali, S.R., Junkins, J.L.: Optimal open-loop and stable feedback control of rigid spacecraft maneuvers. J. Astronaut. Sci. **32**(2), 105–122 (1984)
48. White, J.S., Shigemoto, F.H., Bourquin, K.: Satellite attitude control utilizing the Earth's magnetic field. Tech. Rep. NASA-TN-D-1068, A-474, NASA Ames Research Center, Moffett Field (1961)
49. Wie, B.: Space Vehicle Dynamics and Control, 2nd edn. American Institute of Aeronautics and Astronautics, Reston (2008)
50. Wie, B., Barba, P.M.: Quaternion feedback for spacecraft large angle maneuvers. J. Guid. Contr. Dynam. **8**(3), 360–365 (1985)

Chapter 8
Quaternion Identities

The purpose of this chapter is to present a collection of vector and quaternion identities that are useful for control and estimation computations. Many of them used throughout this text. Several appear in Chap. 2 but are repeated here for convenience.

8.1 Cross Product Identities

The 3×3 skew-symmetric *cross-product matrix* is defined as

$$[\mathbf{u}\times] \equiv \begin{bmatrix} 0 & -u_3 & u_2 \\ u_3 & 0 & -u_1 \\ -u_2 & u_1 & 0 \end{bmatrix} \tag{8.1}$$

giving $[\mathbf{u}\times]\mathbf{v} = \mathbf{u} \times \mathbf{v}$. Since $[\mathbf{u}\times]\mathbf{u} = \mathbf{0}$, $[\mathbf{u}\times]$ must be singular. The eigenvalues of $[\mathbf{u}\times]$ are given by $\lambda_1 = 0$ and $\lambda_{2,3} = \pm i\,\|\mathbf{u}\|$. Some useful identities for the cross-product matrix include [6]:

$$[\mathbf{u}\times]^T = -[\mathbf{u}\times] \tag{8.2a}$$

$$\text{adj}\left([\mathbf{u}\times]\right) = \mathbf{u}\,\mathbf{u}^T \tag{8.2b}$$

$$[\mathbf{u}\times]\mathbf{v} = -[\mathbf{v}\times]\mathbf{u} \tag{8.2c}$$

$$[\mathbf{u}\times][\mathbf{v}\times] = -\left(\mathbf{u}^T\mathbf{v}\right) I_3 + \mathbf{v}\,\mathbf{u}^T \tag{8.2d}$$

$$[\mathbf{u}\times][\mathbf{v}\times]\mathbf{w} = \mathbf{u} \times (\mathbf{v} \times \mathbf{w}) = \left(\mathbf{u}^T\mathbf{w}\right)\mathbf{v} - \left(\mathbf{u}^T\mathbf{v}\right)\mathbf{w} \tag{8.2e}$$

$$[\mathbf{u}\times]^3 = -\|\mathbf{u}\|^2[\mathbf{u}\times] \tag{8.2f}$$

$$[\mathbf{u}\times][\mathbf{v}\times] - [\mathbf{v}\times][\mathbf{u}\times] = \mathbf{v}\,\mathbf{u}^T - \mathbf{u}\,\mathbf{v}^T = [(\mathbf{u} \times \mathbf{v})\times] \tag{8.2g}$$

F.L. Markley and J.L. Crassidis, *Fundamentals of Spacecraft Attitude Determination and Control*, Space Technology Library 33, DOI 10.1007/978-1-4939-0802-8_8,
© Springer Science+Business Media New York 2014

$$(I_3 - [\mathbf{u}\times])(I_3 + [\mathbf{u}\times])^{-1} = \frac{1}{1+\|\mathbf{u}\|^2}\left\{(1-\|\mathbf{u}\|^2)I_3 + 2\mathbf{u}\,\mathbf{u}^T - 2[\mathbf{u}\times]\right\} \tag{8.2h}$$

$$\|\mathbf{u}\times\mathbf{v}\|^2 I_3 = \|\mathbf{u}\|^2\mathbf{v}\,\mathbf{v}^T + \|\mathbf{v}\|^2\mathbf{u}\,\mathbf{u}^T - \left(\mathbf{u}^T\mathbf{v}\right)\left(\mathbf{u}\,\mathbf{v}^T + \mathbf{v}\,\mathbf{u}^T\right) + (\mathbf{u}\times\mathbf{v})(\mathbf{u}\times\mathbf{v})^T \tag{8.2i}$$

where I_n is an $n \times n$ identity matrix. Some useful identities involving the cross-product matrix and an arbitrary 3×3 matrix M are:

$$\text{tr}(M)[\mathbf{u}\times] = M[\mathbf{u}\times] + [\mathbf{u}\times]M^T + [(M^T\mathbf{u})\times] \tag{8.3a}$$

$$M[\mathbf{u}\times]M^T = [\{\text{adj}(M^T)\mathbf{u}\}\times] \tag{8.3b}$$

$$[\{(M\mathbf{u})\times(M\mathbf{v})\}\times] = M[(\mathbf{u}\times\mathbf{v})\times]M^T \tag{8.3c}$$

$$(M\mathbf{u})\times(M\mathbf{v}) = \text{adj}(M^T)\,(\mathbf{u}\times\mathbf{v}) \tag{8.3d}$$

$$[\mathbf{u}\times][\text{tr}(M)I_3 - M][\mathbf{u}\times]^T = (\mathbf{u}^T M\mathbf{u})I_3 - \mathbf{u}\,\mathbf{u}^T M^T - M^T\mathbf{u}\,\mathbf{u}^T + \|\mathbf{u}\|^2 M^T \tag{8.3e}$$

where tr denotes the trace operator and adj denotes the adjoint matrix. If we write M in terms of its columns

$$M = \begin{bmatrix}\mathbf{u}_1 & \mathbf{u}_2 & \mathbf{u}_3\end{bmatrix} \tag{8.4}$$

then

$$\det(M) = \mathbf{u}_1^T(\mathbf{u}_2\times\mathbf{u}_3) \tag{8.5}$$

where det denotes the determinant. Also, if A is an orthogonal matrix with determinant 1, then from we have Eq. (8.3b)

$$A[\mathbf{u}\times]A^T = [(A\mathbf{u})\times] \tag{8.6}$$

These cross product relations are useful in proving many of the quaternion identities shown in this chapter.

8.2 Basic Quaternion Identities

The quaternion has a vector part, $\mathbf{q}_{1:3}$, and a scalar part, q_4:

$$\mathbf{q} \equiv \begin{bmatrix} q_1 \\ q_2 \\ q_3 \\ q_4 \end{bmatrix} \equiv \begin{bmatrix} \mathbf{q}_{1:3} \\ q_4 \end{bmatrix} \tag{8.7}$$

We define two 4 × 4 matrices analogous to the 3 × 3 cross-product matrix:

$$[\mathbf{q}\otimes] \equiv \begin{bmatrix} q_4 I_3 - [\mathbf{q}_{1:3}\times] & \mathbf{q}_{1:3} \\ -\mathbf{q}_{1:3}^T & q_4 \end{bmatrix} = \begin{bmatrix} \Psi(\mathbf{q}) & \mathbf{q} \end{bmatrix} \tag{8.8a}$$

$$[\mathbf{q}\odot] \equiv \begin{bmatrix} q_4 I_3 + [\mathbf{q}_{1:3}\times] & \mathbf{q}_{1:3} \\ -\mathbf{q}_{1:3}^T & q_4 \end{bmatrix} = \begin{bmatrix} \Xi(\mathbf{q}) & \mathbf{q} \end{bmatrix} \tag{8.8b}$$

with $\Psi(\mathbf{q})$ and $\Xi(\mathbf{q})$ being the 4 × 3 matrices

$$\Psi(\mathbf{q}) \equiv \begin{bmatrix} q_4 I_3 - [\mathbf{q}_{1:3}\times] \\ -\mathbf{q}_{1:3}^T \end{bmatrix} \tag{8.9a}$$

$$\Xi(\mathbf{q}) \equiv \begin{bmatrix} q_4 I_3 + [\mathbf{q}_{1:3}\times] \\ -\mathbf{q}_{1:3}^T \end{bmatrix} \tag{8.9b}$$

These matrices provide two alternative products of two quaternions $\mathbf{q}$ and $\bar{\mathbf{q}}$:

$$\mathbf{q}\otimes\bar{\mathbf{q}} = [\mathbf{q}\otimes]\,\bar{\mathbf{q}} \tag{8.10a}$$

$$\mathbf{q}\odot\bar{\mathbf{q}} = [\mathbf{q}\odot]\,\bar{\mathbf{q}} \tag{8.10b}$$

The first of these, denoted by $\otimes$, has proved to be more useful in attitude analysis. It follows from these definitions that

$$\mathbf{q}\otimes\bar{\mathbf{q}} = \bar{\mathbf{q}}\odot\mathbf{q} \tag{8.11}$$

Quaternion multiplication is associative, $\mathbf{q}\otimes(\bar{\mathbf{q}}\otimes\bar{\bar{\mathbf{q}}}) = (\mathbf{q}\otimes\bar{\mathbf{q}})\otimes\bar{\bar{\mathbf{q}}}$ and distributive, $\mathbf{q}\otimes(\bar{\mathbf{q}}+\bar{\bar{\mathbf{q}}}) = \mathbf{q}\otimes\bar{\mathbf{q}} + \mathbf{q}\otimes\bar{\bar{\mathbf{q}}}$. Quaternion multiplication is *not* commutative in general, $\mathbf{q}\otimes\bar{\mathbf{q}} \neq \bar{\mathbf{q}}\otimes\mathbf{q}$, paralleling the situation for matrix multiplication. In those cases for which $\mathbf{q}\otimes\bar{\mathbf{q}} = \bar{\mathbf{q}}\otimes\mathbf{q}$, the quaternions $\mathbf{q}$ and $\bar{\mathbf{q}}$ are said to commute. Analogous equations hold for the product $\mathbf{q}\odot\bar{\mathbf{q}}$.

The identity quaternion

$$\mathbf{I}_q \equiv \begin{bmatrix} \mathbf{0}_3 \\ 1 \end{bmatrix} \tag{8.12}$$

obeys $\mathbf{I}_q\otimes\mathbf{q} = \mathbf{q}\otimes\mathbf{I}_q = \mathbf{I}_q\odot\mathbf{q} = \mathbf{q}\odot\mathbf{I}_q = \mathbf{q}$, as required of the identity.

The conjugate quaternion $\mathbf{q}^*$ is obtained, in analogy with the complex conjugate, by changing the sign of the three-vector part:

$$\mathbf{q}^* \equiv \begin{bmatrix} -\mathbf{q}_{1:3} \\ q_4 \end{bmatrix} = T\,\mathbf{q} \tag{8.13}$$

where

$$T \equiv \begin{bmatrix} -I_3 & 0_{3\times1} \\ 0_{1\times3} & 1 \end{bmatrix} \tag{8.14}$$

Note that $T^2 = 1$, so $T^{-1} = T$. The conjugate of the product of two quaternions $\bar{\mathbf{q}}$ and $\mathbf{q}$ is the product of the conjugates in the opposite order:

$$(\bar{\mathbf{q}} \otimes \mathbf{q})^* = \mathbf{q}^* \otimes \bar{\mathbf{q}}^* \tag{8.15}$$

The product of a quaternion with its conjugate is equal to the square of its norm times the identity quaternion

$$\mathbf{q} \otimes \mathbf{q}^* = \mathbf{q}^* \otimes \mathbf{q} = \mathbf{q} \odot \mathbf{q}^* = \mathbf{q}^* \odot \mathbf{q} = \|\mathbf{q}\|^2 \mathbf{I}_q \tag{8.16}$$

These relations and the associativity of quaternion multiplication can be used to show that

$$\|\bar{\mathbf{q}} \otimes \mathbf{q}\| = \|\bar{\mathbf{q}} \odot \mathbf{q}\| = \|\bar{\mathbf{q}}\| \, \|\mathbf{q}\| \tag{8.17}$$

The inverse of any quaternion having nonzero norm is defined by

$$\mathbf{q}^{-1} \equiv \mathbf{q}^* / \|\mathbf{q}\|^2 \tag{8.18}$$

so that $\mathbf{q} \otimes \mathbf{q}^{-1} = \mathbf{q}^{-1} \otimes \mathbf{q} = \mathbf{q} \odot \mathbf{q}^{-1} = \mathbf{q}^{-1} \odot \mathbf{q} = \mathbf{I}_q$, as required by the definition of an inverse. The inverse of the product of two quaternions is the product of the inverses in the opposite order $(\bar{\mathbf{q}} \otimes \mathbf{q})^{-1} = \mathbf{q}^{-1} \otimes \bar{\mathbf{q}}^{-1}$.

Some useful identities are given by

$$[\mathbf{q}^* \otimes] = [\mathbf{q} \otimes]^T = T\,[\mathbf{q} \odot]\,T \tag{8.19a}$$

$$[\mathbf{q}^* \odot] = [\mathbf{q} \odot]^T = T\,[\mathbf{q} \otimes]\,T \tag{8.19b}$$

$$[\mathbf{q} \otimes][\bar{\mathbf{q}} \odot] = [\bar{\mathbf{q}} \odot][\mathbf{q} \otimes] \tag{8.19c}$$

$$[\mathbf{q} \otimes][\mathbf{q}^* \otimes] = [\mathbf{q} \otimes][\mathbf{q} \otimes]^T = \|\mathbf{q}\|^2 I_4 \tag{8.19d}$$

$$[\mathbf{q} \odot][\mathbf{q}^* \odot] = [\mathbf{q} \odot][\mathbf{q} \odot]^T = \|\mathbf{q}\|^2 I_4 \tag{8.19e}$$

$$[\mathbf{q} \otimes]^{-1} = \|\mathbf{q}\|^{-2}[\mathbf{q}^* \otimes] = [\mathbf{q}^{-1} \otimes] \tag{8.19f}$$

$$[\mathbf{q} \odot]^{-1} = \|\mathbf{q}\|^{-2}[\mathbf{q}^* \odot] = [\mathbf{q}^{-1} \odot] \tag{8.19g}$$

$$[\mathbf{I}_q \otimes] = [\mathbf{I}_q \odot] = I_4 \tag{8.19h}$$

$$[\mathbf{q} \otimes]\mathbf{I}_q = [\mathbf{q} \odot]\mathbf{I}_q = \mathbf{q} \tag{8.19i}$$

$$\begin{aligned} [(\mathbf{q} \otimes \bar{\mathbf{q}})\otimes] &= [\mathbf{q} \otimes][\bar{\mathbf{q}} \otimes] = \|\mathbf{q}\|^{-2}[\mathbf{q} \otimes][(\bar{\mathbf{q}} \otimes \mathbf{q})\otimes][\mathbf{q} \otimes]^T \\ &= \|\bar{\mathbf{q}}\|^{-2}[\bar{\mathbf{q}} \otimes]^T[(\bar{\mathbf{q}} \otimes \mathbf{q})\otimes][\bar{\mathbf{q}} \otimes] \end{aligned} \tag{8.19j}$$

$$\begin{aligned} [(\mathbf{q} \otimes \bar{\mathbf{q}})\odot] &= [\bar{\mathbf{q}} \odot][\mathbf{q} \odot] = \|\mathbf{q}\|^{-2}[\mathbf{q} \odot]^T[(\bar{\mathbf{q}} \otimes \mathbf{q})\odot][\mathbf{q} \odot] \\ &= \|\bar{\mathbf{q}}\|^{-2}[\bar{\mathbf{q}} \odot][(\bar{\mathbf{q}} \otimes \mathbf{q})\odot][\bar{\mathbf{q}} \odot]^T \end{aligned} \tag{8.19k}$$

8.3 The Matrices $\Xi(\mathbf{q})$, $\Psi(\mathbf{q})$, $\Omega(\omega)$, and $\Gamma(\omega)$

The matrices $\Xi(\mathbf{q})$ and $\Psi(\mathbf{q})$ satisfy the identities

$$\Xi^T(\mathbf{q})\Xi(\mathbf{q}) = \Psi^T(\mathbf{q})\Psi(\mathbf{q}) = \|\mathbf{q}\|^2 I_3 \tag{8.20a}$$

$$\Xi(\mathbf{q})\Xi^T(\mathbf{q}) = \Psi(\mathbf{q})\Psi^T(\mathbf{q}) = \|\mathbf{q}\|^2 I_4 - \mathbf{q}\mathbf{q}^T \tag{8.20b}$$

$$\Xi^T(\mathbf{q})\,\mathbf{q} = \Psi^T(\mathbf{q})\,\mathbf{q} = \mathbf{0}_3 \tag{8.20c}$$

$$\Xi^T(\mathbf{q})\bar{\mathbf{q}} = -\Xi^T(\bar{\mathbf{q}})\mathbf{q} \tag{8.20d}$$

$$\Psi^T(\mathbf{q})\bar{\mathbf{q}} = -\Psi^T(\bar{\mathbf{q}})\mathbf{q} \tag{8.20e}$$

$$\Xi^T(\mathbf{q})\Psi(\bar{\mathbf{q}}) = \Xi^T(\bar{\mathbf{q}})\Psi(\mathbf{q}) \tag{8.20f}$$

$$\Psi^T(\mathbf{q})\Xi(\bar{\mathbf{q}}) = \Psi^T(\bar{\mathbf{q}})\Xi(\mathbf{q}) \tag{8.20g}$$

$$\Psi(\mathbf{q})\Psi^T(\bar{\mathbf{q}}) + \mathbf{q}\,\bar{\mathbf{q}}^T = [\mathbf{q}\otimes][\bar{\mathbf{q}}^*\otimes] = [(\mathbf{q}\otimes\bar{\mathbf{q}}^*)\otimes] \tag{8.20h}$$

$$\Xi(\mathbf{q})\Xi^T(\bar{\mathbf{q}}) + \mathbf{q}\,\bar{\mathbf{q}}^T = [\mathbf{q}\odot][\bar{\mathbf{q}}^*\odot] = [(\bar{\mathbf{q}}^*\otimes\mathbf{q})\odot] \tag{8.20i}$$

We will overload the quaternion product notation to allow us to multiply a three-component vector $\boldsymbol{\omega}$ and a quaternion, using the definitions

$$\boldsymbol{\omega}\otimes\mathbf{q} \equiv \begin{bmatrix}\boldsymbol{\omega}\\ 0\end{bmatrix}\otimes\mathbf{q} = [\boldsymbol{\omega}\otimes]\,\mathbf{q} = \mathbf{q}\odot\boldsymbol{\omega} = [\mathbf{q}\odot]\boldsymbol{\omega} = \Xi(\mathbf{q})\boldsymbol{\omega} \tag{8.21a}$$

$$\boldsymbol{\omega}\odot\mathbf{q} \equiv \begin{bmatrix}\boldsymbol{\omega}\\ 0\end{bmatrix}\odot\mathbf{q} = [\boldsymbol{\omega}\odot]\mathbf{q} = \mathbf{q}\otimes\boldsymbol{\omega} = [\mathbf{q}\otimes]\boldsymbol{\omega} = \Psi(\mathbf{q})\boldsymbol{\omega} \tag{8.21b}$$

We also use the alternate notation

$$\Omega(\boldsymbol{\omega}) \equiv [\boldsymbol{\omega}\otimes] = \begin{bmatrix}-[\boldsymbol{\omega}\times] & \boldsymbol{\omega}\\ -\boldsymbol{\omega}^T & 0\end{bmatrix} \tag{8.22a}$$

$$\Gamma(\boldsymbol{\omega}) \equiv [\boldsymbol{\omega}\odot] = \begin{bmatrix}[\boldsymbol{\omega}\times] & \boldsymbol{\omega}\\ -\boldsymbol{\omega}^T & 0\end{bmatrix} \tag{8.22b}$$

Note that the matrices $\Omega(\boldsymbol{\omega})$ and $\Gamma(\boldsymbol{\omega})$ are both skew-symmetric with eigenvalues $\lambda_{1,3} = i\,\|\boldsymbol{\omega}\|$ and $\lambda_{2,4} = -i\,\|\boldsymbol{\omega}\|$.

Some useful identities valid for any 3×1 vectors $\mathbf{b}$, $\mathbf{r}$, and $\boldsymbol{\omega}$ are given by [6]

$$\Omega(\mathbf{b})\Gamma(\mathbf{r}) = \Gamma(\mathbf{r})\Omega(\mathbf{b}) \tag{8.23a}$$

$$\Omega(\boldsymbol{\omega})\mathbf{q} = \Xi(\mathbf{q})\boldsymbol{\omega} \tag{8.23b}$$

$$\Gamma(\boldsymbol{\omega})\mathbf{q} = \Psi(\mathbf{q})\boldsymbol{\omega} \tag{8.23c}$$

$$[\mathbf{q}\otimes]\,\Gamma(\boldsymbol{\omega}) = \Gamma(\boldsymbol{\omega})\,[\mathbf{q}\otimes] \tag{8.23d}$$

$$[\mathbf{q}\odot]\,\Omega(\boldsymbol{\omega}) = \Omega(\boldsymbol{\omega})\,[\mathbf{q}\odot] \tag{8.23e}$$

$$\Omega(\boldsymbol{\omega})\Xi(\mathbf{q}) = -\Xi(\mathbf{q})[\boldsymbol{\omega}\times] - \mathbf{q}\,\boldsymbol{\omega}^T \tag{8.23f}$$

$$\Gamma(\boldsymbol{\omega})\Psi(\mathbf{q}) = \Psi(\mathbf{q})[\boldsymbol{\omega}\times] - \mathbf{q}\,\boldsymbol{\omega}^T \tag{8.23g}$$

$$\Omega^2(\boldsymbol{\omega}) = \Gamma^2(\boldsymbol{\omega}) = -\|\boldsymbol{\omega}\|^2 I_4 \tag{8.23h}$$

$$\det[\Omega(\boldsymbol{\omega})] = \det[\Gamma(\boldsymbol{\omega})] = \|\boldsymbol{\omega}\|^4 \tag{8.23i}$$

$$\Xi^T(\mathbf{q})\Omega(\boldsymbol{\omega})\Xi(\mathbf{q}) = -\|\mathbf{q}\|^2[\boldsymbol{\omega}\times] \tag{8.23j}$$

$$\Psi^T(\mathbf{q})\Gamma(\boldsymbol{\omega})\Psi(\mathbf{q}) = \|\mathbf{q}\|^2[\boldsymbol{\omega}\times] \tag{8.23k}$$

$$\Omega(\mathbf{b}\times\mathbf{r}) = \frac{1}{2}\,[\Omega(\mathbf{r})\Omega(\mathbf{b}) - \Omega(\mathbf{b})\Omega(\mathbf{r})] \tag{8.23l}$$

$$\Gamma(\mathbf{b}\times\mathbf{r}) = \frac{1}{2}\,[\Gamma(\mathbf{b})\Gamma(\mathbf{r}) - \Gamma(\mathbf{r})\Gamma(\mathbf{b})] \tag{8.23m}$$

8.4 Identities Involving the Attitude Matrix

The four-component quaternion representation of attitude is related to the Euler axis/angle representation by

$$\frac{\mathbf{q}}{\|\mathbf{q}\|} = \frac{1}{\sqrt{q_4^2 + \|\mathbf{q}_{1:3}\|^2}}\begin{bmatrix}\mathbf{q}_{1:3}\\ q_4\end{bmatrix} = \begin{bmatrix}\mathbf{e}\sin(\vartheta/2)\\ \cos(\vartheta/2)\end{bmatrix} \tag{8.24}$$

where $\mathbf{e}$ is a unit vector corresponding to the axis of rotation and ϑ is the angle of rotation. Then the attitude matrix is given by

$$\begin{aligned} A(\mathbf{q}) &\equiv \|\mathbf{q}\|^{-2}\,\Xi^T(\mathbf{q})\Psi(\mathbf{q}) = \|\mathbf{q}\|^{-2}\left\{(q_4 I_3 - [\mathbf{q}_{1:3}\times])^2 + \mathbf{q}_{1:3}\,\mathbf{q}_{1:3}^T\right\} \\ &= \|\mathbf{q}\|^{-2}\left\{\left(q_4^2 - \|\mathbf{q}_{1:3}\|^2\right) I_3 + 2\mathbf{q}_{1:3}\,\mathbf{q}_{1:3}^T - 2q_4[\mathbf{q}_{1:3}\times]\right\} \end{aligned} \tag{8.25}$$

which agrees with the Euler axis/angle representation of Eq. (2.108).

The rotation group has only three degrees of freedom, so a quaternion has more components than it needs. In fact, it is clear from Eq. (8.25) that the attitude matrix does not depend on the quaternion norm in any way.[1] For this reason, the attitude quaternion is usually defined with unit norm, $\|\mathbf{q}\| = 1$. This is analogous to

[1]Some authors define the attitude matrix without the factor of $\|\mathbf{q}\|^{-2}$ even if the quaternion is not normalized, with the result that the attitude matrix is not guaranteed to be orthogonal.

requiring that **e** be a unit vector in the Euler axis/angle parameterization, and is the convention adopted in Chap. 2 in this book. For greater generality, however, the identities in this chapter will not assume that the quaternion obeys the norm constraint. If the attitude quaternions are normalized, all the quaternion norms in the following identities have the value unity and can be safely ignored.

Now Eqs. (8.20a)–(8.20c) can be used to show that

$$A^{-1}(\mathbf{q}) = A^T(\mathbf{q}) = A(\mathbf{q}^*) \tag{8.26}$$

and that serial rotations are easily accomplished by multiplying quaternions,

$$A(\bar{\mathbf{q}})A(\mathbf{q}) = A(\bar{\mathbf{q}} \otimes \mathbf{q}) = A(\mathbf{q} \odot \bar{\mathbf{q}}) \tag{8.27}$$

regardless of whether or not the quaternions are normalized. The order of multiplication of the quaternions using the $\otimes$ product is the same as that of the attitude matrix, which is a major advantage of this multiplication convention.

Quaternion multiplication can be used in place of matrix multiplication to transform a three-component vector [6]:

$$\mathbf{q} \otimes \boldsymbol{\omega} \otimes \mathbf{q}^{-1} = \|\mathbf{q}\|^{-2}\,\mathbf{q} \otimes \begin{bmatrix} \boldsymbol{\omega} \\ 0 \end{bmatrix} \otimes \mathbf{q}^* = \begin{bmatrix} A(\mathbf{q})\boldsymbol{\omega} \\ 0 \end{bmatrix} \tag{8.28a}$$

$$\mathbf{q}^{-1} \otimes \boldsymbol{\omega} \otimes \mathbf{q} = \|\mathbf{q}\|^{-2}\,\mathbf{q}^* \otimes \begin{bmatrix} \boldsymbol{\omega} \\ 0 \end{bmatrix} \otimes \mathbf{q} = \begin{bmatrix} A^T(\mathbf{q})\boldsymbol{\omega} \\ 0 \end{bmatrix} \tag{8.28b}$$

Some useful identities are [6]

$$[\mathbf{q}\otimes][\mathbf{q}^{-1}\odot] = [\mathbf{q}^{-1}\odot][\mathbf{q}\otimes] = \begin{bmatrix} A(\mathbf{q}) & 0_{3\times1} \\ 0_{1\times3} & 1 \end{bmatrix} \tag{8.29a}$$

$$[\mathbf{q}\otimes]\Omega(\boldsymbol{\omega})[\mathbf{q}^{-1}\otimes] = \Omega\left(A(\mathbf{q})\boldsymbol{\omega}\right) \tag{8.29b}$$

$$[\mathbf{q}^{-1}\otimes]\Omega(\boldsymbol{\omega})[\mathbf{q}\otimes] = \Omega\left(A^T(\mathbf{q})\boldsymbol{\omega}\right) \tag{8.29c}$$

$$[\mathbf{q}^{-1}\odot]\Gamma(\boldsymbol{\omega})[\mathbf{q}\odot] = \Gamma\left(A(\mathbf{q})\boldsymbol{\omega}\right) \tag{8.29d}$$

$$[\mathbf{q}\odot]\Gamma(\boldsymbol{\omega})[\mathbf{q}^{-1}\odot] = \Gamma\left(A^T(\mathbf{q})\boldsymbol{\omega}\right) \tag{8.29e}$$

$$[\mathbf{q}^{-1}\otimes]\Xi(\mathbf{q}) = \begin{bmatrix} A^T(\mathbf{q}) \\ 0_{1\times3} \end{bmatrix} \tag{8.29f}$$

$$[\mathbf{q}^{-1}\odot]\Psi(\mathbf{q}) = \begin{bmatrix} A(\mathbf{q}) \\ 0_{1\times3} \end{bmatrix} \tag{8.29g}$$

Identities involving $[(\mathbf{q} \otimes \bar{\mathbf{q}}^*)\otimes]$ are given by

$$[(\mathbf{q} \otimes \bar{\mathbf{q}}^*)\otimes] = \Xi(\mathbf{q})A(\mathbf{q} \otimes \bar{\mathbf{q}}^*)\Xi^T(\bar{\mathbf{q}}) + \mathbf{q}\,\bar{\mathbf{q}}^T \tag{8.30a}$$

$$[(\mathbf{q} \otimes \bar{\mathbf{q}}^*)\otimes]\Xi(\bar{\mathbf{q}}) = \|\bar{\mathbf{q}}\|^2\,\Xi(\mathbf{q})A(\mathbf{q} \otimes \bar{\mathbf{q}}^*) \tag{8.30b}$$

$$\Xi^T(\mathbf{q})[(\mathbf{q}\otimes\bar{\mathbf{q}}^*)\otimes] = \|\mathbf{q}\|^2 A(\mathbf{q}\otimes\bar{\mathbf{q}}^*)\Xi^T(\bar{\mathbf{q}}) \tag{8.30c}$$

$$\Xi^T(\mathbf{q})[(\mathbf{q}\otimes\bar{\mathbf{q}}^*)\otimes]\Xi(\bar{\mathbf{q}}) = \|\mathbf{q}\|^2\|\bar{\mathbf{q}}\|^2 A(\mathbf{q}\otimes\bar{\mathbf{q}}^*) \tag{8.30d}$$

Identities involving $[(\bar{\mathbf{q}}^*\otimes\mathbf{q})\odot]$ are given by

$$[(\bar{\mathbf{q}}^*\otimes\mathbf{q})\odot] = \Psi(\mathbf{q})A(\mathbf{q}^*\otimes\bar{\mathbf{q}})\Psi^T(\bar{\mathbf{q}}) + \mathbf{q}\,\bar{\mathbf{q}}^T \tag{8.31a}$$

$$[(\bar{\mathbf{q}}^*\otimes\mathbf{q})\odot]\Psi(\bar{\mathbf{q}}) = \|\bar{\mathbf{q}}\|^2\Psi(\mathbf{q})A(\mathbf{q}^*\otimes\bar{\mathbf{q}}) \tag{8.31b}$$

$$\Psi^T(\mathbf{q})[(\bar{\mathbf{q}}^*\otimes\mathbf{q})\odot] = \|\mathbf{q}\|^2 A(\mathbf{q}^*\otimes\bar{\mathbf{q}})\Psi^T(\bar{\mathbf{q}}) \tag{8.31c}$$

$$\Psi^T(\mathbf{q})[(\bar{\mathbf{q}}^*\otimes\mathbf{q})\odot]\Psi(\bar{\mathbf{q}}) = \|\mathbf{q}\|^2\|\bar{\mathbf{q}}\|^2 A(\mathbf{q}^*\otimes\bar{\mathbf{q}}) \tag{8.31d}$$

Equations (8.19j) or (8.19k) can be used to exchange $\bar{\mathbf{q}}^*\otimes\mathbf{q}$ and $\mathbf{q}\otimes\bar{\mathbf{q}}^*$ if needed.

We now derive some useful identities for attitude determination. If $\mathbf{b}$ and $\mathbf{r}$ are any two three-component vectors, then

$$\|\mathbf{q}\|^2\mathbf{b}^T A(\mathbf{q})\mathbf{r} = \mathbf{b}^T\Xi^T(\mathbf{q})\Psi(\mathbf{q})\mathbf{r} = [\Omega(\mathbf{b})\mathbf{q}]^T\,\Gamma(\mathbf{r})\mathbf{q} = -\mathbf{q}^T\Omega(\mathbf{b})\Gamma(\mathbf{r})\mathbf{q} \tag{8.32}$$

where

$$-\Omega(\mathbf{b})\Gamma(\mathbf{r}) = \begin{bmatrix} B + B^T - \mathrm{tr}(B)I_3 & (\mathbf{b}\times\mathbf{r}) \\ (\mathbf{b}\times\mathbf{r})^T & \mathrm{tr}(B) \end{bmatrix} \equiv K \tag{8.33}$$

with

$$B \equiv \mathbf{b}\,\mathbf{r}^T \tag{8.34}$$

This was shown in Sect. 5.3 to lead to Davenport's q method [7]. Equations (8.23a) and (8.23h) can be used to show that

$$K = -\Omega(\mathbf{b})\Gamma(\mathbf{r}) = \frac{1}{2}\left[\left(\|\mathbf{b}\|^2 + \|\mathbf{r}\|^2\right)I_4 - C^T C\right] \tag{8.35}$$

where

$$C \equiv \Gamma(\mathbf{r}) - \Omega(\mathbf{b}) = \begin{bmatrix} [(\mathbf{b}+\mathbf{r})\times] & -(\mathbf{b}-\mathbf{r}) \\ (\mathbf{b}-\mathbf{r})^T & 0 \end{bmatrix} \tag{8.36}$$

This has been used to develop a square-root attitude determination algorithm [5]. Equations (8.23b) and (8.23c) lead to an identity closely related to Eq. (8.35) that is used in a measurement model that is linear in the quaternion [1]:

$$\Psi(\mathbf{q})\mathbf{r} - \Xi(\mathbf{q})\mathbf{b} = C\mathbf{q} \tag{8.37}$$

Also, if $\mathbf{b} = A(\mathbf{q})\mathbf{r}$, then

$$2\,\Xi(\mathbf{q})[\mathbf{b}\times]\Xi^T(\mathbf{q}) = 2\,\Psi(\mathbf{q})[\mathbf{r}\times]\Psi^T(\mathbf{q}) = \|\mathbf{q}\|^2 C \tag{8.38}$$

Other identities are given by

$$\Xi(\mathbf{q})A(\mathbf{q}) = \Psi(\mathbf{q}) \tag{8.39a}$$

$$\Omega\left(A(\mathbf{q})\boldsymbol{\omega}\right)\mathbf{q} = \Gamma(\boldsymbol{\omega})\mathbf{q} \tag{8.39b}$$

$$\Xi^T(\mathbf{q})\Gamma(\boldsymbol{\omega})\Xi(\mathbf{q}) = \|\mathbf{q}\|^2[A(\mathbf{q})\boldsymbol{\omega}\times] \tag{8.39c}$$

$$\Psi^T(\mathbf{q})\Omega(\boldsymbol{\omega})\Psi(\mathbf{q}) = -\|\mathbf{q}\|^2[A^T(\mathbf{q})\boldsymbol{\omega}\times] \tag{8.39d}$$

$$\Xi\left(\Gamma(\boldsymbol{\omega})\mathbf{q}\right) = \Gamma(\boldsymbol{\omega})\Xi(\mathbf{q}) = \left\{\Psi(\mathbf{q})[\boldsymbol{\omega}\times] - \mathbf{q}\,\boldsymbol{\omega}^T\right\}A^T(\mathbf{q}) \tag{8.39e}$$

$$\Psi\left(\Omega(\boldsymbol{\omega})\mathbf{q}\right) = \Omega(\boldsymbol{\omega})\Psi(\mathbf{q}) = -\left\{\Xi(\mathbf{q})[\boldsymbol{\omega}\times] + \mathbf{q}\,\boldsymbol{\omega}^T\right\}A(\mathbf{q}) \tag{8.39f}$$

$$\Xi\left(\Omega(\boldsymbol{\omega})\mathbf{q}\right) = \Gamma\left(A^T(\mathbf{q})\boldsymbol{\omega}\right)\Xi(\mathbf{q}) = \Xi(\mathbf{q})[\boldsymbol{\omega}\times] - \mathbf{q}\,\boldsymbol{\omega}^T \tag{8.39g}$$

$$\Psi\left(\Gamma(\boldsymbol{\omega})\mathbf{q}\right) = \Omega\left(A(\mathbf{q})\boldsymbol{\omega}\right)\Psi(\mathbf{q}) = -\Psi(\mathbf{q})[\boldsymbol{\omega}\times] - \mathbf{q}\,\boldsymbol{\omega}^T \tag{8.39h}$$

$$\left[\left(\Xi^T(\mathbf{q})\,K\,\mathbf{q}\right)\times\right] = \|\mathbf{q}\|^2\left[A(\mathbf{q})B^T - B\,A^T(\mathbf{q})\right] \tag{8.39i}$$

$$\Xi^T(\mathbf{q})\,K\,\mathbf{q} = \|\mathbf{q}\|^2[\mathbf{b}\times]A(\mathbf{q})\mathbf{r} \tag{8.39j}$$

$$\Psi^T(\mathbf{q})\,K\,\mathbf{q} = -\|\mathbf{q}\|^2[\mathbf{r}\times]A^T(\mathbf{q})\mathbf{b} \tag{8.39k}$$

with K and B given by Eqs. (8.33) and (8.34), respectively. All the identities in this section except Eq. (8.38) are valid for any 3×1 vectors $\mathbf{b}$, $\mathbf{r}$, and $\boldsymbol{\omega}$.

We now use Eqs. (8.25) and (8.23c) to derive the measurement sensitivity matrix used in the additive extended Kalman filter [3]. Our goal is to find an expression for

$$H \equiv \frac{\partial}{\partial \mathbf{q}}[A(\mathbf{q})\mathbf{r}] = \frac{\partial}{\partial \mathbf{q}}\left[\|\mathbf{q}\|^{-2}\,\Xi^T(\mathbf{q})\Gamma(\mathbf{r})\mathbf{q}\right] \tag{8.40}$$

It is important to include the factor of $\|\mathbf{q}\|^{-2}$ when evaluating the partial derivatives because the components of $\mathbf{q}$ are varied independently during this process, violating the norm constraint. Evaluating the partials in Eq. (8.40) gives

$$H = \|\mathbf{q}\|^{-2}\left[\frac{\partial}{\partial \mathbf{q}}\Xi^T(\mathbf{q})\right]\Gamma(\mathbf{r})\mathbf{q} + \|\mathbf{q}\|^{-2}\Xi^T(\mathbf{q})\Gamma(\mathbf{r}) - 2\,\|\mathbf{q}\|^{-4}\Xi^T(\mathbf{q})\Gamma(\mathbf{r})\mathbf{q}\,\mathbf{q}^T \tag{8.41}$$

Note in the first term that $\Gamma(\mathbf{r})\mathbf{q} \equiv \bar{\mathbf{q}}$ is an unnormalized quaternion that is assumed constant during the differentiation, so we can use Eq. (8.20d) to get

$$\left[\frac{\partial}{\partial \mathbf{q}}\Xi^T(\mathbf{q})\right]\Gamma(\mathbf{r})\mathbf{q} = \frac{\partial}{\partial \mathbf{q}}\left[\Xi^T(\mathbf{q})\bar{\mathbf{q}}\right] = -\frac{\partial}{\partial \mathbf{q}}\left[\Xi^T(\bar{\mathbf{q}})\mathbf{q}\right] = -\Xi^T(\bar{\mathbf{q}}) = \Xi^T(\mathbf{q})\Gamma(\mathbf{r}) \tag{8.42}$$

where the last equality results from Eq. (8.39e). Collecting terms and using Eq. (8.20b) yields

$$H = 2\|\mathbf{q}\|^{-4}\Xi^T(\mathbf{q})\Gamma(\mathbf{r})\left(\|\mathbf{q}\|^2 I_4 - \mathbf{q}\,\mathbf{q}^T\right) = 2\|\mathbf{q}\|^{-4}\Xi^T(\mathbf{q})\Gamma(\mathbf{r})\Xi(\mathbf{q})\Xi^T(\mathbf{q}) \tag{8.43}$$

Then applying Eq. (8.39c) gives the final result

$$H = 2\|\mathbf{q}\|^{-2}[A(\mathbf{q})\mathbf{r}\times]\,\Xi^T(\mathbf{q}) \tag{8.44}$$

Note that $\|\mathbf{q}\| = 1$ in most applications and that the sensitivity matrix in the multiplicative filter is $[A(\mathbf{q})\mathbf{r}\times]$ [3].

8.5 Error Quaternions

A common quantity used in estimation and control is the error quaternion between two quaternions, denoted by

$$\delta\mathbf{q} \equiv \begin{bmatrix} \delta\mathbf{q}_{1:3} \\ \delta q_4 \end{bmatrix} = \mathbf{q} \otimes \bar{\mathbf{q}}^{-1} = \|\bar{\mathbf{q}}\|^{-2}\mathbf{q} \otimes \bar{\mathbf{q}}^* \tag{8.45}$$

where $\bar{\mathbf{q}}$ is the estimated quaternion in estimation theory or the desired quaternion in control theory. The rules of quaternion multiplication show that $\|\delta\mathbf{q}\| = \|\mathbf{q}\|/\|\bar{\mathbf{q}}\|$, and that $\delta\mathbf{q}_{1:3}$ and δq_4 are given by

$$\delta\mathbf{q}_{1:3} = \|\bar{\mathbf{q}}\|^{-2}\,\Xi^T(\bar{\mathbf{q}})\mathbf{q} = -\|\bar{\mathbf{q}}\|^{-2}\,\Xi^T(\mathbf{q})\bar{\mathbf{q}} \tag{8.46a}$$

$$\delta q_4 = \|\bar{\mathbf{q}}\|^{-2}\,\bar{\mathbf{q}}^T\mathbf{q} \tag{8.46b}$$

If $\delta\mathbf{q}$ is normalized and is close to the identity quaternion, then $\delta\mathbf{q}_{1:3} \approx \boldsymbol{\alpha}/2$ and $\delta q_4 \approx 1$, where $\boldsymbol{\alpha}$ is a vector of small angle rotations. As $\mathbf{q}$ approaches $\bar{\mathbf{q}}$, then $\delta\mathbf{q}_{1:3}$ and $\boldsymbol{\alpha}$ both approach zero.

A "space-referenced error quaternion," $\delta\mathbf{q}_I$, is defined by

$$\delta\mathbf{q}_I \equiv \begin{bmatrix} \delta\mathbf{q}_{1:3I} \\ \delta q_4 \end{bmatrix} = \bar{\mathbf{q}}^{-1} \otimes \mathbf{q} = \bar{\mathbf{q}}^{-1} \otimes \delta\mathbf{q} \otimes \bar{\mathbf{q}} \tag{8.47a}$$

$$\delta\mathbf{q} = \bar{\mathbf{q}} \otimes \delta\mathbf{q}_I \otimes \bar{\mathbf{q}}^{-1} \tag{8.47b}$$

Quaternion multiplication shows that $\delta q_{4I} = \delta q_4$ and that $\delta\mathbf{q}_{1:3I}$ is given by

$$\delta\mathbf{q}_{1:3I} = \|\bar{\mathbf{q}}\|^{-2}\Psi^T(\bar{\mathbf{q}})\mathbf{q} = -\|\bar{\mathbf{q}}\|^{-2}\Psi^T(\mathbf{q})\bar{\mathbf{q}} \tag{8.48}$$

The meaning of the space-referenced error quaternion is clearly shown by[2]

$$\boldsymbol{\delta}\mathbf{q}_{1:3} = A(\mathbf{q})\boldsymbol{\delta}\mathbf{q}_{1:3_I} = A(\bar{\mathbf{q}})\boldsymbol{\delta}\mathbf{q}_{1:3_I} \tag{8.49a}$$

$$\boldsymbol{\delta}\mathbf{q}_{1:3_I} = A^T(\mathbf{q})\boldsymbol{\delta}\mathbf{q}_{1:3} = A^T(\bar{\mathbf{q}})\boldsymbol{\delta}\mathbf{q}_{1:3} \tag{8.49b}$$

Identities involving the error quaternion are given by [2]

$$\Xi^T(\bar{\mathbf{q}})\Xi(\mathbf{q}) = \|\bar{\mathbf{q}}\|^2\{\delta q_4 I_3 + [\boldsymbol{\delta}\mathbf{q}_{1:3}\times]\} \tag{8.50a}$$

$$\left[\Xi^T(\bar{\mathbf{q}})\Xi(\mathbf{q})\right]^{-1} = \|\bar{\mathbf{q}}\|^{-2}\frac{I_3 + A(\boldsymbol{\delta}\mathbf{q})}{2\delta q_4} \tag{8.50b}$$

$$\left[\Xi^T(\bar{\mathbf{q}})\Xi(\mathbf{q})\right]^{-1}\boldsymbol{\delta}\mathbf{q}_{1:3} = \|\bar{\mathbf{q}}\|^{-2}\frac{\boldsymbol{\delta}\mathbf{q}_{1:3}}{\delta q_4} \tag{8.50c}$$

$$\Psi^T(\bar{\mathbf{q}})\Psi(\mathbf{q}) = \|\bar{\mathbf{q}}\|^2\{\delta q_4 I_3 - [\boldsymbol{\delta}\mathbf{q}_{1:3_I}\times]\} \tag{8.50d}$$

$$\left[\Psi^T(\bar{\mathbf{q}})\Psi(\mathbf{q})\right]^{-1} = \|\bar{\mathbf{q}}\|^{-2}\frac{I_3 + A^T(\boldsymbol{\delta}\mathbf{q}_I)}{2\delta q_4} \tag{8.50e}$$

$$\left[\Psi^T(\bar{\mathbf{q}})\Psi(\mathbf{q})\right]^{-1}\boldsymbol{\delta}\mathbf{q}_{1:3_I} = \|\bar{\mathbf{q}}\|^{-2}\frac{\boldsymbol{\delta}\mathbf{q}_{1:3_I}}{\delta q_4} \tag{8.50f}$$

$$A(\boldsymbol{\delta}\mathbf{q}) = A(\mathbf{q})A^T(\bar{\mathbf{q}}) = \left[\Xi^T(\bar{\mathbf{q}})\Xi(\mathbf{q})\right]^{-1}\Xi^T(\mathbf{q})\Xi(\bar{\mathbf{q}}) \tag{8.50g}$$

$$A^T(\boldsymbol{\delta}\mathbf{q}_I) = A^T(\mathbf{q})A(\bar{\mathbf{q}}) = \left[\Psi^T(\bar{\mathbf{q}})\Psi(\mathbf{q})\right]^{-1}\Psi^T(\mathbf{q})\Psi(\bar{\mathbf{q}}) \tag{8.50h}$$

Note that the inverses of $\Xi^T(\bar{\mathbf{q}})\Xi(\mathbf{q})$ and $\Psi^T(\bar{\mathbf{q}})\Psi(\mathbf{q})$ are singular for 180° errors. Equation (8.50) can used to develop a control law that produces linear error dynamics [2, 4]. Also note that both Eqs. (8.50c) and (8.50f) are related to the Rodrigues parameters or Gibbs vector.

8.6 Quaternion Kinematics

The quaternion kinematic equation is given by

$$\dot{\mathbf{q}} = \frac{1}{2}\omega \otimes \mathbf{q} = \frac{1}{2}\Omega(\omega)\mathbf{q} = \frac{1}{2}\Xi(\mathbf{q})\omega \tag{8.51}$$

Note that $\mathbf{q}^T\dot{\mathbf{q}} = 0$, so this equation preserves the quaternion norm. A major advantage of using normalized quaternions is that the kinematic equation is linear in the quaternion and is also free of singularities. Differentiating Eq. (8.51) and using Eq. (8.23h) gives

---

[2]These equations are true even though $A(\mathbf{q}) \neq A(\bar{\mathbf{q}})$ because $A(\boldsymbol{\delta}\mathbf{q})\boldsymbol{\delta}\mathbf{q}_{1:3} = \boldsymbol{\delta}\mathbf{q}_{1:3}$.

$$\ddot{\mathbf{q}} = \frac{1}{2}\Xi(\mathbf{q})\dot{\omega} + \frac{1}{2}\Omega(\omega)\dot{\mathbf{q}} = \frac{1}{2}\Xi(\mathbf{q})\dot{\omega} - \frac{1}{4}\|\omega\|^2\mathbf{q} \tag{8.52}$$

Substituting a dynamics equation for $\dot{\omega}$ relates the quaternion to a torque input.

The inverse kinematic equation is given by multiplying Eq. (8.51) by $\Xi^T(\mathbf{q})$ and applying Eq. (8.20a), yielding

$$\omega = 2\|\mathbf{q}\|^{-2}\,\Xi^T(\mathbf{q})\,\dot{\mathbf{q}} \tag{8.53}$$

If we define the "space-referenced angular velocity," ω_I, by

$$\begin{bmatrix} \omega_I \\ 0 \end{bmatrix} = \mathbf{q}^{-1} \otimes \omega \otimes \mathbf{q} = \begin{bmatrix} A^T(\mathbf{q})\omega \\ 0 \end{bmatrix} \tag{8.54}$$

then the following relationships can be derived:

$$\mathbf{q}^* \otimes \omega = \omega_I \otimes \mathbf{q}^* \tag{8.55a}$$

$$\Gamma(\omega)\mathbf{q}^* = \Psi(\mathbf{q}^*)\omega = \Omega(\omega_I)\mathbf{q}^* = \Xi(\mathbf{q}^*)\omega_I \tag{8.55b}$$

$$\omega \otimes \mathbf{q} = \mathbf{q} \otimes \omega_I \tag{8.55c}$$

$$\Omega(\omega)\mathbf{q} = \Xi(\mathbf{q})\omega = \Gamma(\omega_I)\mathbf{q} = \Psi(\mathbf{q})\omega_I \tag{8.55d}$$

The definition of the attitude matrix in Eq. (8.25) can be used to show that the quaternion kinematic equation in Eq. (8.51) can also be written as

$$\dot{\mathbf{q}} = \frac{1}{2}\Psi(\mathbf{q})\omega_I = \frac{1}{2}\Gamma(\omega_I)\mathbf{q} \tag{8.56a}$$

$$\omega_I = 2\|\mathbf{q}\|^{-2}\,\Psi^T(\mathbf{q})\,\dot{\mathbf{q}} \tag{8.56b}$$

The derivative of the matrix $\Xi(\mathbf{q})$ is given by [6]

$$\begin{aligned}\frac{d}{dt}\Xi(\mathbf{q}) &= \frac{1}{2}\Gamma(\omega_I)\Xi(\mathbf{q}) = \frac{1}{2}\Xi(\mathbf{q})[\omega\times] - \frac{1}{2}\mathbf{q}\,\omega^T \\ &= \frac{1}{2}\Omega(\omega)\Xi(\mathbf{q}) + \Xi(\mathbf{q})[\omega\times]\end{aligned} \tag{8.57}$$

The derivative of the matrix $\Psi(\mathbf{q})$ is given by

$$\begin{aligned}\frac{d}{dt}\Psi(\mathbf{q}) &= \frac{1}{2}\Omega(\omega)\Psi(\mathbf{q}) = -\frac{1}{2}\Psi(\mathbf{q})[\omega_I\times] - \frac{1}{2}\mathbf{q}\,\omega_I^T \\ &= \frac{1}{2}\Gamma(\omega_I)\Psi(\mathbf{q}) - \Psi(\mathbf{q})[\omega_I\times]\end{aligned} \tag{8.58}$$

The identity of the various forms of the right sides of Eqs. (8.57) and (8.58) is proven by Eqs. (8.54), (8.39g), (8.23f), (8.39h), and (8.23g).

Derivatives of the matrices $[\mathbf{q}\otimes]$ and $[\mathbf{q}\odot]$ are given by [6]

$$\frac{d}{dt}[\mathbf{q}\otimes] = \frac{1}{2}\Omega(\boldsymbol{\omega})[\mathbf{q}\otimes] = \frac{1}{2}[\mathbf{q}\otimes]\Omega(\boldsymbol{\omega}_I) \tag{8.59a}$$

$$\frac{d}{dt}[\mathbf{q}\odot] = \frac{1}{2}[\mathbf{q}\odot]\Gamma(\boldsymbol{\omega}) = \frac{1}{2}\Gamma(\boldsymbol{\omega}_I)[\mathbf{q}\odot] \tag{8.59b}$$

where the identities in Eqs. (8.29c) and (8.29e) have been used in Eq. (8.59).

Taking the time derivative of Eq. (8.45) gives the kinematical relationship for the error quaternion $\boldsymbol{\delta}\mathbf{q}$:

$$\begin{aligned}\frac{d}{dt}\boldsymbol{\delta}\mathbf{q} &= \dot{\mathbf{q}}\otimes\bar{\mathbf{q}}^{-1} + \mathbf{q}\otimes\frac{d\,\bar{\mathbf{q}}^{-1}}{dt} = \frac{1}{2}\boldsymbol{\omega}\otimes\mathbf{q}\otimes\bar{\mathbf{q}}^{-1} + \mathbf{q}\otimes\frac{d\,\bar{\mathbf{q}}^{-1}}{dt}\\ &= \frac{1}{2}\boldsymbol{\omega}\otimes\boldsymbol{\delta}\mathbf{q} + \mathbf{q}\otimes\frac{d\,\bar{\mathbf{q}}^{-1}}{dt}\end{aligned} \tag{8.60}$$

We now need to determine an expression for the derivative of $\bar{\mathbf{q}}^{-1}$. The estimated/desired quaternion kinematic model follows

$$\dot{\bar{\mathbf{q}}} = \frac{1}{2}\bar{\boldsymbol{\omega}}\otimes\bar{\mathbf{q}} \tag{8.61}$$

Taking the time derivative of $\mathbf{I}_q = \bar{\mathbf{q}}\otimes\bar{\mathbf{q}}^{-1}$ gives

$$\mathbf{0} = \dot{\bar{\mathbf{q}}}\otimes\bar{\mathbf{q}}^{-1} + \bar{\mathbf{q}}\otimes\frac{d\,\bar{\mathbf{q}}^{-1}}{dt} = \frac{1}{2}\bar{\boldsymbol{\omega}}\otimes\bar{\mathbf{q}}\otimes\bar{\mathbf{q}}^{-1} + \bar{\mathbf{q}}\otimes\frac{d\,\bar{\mathbf{q}}^{-1}}{dt} = \frac{1}{2}\bar{\boldsymbol{\omega}} + \bar{\mathbf{q}}\otimes\frac{d\,\bar{\mathbf{q}}^{-1}}{dt} \tag{8.62}$$

yielding

$$\frac{d\,\bar{\mathbf{q}}^{-1}}{dt} = -\frac{1}{2}\bar{\mathbf{q}}^{-1}\otimes\bar{\boldsymbol{\omega}} \tag{8.63}$$

and thus

$$\frac{d}{dt}\boldsymbol{\delta}\mathbf{q} = \frac{1}{2}\left(\boldsymbol{\omega}\otimes\boldsymbol{\delta}\mathbf{q} - \mathbf{q}\otimes\bar{\mathbf{q}}^{-1}\otimes\bar{\boldsymbol{\omega}}\right) = \frac{1}{2}\left(\boldsymbol{\omega}\otimes\boldsymbol{\delta}\mathbf{q} - \boldsymbol{\delta}\mathbf{q}\otimes\bar{\boldsymbol{\omega}}\right) \tag{8.64}$$

We now define the following error angular velocity: $\boldsymbol{\delta\omega} \equiv \boldsymbol{\omega} - \bar{\boldsymbol{\omega}}$. Substituting $\boldsymbol{\omega} = \bar{\boldsymbol{\omega}} + \boldsymbol{\delta\omega}$ into Eq. (8.64) leads to

$$\frac{d}{dt}\boldsymbol{\delta}\mathbf{q} = \frac{1}{2}\left(\bar{\boldsymbol{\omega}}\otimes\boldsymbol{\delta}\mathbf{q} - \boldsymbol{\delta}\mathbf{q}\otimes\bar{\boldsymbol{\omega}}\right) + \frac{1}{2}\boldsymbol{\delta\omega}\otimes\boldsymbol{\delta}\mathbf{q} \tag{8.65}$$

Evaluating the quaternion products gives

$$\frac{d}{dt}\boldsymbol{\delta}\mathbf{q}_{1:3} = -\bar{\omega} \times \boldsymbol{\delta}\mathbf{q}_{1:3} + \frac{1}{2}\left(\delta q_4\, \boldsymbol{\delta\omega} - \boldsymbol{\delta\omega} \times \boldsymbol{\delta}\mathbf{q}_{1:3}\right) \tag{8.66a}$$

$$\frac{d}{dt}\delta q_4 = -\frac{1}{2}\boldsymbol{\delta\omega}^T \boldsymbol{\delta}\mathbf{q}_{1:3} \tag{8.66b}$$

Equation (8.66a) can also be derived by differentiating Eq. (8.46a), leading to

$$\frac{d}{dt}\boldsymbol{\delta}\mathbf{q}_{1:3} = \frac{1}{2\|\bar{\mathbf{q}}\|^2}\left[\Xi^T(\bar{\mathbf{q}})\Xi(\mathbf{q})\omega - \Xi^T(\mathbf{q})\Xi(\bar{\mathbf{q}})\bar{\omega}\right] \tag{8.67}$$

Equation (8.50a) can used to prove that Eq. (8.66a) is equivalent to Eq. (8.67).

We define the "space-referenced error angular velocity," $\boldsymbol{\delta\omega}_I$, by

$$\boldsymbol{\delta\omega}_I \equiv \omega_I - \bar{\omega}_I = A^T(\mathbf{q})\omega - A^T(\bar{\mathbf{q}})\bar{\omega} \tag{8.68}$$

Note that $\boldsymbol{\delta\omega}_I \neq A^T\boldsymbol{\delta\omega}$ for either $A(\mathbf{q})$ or $A(\bar{\mathbf{q}})$,[3] but that Eq. (8.49b) gives

$$\boldsymbol{\delta}\mathbf{q}_{1:3_I}^T\boldsymbol{\delta\omega}_I = \left[A^T(\mathbf{q})\boldsymbol{\delta}\mathbf{q}_{1:3}\right]^T A^T(\mathbf{q})\omega - \left[A^T(\bar{\mathbf{q}})\boldsymbol{\delta}\mathbf{q}_{1:3}\right]^T A^T(\bar{\mathbf{q}})\bar{\omega} = \boldsymbol{\delta}\mathbf{q}_{1:3}^T\boldsymbol{\delta\omega} \tag{8.69}$$

Following the derivation of Eq. (8.65), we can evaluate the derivative of $\boldsymbol{\delta}\mathbf{q}_I$ as

$$\begin{aligned}\frac{d}{dt}\boldsymbol{\delta}\mathbf{q}_I &= \frac{1}{2}\left(\boldsymbol{\delta}\mathbf{q}_I \otimes \omega_I - \bar{\omega}_I \otimes \boldsymbol{\delta}\mathbf{q}_I\right)\\ &= \frac{1}{2}\left(\boldsymbol{\delta}\mathbf{q}_I \otimes \bar{\omega}_I - \bar{\omega}_I \otimes \boldsymbol{\delta}\mathbf{q}_I\right) + \frac{1}{2}\boldsymbol{\delta}\mathbf{q}_I \otimes \boldsymbol{\delta\omega}_I\end{aligned} \tag{8.70}$$

which has the vector and scalar parts

$$\frac{d}{dt}\boldsymbol{\delta}\mathbf{q}_{1:3_I} = \bar{\omega}_I \times \boldsymbol{\delta}\mathbf{q}_{1:3_I} + \frac{1}{2}\left(\delta q_4\boldsymbol{\delta\omega}_I + \boldsymbol{\delta\omega}_I \times \boldsymbol{\delta}\mathbf{q}_{1:3_I}\right) \tag{8.71a}$$

$$\frac{d}{dt}\delta q_4 = -\frac{1}{2}\boldsymbol{\delta\omega}_I^T\boldsymbol{\delta}\mathbf{q}_{1:3_I} \tag{8.71b}$$

Equation (8.71a) can also be derived by differentiating Eq. (8.48), leading to

$$\frac{d}{dt}\boldsymbol{\delta}\mathbf{q}_{1:3_I} = \frac{1}{2\|\bar{\mathbf{q}}\|^2}\left[\Psi^T(\bar{\mathbf{q}})\Psi(\mathbf{q})\omega_I - \Psi^T(\mathbf{q})\Psi(\bar{\mathbf{q}})\bar{\omega}_I\right] \tag{8.72}$$

Equation (8.50d) can used to prove that Eq. (8.71a) is equivalent to Eq. (8.72).

[3] An orthogonal transformation would preserve the norm, giving $\|\boldsymbol{\delta\omega}_I\| = \|\boldsymbol{\delta\omega}\|$, and it is not difficult to show that $\|\boldsymbol{\delta\omega}_I\|^2 = \|\boldsymbol{\delta\omega}\|^2 + 2\omega^T[I_3 - A(\boldsymbol{\delta\omega})]\bar{\omega}$.

Some kinematic identities involving $\left[\Xi^T(\bar{\mathbf{q}})\Xi(\mathbf{q})\right]^{-1}$ are given by

$$2\left[\Xi^T(\bar{\mathbf{q}})\Xi(\mathbf{q})\right]^{-1}\Xi^T(\mathbf{q})\dot{\bar{\mathbf{q}}} = A(\delta\mathbf{q})\,\bar{\omega} \tag{8.73a}$$

$$2\left[\Xi^T(\bar{\mathbf{q}})\Xi(\mathbf{q})\right]^{-1}\Xi^T(\mathbf{q})\ddot{\bar{\mathbf{q}}} = A(\delta\mathbf{q})\,\dot{\bar{\omega}} + \frac{\|\bar{\omega}\|^2}{2\delta q_4}\delta\mathbf{q}_{1:3} \tag{8.73b}$$

$$2\left[\Xi^T(\bar{\mathbf{q}})\Xi(\mathbf{q})\right]^{-1}\Xi^T(\dot{\bar{\mathbf{q}}})\Omega(\omega)\mathbf{q} = \left([\omega\times] + \frac{\delta\mathbf{q}_{1:3}\,\omega^T}{\delta q_4}\right)A(\delta\mathbf{q})\,\bar{\omega} \tag{8.73c}$$

References

1. Choukroun, D., Bar-Itzhack, I.Y., Oshman, Y.: Novel quaternion Kalman filter. IEEE Trans. Aerospace Electron. Syst. **42**(1), 174–190 (2006)
2. Crassidis, J.L., Junkins, J.L.: Optimal Estimation of Dynamic Systems, 2nd edn. CRC Press, Boca Raton (2012)
3. Lefferts, E.J., Markley, F.L., Shuster, M.D.: Kalman filtering for spacecraft attitude estimation. J. Guid. Contr. Dynam. **5**(5), 417–429 (1982)
4. Paielli, R.A., Bach, R.E.: Attitude control with realization of linear error dynamics. J. Guid. Contr. Dynam. **16**(1), 182–189 (1993)
5. Pittelkau, M.E.: Square root quaternion estimation. In: AIAA/AAS Astrodynamics Specialist Conference and Exhibit. Monterey, CA (2002). AIAA 2002-4914
6. Shuster, M.D.: A survey of attitude representations. J. Astronaut. Sci. **41**(4), 439–517 (1993)
7. Shuster, M.D., Oh, S.D.: Attitude determination from vector observations. J. Guid. Contr. **4**(1), 70–77 (1981)

Chapter 9
Euler Angles

This chapter presents explicit expressions for the attitude matrices and kinematic matrices for all the 12 Euler and Tait-Bryan angle attitude representations. We first present the attitude matrices for the symmetric and asymmetric sets. Then the kinematic matrices $B(\theta, \psi)$, defined by Eq. (3.39), and their inverses are shown for all the symmetric and asymmetric sets. These matrices are used in the kinematic relations:

$$\begin{bmatrix} \dot{\phi} \\ \dot{\theta} \\ \dot{\psi} \end{bmatrix} = B(\theta, \psi)\,\boldsymbol{\omega} \tag{9.1}$$

where $\boldsymbol{\omega} \equiv [\omega_1 \ \omega_2 \ \omega_3]^T$ is the angular velocity vector, and

$$\boldsymbol{\omega} = B^{-1}(\theta, \psi) \begin{bmatrix} \dot{\phi} \\ \dot{\theta} \\ \dot{\psi} \end{bmatrix} \tag{9.2}$$

The abbreviated notation c $\equiv$ cos and s $\equiv$ sin is employed in Tables 9.1, 9.2, 9.3, 9.4.

Table 9.5 gives equations for converting the symmetric and asymmetric Euler angle sets directly to quaternions, without the need to compute the attitude matrix as an intermediate step.

F.L. Markley and J.L. Crassidis, *Fundamentals of Spacecraft Attitude Determination and Control*, Space Technology Library 33, DOI 10.1007/978-1-4939-0802-8_9,
© Springer Science+Business Media New York 2014

Table 9.1 Attitude matrix: six symmetric sets

Axes	Attitude matrix
1–2–1	$\begin{bmatrix} c\theta & s\phi\, s\theta & -c\phi\, s\theta \\ s\theta\, s\psi & c\phi\, c\psi - s\phi\, c\theta\, s\psi & s\phi\, c\psi + c\phi\, c\theta\, s\psi \\ s\theta\, c\psi & -c\phi\, s\psi - s\phi\, c\theta\, c\psi & -s\phi\, s\psi + c\phi\, c\theta\, c\psi \end{bmatrix}$
1–3–1	$\begin{bmatrix} c\theta & c\phi\, s\theta & s\phi\, s\theta \\ -s\theta\, c\psi & -s\phi\, s\psi + c\phi\, c\theta\, c\psi & c\phi\, s\psi + s\phi\, c\theta\, c\psi \\ s\theta\, s\psi & -s\phi\, c\psi - c\phi\, c\theta\, s\psi & c\phi\, c\psi - s\phi\, c\theta\, s\psi \end{bmatrix}$
2–1–2	$\begin{bmatrix} c\phi\, c\psi - s\phi\, c\theta\, s\psi & s\theta\, s\psi & -s\phi\, c\psi - c\phi\, c\theta\, s\psi \\ s\phi\, s\theta & c\theta & c\phi\, s\theta \\ c\phi\, s\psi + s\phi\, c\theta c\psi & -s\theta\, c\psi & -s\phi\, s\psi + c\phi\, c\theta\, c\psi \end{bmatrix}$
2–3–2	$\begin{bmatrix} -s\phi\, s\psi + c\phi\, c\theta\, c\psi & s\theta\, c\psi & -c\phi\, s\psi - s\phi\, c\theta\, c\psi \\ -c\phi\, s\theta & c\theta & s\phi\, s\theta \\ s\phi\, c\psi + c\phi\, c\theta\, s\psi & s\theta\, s\psi & c\phi\, c\psi - s\phi\, c\theta\, s\psi \end{bmatrix}$
3–1–3	$\begin{bmatrix} c\phi\, c\psi - s\phi\, c\theta\, s\psi & s\phi\, c\psi + c\phi\, c\theta\, s\psi & s\theta\, s\psi \\ -c\phi\, s\psi - s\phi\, c\theta\, c\psi & -s\phi\, s\psi + c\phi\, c\theta\, c\psi & s\theta\, c\psi \\ s\phi\, s\theta & -c\phi\, s\theta & c\theta \end{bmatrix}$
3–2–3	$\begin{bmatrix} -s\phi\, s\psi + c\phi\, c\theta\, c\psi & c\phi\, s\psi + s\phi\, c\theta\, c\psi & -s\theta\, c\psi \\ -s\phi\, c\psi - c\phi\, c\theta\, s\psi & c\phi\, c\psi - s\phi\, c\theta\, s\psi & s\theta\, s\psi \\ c\phi\, s\theta & s\phi\, s\theta & c\theta \end{bmatrix}$

Table 9.2 Attitude matrix: six asymmetric sets

Axes	Attitude matrix
1–2–3	$\begin{bmatrix} c\theta\, c\psi & c\phi\, s\psi + s\phi\, s\theta\, c\psi & s\phi\, s\psi - c\phi\, s\theta\, c\psi \\ -c\theta\, s\psi & c\phi\, c\psi - s\phi\, s\theta\, s\psi & s\phi\, c\psi + c\phi\, s\theta\, s\psi \\ s\theta & -s\phi\, c\theta & c\phi\, c\theta \end{bmatrix}$
1–3–2	$\begin{bmatrix} c\theta\, c\psi & s\phi\, s\psi + c\phi\, s\theta\, c\psi & -c\phi\, s\psi + s\phi\, s\theta\, c\psi \\ -s\theta & c\phi\, c\theta & s\phi\, c\theta \\ c\theta\, s\psi & -s\phi\, c\psi + c\phi\, s\theta\, s\psi & c\phi\, c\psi + s\phi\, s\theta\, s\psi \end{bmatrix}$
2–1–3	$\begin{bmatrix} c\phi\, c\psi + s\phi\, s\theta\, s\psi & c\theta\, s\psi & -s\phi\, c\psi + c\phi\, s\theta\, s\psi \\ -c\phi\, s\psi + s\phi\, s\theta\, c\psi & c\theta\, c\psi & s\phi\, s\psi + c\phi\, s\theta\, c\psi \\ s\phi\, c\theta & -s\theta & c\phi\, c\theta \end{bmatrix}$
2–3–1	$\begin{bmatrix} c\phi\, c\theta & s\theta & -s\phi\, c\theta \\ s\phi\, s\psi - c\phi\, s\theta\, c\psi & c\theta\, c\psi & c\phi\, s\psi + s\phi\, s\theta\, c\psi \\ s\phi\, c\psi + c\phi\, s\theta\, s\psi & -c\theta\, s\psi & c\phi\, c\psi - s\phi\, s\theta\, s\psi \end{bmatrix}$
3–1–2	$\begin{bmatrix} c\phi\, c\psi - s\phi\, s\theta\, s\psi & s\phi\, c\psi + c\phi\, s\theta\, s\psi & -c\theta\, s\psi \\ -s\phi\, c\theta & c\phi\, c\theta & s\theta \\ c\phi\, s\psi + s\phi\, s\theta\, c\psi & s\phi\, s\psi - c\phi\, s\theta\, c\psi & c\theta\, c\psi \end{bmatrix}$
3–2–1	$\begin{bmatrix} c\phi\, c\theta & s\phi\, c\theta & -s\theta \\ -s\phi\, c\psi + c\phi\, s\theta\, s\psi & c\phi\, c\psi + s\phi\, s\theta\, s\psi & c\theta\, s\psi \\ s\phi\, s\psi + c\phi\, s\theta\, c\psi & -c\phi\, s\psi + s\phi\, s\theta\, c\psi & c\theta\, c\psi \end{bmatrix}$

Table 9.3 The matrix $B(\theta, \psi)$

Axes	Symmetric sets	Axes	Asymmetric sets
1–2–1	$\frac{1}{s\theta}\begin{bmatrix} 0 & s\psi & c\psi \\ 0 & s\theta\, c\psi & -s\theta\, s\psi \\ s\theta & -c\theta\, s\psi & -c\theta\, c\psi \end{bmatrix}$	1–2–3	$\frac{1}{c\theta}\begin{bmatrix} c\psi & -s\psi & 0 \\ c\theta\, s\psi & c\theta\, c\psi & 0 \\ -s\theta\, c\psi & s\theta\, s\psi & c\theta \end{bmatrix}$
1–3–1	$\frac{1}{s\theta}\begin{bmatrix} 0 & -c\psi & s\psi \\ 0 & s\theta\, s\psi & s\theta\, c\psi \\ s\theta & c\theta\, c\psi & -c\theta\, s\psi \end{bmatrix}$	1–3–2	$\frac{1}{c\theta}\begin{bmatrix} c\psi & 0 & s\psi \\ -c\theta\, s\psi & 0 & c\theta\, c\psi \\ s\theta\, c\psi & c\theta & s\theta\, s\psi \end{bmatrix}$
2–1–2	$\frac{1}{s\theta}\begin{bmatrix} s\psi & 0 & -c\psi \\ s\theta\, c\psi & 0 & s\theta\, s\psi \\ -c\theta\, s\psi & s\theta & c\theta\, c\psi \end{bmatrix}$	2–1–3	$\frac{1}{c\theta}\begin{bmatrix} s\psi & c\psi & 0 \\ c\theta\, c\psi & -c\theta\, s\psi & 0 \\ s\theta\, s\psi & s\theta\, c\psi & c\theta \end{bmatrix}$
2–3–2	$\frac{1}{s\theta}\begin{bmatrix} c\psi & 0 & s\psi \\ -s\theta\, s\psi & 0 & s\theta\, c\psi \\ -c\theta\, c\psi & s\theta & -c\theta\, s\psi \end{bmatrix}$	2–3–1	$\frac{1}{c\theta}\begin{bmatrix} 0 & c\psi & -s\psi \\ 0 & c\theta\, s\psi & c\theta\, c\psi \\ c\theta & -s\theta\, c\psi & s\theta\, s\psi \end{bmatrix}$
3–1–3	$\frac{1}{s\theta}\begin{bmatrix} s\psi & c\psi & 0 \\ s\theta\, c\psi & -s\theta\, s\psi & 0 \\ -c\theta\, s\psi & -c\theta\, c\psi & s\theta \end{bmatrix}$	3–1–2	$\frac{1}{c\theta}\begin{bmatrix} -s\psi & 0 & c\psi \\ c\theta\, c\psi & 0 & c\theta\, s\psi \\ s\theta\, s\psi & c\theta & -s\theta\, c\psi \end{bmatrix}$
3–2–3	$\frac{1}{s\theta}\begin{bmatrix} -c\psi & s\psi & 0 \\ s\theta\, s\psi & s\theta\, c\psi & 0 \\ c\theta\, c\psi & -c\theta\, s\psi & s\theta \end{bmatrix}$	3–2–1	$\frac{1}{c\theta}\begin{bmatrix} 0 & s\psi & c\psi \\ 0 & c\theta\, c\psi & -c\theta\, s\psi \\ c\theta & s\theta\, s\psi & s\theta\, c\psi \end{bmatrix}$

Table 9.4 The matrix $B^{-1}(\theta, \psi)$

Axes	Symmetric sets	Axes	Asymmetric sets
1–2–1	$\begin{bmatrix} c\theta & 0 & 1 \\ s\theta\, s\psi & c\psi & 0 \\ s\theta\, c\psi & -s\psi & 0 \end{bmatrix}$	1–2–3	$\begin{bmatrix} c\theta\, c\psi & s\psi & 0 \\ -c\theta\, s\psi & c\psi & 0 \\ s\theta & 0 & 1 \end{bmatrix}$
1–3–1	$\begin{bmatrix} c\theta & 0 & 1 \\ -s\theta\, c\psi & s\psi & 0 \\ s\theta\, s\psi & c\psi & 0 \end{bmatrix}$	1–3–2	$\begin{bmatrix} c\theta\, c\psi & -s\psi & 0 \\ -s\theta & 0 & 1 \\ c\theta\, s\psi & c\psi & 0 \end{bmatrix}$
2–1–2	$\begin{bmatrix} s\theta\, s\psi & c\psi & 0 \\ c\theta & 0 & 1 \\ -s\theta\, c\psi & s\psi & 0 \end{bmatrix}$	2–1–3	$\begin{bmatrix} c\theta\, s\psi & c\psi & 0 \\ c\theta\, c\psi & -s\psi & 0 \\ -s\theta & 0 & 1 \end{bmatrix}$
2–3–2	$\begin{bmatrix} s\theta\, c\psi & -s\psi & 0 \\ c\theta & 0 & 1 \\ s\theta\, s\psi & c\psi & 0 \end{bmatrix}$	2–3–1	$\begin{bmatrix} s\theta & 0 & 1 \\ c\theta\, c\psi & s\psi & 0 \\ -c\theta\, s\psi & c\psi & 0 \end{bmatrix}$
3–1–3	$\begin{bmatrix} s\theta\, s\psi & c\psi & 0 \\ s\theta\, c\psi & -s\psi & 0 \\ c\theta & 0 & 1 \end{bmatrix}$	3–1–2	$\begin{bmatrix} -c\theta\, s\psi & c\psi & 0 \\ s\theta & 0 & 1 \\ c\theta\, c\psi & s\psi & 0 \end{bmatrix}$
3–2–3	$\begin{bmatrix} -s\theta\, c\psi & s\psi & 0 \\ s\theta\, s\psi & c\psi & 0 \\ c\theta & 0 & 1 \end{bmatrix}$	3–2–1	$\begin{bmatrix} -s\theta & 0 & 1 \\ c\theta\, s\psi & c\psi & 0 \\ c\theta\, c\psi & -s\psi & 0 \end{bmatrix}$

Table 9.5 Euler angle to quaternion conversions

Axes	Quaternion	Axes	Quaternion
1–2–1	$\begin{bmatrix} \bar{c}\theta\,\bar{s}(\phi+\psi) \\ \bar{s}\theta\,\bar{c}(\phi-\psi) \\ \bar{s}\theta\,\bar{s}(\phi-\psi) \\ \bar{c}\theta\,\bar{c}(\phi+\psi) \end{bmatrix}$	1–2–3	$\begin{bmatrix} \bar{s}\phi\,\bar{c}\theta\,\bar{c}\psi+\bar{c}\phi\,\bar{s}\theta\,\bar{s}\psi \\ \bar{c}\phi\,\bar{s}\theta\,\bar{c}\psi-\bar{s}\phi\,\bar{c}\theta\,\bar{s}\psi \\ \bar{c}\phi\,\bar{c}\theta\,\bar{s}\psi+\bar{s}\phi\,\bar{s}\theta\,\bar{c}\psi \\ \bar{c}\phi\,\bar{c}\theta\,\bar{c}\psi-\bar{s}\phi\,\bar{s}\theta\,\bar{s}\psi \end{bmatrix}$
1–3–1	$\begin{bmatrix} \bar{c}\theta\,\bar{s}(\phi+\psi) \\ \bar{s}\theta\,\bar{s}(\psi-\phi) \\ \bar{s}\theta\,\bar{c}(\psi-\phi) \\ \bar{c}\theta\,\bar{c}(\phi+\psi) \end{bmatrix}$	1–3–2	$\begin{bmatrix} \bar{s}\phi\,\bar{c}\theta\,\bar{c}\psi-\bar{c}\phi\,\bar{s}\theta\,\bar{s}\psi \\ \bar{c}\phi\,\bar{c}\theta\,\bar{s}\psi-\bar{s}\phi\,\bar{s}\theta\,\bar{c}\psi \\ \bar{c}\phi\,\bar{s}\theta\,\bar{c}\psi+\bar{s}\phi\,\bar{c}\theta\,\bar{s}\psi \\ \bar{c}\phi\,\bar{c}\theta\,\bar{c}\psi+\bar{s}\phi\,\bar{s}\theta\,\bar{s}\psi \end{bmatrix}$
2–1–2	$\begin{bmatrix} \bar{s}\theta\,\bar{c}(\psi-\phi) \\ \bar{c}\theta\,\bar{s}(\phi+\psi) \\ \bar{s}\theta\,\bar{s}(\psi-\phi) \\ \bar{c}\theta\,\bar{c}(\phi+\psi) \end{bmatrix}$	2–1–3	$\begin{bmatrix} \bar{c}\phi\,\bar{s}\theta\,\bar{c}\psi+\bar{s}\phi\,\bar{c}\theta\,\bar{s}\psi \\ \bar{s}\phi\,\bar{c}\theta\,\bar{c}\psi-\bar{c}\phi\,\bar{s}\theta\,\bar{s}\psi \\ \bar{c}\phi\,\bar{c}\theta\,\bar{s}\psi-\bar{s}\phi\,\bar{s}\theta\,\bar{c}\psi \\ \bar{c}\phi\,\bar{c}\theta\,\bar{c}\psi+\bar{s}\phi\,\bar{s}\theta\,\bar{s}\psi \end{bmatrix}$
2–3–2	$\begin{bmatrix} \bar{s}\theta\,\bar{s}(\phi-\psi) \\ \bar{c}\theta\,\bar{s}(\phi+\psi) \\ \bar{s}\theta\,\bar{c}(\phi-\psi) \\ \bar{c}\theta\,\bar{c}(\phi+\psi) \end{bmatrix}$	2–3–1	$\begin{bmatrix} \bar{c}\phi\,\bar{c}\theta\,\bar{s}\psi+\bar{s}\phi\,\bar{s}\theta\,\bar{c}\psi \\ \bar{s}\phi\,\bar{c}\theta\,\bar{c}\psi+\bar{c}\phi\,\bar{s}\theta\,\bar{s}\psi \\ \bar{c}\phi\,\bar{s}\theta\,\bar{c}\psi-\bar{s}\phi\,\bar{c}\theta\,\bar{s}\psi \\ \bar{c}\phi\,\bar{c}\theta\,\bar{c}\psi-\bar{s}\phi\,\bar{s}\theta\,\bar{s}\psi \end{bmatrix}$
3–1–3	$\begin{bmatrix} \bar{s}\theta\,\bar{c}(\phi-\psi) \\ \bar{s}\theta\,\bar{s}(\phi-\psi) \\ \bar{c}\theta\,\bar{s}(\phi+\psi) \\ \bar{c}\theta\,\bar{c}(\phi+\psi) \end{bmatrix}$	3–1–2	$\begin{bmatrix} \bar{c}\phi\,\bar{s}\theta\,\bar{c}\psi-\bar{s}\phi\,\bar{c}\theta\,\bar{s}\psi \\ \bar{c}\phi\,\bar{c}\theta\,\bar{s}\psi+\bar{s}\phi\,\bar{s}\theta\,\bar{c}\psi \\ \bar{c}\phi\,\bar{s}\theta\,\bar{s}\psi+\bar{s}\phi\,\bar{c}\theta\,\bar{c}\psi \\ \bar{c}\phi\,\bar{c}\theta\,\bar{c}\psi-\bar{s}\phi\,\bar{s}\theta\,\bar{s}\psi \end{bmatrix}$
3–2–3	$\begin{bmatrix} \bar{s}\theta\,\bar{s}(\psi-\phi) \\ \bar{s}\theta\,\bar{c}(\psi-\phi) \\ \bar{c}\theta\,\bar{s}(\phi+\psi) \\ \bar{c}\theta\,\bar{c}(\phi+\psi) \end{bmatrix}$	3–2–1	$\begin{bmatrix} \bar{c}\phi\,\bar{c}\theta\,\bar{s}\psi-\bar{s}\phi\,\bar{s}\theta\,\bar{c}\psi \\ \bar{c}\phi\,\bar{s}\theta\,\bar{c}\psi+\bar{s}\phi\,\bar{c}\theta\,\bar{s}\psi \\ \bar{s}\phi\,\bar{c}\theta\,\bar{c}\psi-\bar{c}\phi\,\bar{s}\theta\,\bar{s}\psi \\ \bar{c}\phi\,\bar{c}\theta\,\bar{c}\psi+\bar{s}\phi\,\bar{s}\theta\,\bar{s}\psi \end{bmatrix}$

where $\bar{c}(\cdot) \equiv \cos(\cdot/2)$ and $\bar{s}(\cdot) \equiv \sin(\cdot/2)$

Chapter 10
Orbital Dynamics

The study of bodies in orbit has attracted the world's greatest mathematicians in the past, and remains a flourishing subject area in the present. In fact many useful mathematical concepts, such as Bessel functions and nonlinear least squares, can be directly traced back to the study of orbital motion. Here the basic equations and concepts of orbital dynamics are introduced. More details can be found in the references herein.

In the seventeenth century, Johannes Kepler propounded his three laws of planetary motion defining the shape of planetary orbits, the velocity at which planets travel around the Sun, and the time required for a planet to complete an orbit. They state that

1. The orbit of each planet is an ellipse, with the Sun at a focus.
2. The line joining the planet to the Sun sweeps out equal areas in equal times.
3. The square of the period of a planet is proportional to the cube of its mean distance from the Sun.

Since ellipses are a basic element in the study of orbital motion, we begin this chapter with a review of the geometry of ellipses.

In this chapter all position, $\mathbf{r}$, and velocity, $\mathbf{v}$, vectors are expressed in inertial coordinates, unless a different frame is explicitly indicated. Thus, the subscript I is removed from these vectors for brevity.

10.1 Geometry of Ellipses

This section provides a review of elliptical geometry, which will later be used to describe the motion of a spacecraft in orbit. Figure 10.1 shows a basic ellipse using a two-center bipolar coordinate system, centered at (x_0, y_0). The variable a is known as the *semimajor* axis and the variable b is the *semiminor* axis. By definition the two-center bipolar coordinate equation is given by

F.L. Markley and J.L. Crassidis, *Fundamentals of Spacecraft Attitude Determination and Control*, Space Technology Library 33, DOI 10.1007/978-1-4939-0802-8_10,
© Springer Science+Business Media New York 2014

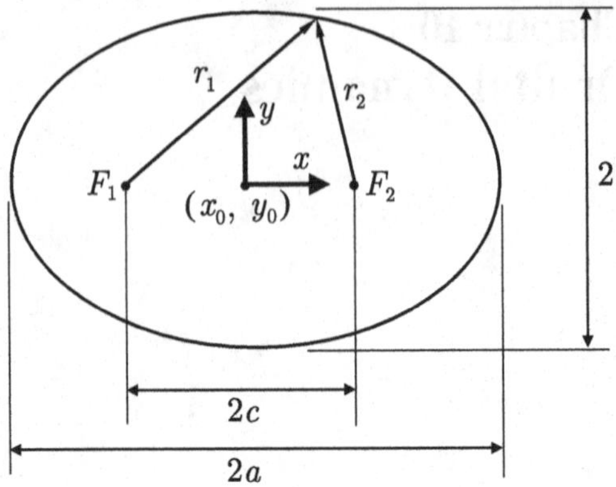

Fig. 10.1 Two-center bipolar ellipse coordinate system

$$r_1 + r_2 = 2a \tag{10.1}$$

Assuming that $x_0 = y_0 = 0$, the first focus F_1 is at $(-c, 0)$ and the second focus F_2 is at $(c, 0)$. We wish to describe a point given by both r_1 and r_2 using the coordinates x and y. It is easy to show that $r_1 = \sqrt{(x+c)^2 + y^2}$ and $r_2 = \sqrt{(x-c)^2 + y^2}$. Substituting these quantities into Eq. (10.1) gives

$$\sqrt{x^2 + y^2 + c^2 + 2cx} + \sqrt{x^2 + y^2 + c^2 - 2cx} = 2a \tag{10.2}$$

Squaring both sides of Eq. (10.2) gives

$$2(x^2 + y^2 + c^2) + 2\sqrt{(x^2 + y^2 + c^2)^2 - 4c^2x^2} = 4a^2 \tag{10.3}$$

Rearranging terms and squaring again leads to

$$\begin{aligned}(x^2 + y^2 + c^2)^2 - 4c^2x^2 &= [2a^2 - (x^2 + y^2 + c^2)]^2 \\ &= 4a^4 - 4a^2(x^2 + y^2 + c^2) + (x^2 + y^2 + c^2)^2\end{aligned} \tag{10.4}$$

Grouping terms and dividing by $4a^2(a^2 - c^2)$ then leads to

$$\frac{x^2}{a^2} + \frac{y^2}{a^2 - c^2} = 1 \tag{10.5}$$

The points on the ellipse at $x = 0$ have $y = \pm b$, so $r_1 = r_2 = \sqrt{c^2 + b^2} = a$ at these points. Substituting $a^2 - c^2 = b^2$ into Eq. (10.5) gives the basic equation for an ellipse:

$$\frac{x^2}{a^2} + \frac{y^2}{b^2} = 1 \tag{10.6}$$

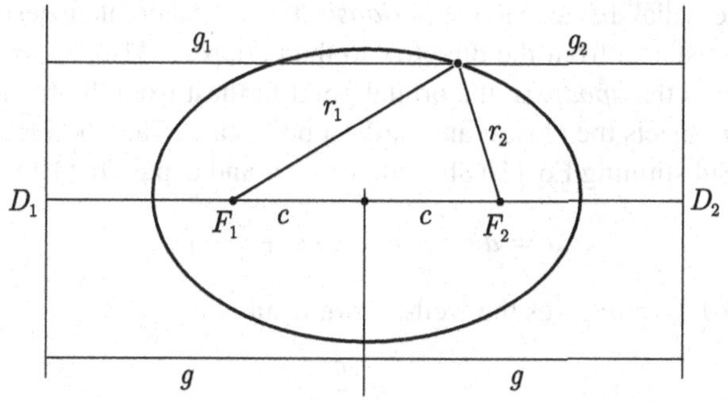

Fig. 10.2 Focal definition of ellipse

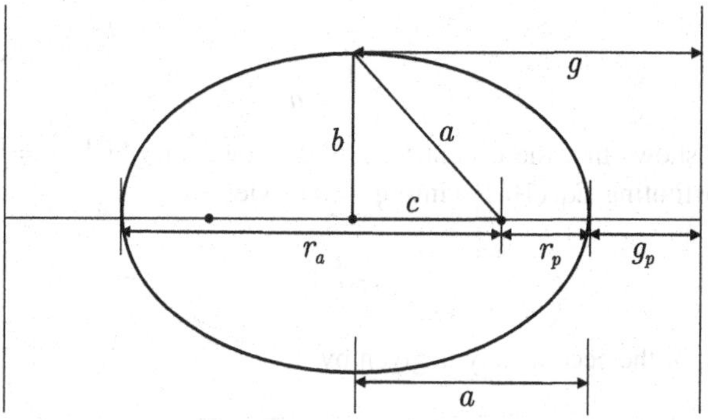

Fig. 10.3 Relationships in focal definition of ellipse

Another way to describe an ellipse is to use the focal definition, which states that an ellipse is the locus of points the ratio of whose distances from a fixed line, the directrix, to a fixed point, the focus, is a constant e, called the eccentricity. This definition is depicted in Fig. 10.2, where the vertical line through D_1 is the directrix associated with F_1 and the vertical line through D_2 is the directrix associated with F_2. The eccentricity is $e = r_1/g_1 = r_2/g_2$, where g_1 and g_2 are the distances from the comparative directrices to the desired point on the ellipse. Adding $r_1 = e\, g_1$ and $r_2 = e\, g_2$ gives

$$r_1 + r_2 = e(g_1 + g_2) \tag{10.7}$$

Comparing Eqs. (10.1) and (10.7) gives $a = e\, g$, where g is the distance shown in Fig. 10.3.

Referring to Fig. 10.3 the following relationships can be seen:

$$r_p = a - c = e\, g_p \tag{10.8a}$$

$$g = a + g_p \tag{10.8b}$$

where r_p is the radial distance to the *periapsis*, the orbital point closest to the focus, and g_p is the distance from the directrix to the periapsis. Also, $r_a = a + c$ is the radial distance to the *apoapsis*, the orbital point farthest from the focus. For Earth-orbiting space objects the closest and farthest points are called perigee and apogee, respectively. Substituting Eq. (10.8b) into $a = e\,g$ and using Eq. (10.8a) gives

$$a = ae + e\,g_p = ae + a - c \tag{10.9}$$

Solving Eq. (10.9) for e gives the well-known result:

$$e = \frac{c}{a} \tag{10.10}$$

Substituting $c = \sqrt{a^2 - b^2}$ into this yields

$$e = \sqrt{1 - \frac{b^2}{a^2}} \tag{10.11}$$

which clearly shows that the eccentricity of a circle, for which $a = b$, is zero as expected. Substituting Eq. (10.10) into $g = a/e$ yields

$$g = \frac{a^2}{c} \tag{10.12}$$

Another form for the eccentricity is given by

$$e = \frac{a - c}{g - a} = \frac{\sqrt{b^2 + c^2}}{g} \tag{10.13}$$

Substituting $c = e\,a$ into $r_p = a - c$ and $r_a = a + c$ gives

$$r_p = a(1 - e) \tag{10.14a}$$

$$r_a = a(1 + e) \tag{10.14b}$$

Equations (10.6) and (10.10) give $y = b\sqrt{1 - e^2}$ when $x = c$. This distance is known as the *semilatus rectum*, a combination of the Latin words "semi," meaning half, "latus," meaning side, and "rectum," meaning straight. It is shown as p in Fig. 10.4, so we have

$$p = a(1 - e^2) \tag{10.15}$$

It is often more convenient to represent an ellipse in polar coordinate form rather than the Cartesian form of Eq. (10.6). Referring to Fig. 10.5 the x and y components are related to the polar coordinates r and ν by

$$x = c + r\cos\nu \tag{10.16a}$$

$$y = r\sin\nu \tag{10.16b}$$

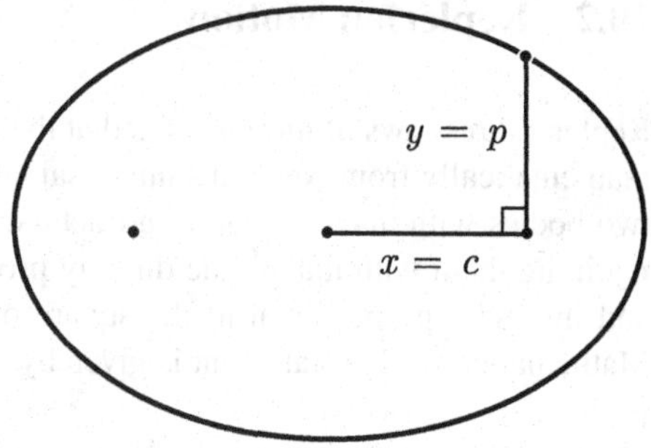

Fig. 10.4 Semilatus rectum definition

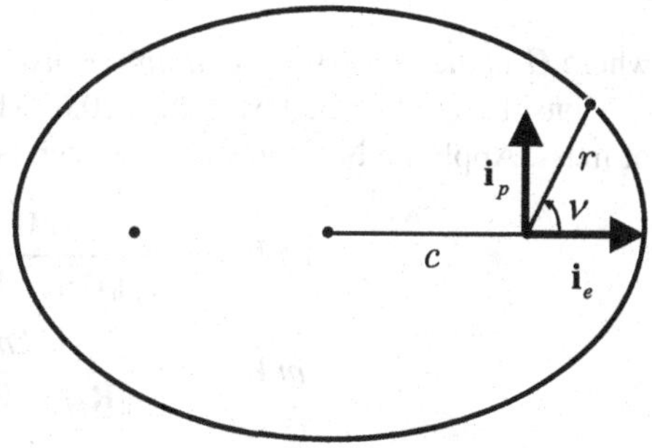

Fig. 10.5 Polar coordinate definition of an ellipse

where ν is known as the *true anomaly*, which is defined as a positive rotation from the periapsis direction. Substituting Eq. (10.16) into Eq. (10.6) and multiplying both sides by b^2 gives

$$(b^2/a^2)(c^2 + 2\,c\,r\cos\nu + r^2\cos^2\nu) + r^2\sin^2\nu = b^2 \tag{10.17}$$

Substituting $\sin^2\nu = 1-\cos^2\nu$, using $b^2 = a^2(1-e^2)$ from Eq. (10.11) and $c = e\,a$ from Eq. (10.10), and collecting terms leads to

$$\begin{aligned} 0 &= r^2 - e^2r^2\cos^2\nu + 2a(1-e^2)\,e\,r\cos\nu - a^2(1-e^2)^2 \\ &= r^2 - [a(1-e^2) - e\,r\cos\nu]^2 \end{aligned} \tag{10.18}$$

so r is given by

$$r = \pm[a(1-e^2) - e\,r\cos\nu] \tag{10.19}$$

The plus sign must be used because r is positive. This leads to

$$r = \frac{a(1-e^2)}{1 + e\cos\nu} \tag{10.20}$$

Using Eq. (10.15) expresses this in terms of the semilatus rectum

$$r = \frac{p}{1 + e\cos\nu} \tag{10.21}$$

Note that when $e = 0$ then $r = p$, which represents a circle.

10.2 Keplerian Motion

Kepler's three laws of motion, stated at the beginning of this chapter, can be proven mathematically from Newton's universal law of gravitation. Newton stated that any two bodies with mass M and m attract each other by force that acts along the line $\mathbf{r}$ joining them with magnitude directly proportional to the product of their masses and inversely proportional to the square of the distance $r = \|\mathbf{r}\|$ between them. Mathematically, this statement is given by

$$\mathbf{F}_g = -GMm\,\frac{\mathbf{r}}{r^3} \tag{10.22}$$

where G is the *universal gravitation constant* [3,4].

Consider the two bodies in Fig. 10.6, where $\mathbf{R}_c$ denotes the vector to the center of mass. Applying Newton's law for each body we obtain

$$M\,\ddot{\mathbf{R}}_M = -\frac{GMm}{\|\mathbf{R}_M - \mathbf{R}_m\|^3}(\mathbf{R}_M - \mathbf{R}_m) \tag{10.23a}$$

$$m\,\ddot{\mathbf{R}}_m = -\frac{GMm}{\|\mathbf{R}_M - \mathbf{R}_m\|^3}(\mathbf{R}_m - \mathbf{R}_M) \tag{10.23b}$$

Adding Eqs. (10.23a) and (10.23b) gives

$$M\ddot{\mathbf{R}}_M + m\ddot{\mathbf{R}}_m = \mathbf{0} \tag{10.24}$$

The center of mass is given by

$$\mathbf{R}_c = \frac{M\mathbf{R}_M + m\mathbf{R}_m}{M + m} \tag{10.25}$$

so Eq. (10.24) gives $\ddot{\mathbf{R}}_c = \mathbf{0}$ at all time. This states that the center of mass moves in a straight line at constant velocity. We are also interested in determining the relative position, $\mathbf{r} = \mathbf{R}_m - \mathbf{R}_M$, of the smaller mass (e.g. a satellite) with respect to the large mass (e.g. the Earth). Subtracting Eq. (10.23a) from Eq. (10.23b) leads to

$$\ddot{\mathbf{r}} = -\frac{\mu}{r^3}\mathbf{r} \tag{10.26}$$

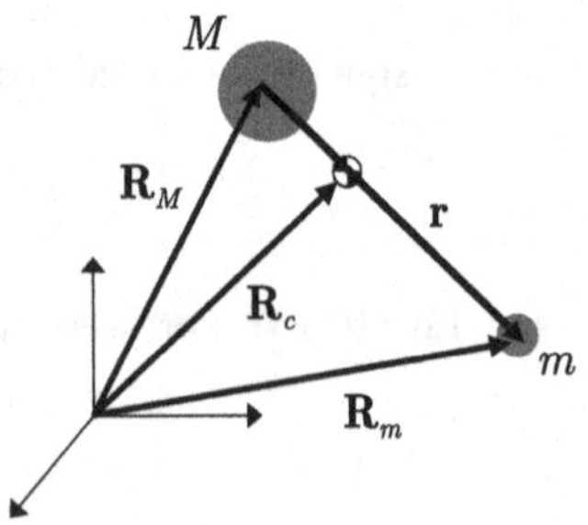

Fig. 10.6 Relative motion of two bodies

where $\mu = G(M+m)$ is the *gravitational parameter*. The approximation $\mu \approx GM$ is obviously a good one for most orbiting bodies since $M \gg m$. Perturbations due to conservative forces, such as the gravity differential force due to the Earth's oblateness, and non-conservative forces, such as drag and solar radiation pressure, are often added to the right hand side of Eq. (10.26). These perturbations will be discussed in Sect. 10.3.

Equation (10.26) can be used to show that the energy and angular momentum of orbital motion are conserved. The *specific kinetic energy*, i.e. the kinetic energy per unit mass, is given by

$$T \equiv \frac{1}{2}v^2 \tag{10.27}$$

The time derivatives of T, using Eq. (10.26), is

$$\dot{T} = \frac{d}{dt}\left(\frac{\dot{\mathbf{r}} \cdot \dot{\mathbf{r}}}{2}\right) = \dot{\mathbf{r}} \cdot \ddot{\mathbf{r}} = -\frac{\mu \dot{\mathbf{r}} \cdot \mathbf{r}}{r^3} \tag{10.28}$$

The *specific potential energy* is defined by

$$V \equiv -\frac{\mu}{r} \tag{10.29}$$

The time derivative of V is

$$\dot{V} = -\mu \frac{d}{dt}(\mathbf{r} \cdot \mathbf{r})^{-1/2} = \frac{\mu \dot{\mathbf{r}} \cdot \mathbf{r}}{r^3} \tag{10.30}$$

Adding Eqs. (10.28) and (10.30) proves that the *specific energy*

$$\mathscr{E} = T + V \tag{10.31}$$

has time derivative zero, i.e. that it is conserved.

The *specific angular momentum* is given by

$$\mathbf{h} \equiv \mathbf{r} \times \mathbf{v} = \mathbf{r} \times \dot{\mathbf{r}} \tag{10.32}$$

Its time derivative is

$$\dot{\mathbf{h}} = \dot{\mathbf{r}} \times \dot{\mathbf{r}} + \mathbf{r} \times \ddot{\mathbf{r}} = -\frac{\mu}{r^3}\mathbf{r} \times \mathbf{r} = \mathbf{0} \tag{10.33}$$

showing that the angular momentum is conserved. Since $\mathbf{r}$ and $\mathbf{v}$ are both perpendicular to the constant vector $\mathbf{h}$, the position and velocity vectors must remain in a fixed plane, called the *orbital plane*, perpendicular to $\mathbf{h}$.

We now use Eq. (10.26) to prove Kepler's three laws. Figure 10.7 depicts the area swept out by a line joining the primary mass and its satellite. This area is half the area of a parallelogram formed by $\mathbf{r}$ and $d\mathbf{r}$:

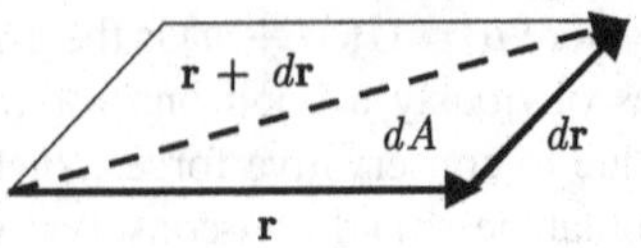

Fig. 10.7 Area sweep of primary mass and satellite

$$\begin{aligned} dA &= \frac{1}{2}\|\mathbf{r} \times d\mathbf{r}\| = \frac{1}{2}\|\mathbf{r} \times \frac{d\mathbf{r}}{dt} dt\| \\ &= \frac{1}{2}\|\mathbf{r} \times \dot{\mathbf{r}}\| dt = \frac{1}{2} h\, dt \end{aligned} \tag{10.34}$$

Since $\mathbf{h}$ is constant then Eq. (10.34) indicates that the line sweeps out equal areas in equal times, which proves Kepler's second law.

Since $\mathbf{h}$ is constant, then

$$\frac{d}{dt}(\dot{\mathbf{r}} \times \mathbf{h}) = \ddot{\mathbf{r}} \times \mathbf{h} = -\frac{\mu}{r^3}\mathbf{r} \times \mathbf{h} = -\frac{\mu}{r^3}\mathbf{r} \times (\mathbf{r} \times \dot{\mathbf{r}}) = -\frac{\mu}{r^3}\left[(\mathbf{r} \cdot \dot{\mathbf{r}})\mathbf{r} - r^2\dot{\mathbf{r}}\right] \tag{10.35}$$

where Eq. (8.2e), the "bac−cab" rule, has been used. Combining this with Eq. (10.30) gives

$$\frac{d}{dt}\left(\dot{\mathbf{r}} \times \mathbf{h} - \frac{\mu}{r}\mathbf{r}\right) = \mathbf{0} \tag{10.36}$$

Integrating this equation gives

$$\dot{\mathbf{r}} \times \mathbf{h} - \frac{\mu}{r}\mathbf{r} = \mu\, \mathbf{e} \tag{10.37}$$

where the integration constant $\mu\,\mathbf{e}$ is commonly referred to as the *Laplace vector*. Clearly $\dot{\mathbf{r}} \times \mathbf{h}$ is in the orbit plane, so $\mathbf{e}$ must also be in this plane. Forming the dot product of (10.37) with $\mathbf{r}$ gives

$$\mu\,\mathbf{r} \cdot \mathbf{e} = \mathbf{r} \cdot (\dot{\mathbf{r}} \times \mathbf{h}) - \mu r = (\mathbf{r} \times \dot{\mathbf{r}}) \cdot \mathbf{h} - \mu r = h^2 - \mu r \tag{10.38}$$

where we have used Eq. (2.56a). Let ψ be the angle between $\mathbf{r}$ and $\mathbf{e}$, so that Eq. (10.38) becomes $h^2 = \mu\, r + \mu\, r\, e\, \cos\psi$, or

$$r = \frac{h^2/\mu}{1 + e\, \cos\psi} \tag{10.39}$$

This is just Eq. (10.21) for an ellipse in polar coordinates, where h^2/μ is the semilatus rectum, $\psi \equiv \nu$ is the true anomaly shown in Fig. 10.5, and e is the eccentricity, proving Kepler's first law. It also requires the vector $\mathbf{e}$, which is known as the *eccentricity vector*, to point along the periapsis direction because ν is the angle between $\mathbf{r}$ and $\mathbf{e}$. The magnitude of the eccentricity vector is the eccentricity of the elliptical orbit, so **e** is not a unit vector as its symbol might mislead one into thinking. The equation $p \equiv h^2/\mu$ gives the following useful expression for h:

$$h = \sqrt{\mu p} \tag{10.40}$$

Note that Eq. (10.39) is actually more general than just for ellipses. It is a polar equation of a conic section, where the value of e dictates the type of orbit. If $e = 0$ then the orbit is circular with $a = r$; if $e < 1$ then the orbit is elliptical with $a > 0$; if $e = 1$ then the orbit is parabolic with infinite semimajor axis; and if $e > 1$ then the orbit is hyperbolic with $a < 0$. Elliptic orbits are the most common type because they are used for planets and planetary satellites. Parabolic orbits are rarely found, but orbits of some comets approximate a parabola. These orbits are a borderline case between an open and closed orbit. The orbit of an interplanetary probe sent from the Earth must be a hyperbolic orbit if the probe is to escape the Earth's gravitational field with finite speed. Parabolic and hyperbolic orbits are one-way trips to infinity that will never retrace the same path again.

Equation (10.34) can be written as $dt = (2/h)\,dA$. Integrating this equation over an entire orbit and noting that the area of an ellipse is given by $\pi\, a\, b$ yields

$$\rho = \frac{2\pi ab}{h} \tag{10.41}$$

where ρ is the orbital period. Substituting Eq. (10.40) and $b = \sqrt{ap}$, which follows from Eqs. (10.11) and (10.15), yields

$$\rho = \frac{2\pi}{\sqrt{\mu}}a^{3/2} = \frac{2\pi}{n} \tag{10.42}$$

where n is the *mean motion*, defined by

$$n \equiv \sqrt{\frac{\mu}{a^3}} \tag{10.43}$$

This proves Kepler's third law. For a circular 300 km low-Earth orbit, Eq. (10.42) gives a period of about 90 min. A *geosynchronous orbit* with a period of one sidereal day (approximately 23 h 56 min and 4 s), matching the Earth's sidereal rotation period, requires an altitude of 35,786 km. Communications satellites are often placed in *geostationary orbits*, equatorial circular geosynchronous orbits, because they appear to be stationary with respect to the ground.

Having proven Kepler's three laws from Newton's gravitational law, we now investigate energy further. Since energy is conserved, it can be computed at any convenient point in the orbit. Here, we choose the periapsis point with $v_p = h/r_p$. Using Eqs. (10.14a), (10.40), and (10.15) gives

$$\mathscr{E} = \frac{h^2}{2r_p^2} - \frac{\mu}{r_p} = \frac{\mu a(1-e^2)}{2a^2(1-e)^2} - \frac{\mu}{a(1-e)} = \frac{\mu a(1+e) - 2a\mu}{2a^2(1-e)} = -\frac{\mu}{2a} \tag{10.44}$$

Substituting this expression into Eq. (10.31), substituting the definitions of T and V, and rearranging gives

$$v^2 = \mu \left(\frac{2}{r} - \frac{1}{a} \right) \tag{10.45}$$

Equation (10.45) is known as the *vis viva* equation from the Latin for living force. The escape velocity at a distance r from a center of force is the minimum velocity needed to escape its gravitational pull. This is the velocity needed to achieve a parabolic orbit, and is obtained by setting $a = \infty$ in Eq. (10.45):

$$v_{\text{esc}} = \sqrt{\frac{2\mu}{r}} \tag{10.46}$$

The escape velocity for a body on the Earth's surface, with r being the radius of the Earth, is 11.06 km/s. The escape velocity for a body on the Moon's surface, with r and μ being the radius and gravitational parameter of the Moon, respectively, is 1.68 km/s, which is almost 10 times less than for the Earth.

10.2.1 Classical Orbital Elements

Equation (10.26) requires the initial position, $\mathbf{r}(t_0)$, and velocity, $\dot{\mathbf{r}}(t_0)$, with respect to some inertial coordinate system. Unfortunately $\mathbf{r}(t_0)$ and $\dot{\mathbf{r}}(t_0)$ do not provide the most convenient characterization of the orbit. The six classical Keplerian orbital elements give a more satisfying physical characterization of the orbit than the Cartesian position and velocity vectors. The dimensional elements are given by

- a = semimajor axis (size of the orbit)
- e = eccentricity (shape of the orbit)
- M_0 = initial mean anomaly (related to the spacecraft's initial position in the orbit)

The orientation elements are given by

- i = inclination (angle between orbit plane and reference plane)
- Ω = right ascension of the ascending node (angle between vernal equinox direction and the line of nodes)
- ω = argument of periapsis or perigee (angle between the ascending node direction and periapsis or perigee direction)

Figure 10.8 shows the orientation elements and the unit vectors that define the *perifocal coordinate system*. The vector $\mathbf{i}_e$ points along the periapsis direction, the vector $\mathbf{i}_p$ points along the semilatus rectum direction, and the vector $\mathbf{i}_h$ points along the momentum direction, which is the orbit normal. Note that the perifocal system is specific to a particular orbit, and that $\mathbf{i}_e$ and $\mathbf{i}_p$ are in the orbital plane. The line of nodes vector is given by the intersection of the reference plane (e.g. the Earth's equatorial plane) and the orbital plane.

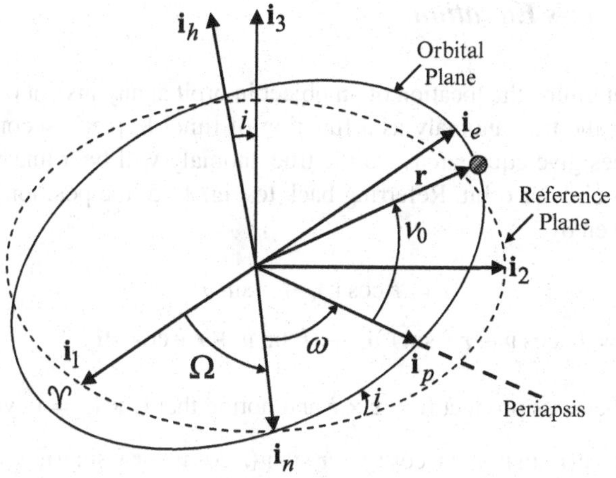

Fig. 10.8 Coordinate system geometry and orbital elements

The transformation from GCI to perifocal coordinates is given by

$$\mathbf{i}_e = A_{11}\mathbf{i}_1 + A_{12}\mathbf{i}_2 + A_{13}\mathbf{i}_3 \tag{10.47a}$$

$$\mathbf{i}_p = A_{21}\mathbf{i}_1 + A_{22}\mathbf{i}_2 + A_{23}\mathbf{i}_3, \tag{10.47b}$$

$$\mathbf{i}_h = A_{31}\mathbf{i}_1 + A_{32}\mathbf{i}_2 + A_{33}\mathbf{i}_3 \tag{10.47c}$$

where A_{ij} are elements of a rotation matrix. The transformation is accomplished using a 3−1−3 Euler rotation, with rotation matrix elements given by

$$A_{11} = \cos\Omega\cos\omega - \sin\Omega\sin\omega\cos i \tag{10.48a}$$

$$A_{12} = \sin\Omega\cos\omega + \cos\Omega\sin\omega\cos i \tag{10.48b}$$

$$A_{13} = \sin\omega\sin i \tag{10.48c}$$

$$A_{21} = -\cos\Omega\sin\omega - \sin\Omega\cos\omega\cos i \tag{10.48d}$$

$$A_{22} = -\sin\Omega\sin\omega + \cos\Omega\cos\omega\cos i \tag{10.48e}$$

$$A_{23} = \cos\omega\sin i \tag{10.48f}$$

$$A_{31} = \sin\Omega\sin i \tag{10.48g}$$

$$A_{32} = -\cos\Omega\sin i \tag{10.48h}$$

$$A_{33} = \cos i \tag{10.48i}$$

The inverse relationships are given in Sect. 10.2.3.

10.2.2 Kepler's Equation

We wish to determine the location of an object in orbit at any instant of time, which means finding the true anomaly as a function of time. Kepler's second law states that equal times give equal areas, so the true anomaly will be a linear function of time only in a circular orbit. Referring back to Fig. 10.5, the position and velocity vectors are given by

$$\mathbf{r} = r\cos\nu\,\mathbf{i}_e + r\sin\nu\,\mathbf{i}_p \tag{10.49a}$$

$$\dot{\mathbf{r}} = (\dot{r}\cos\nu - r\,\dot{\nu}\sin\nu)\mathbf{i}_e + (\dot{r}\sin\nu + r\,\dot{\nu}\cos\nu)\mathbf{i}_p \tag{10.49b}$$

Performing the cross product $\mathbf{h} = \mathbf{r} \times \dot{\mathbf{r}}$ and noting that $\mathbf{i}_e \times \mathbf{i}_p = \mathbf{i}_h$ yields

$$\mathbf{h} = [r\cos\nu(\dot{r}\sin\nu + r\,\dot{\nu}\cos\nu) - r\sin\nu(\dot{r}\cos\nu - r\,\dot{\nu}\sin\nu)]\,\mathbf{i}_h = r^2\dot{\nu}\,\mathbf{i}_h \tag{10.50}$$

Thus $h = r^2\dot{\nu}$, which shows that $r^2\dot{\nu}$ must be a constant. This equation can be written as

$$h\,dt = r^2\,d\nu \tag{10.51}$$

Substituting Eqs. (10.39) and (10.40) into Eq. (10.51) gives

$$\sqrt{\frac{\mu}{p^3}}\,dt = \frac{d\nu}{(1 + e\cos\nu)^2} \tag{10.52}$$

This equation needs to be integrated to determine the true anomaly difference to go from some initial time t_0 to another time t_1:

$$\sqrt{\frac{\mu}{p^3}}(t_1 - t_0) = \int_{\nu_0}^{\nu_1} \frac{d\nu}{(1 + e\cos\nu)^2} \tag{10.53}$$

The right hand side of Eq. (10.53) involves finding a solution to a nonstandard elliptic integral. Unfortunately, no closed-form solution exists. Kepler essentially performed a change of variables to convert the integral equation to an algebraic equation.

Kepler projected the object's position vertically onto a circle circumscribed around the ellipse, as shown in Fig. 10.9. Note that the origin in this figure is at the focus, rather than at the center of the ellipse, as had been assumed in Sect. 10.1. The remainder of this chapter will assume that the origin is at the focus. The angle E is known as the *eccentric anomaly*. It is clear from the figure that

$$x = a\cos E - c = a(\cos E - e) \tag{10.54}$$

Then

$$a\cos E = c + x = ae + \frac{a(1 - e^2)}{1 + e\cos\nu}\cos\nu = \frac{a(e + \cos\nu)}{1 + e\cos\nu} \tag{10.55}$$

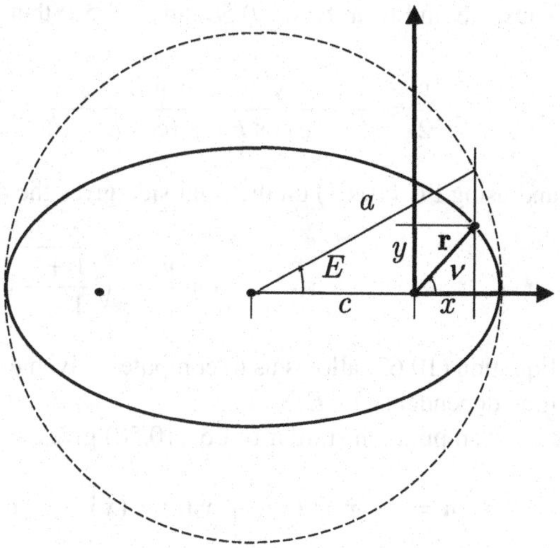

Fig. 10.9 Kepler's geometry

Solving this equation for $\cos \nu$ gives

$$\cos \nu = \frac{\cos E - e}{1 - e \cos E} \tag{10.56}$$

and then substituting this into $r = x/\cos \nu$ gives

$$r = a(1 - e \cos E) \tag{10.57}$$

It follows from Eq. (10.56) that

$$\sin^2 \nu = 1 - \cos^2 \nu = \frac{(1 - e \cos E)^2 - (\cos E - e)^2}{(1 - e \cos E)^2} = \frac{(1 - e^2) \sin^2 E}{(1 - e \cos E)^2} \tag{10.58}$$

Figure 10.9 also shows that $\sin \nu$ has the same sign as $\sin E$, so

$$\sin \nu = \frac{\sqrt{1 - e^2} \sin E}{1 - e \cos E} \tag{10.59}$$

Combining Eqs. (10.57) and (10.59) gives

$$y = r \sin \nu = a\sqrt{1 - e^2} \sin E \tag{10.60}$$

We now use Eqs. (10.54), (10.57), and (10.60) to find a relationship between the eccentric anomaly and true anomaly. Trigonometric half-angle identities for any angle θ give

$$\tan \frac{\theta}{2} = \frac{\sin(\theta/2)}{\cos(\theta/2)} = \frac{2 \sin(\theta/2) \cos(\theta/2)}{2 \cos^2(\theta/2)} = \frac{\sin \theta}{1 + \cos \theta} \tag{10.61}$$

Thus, we find from Eqs. (10.56) and (10.59) that

$$\tan\frac{\nu}{2} = \frac{\sqrt{1-e^2}\sin E}{(1-e\cos E)+(\cos E-e)} = \sqrt{\frac{1+e}{1-e}}\,\frac{\sin E}{1+\cos E} \tag{10.62}$$

and using Eq. (10.61) on the right side gives the desired result

$$\tan\frac{\nu}{2} = \sqrt{\frac{1+e}{1-e}}\tan\frac{E}{2} \tag{10.63}$$

Equation (10.63) allows us to compute ν given e and E, but we still need to find the time dependence of E.

A computation similar to Eq. (10.50) gives

$$\mathbf{h} = \mathbf{r}\times\dot{\mathbf{r}} = (x\,\mathbf{i}_e + y\,\mathbf{i}_p)\times(\dot{x}\,\mathbf{i}_e + \dot{y}\,\mathbf{i}_p) = (x\,\dot{y} - y\,\dot{x})\mathbf{i}_h \tag{10.64}$$

Thus $h = x\,\dot{y} - y\,\dot{x}$. Substituting Eqs. (10.54) and (10.60), along with their time derivatives, into h leads to

$$\begin{aligned} h &= a^2(\cos E - e)\sqrt{1-e^2}\cos E\,\frac{dE}{dt} + a^2\sqrt{1-e^2}\sin^2 E\,\frac{dE}{dt} \\ &= a^2\sqrt{1-e^2}(1-e\cos E)\frac{dE}{dt} \end{aligned} \tag{10.65}$$

Substituting $h = \sqrt{\mu p} = \sqrt{\mu a(1-e^2)}$ into this gives

$$\sqrt{\frac{\mu}{a^3}} = n = (1-e\cos E)\frac{dE}{dt} = \frac{d}{dt}(E - e\sin E) \tag{10.66}$$

where n is the mean motion, defined by Eq. (10.43). Integrating both sides of Eq. (10.66) easily leads to

$$n(t_1 - t_0) = (E - e\sin E)\big|_{E_0}^{E_1} \tag{10.67}$$

Note that if we set $E_0 = 0$ and $E_1 = 2\pi$, then we get Kepler's third law! The mean anomaly at a general time t is given by

$$M(t) = M_0 + n(t - t_0) \tag{10.68}$$

where $M_0 = E_0 - e\sin E_0$ is mean anomaly at time t_0. Then Eq. (10.67) can written as

$$M(t) = E(t) - e\sin E(t) \tag{10.69}$$

which is known as *Kepler's equation*. Given M_0 and some time past t_0, our goal is to determine $E(t)$ using Eqs. (10.68) and (10.69). Unfortunately, Kepler's equation has no closed-form solution. Kepler's equations has intrigued mathematicians for centuries and spawned a multitude of mathematical techniques [7]. Kepler's equation is well suited for Newton-Raphson's iteration. Define the following function:

$$f(E) \equiv M - (E - e\sin E) \tag{10.70}$$

which is zero if the correct E is found. A series expansion of Eq. (10.69) gives the following approximation for E, which is accurate up to third-order in the eccentricity [4]:

$$E = M + \frac{e\sin M}{1 - e\cos M} - \frac{1}{2}\left(\frac{e\sin M}{1 - e\cos M}\right)^3 + \cdots \tag{10.71}$$

Begin Newton's iteration by starting with some initial guess of E, denoted by $\hat{E}$, which can be given by M for small e or by Eq. (10.71). Then compute the Newton correction

$$\Delta E = -\frac{f(\hat{E})}{f'(\hat{E})} = \frac{M - (\hat{E} - e\sin\hat{E})}{1 - e\cos\hat{E}} \tag{10.72}$$

Next update the current estimate using

$$\hat{E} \leftarrow \hat{E} + \Delta E \tag{10.73}$$

where $\leftarrow$ denotes replacement. Continue iterating until ΔE is below some predefined threshold.

10.2.3 Orbit Propagation

The classical solution is the following: given the classical orbital elements at an epoch time t_0, compute the position and velocity vectors at any time t. The classical solution involving E is summarized in Table 10.1. Note that the position in perifocal coordinates is given by $[x,\ y,\ 0]$. Therefore, only the elements A_{11}, A_{21}, A_{12}, A_{22}, A_{13}, and A_{23} from Eq. (10.48) are needed to convert from perifocal coordinates to GCI coordinates, since

$$\mathbf{r} = \begin{bmatrix} A_{11} & A_{12} & A_{13} \\ A_{21} & A_{22} & A_{23} \\ A_{31} & A_{32} & A_{33} \end{bmatrix}^T \begin{bmatrix} x \\ y \\ 0 \end{bmatrix} \tag{10.74a}$$

$$\dot{\mathbf{r}} = \begin{bmatrix} A_{11} & A_{12} & A_{13} \\ A_{21} & A_{22} & A_{23} \\ A_{31} & A_{32} & A_{33} \end{bmatrix}^T \begin{bmatrix} \dot{x} \\ \dot{y} \\ 0 \end{bmatrix} \tag{10.74b}$$

Table 10.1 Classical orbital solution

$\{a,\ e,\ i,\ \Omega,\ \omega,\ M_0,\ (t-t_0)\} \Rightarrow \{\mathbf{r}(t),\ \dot{\mathbf{r}}(t)\}$	
Mean Anomaly	$n = \sqrt{\mu/a^3}$
	$M = M_0 + n(t - t_0)$
Kepler's Equation	$M = E - e\sin E$
	Solve for E
Compute r	$r = a(1 - e\cos E)$
Compute $x,\ y$	$x = a(\cos E - e)$
	$y = a\sqrt{1-e^2}\sin E$
Compute $\dot{x},\ \dot{y}$	$\dot{x} = -(na^2/r)\sin E$
	$\dot{y} = (na^2/r)\sqrt{1-e^2}\cos E$
Position	$\mathbf{r} = \begin{bmatrix} A_{11} & A_{21} \\ A_{12} & A_{22} \\ A_{13} & A_{23} \end{bmatrix} \begin{bmatrix} x \\ y \end{bmatrix}$
Velocity	$\dot{\mathbf{r}} = \begin{bmatrix} A_{11} & A_{21} \\ A_{12} & A_{22} \\ A_{13} & A_{23} \end{bmatrix} \begin{bmatrix} \dot{x} \\ \dot{y} \end{bmatrix}$

Note that these rotation matrix elements are time-invariant, which means that they only need to be computed once and then stored. First, the mean motion, n, and mean anomaly, M, are computed. Then, Kepler's equation is solved for E. Next, the variables $x,\ y,\ \dot{x},\ \dot{y}$, and r are computed. Finally, the position and velocity vectors at time t are determined.

Another form of the classical solution is possible by substituting $x = r\cos\nu$ and $y = r\sin\nu$ into Eq. (10.74a), which gives

$$\mathbf{r} = r \begin{bmatrix} \cos\Omega\cos\theta - \sin\Omega\sin\theta\cos i \\ \sin\Omega\cos\theta + \cos\Omega\sin\theta\cos i \\ \sin\theta\sin i \end{bmatrix} \tag{10.75}$$

where $\theta = \omega + \nu$. The time derivative of Eq. (10.75) can be shown to be

$$\dot{\mathbf{r}} = -\sqrt{\frac{\mu}{p}} \begin{bmatrix} (\sin\theta + e\sin\omega)\cos\Omega + (\cos\theta + e\cos\omega)\sin\Omega\cos i \\ (\sin\theta + e\sin\omega)\sin\Omega - (\cos\theta + e\cos\omega)\cos\Omega\cos i \\ -(\cos\theta + e\cos\omega)\sin i \end{bmatrix} \tag{10.76}$$

For this solution once E has been determined using Kepler's equation, then ν is determined using Eq. (10.63). The parameters r can be given by either Eq. (10.21) or Eq. (10.57). Also, the constant p is computed using Eq. (10.15).

The inverse problem involves determining the orbital elements from the spacecraft position $\mathbf{r}(t)$ and velocity $\mathbf{v}(t)$ vectors at some time t. First, the following quantities are computed:

$$r = \|\mathbf{r}(t)\|, \quad v = \|\mathbf{v}(t)\| \tag{10.77a}$$

$$\mathbf{h} = \mathbf{r}(t) \times \mathbf{v}(t), \quad h = \|\mathbf{h}\| \tag{10.77b}$$

$$\frac{1}{a} = \frac{2}{r} - \frac{v^2}{\mu} \tag{10.77c}$$

$$\mathbf{e} = \frac{\mathbf{v}(t) \times \mathbf{h}}{\mu} - \frac{\mathbf{r}(t)}{r}, \quad e = \|\mathbf{e}\| \tag{10.77d}$$

Equation (10.77c), derived directly from the *vis viva* equation, is used to determine a. The perifocal coordinates vectors can now be computed, so that

$$\mathbf{i}_h = \mathbf{h}/h = A_{31}\mathbf{i}_1 + A_{32}\mathbf{i}_2 + A_{33}\mathbf{i}_3 \tag{10.78a}$$

$$\mathbf{i}_e = \mathbf{e}/e = A_{11}\mathbf{i}_1 + A_{12}\mathbf{i}_2 + A_{13}\mathbf{i}_3 \tag{10.78b}$$

$$\mathbf{i}_p = \mathbf{i}_h \times \mathbf{i}_e = A_{21}\mathbf{i}_1 + A_{22}\mathbf{i}_2 + A_{23}\mathbf{i}_3 \tag{10.78c}$$

which can be used to extract any desired A_{ij}. The inclination, i, right ascension of the ascending node, Ω, argument of periapsis, ω, are computed by

$$i = \cos^{-1} A_{33}, \quad 0 \le i < \pi \tag{10.79a}$$

$$\Omega = \text{atan2}(A_{31}, -A_{32}), \quad 0 \le \Omega < 2\pi \tag{10.79b}$$

$$\omega = \text{atan2}(A_{13}, A_{23}), \quad 0 \le \omega < 2\pi \tag{10.79c}$$

Referring to Table 10.1 and Eq. (10.57), we see that

$$\begin{aligned} \mathbf{r}(t) \cdot \mathbf{v}(t) &= x\,\dot{x} + y\,\dot{y} = (na^3/r)[-(\cos E - e)\sin E + (1-e^2)\sin E \cos E] \\ &= (na^3/r)e(1 - e\cos E)\sin E = na^2 e \sin E(t) \end{aligned} \tag{10.80}$$

where many time arguments of $E(t)$ have been omitted to save space. Solving Eq. (10.57) for $e\cos E(t)$ gives

$$e\cos E(t) = 1 - \frac{r}{a} \tag{10.81}$$

Equations (10.80) and (10.81) then give the eccentric anomaly at time t as

$$E(t) = \text{atan2}\left(\frac{\mathbf{r}(t)\cdot\mathbf{v}(t)}{na^2}, 1 - \frac{r}{a}\right), \quad 0 \le E(t) < 2\pi \tag{10.82}$$

The mean anomaly at time t is determined by Kepler's equation

$$M(t) = E(t) - e \sin E(t) \tag{10.83}$$

and the mean anomaly at the epoch time t_0 is given by

$$M_0 = M(t) - n(t - t_0) \tag{10.84}$$

Note that the Kepler elements a, e, i, Ω, ω, and M_0 are independent of the time t at which $\mathbf{r}(t)$ and $\mathbf{v}(t)$ are determined. Note also that quadrants are important, which is why the atan2 function is used in Eqs. (10.78) and (10.82).

10.3 Disturbing Forces[1]

A spacecraft is affected by many disturbing accelerations, or perturbations. Four important disturbance sources are gravity due to a non-spherical Earth,[2] denoted by $\mathbf{a}_{\text{grav}}$; forces due to bodies other than the central body, denoted by $\mathbf{a}_{\text{third}}$; aerodynamic drag, denoted by $\mathbf{a}_{\text{aero}}$; and solar radiation pressure (SRP), denoted by $\mathbf{a}_{\text{SRP}}$. The modification of Eq. (10.26) include these disturbances is

$$\ddot{\mathbf{r}} = -\frac{\mu}{r^3}\mathbf{r} + \mathbf{a}_{\text{perturb}} = -\frac{\mu}{r^3}\mathbf{r} + \mathbf{a}_{\text{grav}} + \mathbf{a}_{\text{third}} + \mathbf{a}_{\text{aero}} + \mathbf{a}_{\text{SRP}} \tag{10.85}$$

Note that gravity and third-body disturbances are conservative, meaning that they can be computed as gradients of potential functions, while drag and SRP are not.

The most straightforward way to compute perturbed orbital motion is to integrate Eq. (10.85) numerically. This is known as the method of special perturbations, and can be the most accurate method, but also the most computationally expensive. The method of general perturbations, which is actually less general, was developed to speed the computation [10]. We provide a brief discussion of general perturbation theory in Sect. 10.4.1.

10.3.1 Non-Spherical Gravity

The Earth is not a perfect sphere, and mass is distributed nonuniformly throughout the Earth. Since gravity depends directly on mass, the gravity field will reflect this nonuniformity. The most common approach to model non-spherical gravity uses a

---

[1] The authors would like to thank Christopher K. Nebelecky for the contributions in this section.

[2] The generalization to any other central body is straightforward; we choose the Earth for specificity.

spherical harmonic expansion, which we will now develop. We consider a body such as the Earth to be built up of a large number, N, of points with mass m_i located at points $\mathbf{r}_i$ These give rise to a gravitational potential at a point $\mathbf{r}$ of

$$U(\mathbf{r}) = \sum_{i=1}^{N} \frac{Gm_i}{\|\mathbf{r} - \mathbf{r}_i\|} \tag{10.86}$$

where G is Newton's universal gravitational constant. In the Earth-Centered/Earth-Fixed (ECEF) frame

$$\mathbf{r} = r \begin{bmatrix} \cos\lambda' \cos\phi \\ \cos\lambda' \sin\phi \\ \sin\lambda' \end{bmatrix} \quad \text{and} \quad \mathbf{r}_i = r_i \begin{bmatrix} \cos\lambda'_i \cos\phi_i \\ \cos\lambda'_i \sin\phi_i \\ \sin\lambda'_i \end{bmatrix} \tag{10.87}$$

where λ' is the geocentric latitude[3] and ϕ is the longitude, so

$$\begin{aligned} \|\mathbf{r} - \mathbf{r}_i\|^{-1} &= (r^2 + r_i^2 - 2\mathbf{r} \cdot \mathbf{r}_i)^{-1/2} \\ &= \frac{1}{r}\left\{1 + \left(\frac{r_i}{r}\right)^2 - \frac{2r_i}{r}\left[\cos\lambda' \cos\lambda'_i \cos(\phi - \phi_i) + \sin\lambda' \sin\lambda'_i\right]\right\}^{-1/2} \end{aligned} \tag{10.88}$$

We separate the dependence on $\mathbf{r}$ from that on $\mathbf{r}_i$ by means of the identity [2]

$$\begin{aligned} &\left\{1 + \left(\frac{r_i}{r}\right)^2 - \frac{2r_i}{r}\left[\cos\lambda' \cos\lambda'_i \cos(\phi - \phi_i) + \sin\lambda' \sin\lambda'_i\right]\right\}^{-1/2} \\ &= \sum_{n=0}^{\infty} \left(\frac{r_i}{r}\right)^n \sum_{m=0}^{n} (2 - \delta_{0m}) \frac{(n-m)!}{(n+m)!} P_n^m(\sin\lambda') P_n^m(\sin\lambda'_i) \cos\left(m(\phi - \phi_i)\right) \end{aligned} \tag{10.89}$$

where δ_{0m} is the Kronecker delta and P_n^m, the associated Legendre function of degree n and order m, is defined in terms of the Legendre polynomials[4]

$$P_n(x) = \frac{1}{2^n n!} \frac{d^n}{dx^n}(x^2 - 1)^n \tag{10.90}$$

by[5]

$$P_n^m(x) = (1 - x^2)^{m/2} \frac{d^m}{dx^m} P_n(x) \tag{10.91}$$

---

[3] We use λ' for latitude to avoid confusion with the geodetic latitude of Sect. 2.6.3.

[4] Equation (10.90) is the well-known Rodrigues equation for the Legendre polynomials.

[5] We follow Vallado's notation [21]. Montenbruck and Gill [14] and Abramowitz and Stegun [1] denote these functions by P_{nm} and define $P_n^m = (-1)^m P_{nm}$.

The associated Legendre functions for degrees 0 to 4 are [21]

$$
\begin{aligned}
P_0^0(\sin\lambda') &= 1 & P_3^2(\sin\lambda') &= 15\cos^2\lambda'\sin\lambda' \\
P_1^0(\sin\lambda') &= \sin\lambda' & P_3^3(\sin\lambda') &= 15\cos^3\lambda' \\
P_1^1(\sin\lambda') &= \cos\lambda' & P_4^0(\sin\lambda') &= \frac{1}{8}(35\sin^4\lambda' - 30\sin^2\lambda' + 3) \\
P_2^0(\sin\lambda') &= \frac{1}{2}(3\sin^2\lambda' - 1) & P_4^1(\sin\lambda') &= \frac{5}{2}\cos\lambda'(7\sin^3\lambda' - 3\sin\lambda') \\
P_2^1(\sin\lambda') &= 3\sin\lambda'\cos\lambda' & P_4^2(\sin\lambda') &= \frac{15}{2}\cos^2\lambda'(7\sin^2\lambda' - 1) \\
P_2^2(\sin\lambda') &= 3\cos^2\lambda' & P_4^3(\sin\lambda') &= 105\cos^3\lambda'\sin\lambda' \\
P_3^0(\sin\lambda') &= \frac{1}{2}(5\sin^3\lambda' - 3\sin\lambda') & P_4^4(\sin\lambda') &= 105\cos^4\lambda' \\
P_3^1(\sin\lambda') &= \frac{3}{2}\cos\lambda'(5\sin^2\lambda' - 1) & &
\end{aligned}
\tag{10.92}
$$

Equations (10.90) and (10.91) are not useful for computation. The customary procedure is to compute P_0^0, P_1^0, and P_1^1 from their explicit expressions and then to find the associated Legendre functions of higher degree recursively by

$$P_n^0(\sin\lambda') = \frac{1}{n}\left[(2n-1)\sin\lambda' P_{n-1}^0(\sin\lambda') - (n-1)P_{n-2}^0(\sin\lambda')\right] \tag{10.93a}$$

$$P_n^m(\sin\lambda') = \sin\lambda' P_{n-1}^m(\sin\lambda') + (n+m-1)\cos\lambda' P_{n-1}^{m-1}(\sin\lambda'), \quad \text{for } 0 < m < n \tag{10.93b}$$

$$P_n^n(\sin\lambda') = (2n-1)\cos\lambda' P_{n-1}^{n-1}(\sin\lambda') \tag{10.93c}$$

or some equivalent recursion relations.

Putting Eqs. (10.86), (10.88), and (10.89) together, and noting that $P_0^0(x) = 1$, gives the *spherical harmonics expansion* for the gravity potential

$$U(\mathbf{r}) = \frac{\mu}{r}\left\{1 + \sum_{n=1}^{\infty}\left(\frac{R}{r}\right)^n \sum_{m=0}^{n} P_n^m(\sin\lambda')[C_n^m\cos(m\phi) + S_n^m\sin(m\phi)]\right\} \tag{10.94}$$

where $\mu \equiv GM$, with $M = \sum_{i=1}^{N} m_i$ being the total mass of all the particles in the central body and, for $n \geq 1$ and $0 \leq m \leq n$,

$$C_n^m = \frac{2-\delta_{0m}}{M}\frac{(n-m)!}{(n+m)!}\sum_{i=1}^{N} m_i\left(\frac{r_i}{R}\right)^n P_n^m(\sin\lambda_i')\cos(m\phi_i) \tag{10.95a}$$

$$S_n^m = \frac{2-\delta_{0m}}{M}\frac{(n-m)!}{(n+m)!}\sum_{i=1}^{N} m_i\left(\frac{r_i}{R}\right)^n P_n^m(\sin\lambda_i')\sin(m\phi_i) \tag{10.95b}$$

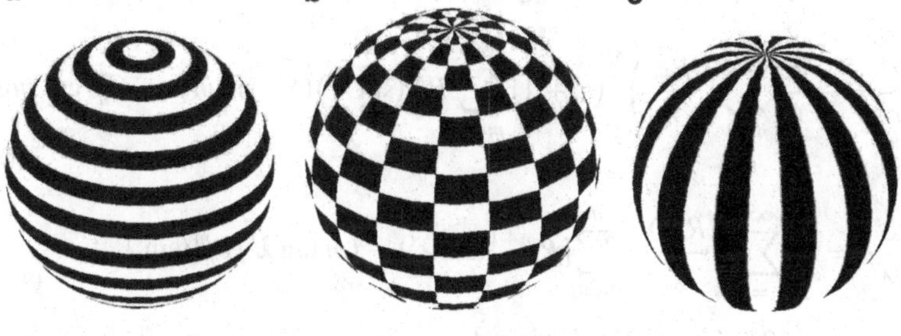

Fig. 10.10 Types of spherical harmonics. (**a**) Zonal. (**b**) Tesseral. (**c**) Sectoral

The parameter R characterizes the size of the mass distribution; for Earth gravity modeling, $M = M_\oplus$ and $R = R_\oplus$. Note that Eq. (10.95) is not used to compute C_n^m and S_n^m; they are found by fitting gravity measurement data, including tracking of satellite orbits.

The spherical harmonics fall into three classes: zonal harmonics for $m = 0$, tesseral harmonics for $0 < m < n$, and sectoral harmonics for $m = n$. These are illustrated in Fig. 10.10.

Note that Eq. (10.95) implies that $S_n^0 = 0$, so there are only $2n + 1$ nonzero terms in the spherical harmonics expansion for any degree n. We can eliminate some other coefficients by choosing our coordinate system wisely. Equations (10.95) and (10.87) say that

$$\begin{bmatrix} C_1^1 \\ S_1^1 \\ C_1^0 \end{bmatrix} = \frac{1}{MR} \sum_{i=1}^{N} m_i \mathbf{r}_i \tag{10.96}$$

Thus C_1^0, C_1^1, and S_1^1 will all vanish by Eq. (3.55) if we place the origin of our coordinate system at the central body's center of mass, as is done for ECEF coordinates. Furthermore, Eqs. (10.95), (10.87), and (3.69) say that $C_2^1 = -J_{13}/MR^2$ and $S_2^1 = -J_{23}/MR^2$, where J is the central body's moment of inertia tensor. Thus these coefficients also vanish if we choose the z axis of our coordinate frame to be a principal axis of inertia of the central body.

The acceleration due to gravity is determined by taking the gradient of the potential function in Eq. (10.94). The leading term μ/r in the potential function gives the first term on the right side of Eq. (10.85), so we denote the potential with this term omitted by U'. We also set the origin of the coordinate frame at the Earth's center of mass so that the $n = 1$ terms in Eq. (10.94) vanish by Eq. (10.96). The geopotential U' is expressed in terms of radial distance, geocentric latitude, and longitude, so we apply the chain rule to obtain acceleration due to the non-spherical Earth as

$$\mathbf{a}_{\text{grav}} = \nabla U'(\mathbf{r}) = \frac{\partial U'}{\partial r}\nabla r + \frac{\partial U'}{\partial \lambda'}\nabla \lambda' + \frac{\partial U'}{\partial \phi}\nabla \phi \tag{10.97}$$

The partials of the geopotential are given by

$$\frac{\partial U'}{\partial r} = -\frac{\mu}{r^2}\sum_{n=2}^{\infty}\left(\frac{R_\oplus}{r}\right)^n (n+1)\sum_{m=0}^{n} P_n^m(\sin\lambda')(S_n^m \sin m\phi + C_n^m \cos m\phi) \tag{10.98a}$$

$$\frac{\partial U'}{\partial \lambda'} = \frac{\mu}{r}\sum_{n=2}^{\infty}\left(\frac{R_\oplus}{r}\right)^n \sum_{m=0}^{n}[P_n^{m+1}(\sin\lambda') - (m\tan\lambda')P_n^m(\sin\lambda')] \times (S_n^m \sin m\phi + C_n^m \cos m\phi) \tag{10.98b}$$

$$\frac{\partial U'}{\partial \phi} = \frac{\mu}{r}\sum_{n=2}^{\infty}\left(\frac{R_\oplus}{r}\right)^n \sum_{m=0}^{n} m P_n^m(\sin\lambda')(S_n^m \cos m\phi - C_n^m \sin m\phi) \tag{10.98c}$$

The gradients of the radius, latitude, and longitude are given by

$$\nabla r = \frac{\mathbf{r}}{r} \tag{10.99a}$$

$$\nabla \lambda' = \frac{1}{\sqrt{x^2+y^2}}\left(-z\frac{\mathbf{r}}{r^2} + \nabla z\right) \tag{10.99b}$$

$$\nabla \phi = \frac{1}{x^2+y^2}(x\,\nabla y - y\,\nabla x) \tag{10.99c}$$

where

$$\nabla x = \begin{bmatrix}1\\0\\0\end{bmatrix}, \quad \nabla y = \begin{bmatrix}0\\1\\0\end{bmatrix}, \quad \nabla z = \begin{bmatrix}0\\0\\1\end{bmatrix} \tag{10.100}$$

The Cartesian coordinates of the acceleration are then given by

$$a_{\text{grav}_x} = \left(\frac{1}{r}\frac{\partial U'}{\partial r} - \frac{z}{r^2\sqrt{x^2+y^2}}\frac{\partial U'}{\partial \lambda'}\right)x - \left(\frac{1}{x^2+y^2}\frac{\partial U'}{\partial \phi}\right)y \tag{10.101a}$$

$$a_{\text{grav}_y} = \left(\frac{1}{r}\frac{\partial U'}{\partial r} - \frac{z}{r^2\sqrt{x^2+y^2}}\frac{\partial U'}{\partial \lambda'}\right)y + \left(\frac{1}{x^2+y^2}\frac{\partial U'}{\partial \phi}\right)x \tag{10.101b}$$

$$a_{\text{grav}_z} = \left(\frac{1}{r}\frac{\partial U'}{\partial r}\right)z + \frac{\sqrt{x^2+y^2}}{r^2}\frac{\partial U'}{\partial \lambda'} \tag{10.101c}$$

As an example we consider a 6th-order spherical harmonic geopotential model including only the zonal harmonics. Zonal harmonics are especially simple because they are symmetrical about the polar axis, i.e. there is no dependence on the longitude or on the Greenwich hour angle. A specific notation, $J_n \equiv -C_n^0$, is used

Table 10.2 Zonal coefficients

i	J_i	$\times 10$
2	1.08262668355	−3
3	−2.53265648533	−6
4	−1.61962159137	−6
5	−2.27296082869	−7
6	5.40681239107	−7

for these coefficients. Values for the zonal coefficients of the Earth can be found in Table 10.2 [21]. By far the strongest perturbation due to the Earth's shape arises from J_2, with J_3 being more than 400 times smaller.

The perturbing acceleration is given by

$$\mathbf{a}_{\text{grav}} = \sum_{i=2}^{6} \mathbf{a}_{J_i} \tag{10.102}$$

where the individual terms are [17]

$$\mathbf{a}_{J_2} = -\frac{3}{2} J_2 \left(\frac{\mu}{r^2}\right) \left(\frac{R_\oplus}{r}\right)^2 \begin{bmatrix} \left(1 - 5\left(\frac{z}{r}\right)^2\right) \frac{x}{r} \\ \left(1 - 5\left(\frac{z}{r}\right)^2\right) \frac{y}{r} \\ \left(3 - 5\left(\frac{z}{r}\right)^2\right) \frac{z}{r} \end{bmatrix} \tag{10.103a}$$

$$\mathbf{a}_{J_3} = -\frac{1}{2} J_3 \left(\frac{\mu}{r^2}\right) \left(\frac{R_\oplus}{r}\right)^3 \begin{bmatrix} 5\left(7\left(\frac{z}{r}\right)^3 - 3\left(\frac{z}{r}\right)\right) \frac{x}{r} \\ 5\left(7\left(\frac{z}{r}\right)^3 - 3\left(\frac{z}{r}\right)\right) \frac{y}{r} \\ 3\left(10\left(\frac{z}{r}\right)^2 - \frac{35}{3}\left(\frac{z}{r}\right)^4 - 1\right) \end{bmatrix} \tag{10.103b}$$

$$\mathbf{a}_{J_4} = -\frac{5}{8} J_4 \left(\frac{\mu}{r^2}\right) \left(\frac{R_\oplus}{r}\right)^4 \begin{bmatrix} \left(3 - 42\left(\frac{z}{r}\right)^2 + 63\left(\frac{z}{r}\right)^4\right) \frac{x}{r} \\ \left(3 - 42\left(\frac{z}{r}\right)^2 + 63\left(\frac{z}{r}\right)^4\right) \frac{y}{r} \\ -\left(15 - 70\left(\frac{z}{r}\right)^2 + 63\left(\frac{z}{r}\right)^4\right) \frac{z}{r} \end{bmatrix} \tag{10.103c}$$

$$\mathbf{a}_{J_5} = -\frac{1}{8} J_5 \left(\frac{\mu}{r^2}\right) \left(\frac{R_\oplus}{r}\right)^5 \begin{bmatrix} 3\left(35\left(\frac{z}{r}\right) - 210\left(\frac{z}{r}\right)^3 + 231\left(\frac{z}{r}\right)^5\right) \frac{x}{r} \\ 3\left(35\left(\frac{z}{r}\right) - 210\left(\frac{z}{r}\right)^3 + 231\left(\frac{z}{r}\right)^5\right) \frac{y}{r} \\ \left(15 - 315\left(\frac{z}{r}\right)^2 + 945\left(\frac{z}{r}\right)^4 - 693\left(\frac{z}{r}\right)^6\right) \end{bmatrix} \tag{10.103d}$$

$$\mathbf{a}_{J_6} = \frac{1}{16} J_6 \left(\frac{\mu}{r^2}\right) \left(\frac{R_\oplus}{r}\right)^6 \begin{bmatrix} \left(35 - 945\left(\frac{z}{r}\right)^2 + 3465\left(\frac{z}{r}\right)^4 - 3003\left(\frac{z}{r}\right)^6\right) \frac{x}{r} \\ \left(35 - 945\left(\frac{z}{r}\right)^2 + 3465\left(\frac{z}{r}\right)^4 - 3003\left(\frac{z}{r}\right)^6\right) \frac{y}{r} \\ \left(2205\left(\frac{z}{r}\right)^2 - 4851\left(\frac{z}{r}\right)^4 + 3003\left(\frac{z}{r}\right)^6 - 315\right) \frac{z}{r} \end{bmatrix} \tag{10.103e}$$

The gravitational parameter μ and mean equatorial Earth radius, $R_\oplus$ are given by

$$\mu = 3.986004418 \times 10^{14} \ \frac{\text{m}^3}{\text{s}^2} \tag{10.104a}$$

$$R_\oplus = 6.378\ 137 \times 10^6 \ \text{m} \tag{10.104b}$$

10.3.2 Third-Body Forces

The term *third-body* is usually applied to any body other than the two bodies producing two-body Keplerian motion. Consider the motion of a body with mass m_2 at position $\mathbf{r}_2$ about a body with mass m_1 at position $\mathbf{r}_1$, under the influence of $n-2$ other bodies with masses m_i at positions $\mathbf{r}_i$

$$\ddot{\mathbf{r}}_1 = -\frac{Gm_2}{\|\mathbf{r}_1 - \mathbf{r}_2\|^3}(\mathbf{r}_1 - \mathbf{r}_2) - \sum_{i=3}^{N} \frac{Gm_i}{\|\mathbf{r}_1 - \mathbf{r}_i\|^3}(\mathbf{r}_1 - \mathbf{r}_i) \tag{10.105a}$$

$$\ddot{\mathbf{r}}_2 = -\frac{Gm_1}{\|\mathbf{r}_1 - \mathbf{r}_2\|^3}(\mathbf{r}_2 - \mathbf{r}_1) - \sum_{i=3}^{N} \frac{Gm_i}{\|\mathbf{r}_2 - \mathbf{r}_i\|^3}(\mathbf{r}_2 - \mathbf{r}_i) \tag{10.105b}$$

As in the two-body case, subtracting Eq. (10.105a) from Eq. (10.105b) gives, with $\mathbf{r} \equiv \mathbf{r}_2 - \mathbf{r}_1$, $r \equiv \|\mathbf{r}\|$, $\mu \equiv G(m_1 + m_2)$, and $\mu_i \equiv Gm_i$ for $i \geq 3$,

$$\ddot{\mathbf{r}} = -\frac{\mu}{r^3}\mathbf{r} - \sum_{i=3}^{N} \mu_i \left(\frac{\mathbf{r}_1 - \mathbf{r}_i + \mathbf{r}}{\|\mathbf{r}_1 - \mathbf{r}_i + \mathbf{r}\|^3} - \frac{\mathbf{r}_1 - \mathbf{r}_i}{\|\mathbf{r}_1 - \mathbf{r}_i\|^3} \right) \tag{10.106}$$

The terms in the sum can be treated as small perturbations if all the other bodies are much farther away from m_1 than is m_2, i.e. if $\|\mathbf{r}_1 - \mathbf{r}_i\| \ll r$ for all $i \geq 3$. That is why the gravitational force of the Sun can be treated as a perturbation when analyzing the motion of a spacecraft about the Earth.

Let us now expand the first term in the parentheses of Eq. (10.106) in powers of $\mathbf{r}/\|\mathbf{r}_1 - \mathbf{r}_i\|$ and retain only the first-order term. With

$$\begin{aligned} \|\mathbf{r}_1 - \mathbf{r}_i + \mathbf{r}\|^{-3} &= \left[\|\mathbf{r}_1 - \mathbf{r}_i\|^2 + 2\mathbf{r} \cdot (\mathbf{r}_1 - \mathbf{r}_i) + r^2\right]^{-3/2} \\ &\approx \|\mathbf{r}_1 - \mathbf{r}_i\|^{-3}\left[1 - 3\mathbf{r} \cdot (\mathbf{r}_1 - \mathbf{r}_i)/\|\mathbf{r}_1 - \mathbf{r}_i\|^2\right] \end{aligned} \tag{10.107}$$

this gives

$$\ddot{\mathbf{r}} = -\frac{\mu}{r^3}\mathbf{r} - \sum_{i=3}^{N} \mu_i \frac{\mathbf{r} - 3(\mathbf{u}_i \cdot \mathbf{r})\mathbf{u}_i}{\|\mathbf{r}_1 - \mathbf{r}_i\|^3} \tag{10.108}$$

where $\mathbf{u}_i \equiv (\mathbf{r}_1 - \mathbf{r}_i)/\|\mathbf{r}_1 - \mathbf{r}_i\|$. However, it is often better to use Eq. (10.106) than this approximation. For instance, higher order terms in the expansion for the perturbation of the Moon on a spacecraft in geosynchronous orbit are not negligible, because $r/\|\mathbf{r}_1 - \mathbf{r}_i\| \approx 1/9$ in that case.

10.3.3 Atmospheric Drag

For objects in low-Earth orbit, atmospheric drag represents a significant perturbing force. The drag force that a spacecraft experiences is given by

$$\mathbf{F}_{\text{aero}} = -\frac{1}{2}\rho\, C_D S \|\mathbf{v}_{\text{rel}}\| \mathbf{v}_{\text{rel}} \tag{10.109}$$

where ρ is the local atmospheric density, which is discussed in Sect. 11.2, C_D is a dimensionless drag coefficient, S is the spacecraft area projected along the direction of motion, and $\mathbf{v}_{\text{rel}}$ is the relative velocity of the spacecraft with respect to the atmosphere. As discussed in Sect. 3.3.6.3, the relative velocity, $\mathbf{v}_{\text{rel}}$, is different from the GCI velocity of the spacecraft because the atmosphere is not stationary in the GCI frame. The equations accounting for atmospheric motion can be found in that section.

The drag coefficient, C_D, is a dimensionless parameter that quantifies how a spacecraft interacts with the surrounding medium to retard its motion. The drag coefficient is a function of the shape of the spacecraft, its orientation with respect to $\mathbf{v}_{\text{rel}}$, its surface properties, and the composition of the atmosphere. Often, drag coefficients are determined via experimental and/or finite element analysis. Experiments are in their own right situation-dependent as they consider only a single composition for the surrounding medium. Vallado [23] points out that the drag coefficient for several common shapes can increase significantly simply by changing the orbital altitude. This is because the composition of the atmosphere changes with altitude and the surface materials of the spacecraft will, in general, interact with the different compositions differently.

The projected area of a spherical spacecraft is unchanging, but S depends on the spacecraft attitude for all other shapes. It is typical to model the geometry of non-spherical spacecraft as a collection of N flat plates of area S_i and outward normal unit vector $\mathbf{n}_B^i$ expressed in the spacecraft body-fixed coordinate system. The inclination of the ith plate to the relative velocity is given by

$$\cos\theta_{\text{aero}}^i = \left(A^T \mathbf{n}_B^i\right)^T \left(\frac{\mathbf{v}_{\text{rel}}}{\|\mathbf{v}_{\text{rel}}\|}\right) \tag{10.110}$$

where A is the attitude matrix that rotates the GCI frame to the spacecraft body-fixed frame. The drag force in this model is

$$\mathbf{F}_{\text{aero}} = -\frac{1}{2}\rho\, C_D \|\mathbf{v}_{\text{rel}}\| \mathbf{v}_{\text{rel}} \sum_{i=1}^{N}{}' S_i \cos\theta_{\text{aero}}^i \tag{10.111}$$

where the prime on the sum indicates that only plates with $\cos\theta^i_{\text{aero}} > 0$ are included in the summation. Note that this algorithm does not account for potential self-shielding that would exist on concave spacecraft.

It is typical to group the drag coefficient, projected frontal area, and mass into a single term called the ballistic coefficient, B:

$$B = \frac{m}{C_D S} \tag{10.112}$$

The ballistic coefficient describes the relative magnitude of the effect of drag to inertial forces and gravity. A spacecraft with a low ballistic coefficient will be more susceptible to effects caused by drag than a spacecraft with a high ballistic coefficient. Most orbit determination algorithms can also refine an initial estimate of the ballistic coefficient. Note that some texts define the ballistic coefficient as the inverse of that in Eq. (10.112).

10.3.4 Solar Radiation Pressure

Solar radiation pressure (SRP) is another non-conservative force acting on spacecraft. It is dominated by drag for spacecraft in low-Earth orbit, but SRP will generally outweigh drag in higher altitude orbits ($\geq$ 800 km). Like drag, SRP can be characterized using either a simple or high fidelity model depending upon the level of accuracy needed as well as a priori knowledge of the spacecraft. The mechanism by which SRP affects the orbit of a spacecraft is through momentum exchange between the spacecraft and photons incident on the spacecraft. Because of this, SRP is a fundamentally different perturbation from that of drag. Whereas drag acts throughout the entire orbit, SRP only contributes as times when the spacecraft is not in the shadow of the Earth or another body.

The simplest of SRP models assumes that the force on the spacecraft due to solar radiation can be characterized as [21]

$$\mathbf{F}_{\text{SRP}} = -P_{\odot}\, c_{\text{SRP}}\, S\, \mathbf{e}_{\text{sat}\odot} \tag{10.113}$$

where $P_{\odot}$ is the pressure of solar radiation, S is the Sun-facing area of the spacecraft, and $\mathbf{e}_{\text{sat}\odot}$ is a unit vector directed from the spacecraft to the center of the Sun, expressed in inertial coordinates. Section 11.3 presents methods for computing the Sun position, solar pressure, and conditions for shadowing. The constant c_{SRP} defines how the incident radiation interacts with the spacecraft. A value of $c_{\text{SRP}} = 0.0$ implies that the spacecraft is transparent, so not affected by any incoming radiation. When $c_{\text{SRP}} = 2.0$, the spacecraft acts like a mirror with all incident radiation perfectly reflected directly back to the Sun. While c_{SRP} can vary for a single spacecraft depending on its attitude, for most practical problems a value of c_{SRP} between 1.0 and 2.0 is sufficient for preliminary simulations.

More sophisticated methods can be used to obtain more accurate estimates of SRP if detailed knowledge of the spacecraft is available. These methods decompose the SRP into contributions due to specular reflection, diffuse reflection, absorption, and emission. Suppose that spacecraft's surface can be modeled as a collection of N flat plates of area S_i, outward normal $\mathbf{n}_B^i$ in the body coordinate frame, specular reflection coefficient R_{spec}^i, diffuse reflection coefficient R_{diff}^i, and absorption coefficient R_{abs}^i. The coefficients sum to unity; $R_{\text{spec}}^i + R_{\text{diff}}^i + R_{\text{abs}}^i = 1$. Diffuse reflection is assumed to be Lambertian, which means that the intensity of the reflected light in any direction is proportional to the cosine of the angle between the reflection direction and the normal.

The inclination of the ith plate to the spacecraft-to-Sun vector is given by

$$\cos\theta_{\text{SRP}}^i = \left(\mathbf{n}_I^i\right)^T \mathbf{e}_{\text{sat}\odot} = \left(A^T \mathbf{n}_B^i\right)^T \mathbf{e}_{\text{sat}\odot} \tag{10.114}$$

where A is the attitude matrix that rotates the GCI frame to the spacecraft body-fixed frame. The force due to SRP can then be expressed as [21]

$$\mathbf{F}_{\text{SRP}} = -P_\odot \sum_{i=1}^{N}{}' S_i \cos\theta_{\text{SRP}}^i \left[2\left(\frac{R_{\text{diff}}^i}{3} + R_{\text{spec}}^i \cos\theta_{\text{SRP}}^i \right) \mathbf{n}_I^i + (1 - R_{\text{spec}}^i)\mathbf{e}_{\text{sat}\odot} \right] \tag{10.115}$$

where the prime on the sum indicates that only plates with $\cos\theta_{\text{SRP}}^i > 0$ are included in the summation.

Equation (10.115) provides a good approximation for the SRP acting on a spacecraft of basic geometry, but it has several limitations. First, it should be mentioned that the Sun is not the only source of radiation, although it is by far the largest for Earth-orbiting spacecraft. Reflected light from the Earth or the Moon, called *albedo*, can be significant if very precise dynamical modeling is required; and models incorporating this effect have been developed [5].

Secondly, the force due to thermal radiation emitted from the spacecraft has been ignored. A spacecraft is usually in a long-term energy balance, so all the absorbed radiation is emitted as thermal radiation, although not necessarily at the same time or from the same surface as its absorption. A simple way to model thermal radiation is to consider a portion of the absorbed radiation to be diffusely reflected, by increasing R_{diff}^i and decreasing R_{abs}^i. However, this assumes that the energy is re-radiated from the same surface and at the same time as its absorption. A more accurate computation requires knowledge of the absolute temperature T_i and emissivity ϵ^i (a dimensionless constant between 0 and 1) of each surface. Then the thermal radiation flux from the surface is given by the Stefan-Boltzmann law

$$\mathscr{F}_{\text{thermal}}^i = \epsilon^i \sigma T_i^4 \tag{10.116}$$

where $\sigma = 5.67 \times 10^{-8}\ \text{Wm}^{-2}\text{K}^{-4}$ is the Stefan-Boltzmann constant. If the thermal radiation from every surface is Lambertian, it gives rise to a net force

$$\mathbf{F}_{\text{thermal}} = -\frac{2}{3} \sum_{i=1}^{N} \mathscr{F}_{\text{thermal}}^i S_i \mathbf{n}_I^i \tag{10.117}$$

The reason that thermal radiation is usually negligible is that it is emitted with roughly equal flux in all directions, so that the net radiation force is generally small. A careful treatment of thermal radiation has been used to explain the anomalous acceleration of the Pioneer 10 and 11 spacecraft, though [18, 20]. This analysis had to account for onboard nuclear energy sources on these spacecraft, which resulted in more radiation energy being emitted than absorbed.

Finally, Eq. (10.115) does not explicitly account for potential self-shadowing of concave spacecraft. If the configuration of the spacecraft is known a priori, self-shadowing can be taken into account by replacing S_i with the area of the flat plate that is visible to the Sun after accounting for self-shadowing. Modeling the effects of reflected radiation or thermal radiation from one surface striking another surface is an additional complication. Another drawback to Eq. (10.115) is that it is only valid for a collection of flat surfaces with uniquely defined outward normals. Most real spacecraft have some curved surfaces, and accurately approximating these surfaces by a collection of flat plates causes the size of the model to grow, increasing the computational burden.

10.4 Perturbation Methods

Perturbation methods find solutions to complex problems by relating them to solutions of simpler problems. In our case, the solution of Eq. (10.85) is close to the solution of the Kepler problem with all but the first term on the right side absent. We will only give a very brief discussion of one of these methods; much fuller treatments can be found in many books [4, 8, 16, 22].

10.4.1 Variation of Parameters

In the variation of parameters (VOP) method, constant parameters that characterize the unperturbed motion of a system are treated as slowly-varying parameters in a representation of the perturbed motion. The six Kepler orbit elements are well suited for the application of the VOP method to perturbed orbital motion. Consider that we have a solution of Eq. (10.85) that gives a spacecraft's position and velocity as functions of time. Then at any particular time t, we can compute a set of six Kepler elements using Eqs. (10.77)–(10.84). These are called *osculating elements*, and the Keplerian orbit with these elements is the *osculating orbit*, after the Latin word "osculum," meaning kiss, because the osculating orbit is tangent to, or kisses, the perturbed orbit at time t.

The osculating Keplerian elements $a(t)$, $e(t)$, $i(t)$, $\Omega(t)$, $\omega(t)$, and $M_0(t)$ that represent the perturbed orbit are functions of time, and obey a set of nonlinear coupled differential equations. These equations for the case of conservative perturbations, i.e. those derivable from a potential function, were developed by

Lagrange; and Gauss found the equivalent equations in terms of the perturbing acceleration $\mathbf{a}_{\text{perturb}}$. Exact integration of these equations has no real advantage over direct integration of Eq. (10.85), but they are very useful as a starting point for approximations. Their widespread application to the dynamics of Earth-orbiting spacecraft began with the work of Brouwer [6] and Kozai [12], who analyzed the effect of the zonal gravity harmonics on the osculating elements. Hoots has summarized the history of these developments [10].

The variations of the osculating elements can be separated into periodic terms and secular terms. The orbit elements obtained after removing the periodic terms (in some analytic model and to some degree of approximation) are called *mean elements*. The mean elements are smoother functions of time than the osculating elements, and they obey simpler differential equations. The basic idea of the VOP method is to propagate the mean elements and then add the periodic terms analytically. Care must be taken, because different analytic methods result in somewhat different mean elements.

10.4.2 Two Line Elements

The U.S. Space Surveillance System has been tracking and maintaining a catalog of man-made Earth-orbiting satellites since the dawn of the space age; by 2004 the catalog contained more than 10,000 objects [10]. The catalog is in the form of North American Air Defense Command (NORAD) two line element (TLE) sets, which provide the basic parameters to predict the position and velocity of a spacecraft. Tables 10.3 and 10.4 describe the information in the TLE set.[6] The orbital elements in Line 2 are mean values obtained by removing periodic variations in a particular way. Accurate orbit predictions are obtained only if the prediction model computes these periodic variations in exactly the same way that they were removed in computing the elements. Using any other prediction model, even one that is theoretically more accurate, will produce inferior predictions. The model currently used for this purpose is the simplified general perturbations model SGP4, which includes low-order zonal gravity terms and an approximate aerodynamic drag model [10, 22].

An example for a TLE from the Tropical Rainfall Measuring Mission (TRMM) spacecraft is given by

```
TRMM
1 25063U 97074A   11130.2059828   .00013273  00000-0  18660-3 0  6592
2 25063 034.9640 081.2155 0001042 240.3761 119.6798 15.55777853 76795 4
```

The title line contains the spacecraft name. The other two lines are described in Tables 10.3 and 10.4. Note that there is an assumed decimal point in the B^* value, so the value in the example is by 1.8660×10^{-4}. The term B^* is given by [21]

[6] The format is also described in http://en.wikipedia.org/wiki/Two-line_element_set.

Table 10.3 Line 1 TLE information

Field	Columns	Content	Example
1	01–01	Line number	1
2	03–07	Satellite number	25063
3	08–08	Classification (U=Unclassified)	U
4	10–11	International Designator (Last two digits of launch year)	97
5	12–14	International Designator (Launch number of the year)	074
6	15–17	International Designator (Piece of the launch)	A
7	19–20	Epoch Year (Last two digits of year)	11
8	21–32	Epoch (Day of the year and fractional portion of the day)	130.20598286
9	34–43	First Time Derivative of the Mean Motion divided by two	.00013273
10	45–52	Second Time Derivative of Mean Motion divided by six	00000–0
11	54–61	B^* drag term (decimal point assumed)	18660–3
12	63–63	The number 0 (Originally "Ephemeris type")	0
13	65–68	Element number	659
14	69–69	Checksum (Modulo 10)	2

Table 10.4 Line 2 TLE information

Field	Columns	Content	Example
1	01–01	Line number	2
2	03–07	Satellite number	25063
3	09–16	Inclination (Deg)	34.9640
4	18–25	Right Ascension (Deg)	81.2155
5	27–33	Eccentricity (decimal point assumed)	0001042
6	35–42	Argument of Perigee (Deg)	240.3761
7	44–51	Mean Anomaly (Deg)	119.6798
8	53–63	Mean Motion (Revs per Day)	15.55777853
9	64–68	Revolution number at epoch (Revs)	76795
10	69–69	Checksum (Modulo 10)	4

$$B^* = \frac{1}{2}\frac{C_D S}{m}\rho_0 R_\oplus \tag{10.118}$$

where C_D is the drag coefficient, S is the cross-sectional area, m is the mass, ρ_0 is the atmospheric density at perigee (assumed to be 2.461×10^{-5} kg/m^2/ER), and $R_\oplus$ is the Earth radius (ER), typically given by 6378.135 km. The ballistic coefficient, B, defined by Eq. (10.112), is related to B^* by

$$B = \frac{R_\oplus \rho_0}{2B^*} \tag{10.119}$$

Using $R_\oplus = 6378.135$ km the constant conversion is given by

$$B = \frac{1}{12.741621\, B^*}\ \frac{\text{kg}}{\text{m}^2} \tag{10.120}$$

10.4.3 A Useful Approximation, Secular J_2 Effects Only

We noted in Sect. 10.3.1 that the J_2 zonal term gave by far the largest perturbation due to non-spherical gravitational field of the Earth. In many cases, a simple Keplerian propagation including only secular J_2 perturbations is adequate for the orbital accuracy required. This is the case, for example, when the output is used for graphical visualization. In such instances, we can tolerate errors of order J_2, but want to avoid errors of order $J_2(t - t_0)$.

Analysis shows that the mean elements $\bar{a}$, $\bar{e}$, and $\bar{i}$ are constant, and the other mean elements have the following secular variation to first order in J_2 [16]:

$$\bar{\omega}(t) = \bar{\omega}(t_0) + \frac{3}{2}J_2\left(\frac{R_\oplus}{\bar{a}}\right)^2 (1-\bar{e}^2)^{-2}\left(2 - \frac{5}{2}\sin^2\bar{i}\right)\bar{n}(t-t_0) \tag{10.121a}$$

$$\bar{\Omega}(t) = \bar{\Omega}(t_0) - \frac{3}{2}J_2\left(\frac{R_\oplus}{\bar{a}}\right)^2 (1-\bar{e}^2)^{-2}\left(\cos\bar{i}\right)\bar{n}(t-t_0) \tag{10.121b}$$

$$\bar{M}_0(t) = \bar{M}_0(t_0) + \frac{3}{2}J_2\left(\frac{R_\oplus}{\bar{a}}\right)^2 (1-\bar{e}^2)^{-3/2}\left(1 - \frac{3}{2}\sin^2\bar{i}\right)\bar{n}(t-t_0) \tag{10.121c}$$

where the overbars indicate mean elements, and $\bar{n} = \sqrt{\mu/\bar{a}^3}$. We can combine the last of these equations with Eq. (10.68) to get

$$\bar{M}(t) = \bar{M}_0(t_0) + \hat{n}(t-t_0) \tag{10.122}$$

where

$$\hat{n} = \bar{n}\left[1 + \frac{3}{2}J_2\left(\frac{R_\oplus}{\bar{a}}\right)^2 (1-\bar{e}^2)^{-3/2}\left(1 - \frac{3}{2}\sin^2\bar{i}\right)\right] = \sqrt{\frac{\mu}{\hat{a}^3}} \tag{10.123}$$

with

$$\hat{a} = \bar{a}\left[1 - J_2\left(\frac{R_\oplus}{\bar{a}}\right)^2 (1-\bar{e}^2)^{-3/2}\left(1 - \frac{3}{2}\sin^2\bar{i}\right)\right] \tag{10.124}$$

If we can tolerate errors of order J_2, we only need to add the first-order corrections of Eqs. (10.121) and (10.122), and can otherwise propagate the orbit as if it were an unperturbed orbit. It is unnecessary to compute the periodic corrections for many orbit and attitude analyses, so we can treat the propagated mean elements as if they were osculating elements. The initial values of the mean elements can be obtained from a TLE set, or they can simply be postulated for a mission design study.

We often want to initialize an orbit propagation with position and velocity vectors at some epoch time. In order to use the simple Keplerian propagator incorporating secular J_2 effects, we first convert the position and velocity to osculating orbital elements. In principle, we should then remove the periodic variations to obtain mean elements. Since we are willing to tolerate errors of order J_2, however, this osculating-to-mean conversion is unnecessary for five of the six Keplerian orbit elements. We must compute the mean value of the semimajor axis, however, because Eq. (10.122) shows that an error of order J_2 in the semimajor axis will lead to an error of the same order of magnitude in the mean motion and thus to propagation errors growing like $J_2\hat{n}(t-t_0)$, which we are not willing to tolerate [13]. The mean-to-osculating transformation of the semimajor axis is given by [16]

$$a(t) = \bar{a} + \frac{J_2 R_\oplus^2}{\bar{a}}\left\{\left(\frac{\bar{a}}{r}\right)^3\left[1-3\sin^2\bar{i}\,\sin^2(\bar{\omega}+\nu)\right]-(1-\bar{e}^2)^{-3/2}\left(1-\frac{3}{2}\sin^2\bar{i}\right)\right\} \tag{10.125}$$

After propagating the orbit to times of interest, we apply the inverse, mean-to-osculating, transformation of the semimajor axis. This is not strictly necessary, because ignoring it leads to tolerable errors of order J_2, but it ensures that the output position and velocity agree with the input position and velocity at the epoch time. Since we convert back to the osculating semimajor axis, we are free to use the quantity $\hat{a}$ of Eq. (10.124) as our mean semimajor axis on the right side of Eqs. (10.121a) and (10.121b), because that introduces differences only of order J_2^2. Combining Eqs. (10.124) and (10.125) and using Eq. (10.20) to express $r(t)$ in terms of the mean elements and the true anomaly $\nu(t)$ gives

$$a(t) = \hat{a} + \left(\frac{J_2 R_\oplus^2}{\hat{a}}\right) g(t) \tag{10.126}$$

where

$$g(t) \equiv \left(\frac{1+\bar{e}\cos\nu(t)}{1-\bar{e}^2}\right)^3\left[1-3\sin^2\bar{i}\,\sin^2(\bar{\omega}(t)+\nu(t))\right] \tag{10.127}$$

We use the exact inverse of this equation

$$\hat{a} = \frac{1}{2}\left[a(t_0)+\sqrt{a^2(t_0)-4J_2R_\oplus^2 g(t_0)}\right] \tag{10.128}$$

to initialize $\hat{a}$ at the epoch time, so that the output position and velocity agree with the input to computer precision. Thus the steps in this orbit propagator are

- Convert $\mathbf{r}(t_0)$ and $\mathbf{v}(t_0)$ to osculating Keplerian orbit elements
- Use Eq. (10.128) to compute $\hat{a}$
- Treat $\hat{a}$ and the other five osculating elements as mean elements
- Compute the mean motion $\hat{n} = \sqrt{\mu/\hat{a}^3}$
- Use Eqs. (10.121a), (10.121b), and (10.122) to propagate to time t

- Use Eq. (10.126) to compute $a(t)$
- Convert the Keplerian orbit elements to $\mathbf{r}(t)$ and $\mathbf{v}(t)$

This orbit propagator was used in Goddard Space Flight Center's Mission Planning Graphics Tool (MPGT), and also to drive a wall display at the Smithsonian Air and Space Museum.

10.4.4 Sun-Synchronous Orbits

The right ascension of the ascending node in a *Sun-synchronous* orbit increases at a rate of 1 revolution per year, or 1.991×10^{-7} rad/s, so that the orbit plane has a nearly fixed orientation relative to the Sun. Substituting $\dot{\bar{\Omega}} = 1.991 \times 10^{-7}$ rad/s and the numerical values for J_2, μ, and $R_\oplus$ into Eq. (10.121b) shows that the inclination of a Sun-synchronous orbit must obey the relationship

$$\cos \bar{i} = -0.09892\,(1-\bar{e}^2)^2 \left(\frac{\bar{a}}{R_\oplus}\right)^{7/2} = -(1-\bar{e}^2)^2 \left(\frac{\rho}{3.975\,\mathrm{h}}\right)^{7/3} \qquad (10.129)$$

where ρ is the orbit period. Because the cosine is negative, the inclination will be greater than 95.68°, which means that these are *retrograde* orbits.

Sun-synchronous orbits are labeled by the mean local time of the ascending node, i.e. the mean local time at the sub-satellite point when the spacecraft crosses the equator from South to North. These orbits have many useful applications, particularly for observing the Earth's surface, because the lighting conditions at any latitude change only seasonally, so very slowly from orbit to orbit. The inclination of circular Sun-synchronous orbits varies from 95.68° for $a = R_\oplus$ to 99.48° for $a = R_\oplus + 1{,}000$ km, and these high-inclination orbits are well-suited to observing high-latitude regions of the Earth.

The *A Train*, or Afternoon Train, of Earth-observation satellites provides a good example of the use of such orbits. The spacecraft in this constellation are in 1:30 p.m. Sun-synchronous orbits with $i = 98.14°$ and $a = 7{,}078$ km.[7] They are spaced a few minutes apart from each other so their collective observations may be used to build high-definition three-dimensional images of the Earth's atmosphere and surface. There are currently five active satellites in the A Train: the Global Change Observing Mission (GCOM-W1, also known as SHIZUKU), the lead spacecraft in the constellation, launched by JAXA on May 18, 2012: Aqua, 4 min behind GCOM-W1, launched by NASA on May 4, 2002; CloudSat, a cooperative effort between NASA and the Canadian Space Agency, 2 min and 30 s behind Aqua; the Cloud-Aerosol Lidar and Infrared Pathfinder Satellite Observations (CALIPSO) spacecraft, a joint effort of CNES and NASA, launched with Cloudsat on April 28,

[7] The lack of precise agreement with Eq. (10.129) is due to higher-order zonal perturbations.

2006 and following it by no more than 15 s; and Aura, a multi-national satellite, 15 min behind Aqua but crossing the equator 8 min behind it due to a different ascending node, launched by NASA on July 15, 2004.

Several solar observing scientific spacecraft have employed *dawn/dusk* Sun-synchronous orbits, with ascending nodes at 6 a.m. or 6 p.m., to afford a nearly continuous view of the Sun. The near-absence of eclipses in dawn/dusk orbits is also useful for providing photoelectric power and for minimizing thermal transients on sensitive scientific instruments that occur at eclipse entry and exit. The eclipse analysis of Sect. 11.3 of Chap. 11 shows that eclipses cannot be completely avoided unless $a|\sin(\varepsilon + i)| > R_\oplus$, where $\varepsilon = 23.44°$ is the obliquity of the ecliptic. This condition is satisfied only if a is between $1.218R_\oplus$ and $1.522R_\oplus$. A spacecraft in a lower or higher 6 a.m. orbit will suffer eclipses around the northernmost point of its orbit for several weeks near the winter solstice, and a spacecraft in a 6 p.m. orbit outside this altitude range will be eclipsed near the southernmost point of its orbit around the time of the summer solstice. The Canadian RADARSAT-2 spacecraft, launched on December 14, 2007, is an example of a 6 p.m. Sun-synchronous orbit. Its altitude is 798 km, so it has an 11-week eclipse period around the summer solstice, with the longest eclipse duration being 18 min [15].

10.5 Lagrange Points

After Newton solved the equations of motion of two bodies under their gravitational attraction, the famous *two-body problem*, attention turned to the three-body problem. This led to a great deal of deep mathematics that is beyond the scope of this book [4, 8, 16]. We will only consider the *circular restricted three-body problem* (CR3BP). *Restricted* means that the mass of the third body (generally a spacecraft or an asteroid) is so small compared to the masses of the other two bodies (usually the Sun and the Earth,[8] the Earth and the Moon, or the Sun and Jupiter) that ignoring its influence on their motion is an excellent approximation. *Circular* means that the two other bodies, called the *primaries*, move in circular orbits around each other. We will assume that the first primary, with mass m_1, is at least as heavy as the second, with mass m_2. The motion of the third body is most naturally described in a *co-rotating* coordinate system, in which the primaries appear to be at rest. The origin of this frame is at the center of mass of the two primaries, its x axis is in the direction from m_1 to m_2, its z axis is along the direction of the orbital angular velocity of the two primaries, and its y axis completes the right-handed orthogonal triad. Figure 10.11 illustrates the frame and axis definitions.

Our analysis will follow that of Danby [8], and will employ kinematic equations in a rotating frame developed in Sect. 3.1.3. We use Eq. (3.15) for the position vector

[8] In discussing the Sun/Earth Lagrange points, "Earth" means the system of the Earth and the Moon, the Sun is the first body, and the mass and location of the second body are the summed Earth/Moon mass and the location of the Earth/Moon center of mass.

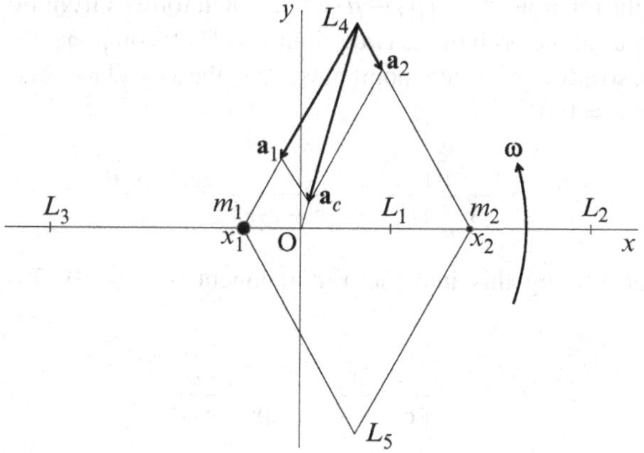

Fig. 10.11 The five Lagrange points

of the spacecraft with respect to the center of mass of the two primaries, letting F denote the inertial frame and G denote the co-rotating frame, which differ only by a rotation about the z axis. Because $\boldsymbol{\omega}$ is in the positive z direction, it has the same representation in both frames, so we can omit its frame-specifying subscript. The angular velocity is also constant, so the $\dot{\boldsymbol{\omega}}$ term vanishes, giving

$$\ddot{\mathbf{r}}_G = A_{GF}\ddot{\mathbf{r}}_F - \boldsymbol{\omega} \times (\boldsymbol{\omega} \times \mathbf{r}_G) - 2\boldsymbol{\omega} \times \dot{\mathbf{r}}_G \tag{10.130}$$

The acceleration in the inertial frame is due to the gravitational attraction of the two primaries,

$$\ddot{\mathbf{r}}_F = -\mu_1 \frac{(\mathbf{r} - \mathbf{r}_1)_F}{\|\mathbf{r} - \mathbf{r}_1\|^3} - \mu_2 \frac{(\mathbf{r} - \mathbf{r}_2)_F}{\|\mathbf{r} - \mathbf{r}_2\|^3} \tag{10.131}$$

where $\mu_i = GM_i$, so the spacecraft dynamics is governed by the equation

$$\ddot{\mathbf{r}}_G = -\mu_1 \frac{(\mathbf{r} - \mathbf{r}_1)_G}{\|\mathbf{r} - \mathbf{r}_1\|^3} - \mu_2 \frac{(\mathbf{r} - \mathbf{r}_2)_G}{\|\mathbf{r} - \mathbf{r}_2\|^3} - \boldsymbol{\omega} \times (\boldsymbol{\omega} \times \mathbf{r}_G) - 2\boldsymbol{\omega} \times \dot{\mathbf{r}}_G \tag{10.132}$$

All vectors have now been expressed in the co-rotating frame, so we will omit the subscript G for the remainder of the discussion.

We want to prove the existence of five *Lagrange points*, or *libration points*, which are points of static equilibrium where $\ddot{\mathbf{r}}$ and $\dot{\mathbf{r}}$ are zero. Thus we are looking for solutions of

$$\mathbf{0} = -\frac{\mu_1}{\|\mathbf{r} - \mathbf{r}_1\|^3} \begin{bmatrix} x - x_1 \\ y \\ z \end{bmatrix} - \frac{\mu_2}{\|\mathbf{r} - \mathbf{r}_2\|^3} \begin{bmatrix} x - x_2 \\ y \\ z \end{bmatrix} + \frac{\mu_1 + \mu_2}{a^3} \begin{bmatrix} x \\ y \\ 0 \end{bmatrix} \tag{10.133}$$

We have used the relation $\omega^2 = (\mu_1 + \mu_2)/a^3$, which follows from Eq. (10.43) with a being the radius of the orbit of the two primaries. The z component of Eq. (10.133) requires $z = 0$, so any equilibrium point must lie in the x–y plane. The y component requires either $y = 0$ or

$$0 = -\frac{\mu_1}{\|\mathbf{r} - \mathbf{r}_1\|^3} - \frac{\mu_2}{\|\mathbf{r} - \mathbf{r}_2\|^3} + \frac{\mu_1 + \mu_2}{a^3} \tag{10.134}$$

If $y \neq 0$, substituting this into the x component of Eq. (10.133) leads to the requirement

$$0 = \frac{\mu_1 x_1}{\|\mathbf{r} - \mathbf{r}_1\|^3} + \frac{\mu_2 x_2}{\|\mathbf{r} - \mathbf{r}_2\|^3} \tag{10.135}$$

The definition of the center of mass means that $\mu_1 x_1 = -\mu_2 x_2$, so it follows from Eq. (10.135) that $\|\mathbf{r} - \mathbf{r}_1\| = \|\mathbf{r} - \mathbf{r}_2\|$. Substituting this into Eq. (10.134) then gives $\|\mathbf{r} - \mathbf{r}_1\| = \|\mathbf{r} - \mathbf{r}_2\| = a$. This defines the two Lagrange points L_4 and L_5 at the vertices of two equilateral triangles in the orbital plane of the two primaries, whose other vertices are the locations of the primaries. These are illustrated for the case that $m_1 = 3m_2$ in Fig. 10.11. The figure shows that the resultant of the acceleration vectors $\mathbf{a}_1$ due to the attraction of m_1 and $\mathbf{a}_2$ due to the attraction of m_2 provides the centripetal acceleration $\mathbf{a}_c = \boldsymbol{\omega} \times (\boldsymbol{\omega} \times \mathbf{r})$ needed for equilibrium at L_4.

There is no simple expression for the location of the collinear Lagrange points, the points with $y = z = 0$. Assume that $m_2 \leq m_1$ and let

$$\beta \equiv \frac{m_2}{m_1 + m_2}, \quad \text{so that} \quad x_1 = -\beta a \quad \text{and} \quad x_2 = (1 - \beta)a \tag{10.136}$$

We also let $x = \xi a$, so that Eq. (10.132) for a body at rest on the x axis gives

$$\frac{\ddot{\xi}}{\omega^2} = -(1 - \beta)\frac{\xi + \beta}{|\xi + \beta|^3} - \beta\frac{\xi + \beta - 1}{|\xi + \beta - 1|^3} + \xi \tag{10.137}$$

The right side of this equation is plotted in Fig. 10.12 for $|\xi| \leq 1.5$ for the case that $m_1 = 3m_2$. The function will have this general shape for any mass ratios. We notice two things. First, it is easy to see that there are three collinear Lagrange points: L_1 between m_1 and m_2, L_2 on the side of m_2 away from m_1, and L_3 on the side of m_1 away from m_2.[9] The second point is that the acceleration in the vicinity of a collinear Lagrange point is always *away* from the Lagrange point, meaning that the collinear Lagrange points are unstable. If $m_2 \ll m_1$, the usual case, it is not too difficult to show that L_2 and L_1 are at $x \approx x_2 \pm (\beta/3)^{1/3}a$, respectively, and that L_3 is at $x \approx -(1 + 5\beta/12)a$. The Sun/Earth L_1 and L_2 points are about 0.01 AU, or about 1.5×10^6 km, from the Earth, because $\beta \approx 3 \times 10^{-6}$ for that system.

[9] This labeling convention is not universally followed.

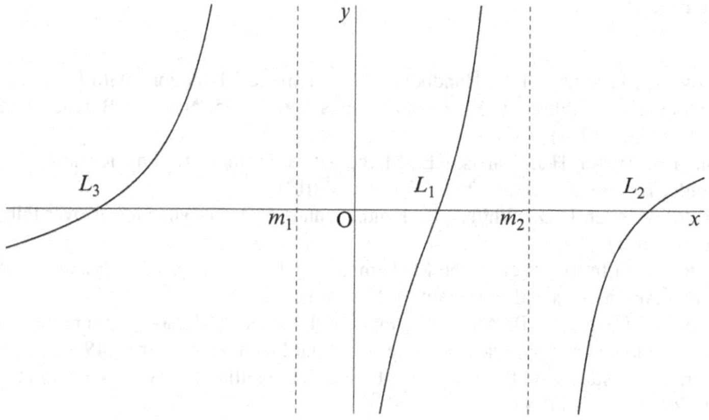

Fig. 10.12 x-axis force, and collinear Lagrange points

A complete analysis of the stability of the Lagrange points requires consideration of the Coriolis acceleration, the last term in Eq. (10.132). We will only state the result obtained by linearizing this equation in the vicinity of L_4 and L_5, Danby shows that these points are stable if $m_1/m_2 > (25 + 3\sqrt{69})/2 = 24.96$. Stable points at L_4 or L_5 are called *Trojan points*, after the Trojan asteroids at the Sun-Jupiter L_4 and L_5 points. The Earth/Moon Trojan points are stable, but solar gravity leads to large motions about these points [11, 19].

Although the collinear Lagrange points are unstable, relatively small control forces can keep spacecraft in orbits around these points, and a variety of uses have been found for spacecraft in such orbits [9]. The International Sun-Earth Explorer 3 (ISEE-3), the Solar and Heliospheric Observatory (SOHO), the Advanced Composition Explorer (ACE), and WIND spacecraft have all been stationed near the Sun/Earth L_1 point to observe the Sun and the solar wind. Radio emissions from the Sun would complicate communication with a spacecraft exactly at L_1. The Sun/Earth L_2 point is useful for stationing an instrument that needs to be shielded from thermal energy emitted by both the Sun and the Earth, which are both on the same side of the spacecraft. Spacecraft are almost never placed exactly at L_2, however, since they would not be able to use Solar power at that point. The WMAP, Herschel, and Planck spacecraft have all been placed in orbits around L_2, and the James Webb Space Telescope (JWST) is planned to be placed there. An orbit about the Earth/Moon L_2 point has been proposed as the location for a data relay satellite to communicate with human or robotic explorers on the back side of the Moon.

References

1. Abramowitz, M., Stegun, I.A.: Handbook of Mathematical Functions with Formulas, Graphs and Mathematical Tables. Applied Mathematics Series - 55. National Bureau of Standards, Washington, DC (1964)
2. Arfken, G.B., Weber, H.J., Harris, F.E.: Mathematical Methods for Physicists: A Comprehensive Guide, 7th edn. Academic Press, Waltham (2013)
3. Bate, R.R., Mueller, D.D., White, J.E.: Fundamentals of Astrodynamics. Dover Publications, New York (1971)
4. Battin, R.H.: An Introduction to the Mathematics and Methods of Astrodynamics. American Institute of Aeronautics and Astronautics, New York (1987)
5. Borderies, N., Longaretti, P.: A new treatment of the albedo radiation pressure in the case of a uniform albedo and of a spherical satellite. Celestial Mech. Dyn. Astron. **49**(1), 69–98 (1990)
6. Brouwer, D.: Solution of the problem of artificial satellite theory without drag. Astron. J. **64**(1274), 378–397 (1959)
7. Colwell, P.: Solving Kepler's Equation over Three Centuries. Willmann-Bell, Richmond (1993)
8. Danby, J.M.A.: Fundamentals of Celestial Mechanics, 2nd edn., 3rd Printing. Willman-Bell, Richmond (1992)
9. Farquhar, R.W.: Fifty Years on the Space Frontier: Halo Orbits, Comets, Asteroids, and More. Outskirts Press, Parker (2011)
10. Hoots, F.R., Schumacher Jr., P.W., Glover, R.A.: History of analytical orbit modeling in the U.S. space surveillance system. J. Guid. Contr. Dynam. **27**(2), 174–185 (2004)
11. Kolenkiewicz, R., Carpenter, L.: Stable periodic orbits about the Sun perturbed Earth-Moon triangular points. AIAA J. **6**(7), 1301–1304 (1968)
12. Kozai, Y.: The motion of a close Earth satellite. Astron. J. **64**(1274), 367–377 (1959)
13. Markley, F.L., Jeletic, J.F.: Fast orbit propagator for graphical display. J. Guid. Contr. Dynam. **14**(2), 473–475 (1991)
14. Montenbruck, O., Gill, E.: Satellite Orbits: Models, Methods, and Applications. Springer, Berlin/Heidelberg/New York (2000)
15. Morena, L.C., James, K.V., Beck, J.: An introduction to the RADARSAT-2 mission. Can. J. Rem. Sens. **30**(3), 221–234 (2004)
16. Roy, A.E.: Orbital Motion, 4th edn. IOP Publishing, Bristol (2005)
17. Schaub, H., Junkins, J.L.: Analytical Mechanics of Aerospace Systems, 2nd edn. American Institute of Aeronautics and Astronautics, New York (2009)
18. Scheffer, L.K.: Conventional forces can explain the anomalous acceleration of Pioneer 10. Phys. Rev. **D67**(8), 8402-1–8402-11 (2003)
19. Tapley, B.D., Lewallen, J.M.: Solar influence on satellite motion near the stable Earth-Moon libration points. AIAA J. **2**(4), 728–732 (1964)
20. Turyshev, S.G., Toth, V.T., Kinsella, G., Lee, S.-C., Lok, S.M., Ellis, J.: Support for the thermal origin of the Pioneer anomaly. Phys. Rev. Lett. **108**(24), 241101 (2012)
21. Vallado, D.A.: Fundamentals of Astrodynamics and Applications, 3rd edn. Microcosm Press, Hawthorne and Springer, New York (2007)
22. Vallado, D.A., Crawford, P., Hujsak, R., Kelso, T.S.: Revisiting Spacetrack Report #3: Rev 2. In: AIAA/AAS Astrodynamics Specialist Conference. Keystone (2006)
23. Vallado, D.A., Finkleman, D.: A critical assessment of satellite drag and atmospheric density modelling. In: AIAA/AAS Astrodynamics Specialist Conference. Honolulu (2008)

Chapter 11
Environment Models

The analysis of spacecraft attitude and trajectory motion depends on several environmental models, as does the analysis of sensor measurements. This chapter presents models of the Earth's magnetic field and atmosphere and of the motion of the Sun and its planets as seen from the Earth.

11.1 Magnetic Field Models

In a region with no appreciable electric currents, a magnetic field can be computed as the negative gradient of a magnetic scalar potential arising from a collection of magnetic poles, each providing a $1/r$ contribution. This is similar to the gravitational potential contributed by a collection of point masses, so magnetic field models use the same mathematical apparatus as gravity field models. Following the same path as in Sect. 10.3.1 leads to a spherical harmonics expansion for the magnetic potential. There is one significant difference between the magnetic and gravitational potentials. Gravitational forces are always attractive, while magnetic forces can be attractive or repulsive, represented by North and South magnetic poles with pole strengths of opposite signs. In fact, isolated poles are never found in nature: poles always appear in pairs, North and South poles of equal and opposite magnitude, comprising a magnetic dipole.[1] As a result, the first term on the right side of the magnetic analog of Eq. (10.94) is absent, leading to

$$V(\mathbf{r}) = a \sum_{n=1}^{\infty} \left(\frac{a}{r}\right)^{n+1} \sum_{m=0}^{n} \bar{P}_n^m(\sin\lambda')[g_n^m(t)\cos(m\phi) + h_n^m(t)\sin(m\phi)] \quad (11.1)$$

[1] Some physical theories predict the occurrence of isolated magnetic monopoles, but none has of yet been detected.

F.L. Markley and J.L. Crassidis, *Fundamentals of Spacecraft Attitude Determination and Control*, Space Technology Library 33, DOI 10.1007/978-1-4939-0802-8_11,
© Springer Science+Business Media New York 2014

This equation is presented in the form used for the International Geomagnetic Reference Field (IGRF) [4]. The parameter $a = 6371.2$ km is the magnetic spherical reference radius, and the coefficients $g_n^m(t)$ and $h_n^m(t)$ are conventionally given in units of nanotesla (nT). The bars over the associated Legendre functions indicate that they are the Schmidt semi- (or quasi-) normalized functions, defined as [20]

$$\bar{P}_n^m = \sqrt{(2 - \delta_{0m})\frac{(n-m)!}{(n+m)!}}\, P_n^m \tag{11.2}$$

The IGRF is produced and maintained under the auspices of the International Association of Geomagnetism and Aeronomy (IAGA). It represents the internal part of the Earth's magnetic field, which is almost entirely generated in the Earth's core and is slowly varying on time scales of years to decades [12]. In addition to the internal field, magnetic fields with time scales ranging mostly from seconds to hours are generated by electric currents in the ionosphere and magnetosphere. These fields, which are not included in the IGRF, are as small as 20 nT during magnetically quiet periods, but can be as large as 1,000 nT or greater during a magnetic storm.

Because of the time variation of the Earth's field, the IGRF provides coefficients $g_n^m(T_i)$ and $h_n^m(T_i)$ at 5-year intervals. The eleventh generation IGRF has coefficients of degrees $n \leq 10$ for $1900 \leq T_i < 2000$ and $n \leq 13$ for $2000 \leq T_i \leq 2010$. The coefficients for times between T_i and T_{i+1} are found by linear interpolation. This IGRF also includes a predictive model for linear extrapolation to years after 2010, comprising secular variations $\dot{g}_n^m(2010)$ and $\dot{h}_n^m(2010)$ (in nT/year) only for degrees $n \leq 8$. When one of the 5-year constituent models is considered definitive, it is called a Definitive Geomagnetic Reference Field (DGRF), and its coefficients are frozen. The eleventh generation IGRF is considered definitive for the years 1945–2005. The RMS errors in the more recent models are estimated to be about 10 nT.

11.1.1 Dipole Model

Keeping only the $n = 1$ terms in Eq. (11.1) gives the dipole approximation

$$\begin{aligned} V(\mathbf{r}) &= \frac{a^3}{r^2}\left[\bar{P}_1^0(\sin\lambda')g_1^0 + \bar{P}_1^1(\sin\lambda')(g_1^1\cos\phi + h_1^1\sin\phi)\right] \\ &= \frac{a^3}{r^2}\left(g_1^0\sin\lambda' + g_1^1\cos\lambda'\cos\phi + h_1^1\cos\lambda'\sin\phi\right) = \frac{\mathbf{m}\cdot\mathbf{r}}{r^3} \end{aligned} \tag{11.3}$$

where

$$\mathbf{m} = a^3\begin{bmatrix} g_1^1 \\ h_1^1 \\ g_1^0 \end{bmatrix} = m\begin{bmatrix} \sin\theta_m\cos\alpha_m \\ \sin\theta_m\sin\alpha_m \\ \cos\theta_m \end{bmatrix} \tag{11.4}$$

is the magnetic dipole in the ECEF frame. The coefficients from the 2005 DGRF: $g_1^0 = -29554.63$ nT, $g_1^1 = -1669.05$ nT, and $h_1^1 = 5077.99$ nT give the dipole magnitude[2] $m = a^3 \times 30{,}034$ nT $= 7.77 \times 10^{22}$ Am^2 as well as the orientation $\theta_m = 169.7°$ and $\alpha_m = 108.2°$. Note that the Earth's North magnetic pole is near the South geographic pole, where it must be because opposite poles attract. The magnetic field is given by

$$\mathbf{B}(\mathbf{r}) = -\nabla V(\mathbf{r}) = \frac{3(\mathbf{m}\cdot\mathbf{r})\mathbf{r} - r^2\mathbf{m}}{r^5} \tag{11.5}$$

The magnetic field strength falls off with distance as $1/r^3$, and it is twice as great at the magnetic poles, $\mathbf{r} \parallel \mathbf{m}$, as at the magnetic equator, $\mathbf{r} \perp \mathbf{m}$.

The Earth's magnetic field is often used as a reference for attitude determination or control, so it is of interest to know how the direction of the ambient magnetic field varies as a spacecraft moves around its orbit. The time derivative of the magnetic field at the location of the spacecraft can be separated into components parallel and perpendicular to the field by

$$\dot{\mathbf{B}} = \|\mathbf{B}\|^{-2}\left[(\mathbf{B}\cdot\dot{\mathbf{B}})\mathbf{B} + (\mathbf{B}\times\dot{\mathbf{B}})\times\mathbf{B}\right] = \|\mathbf{B}\|^{-2}(\mathbf{B}\cdot\dot{\mathbf{B}})\mathbf{B} + \boldsymbol{\Omega}\times\mathbf{B} \tag{11.6}$$

where $\boldsymbol{\Omega} \equiv \|\mathbf{B}\|^{-2}(\mathbf{B}\times\dot{\mathbf{B}})$ is the rotation rate of the magnetic field. The first term on the right side of Eq. (11.6) gives the rate of change of the magnitude of the magnetic field, which is of less interest for attitude estimation and control. For a spacecraft in a circular orbit, r is constant and the spacecraft velocity in the GCI frame is $\dot{\mathbf{r}} = \boldsymbol{\omega}_o \times \mathbf{r}$, where $\boldsymbol{\omega}_o$ is the orbital angular velocity, so

$$\dot{\mathbf{B}} = \frac{3(\mathbf{m}\cdot\mathbf{r})(\boldsymbol{\omega}_o\times\mathbf{r}) + 3[\mathbf{m}\cdot(\boldsymbol{\omega}_o\times\mathbf{r})]\mathbf{r}}{r^5} \tag{11.7}$$

The derivation of Eq. (11.7) assumes that the magnetic dipole is constant in GCI, which is not exactly true due to the rotation of the Earth. The orbital angular velocity of a near-Earth spacecraft is about 15 times greater than the rotation rate of the Earth, though, so ignoring the motion of the dipole in the GCI frame is a reasonable approximation. Then, after a significant amount of vector algebra, we find that

$$\boldsymbol{\Omega} = 3\frac{[(\mathbf{m}\cdot\mathbf{r})^2 + m^2r^2]\boldsymbol{\omega}_o + (\mathbf{m}\cdot\boldsymbol{\omega}_o)[2(\mathbf{m}\cdot\mathbf{r})\mathbf{r} - r^2\mathbf{m}]}{3(\mathbf{m}\cdot\mathbf{r})^2 + m^2r^2} \tag{11.8}$$

The derivation of Eq. (11.8) makes use of the fact that $\mathbf{r}\cdot\boldsymbol{\omega}_o = 0$, because $\boldsymbol{\omega}_o$ is along the orbit normal. This result has two interesting limits. If the orbit plane is

[2]An extra factor of $4\pi/\mu_0 = 10^7$ is needed to convert to Am^2.

the magnetic equatorial plane, then $m\boldsymbol{\omega}_o = \pm\omega_o\mathbf{m}$ and $\mathbf{m}\cdot\mathbf{r} = 0$, so $\boldsymbol{\Omega} = \mathbf{0}$ as expected. In the more interesting case that the orbit plane contains the magnetic pole directions, $\mathbf{m}\cdot\boldsymbol{\omega}_o = 0$, and

$$\boldsymbol{\Omega} = 3\left[\frac{(\mathbf{m}\cdot\mathbf{r})^2 + m^2r^2}{3(\mathbf{m}\cdot\mathbf{r})^2 + m^2r^2}\right]\boldsymbol{\omega}_o, \quad \text{for } \mathbf{m} \text{ in the orbit plane} \tag{11.9}$$

This has the maximum and minimum magnitudes $\boldsymbol{\Omega} = 3\boldsymbol{\omega}_o$ at the magnetic equator and $\boldsymbol{\Omega} = 3\boldsymbol{\omega}_o/2$ at the magnetic poles.

11.2 Atmospheric Density

Modeling the local atmospheric density, ρ, is quite difficult, and many decades of work have resulted in many different models. Vallado [18] provides an excellent description of the development of atmospheric models, including many of the common assumptions built into them. This section is devoted to the development of atmospheric modeling including the numerous existing models with their benefits and shortcomings.

11.2.1 Exponentially Decaying Model Atmosphere

The simplest of all the models is a fully static, exponentially decaying model. This model [19] assumes the atmospheric density decays exponentially with increasing height. This model is fully static in the sense that the densities are independent of time. Also, the exponential model assumes an axially symmetric atmosphere about the polar axis. The exponential model is given by

$$\rho = \rho_0 \exp\left(-\frac{h - h_0}{H}\right) \tag{11.10}$$

where ρ_0 and h_0 are reference density and height, respectively, h is the height above the ellipsoid, and H is the scale height, which is the fractional change in density with height. Table 11.1 gives values of h_0, ρ_0, and H for various ranges of h [19]. The exponential atmosphere is a good choice for preliminary simulations because it yields decent results and is computationally efficient. However, this model is not usually sufficient for high fidelity simulations.

Table 11.1 Exponential atmospheric model

h (km)	h_0 (km)	ρ_0 (kg/m^3)	H (km)
0–25	0	1.225	8.44
25–30	25	3.899×10^{-2}	6.49
30–35	30	1.774×10^{-2}	6.75
35–40	35	8.279×10^{-3}	7.07
40–45	40	3.972×10^{-3}	7.47
45–50	45	1.995×10^{-3}	7.83
50–55	50	1.057×10^{-3}	7.95
55–60	55	5.821×10^{-4}	7.73
60–65	60	3.206×10^{-4}	7.29
65–70	65	1.718×10^{-4}	6.81
70–75	70	8.770×10^{-5}	6.33
75–80	75	4.178×10^{-5}	6.00
80–85	80	1.905×10^{-5}	5.70
85–90	85	8.337×10^{-6}	5.41
90–95	90	3.396×10^{-6}	5.38
95–100	95	1.343×10^{-6}	5.74
100–110	100	5.297×10^{-7}	6.15
110–120	110	9.661×10^{-8}	8.06
120–130	120	2.438×10^{-8}	11.6
130–140	130	8.484×10^{-9}	16.1
140–150	140	3.845×10^{-9}	20.6
150–160	150	2.070×10^{-9}	24.6
160–180	160	1.224×10^{-9}	26.3
180–200	180	5.464×10^{-10}	33.2
200–250	200	2.789×10^{-10}	38.5
250–300	250	7.248×10^{-11}	46.9
300–350	300	2.418×10^{-11}	52.5
350–400	350	9.158×10^{-12}	56.4
400–450	400	3.725×10^{-12}	59.4
450–500	450	1.585×10^{-12}	62.2
500–600	500	6.967×10^{-13}	65.8
600–700	600	1.454×10^{-13}	79.0
700–800	700	3.614×10^{-14}	109.0
800–900	800	1.170×10^{-14}	164.0
900–1,000	900	5.245×10^{-15}	225.0
>1,000	1,000	3.019×10^{-15}	268.0

11.2.2 Harris-Priester Model Atmosphere

The Harris-Priester atmosphere [6] was one of the first attempts to model density in terms of atmospheric temperature. Harris and Priester noted that the atmosphere is mainly governed by the laws of thermodynamics and equilibrium in diffusion, conduction and the absorption of thermal energy from the Sun. Density for the Harris-Priester atmosphere is determined by simultaneous integration of the heat conduction and diffusion equations for each of the molecular constituents in the atmosphere, yielding the number density for each constituent. Using the ideal gas law, the density can then be calculated. The Harris-Priester model has resulted in several lookup tables where one can determine the density as a function of the local time and altitude for a prescribed solar-flux value. These tables are convenient for determining the density variation within single day and may be useful in short term simulations. While the Harris-Priester model accounts for many of the phenomena affecting the atmosphere, it lacks the ability to model seasonal, latitudinal, and geomagnetic variations with the accuracy needed for high fidelity long term simulations.

11.2.3 Jacchia and GOST Model Atmospheres

Throughout the late 1960s and 1970s, Jacchia produced a series of models [7–9] and updates that explicitly handle each of the observed atmospheric phenomena. The basis for each of the Jacchia models is essentially the same as the Harris-Priester model with one major exception. Whereas the Harris-Priester model determines the temperature by integrating the heat conduction equation, Jacchia uses an empirically derived temperature profile to integrate the diffusion equation for each constituent. Another difference between the Harris-Priester and Jacchia models is that the Harris-Priester model accounts for all of the solar phenomena within the integration of the conduction equation whereas Jacchia takes a different approach and adds temperature and/or density corrections to account for each of the phenomena. A particular advantage to Jacchia's models is that he and his colleagues used real spacecraft data to curve fit certain phenomena. This allows Jacchia's models to accurately represent actual atmospheric conditions even though all the phenomena are not precisely modeled. One shortcoming of Jacchia's models are that there is no standard set of lookup tables, meaning that the entire temperature/density profile must be determined via numeric integration for each instant, limiting the models' use in real-time applications.

Jacchia's models must be numerically integrated because of the empirical temperature formula that he used. By proposing an alternative temperature function, Roberts [15] was able to analytically integrate the barometric and diffusion equations [17]. The resulting density profile proved to be a very good approximation to Jacchia's densities for all altitudes. The Jacchia-Roberts atmosphere produces very good results and has been implemented in NASA's Goddard Trajectory Determination System (GTDS) [13].

One last model atmosphere of interest is the Russian GOST atmosphere [21]. The GOST model is a purely empirical model constructed using data from the Cosmos spacecraft [17]. The main advantage to the GOST model is that it is extremely computationally efficient. In contrast the other models described, except the static model, the GOST model determines the density directly rather than computing first the temperature profile of the atmosphere. Five multiplicative correction factors then account for variations in the solar flux, diurnal effect, semiannual changes, and geomagnetic activity. Because the GOST model is empirically derived, it requires numerous extensive lookup tables to determine the proper coefficients corresponding to the current atmospheric conditions. The GOST model has been shown to provide good agreement with other semi-analytic models such as Jacchia 1977 [1] and NRLMSIS-00 [22].

11.2.4 Jacchia-Bowman 2008 (JB2008) Model Atmosphere

Because Jacchia's models treat each of the known phenomena individually, they provide a good base that can be refined given new experimental data and advances in research. Such is the case with the Jacchia-Bowman model atmospheres [2, 3]. Bowman uses new spacecraft drag data and new solar indices in order to develop corrections to Jacchia's 1970 model [7]. Bowman has incorporated new solar indices apart from the standard $F_{10.7}$ index in order to determine the exospheric temperature [2], fitted new semiannual and seasonal-latitudinal corrections, added high altitude corrections, and provided a new means to account for geomagnetic activity. The Jacchia-Bowman model atmosphere uses Jacchia's original temperature function, which means that numeric integration is required in order to determine the density distribution. While the model is slow, Bowman's corrections result in a substantial reduction in density error over other atmospheric models [2].

Because of the high quality of results produced by the Jacchia-Bowman model, it is the standard model used by the Joint Space Operations Center (JSpOC) when tracking objects. The most current version of the Jacchia-Bowman model, JB2008, can be found at the website http://sol.spacenvironment.net/~JB2008/. Along with Fortran source code, the website also offers many relevant publications relating to the JB2008 model and its previous versions, histories of solar flux data and magnetic storm indices. In this section the JB2008 model is described in detail.

11.2.4.1 Temperature Profile

Jacchia's models, from which the JB2008 model is constructed, begin by determining the temperature profile of the atmosphere at a specific time and location. This temperature profile is subsequently used to integrate the diffusion equations in order to determine the local density. Defining the temperature profiles begins by determining the exospheric temperature. The exospheric temperature is influenced by many factors including solar activity, local time and geomagnetic activity. The effect of solar activity on the exospheric temperature can be represented by the uncorrected exospheric temperature, T_c [2]

$$T_c = 392.4 + 3.227\ \bar{F}_S + 0.298\ \Delta F_{10} + 2.259\ \Delta S_{10} + 0.312\ \Delta M_{10} + 0.178\ \Delta Y_{10} \tag{11.11}$$

with

$$\bar{F}_S = W_T \bar{F}_{10} + (1 - W_T)\bar{S}_{10} \tag{11.12a}$$

$$W_T = \min\left(\left(\frac{\bar{F}_{10}}{240}\right)^{1/4}, 1\right) \tag{11.12b}$$

and T_c expressed in degrees Kelvin. The Δ values are given by $\Delta X = X - \bar{X}$ where X is the current value of the solar index and $\bar{X}$ is the 81-day centered average value of the index, centered at the desired time. F_{10}, S_{10}, M_{10}, and Y_{10} are indices which are used to describe solar activity, as described in detail in [2]. These indices are typically tabulated as daily values. However, care must be taken when reading values from a lookup table. Each index has its own associated time lag for implementation. Values for F_{10} and S_{10} have a time lag of 1 day, M_{10} has a 2 day time lag and Y_{10} has a 5 day time lag. To clarify this point, suppose we want to determine the exospheric temperature on 11-February-2011. The values F_{10}, S_{10}, and their centered averages would correspond to those tabulated for 10-February-2011. The M_{10} and $\bar{M}_{10}$ values would come from the tabulated values for 9-February-2011, and the values for Y_{10} would be read from 6-February-2011.

Once the uncorrected exospheric temperature is determined, correction factors corresponding to diurnal and geomagnetic effects are added in. The diurnal variation accounts for the day-to-night variation in density of the atmosphere. Using spacecraft drag data to derive the atmospheric density has shown that the maximum density occurs around 14h00 local solar time while a minimum in density occurs around 04h00 local solar time [10]. Accounting for the diurnal variation, the local atmospheric temperature is given by

$$T_\ell = T_c \left\{1 + R\left[\sin^m(\theta) + (\cos^m \eta - \sin^m \theta)\left|\cos^n(\tau/2)\right|\right]\right\} \tag{11.13}$$

with

$$\theta = \frac{1}{2}\left|\lambda + \delta_\odot\right| \tag{11.14a}$$

$$\eta = \frac{1}{2}\left|\lambda - \delta_\odot\right| \tag{11.14b}$$

$$\tau = H + \beta + p\sin(H + \gamma) \tag{11.14c}$$

$$H = \phi - \phi_\odot \tag{11.14d}$$

where λ is the geodetic latitude, ϕ is the longitude, $\delta_\odot$ is the declination of the Sun, and $\phi_\odot$ is the right ascension (longitude) of the Sun. The constants R, m, n β, p, and γ are derived so that the model given by Eq. (11.13) best fit spacecraft drag data;

$$R = 0.31 \qquad \beta = +37^\circ$$
$$m = 2.5 \qquad p = +6^\circ$$
$$n = 3.0 \qquad \gamma = +43^\circ$$

In addition to modeling of the diurnal bulge, Bowman has introduced an additional diurnal correction [3]. The temperature correction can be determined by first defining

$$F = 0.1(F_{10} - 100) \tag{11.15a}$$

$$T = \frac{LST}{24} \tag{11.15b}$$

$$y = \cos\lambda \tag{11.15c}$$

where LST is the local spacecraft time give by $LST = (H + \pi)\frac{180^\circ}{\pi}\frac{24}{360}$ which is then modulated so that $0 < LST \leq 24$. The temperature corrections were determined by curve-fitting data over certain altitude intervals. With h being the altitude in the ECEF frame given by Eq. (2.77), the results of these fits are given by

$120 \leq h \leq 200$ km

$$\begin{aligned}
H &= (h - 120)/80 \\
A &= C_{17} + C_{18}yT + C_{19}yT^2 + C_{20}yT^3 + C_{21}yF + C_{22}yFT + C_{23}yFT^2 \\
B &= C_1 + C_2F + C_3FT + C_4FT^2 + C_5FT^3 + C_6FT^4 + C_7FT^5 \\
&\quad + C_8yT + C_9yT^2 + C_{10}yT^3 + C_{11}yT^4 + C_{12}yT^5 + C_{13}y + C_{14}Fy \\
&\quad + C_{15}FyT + C_{16}FyT^2 \\
\Delta T_c &= (3A - B)\,H^2 + (B - 2A)\,H^3
\end{aligned} \tag{11.16}$$

$200 < h \leq 240$ km

$$\begin{aligned}
H &= 4/5 \\
\Delta T_c &= C_1H + C_2FH + C_3FHT + C_4FHT^2 + C_5FHT^3 + C_6FHT^4 + C_7FHT^5 \\
&\quad + C_8yHT + C_9yHT^2 + C_{10}yHT^3 + C_{11}yHT^4 + C_{12}yHT^5 + C_{13}yH \\
&\quad + C_{14}FyH + C_{15}FyHT + C_{16}FyHT^2 + C_{17} + C_{18}yT + C_{19}yT^2 + C_{20}yT^3 \\
&\quad + C_{21}Fy + C_{22}FyT + C_{23}FyT^2
\end{aligned} \tag{11.17}$$

$240 < h \le 300$ km

$$\begin{aligned}
H &= (h-200)/50,\ \bar{h} = 3,\ h_p = (h-240)/6 \\
A &= C_1H + C_2FH + C_3FHT + C_4FHT^2 + C_5FHT^3 + C_6FHT^4 \\
&\quad + C_7FHT^5 + C_8yHT \quad + C_9yHT^2 + C_{10}yHT^3 + C_{11}yHT^4 + C_{12}yHT^5 \\
&\quad + C_{13}yH + C_{14}FyH + C_{15}FyHT + C_{16}FyHT^2 + C_{17} + C_{18}yT \\
&\quad + C_{19}yT^2 + C_{20}yT^3 + C_{21}Fy + C_{22}FyT + C_{23}FyT^2 \\
B &= C_1 + C_2F + C_3FT + C_4FT^2 + C_5FT^3 + C_6FT^4 + C_7FT^5 + C_8yT + C_9yT^2 \\
&\quad + C_{10}yT^3 + C_{11}yT^4 + C_{12}yT^5 + C_{13}y + C_{14}Fy + C_{15}FyT + C_{16}FyT^2 \\
X &= B_1 + B_2F + B_3FT + B_4FT^2 + B_5FT^3 + B_6FT^4 + B_7FT^5 + B_8yT \\
&\quad + B_9yT^2 + B_{10}yT^3 + B_{11}yT^4 + B_{12}yT^5 + B_{13}y\bar{h} + B_{14}y\bar{h}T + B_{15}y\bar{h}T^2 \\
&\quad + B_{16}y\bar{h}T^3 + B_{17}y\bar{h}T^4 + B_{18}y\bar{h}T^5 + B_{19}y \\
Y &= B_{13}y + B_{14}yT + B_{15}yT^2 + B_{16}yT^3 + B_{17}yT^4 + B_{18}yT^5 \\
C &= 3X - Y - 3A - 2B \\
D &= X - (A + B + C) \\
\Delta T_c &= A + Bh_p + Ch_p^2 + Dh_p^4
\end{aligned} \tag{11.18}$$

$300 < h \le 600$ km

$$\begin{aligned}
H &= h/100 \\
\Delta T_c &= B_1 + B_2F + B_3FT + B_4FT^2 + B_5FT^3 + B_6FT^4 + B_7FT^5 \\
&\quad + B_8yT + B_9yT^2 + B_{10}yT^3 B_{11}yT^4 + B_{12}yT^5 + B_{13}yH \\
&\quad + B_{14}yHT + B_{15}yHT^2 + B_{16}yHT^3 + B_{17}yHT^4 \\
&\quad + B_{18}yHT^5 + B_{19}y
\end{aligned} \tag{11.19}$$

$600 < h \le 800$ km

$$\begin{aligned}
H &= 6,\ h_p = (h-600)/100 \\
A &= B_1 + B_2F + B_3FT + B_4FT^2 + B_5FT^3 + B_6FT^4 + B_7FT^5 \\
&\quad + B_8yT + B_9yT^2 + B_{10}yT^3 + B_{11}yT^4 + B_{12}yT^5 + B_{13}yH \\
&\quad + B_{14}yHT + B_{15}yHT^2 + B_{16}yHT^3 + B_{17}yHT^4 + B_{18}yHT^5 + B_{19}y \\
B &= B_{13}y + B_{14}yT + B_{15}yT^2 + B_{16}yT^4 + B_{17}yT^4 + B_{18}yT^5 \\
C &= -(3 + 4B)/4 \\
D &= (A + B)/4 \\
\Delta T_c &= A + Bh_p + Ch_p^2 + Dh_p^3
\end{aligned} \tag{11.20}$$

Table 11.2 Diurnal temperature correction factors

i	B_i	C_i
1	-0.457512297×10^1	-0.155986211×10^2
2	-0.512114909×10^1	-0.512114909×10^1
3	-0.693003609×10^2	-0.693003609×10^2
4	0.203716701×10^3	0.203716701×10^3
5	0.703316291×10^3	0.703316291×10^3
6	-0.194349234×10^4	-0.194349234×10^4
7	0.110651308×10^4	0.110651308×10^4
8	-0.174378996×10^3	-0.220835117×10^3
9	0.188594601×10^4	0.143256989×10^4
10	-0.709371517×10^4	-0.318481844×10^4
11	0.922454523×10^4	0.328981513×10^4
12	-0.384508073×10^4	-0.135332119×10^4
13	-0.645841789×10^1	0.199956489×10^2
14	0.409703319×10^2	-0.127093998×10^2
15	-0.482006560×10^3	0.212825156×10^2
16	0.181870931×10^4	-0.275555432×10^1
17	-0.237389204×10^4	0.110234982×10^2
18	0.996703815×10^3	0.148881951×10^3
19	0.361416936×10^2	-0.751640284×10^3
20		0.637876542×10^3
21		0.127093998×10^2
22		-0.212825156×10^2
23		0.275555432×10^1

$h < 120$ km or 800 km $< h$

$$\Delta T_c = 0 \tag{11.21}$$

The coefficients B_i and C_i can be found in Table 11.2.

During calculation of the diurnal temperature variation, Eq. (11.13), it was specifically assumed that this temperature occurs during times of no geomagnetic activity [8]. There are several means to account for the effect of geomagnetic activity on the density. The first method, proposed by Jacchia et al. [10] uses an empirically derived temperature correction based on the 3-h geomagnetic planetary index K_p. The empirical formulation was later updated to incorporate a latitudinal dependence. This method was found to work well for altitudes greater than 200 km, but fails to accurately match data below this height. To overcome this shortcoming a hybrid formula that includes both a temperature and density correction can be applied for altitudes lower than 200 km. A new method proposed by Bowman uses the DST (Disturbance Storm Time) index in order to calculate a temperature correction, ΔT_{DST} [2].

Given T_ℓ, the diurnal correction factor ΔT_c and the geomagnetic correction factor ΔT_{DST}, the local exospheric temperature, T_∞ is calculated as

$$T_\infty = T_\ell + \Delta T_c + \Delta T_{DST} \tag{11.22}$$

The atmospheric temperature profile is assumed to have a constant temperature $T_0 = 183$ K at an altitude of $h_0 = 90$ km. The temperature gradient at h_0 is also assumed equal to zero. From there the temperature rises rapidly to an inflection temperature, T_x at a height of $h_x = 125$ km. Above 125 km the temperature continues to increase in an asymptotic form towards the exospheric temperature T_∞. The inflection temperature has been empirically defined as

$$T_x = a + bT_\infty + c\exp[dT_\infty] \tag{11.23}$$

with the constant coefficients

$$a = 371.6678, \quad b = 0.02385$$
$$c = -392.8292, d = -0.0021357$$

Below h_x the temperature profile is given by a 4th-order polynomial

$$T(h) = T_x + \sum_{n=1}^{4} c_n (h - h_x)^n, \qquad 90 < h < h_x \text{ km} \tag{11.24}$$

where the coefficients, c_n are determined using the following boundary conditions

$$T(h_0) = 183 \tag{11.25a}$$

$$\left.\frac{dT}{dh}\right|_{h_0} = 0 \tag{11.25b}$$

$$G_x \equiv \left.\frac{dT}{dh}\right|_{h_x} = 1.90\frac{T_x - T_0}{h_x - h_0} \tag{11.25c}$$

$$\left.\frac{d^2T}{dh^2}\right|_{h_x} = 0 \tag{11.25d}$$

For $h > h_x$ the temperature profile is asymptotic and the temperature is given by

$$T(h) = T_x + \frac{2(T_\infty - T_x)}{\pi}\tan^{-1}\left(\frac{G_x\pi(h - h_x)}{2(T_\infty - T_x)}\left[1 + 4.5\times 10^{-6}(h - h_x)^{2.5}\right]\right)$$
$$h_x < h \tag{11.26}$$

where G_x is as given in Eq. (11.25c).

11.2.4.2 Barometric Equation

For altitudes between 90 and 105 km, the atmospheric density is computed by integrating the barometric equation. The differential form of the barometric equation is

$$d\ln(\rho) = d\ln\left(\frac{\bar{M}}{T}\right) - \frac{\bar{M}g}{R^*T}dh \tag{11.27}$$

where $R^* = 8.31432$ J/(K-mol) is the ideal gas constant, g is the local acceleration due to gravity given by

$$g = 9.80665\left(1 + \frac{h}{R_e}\right)^{-2} \tag{11.28}$$

where $R_e = 6356.766$ km and $\bar{M}$ is the mean molecular mass of the atmosphere. For $90 < h \leq 105$ km, $\bar{M}$ is approximated by means of a 6th-order polynomial

$$\bar{M}(h) = \sum_{n=0}^{6} c_n(h-90)^n, \qquad 90 < h < 105 \text{ km} \tag{11.29}$$

where the coefficients are given by

$$\begin{aligned} &c_0 = 28.15204, && c_1 = -8.5586 \times 10^{-2}, \; c_2 = 1.2840 \times 10^{-4} \\ &c_3 = -1.0056 \times 10^{-5}, && c_4 = -1.0210 \times 10^{-5}, \; c_5 = 1.5044 \times 10^{-6} \\ &c_6 = 9.9826 \times 10^{-8} \end{aligned}$$

After integrating the barometric equation, the resulting density is given by

$$\rho_{\text{uncorr}}(h) = \rho_0\left(\frac{\bar{M}(h)}{\bar{M}_0}\right)\left(\frac{T_0}{T(h)}\right)\exp\left[-\frac{F(h)}{R^*}\right] \tag{11.30}$$

where

$$F(h) \equiv \int_{h_0}^{h} \frac{\bar{M}(\xi)g(\xi)}{T(\xi)}d\xi \tag{11.31}$$

The density at h_0 is $\rho_0 = 3.46 \times 10^{-6}$ kg/m^3. Note that $T_0 = 183$ K and $\bar{M}_0 \equiv \bar{M}(h_0)$. The subscript "uncorr" has been added to the density in Eq. (11.30) because this density does not include correction factors for the observed semiannual, Eq. (11.39), and seasonal-latitudinal, Eq. (11.42), variations derived from drag data.

These correction factors must be applied to determine the density. The integral given in Eq. (11.31) must be computed numerically. A constant step, 6-point quadrature rule was applied

$$\int_a^b f(x)dx \approx \frac{b-a}{288}[19f(x_0) + 75f(x_1) + 50f(x_2) + 50f(x_3) + 75f(x)_4 + 19f(x_5)] \tag{11.32}$$

where the range from h_0 to h was broken into intervals with $b - a = 10$ m. The JB2008 Fortran code employed a 5-point quadrature rule, also with $b - a = 10$ m, for the $90 < h \leq 105$ km regime.

11.2.4.3 Diffusion Equation

For altitudes between 90 and 105 km, integration of the barometric equation is sufficient to determine the atmospheric density (save for some additional correction factors which will be discussed later). However, altitudes within this range are not typical of spacecraft except during launch and end of life because the drag at this height is high enough such that a spacecraft without thrusting will spiral back to Earth. Above 105 km the atmosphere is assumed to be in diffuse equilibrium. Here the density is computed by determining the number density of each of the individual atmospheric constituents. The atmosphere is assumed to be composed of six components: nitrogen, argon, helium, hydrogen, and atomic and molecular oxygen (O and O_2). The number density for each component is determined by integrating the diffusion equation

$$\frac{dn_i}{n_i} = -\frac{M_i g}{R^* T}dh - \frac{dT}{T}(1 - \alpha_i) \tag{11.33}$$

where n_i, M_i, and α_i are the number density (number of atoms per unit volume), molecular weight, and thermal diffusion coefficient of component i. Before integrating the diffusion equation, we need to know the number density at the boundary, 105 km. Given the density, $\rho(105)$, the number densities for five of the components, excluding hydrogen, are determined as

$$n_i = \frac{q_i N_A \rho(105)}{\bar{M}_{SL}}, \qquad i = \mathrm{N_2, Ar, He} \tag{11.34a}$$

$$n_{\mathrm{O}} = 2N_A \rho(105)\left(\frac{1}{\bar{M}(105)} - \frac{1}{\bar{M}_{SL}}\right) \tag{11.34b}$$

$$n_{\mathrm{O_2}} = N_A \rho(105)\left(\frac{1 + q_{\mathrm{O_2}}}{\bar{M}_{SL}} - \frac{1}{\bar{M}(105)}\right) \tag{11.34c}$$

where q_i is the volumetric fraction of each component in the mixture, N_A is Avogadro's constant, given by $6.02214129 \times 10^{23}$ mol^{-1}, and $\bar{M}_{SL} = 28.960$ is the mean molecular mass of the atmosphere at sea level. The number density

Table 11.3 Atmospheric composition at $h = 90$ km

Species (i)	q_i	M_i	α_i
Nitrogen (N_2)	0.78110	28.0134	0
Oxygen (O_2)	0.20955	31.9988	0
Oxygen (O)	–	15.9994	0
Argon (Ar)	0.0093432	39.948	0
Helium (He)	0.0000061471	4.0026	−0.38
Hydrogen (H)	–	1.00797	0

of hydrogen is not calculated in Eq. (11.34) because hydrogen does not have any impact on the density at altitudes below 500 km. Because of this, altitudes above 105 km must be broken into two distinct regimes, one with $105 < h < 500$ km and one with 500 km $\leq h$. Also note that the density in Eq. (11.34) is the corrected density at 105 km which includes the semiannual and seasonal-latitudinal correction factors alluded to before.

105 < h < 500 km

The volumetric fraction, molecular weight, and thermal diffusion coefficient of each component can be found in Table 11.3.

Integrating Eq. (11.33) results in

$$\ln(n_i(h)) = \ln(n_i(105)) - (1 + \alpha_i)\ln\left(\frac{T(h)}{T(105)}\right) - \frac{M_i G(h)}{R^*} \tag{11.35}$$

where

$$G(h) \equiv \int_{105}^{h} \frac{g(\xi)}{T(\xi)} d\xi \tag{11.36}$$

is again computed numerically by the 6-point quadrature and $b - a = 10$ m. Once the number densities are calculated, the uncorrected density is determined as

$$\rho_{\text{uncorr}}(h) = \frac{1}{N_A} \sum_i M_i n_i(h) \tag{11.37}$$

A discussion of the correction factors which need to be applied to determine the final density will be discussed later.

500 km < h

Above 500 km hydrogen begins to have an impact on the density calculation. To determine the density at any point above 500 km we begin by integrating the barometric equation for all components, except hydrogen, up to an altitude of 500 km. Next we determine the number density of hydrogen at this altitude according to [8]

$$\log_{10}(n_{H(500)}) = 79.13 - 39.40 \log_{10}(T_{500}) + 5.5 \log_{10}(T_{500})^2 \tag{11.38}$$

where T_{500} is the temperature at $h = 500$ km. Now the diffusion equations can be integrated for each component, including hydrogen, up to the desired altitude. The uncorrected density is then determined by Eq. (11.37).

11.2.4.4 Density Correction Factors

The density calculated to this point only includes correction factors accounting for diurnal variations and geomagnetic activity. These are not the only phenomena affecting the density that must be accounted for. The JB2008 atmospheric model includes corrections that account for the observed semiannual variation and seasonal-latitudinal variation of the lower atmosphere. Bowman has also proposed a correction for high altitudes, $h \geq 1{,}000$ k [3].

11.2.4.5 Semiannual Variation

The semiannual variation describes how the density fluctuates throughout the year. Analysis has shown that the period of this variation is about 6 months with density maxima occurring in April and October and density minima occurring in January and July. The semiannual density correction factor has the form

$$\Delta \log_{10} \rho_{SA} = f(h)g(t) \tag{11.39}$$

where $f(h)$ and $g(t)$ are calculated via

$$\bar{F}_{SMJ} = \bar{F}_{10} - 0.7\bar{S}_{10} - 0.04\bar{M}_{10} \tag{11.40a}$$

$$\bar{h} = h/1{,}000 \tag{11.40b}$$

$$f(h) = B_1 + B_2\bar{F}_{SMJ} + B_3\bar{F}_{SMJ}\bar{h} + B_4\bar{F}_{SMJ}\bar{h}^2 + B_5\bar{F}_{SMJ}^2\bar{h} \tag{11.40c}$$

and

$$\bar{F}_{SM} = \bar{F}_{10} - 0.75\bar{S}_{10} - 0.37\bar{M}_{10} \tag{11.41a}$$

$$\omega = 2\pi\frac{DOY - 1}{365} \tag{11.41b}$$

$$\begin{aligned} g(t) = {} & C_1 + C_2\sin(\omega) + C_3\cos(\omega) + C_4\sin(2\omega) + C_5\cos(2\omega) \\ & + \bar{F}_{SM}\left[C_6 + C_7\sin(\omega) + C_8\cos(\omega) + C_9\sin(2\omega) + C_{10}\cos(2\omega)\right] \end{aligned} \tag{11.41c}$$

where DOY is the current day of the year $DOY \in [0, 367)$. The coefficients B_i and C_i can be found in Table 11.4.

Table 11.4 Semiannual variation correction factors

i	B_i	C_i
1	2.689×10^{-1}	-3.633×10^{-1}
2	-1.176×10^{-2}	8.506×10^{-2}
3	2.782×10^{-2}	2.401×10^{-1}
4	-2.782×10^{-2}	-1.897×10^{-1}
5	3.470×10^{-4}	-2.554×10^{-1}
6		-1.790×10^{-2}
7		5.650×10^{-4}
8		-6.407×10^{-4}
9		-3.418×10^{-3}
10		-1.252×10^{-3}

11.2.4.6 Seasonal-Latitudinal Variation

In calculation of the temperature profile, we have assumed the temperature to be a constant 183 K at $h = 90$ km over the entire globe. In reality however, the temperature at this height varies seasonally and over different latitudes. The seasonal-latitudinal correction factor attempts to account for the variation in density arising from our assumption. The correction factor is given by

$$S = 0.2(h - 90)\exp\left[-0.045(h - 90)\right] \tag{11.42a}$$

$$\Phi = \frac{JD - 2400000.5 - 36204}{365.2422} \tag{11.42b}$$

$$P = \sin(2\pi\Phi + 1.72) \tag{11.42c}$$

$$\Delta\log_{10}\rho_{SL} = \text{sign}(\lambda)SP\sin^2(\lambda) \tag{11.42d}$$

where JD is the current Julian Date. The seasonal-latitudinal correction is only applicable to altitudes lower than 200 km. The semiannual variation and seasonal latitudinal variation are multiplicative correction factors to density. Thus the corrected density is given by

$$\log_{10}\rho_{\text{corr}} = \log_{10}\rho_{\text{uncorr}} + \Delta\log_{10}\rho_{SA} + \Delta\log_{10}\rho_{SL} \tag{11.43}$$

11.2.4.7 High Altitude Density Correction

The last correction factor accounts for variations observed at high altitudes. The high altitude correction factor was developed by Bowman after analysis of 25 spacecraft with orbital altitudes between 1,500 and 4,000 km. The high altitude correction factor follows as

$h < 1{,}000$ km

$$F_{HA} = 1 \tag{11.44}$$

$1{,}000 \leq h < 1{,}500$ km

$$H = (h - 1{,}000)/500$$
$$F_{1,500} = D_1 + D_2\bar{F}_{10} + 1{,}500D_3 + 1{,}500D_4\bar{F}_{10}$$
$$\frac{\partial F_{1,500}}{\partial h} = D_3 + D_4\bar{F}_{10}$$
$$F_{HA} = 1 + \left(3F_{1,500} - 500\frac{\partial F_{1,500}}{\partial h} - 3\right)H^2 + \left(500\frac{\partial F_{1,500}}{\partial h} - 2F_{1,500} + 2\right)H^3 \tag{11.45}$$

$1{,}500 \leq h$ km

$$F_{HA} = D_1 + D_2\bar{F}_{10} + D_3h + D_4h\bar{F}_{10} \tag{11.46}$$

where the coefficients are given by

$$D_1 = 2.2 \times 10^{-1}, \quad D_2 = -2.0 \times 10^{-3}$$
$$D_3 = 1.15 \times 10^{-3}, \quad D_4 = -2.11 \times 10^{-6}$$

Once the high altitude correction factor, and corrected density have been determined, the final atmospheric density is given by

$$\rho = F_{HA}\rho_{\text{corr}} \tag{11.47}$$

11.3 Sun Position, Radiation Pressure, and Eclipse Conditions

In order determine solar radiation pressure forces or torques acting on a spacecraft or to process Sun sensor data, we must first determine where the Sun is relative to the spacecraft and whether the spacecraft is shadowed by the Earth or the Moon. The position of the Sun with respect to the Earth can be determined as follows [17]. First, the mean longitude, $\phi_\odot$, and mean anomaly of the Sun, $M_\odot$, are determined in degrees as

$$\phi_\odot = 280.460° + 36{,}000.771\ T_{UT1} \tag{11.48a}$$
$$M_\odot = 357.5277233° + 35999.05034\ T_{UT1} \tag{11.48b}$$

where

$$T_{UT1} = \frac{JD(Y, M, D, h, m, s) - 2{,}451{,}545}{36{,}525} \tag{11.49}$$

with JD computed as in Sect. 2.6.3. Both $\phi_\odot$ and $M_\odot$ are reduced to the range 0° to 360° and the longitude of the ecliptic is determined in degrees as

$$\phi_{\text{ecliptic}} = \phi_\odot + 1.914666471^\circ \sin(M_\odot) + 0.019994643 \sin(2M_\odot) \tag{11.50}$$

The obliquity of the ecliptic is given by

$$\varepsilon = 23.439291^\circ - 0.0130042\, T_{UT1} \tag{11.51}$$

The unit vector in the direction from the Earth to the Sun is then

$$\mathbf{e}_{\oplus\odot} = \begin{bmatrix} \cos(\phi_{\text{ecliptic}}) \\ \cos(\varepsilon)\sin(\phi_{\text{ecliptic}}) \\ \sin(\varepsilon)\sin(\phi_{\text{ecliptic}}) \end{bmatrix} \tag{11.52}$$

We omit subscripts to indicate the coordinate frame because all vectors in this section are expressed in the GCI frame.

The distance, in AU, between the Earth and the Sun can be found by

$$r_{\oplus\odot} = 1.000140612 - 0.016708617\cos(M_\odot) - 0.000139589\cos(2M_\odot) \tag{11.53}$$

and the position vector from the Earth to the Sun is $\mathbf{r}_{\oplus\odot} = r_{\oplus\odot}\mathbf{e}_{\oplus\odot}$. After converting the spacecraft position vector $\mathbf{r}$ to AU, the position vector from the spacecraft to the Sun, expressed in the GCI frame is given

$$\mathbf{r}_{\text{sat}\odot} = \mathbf{r}_{\oplus\odot} - \mathbf{r} \tag{11.54}$$

The distance $r_{\text{sat}\odot}$, between the spacecraft and Sun (in AU) and the unit vector $\mathbf{e}_{\text{sat}\odot}$ are then given by

$$r_{\text{sat}\odot} = \|\mathbf{r}_{\text{sat}\odot}\| \tag{11.55a}$$

$$\mathbf{e}_{\text{sat}\odot} = \frac{\mathbf{r}_{\text{sat}\odot}}{r_{\text{sat}\odot}} \tag{11.55b}$$

The pressure of solar radiation at the position of the spacecraft is then

$$P_\odot = \frac{\mathscr{F}_\odot}{c\, r_{\text{sat}\odot}^2} \tag{11.56}$$

where $\mathscr{F}_\odot$, known as the *solar constant*, is the flux density of solar radiation at a distance of 1 AU from the Sun, and $c = 299{,}792{,}458$ m/s is the speed of light. The solar constant varies over the 11-year solar cycle from 1,361 W/m^2 at solar minimum to 1,363 W/m^2 at solar maximum and is subject to rapid fluctuations as large as 5 W/m^2 at times of high or low solar activity.[3] These fluctuations are very difficult to predict, although daily data does exist [5].

---

[3]These recent measurements [11] are about 5 W/m^2 lower than previous measurements [5].

Given the position vector of the spacecraft and the position vector of the Sun, it can now be determined whether or not the spacecraft is in the shadow of the Earth. There are two main approaches to shadowing. The first assumes that the shadow created by the Earth is a cylindrical projection of the Earth's diameter along the direction of the Sun to the Earth. In the cylindrical approximation, which is generally adequate for spacecraft in lower altitude orbits, the spacecraft is in the Earth's shadow if and only if

$$\mathbf{r} \cdot \mathbf{e}_{\oplus\odot} < -\sqrt{r^2 - R_\oplus^2} \tag{11.57}$$

where $\mathbf{r}$ is the spacecraft position vector and $R_\oplus$ is the equatorial radius of the Earth, which is approximated as a sphere. For a spacecraft in a geosynchronous orbit with radius 42,164 km, Eq. (11.57) gives an eclipse time of 69.6 min for the longest eclipse, which occurs at the equinoxes when the Sun is in the orbital plane of the spacecraft.

The second shadowing approach accounts for the finite diameters of both the Sun and the Earth. This more accurate approach has a conical shadow model with partial shadowing in a region called the penumbra. The more complex eclipse conditions for the conical shadow model, including illumination levels in the penumbra, can be found in Wertz [19]. His equations show that a spacecraft in a geostationary orbit spends 67.5 min in total shadow and 4.3 min in the penumbra during its longest eclipse.

11.4 Orbital Ephemerides of the Sun, Moon, and Planets

The best source for accurate orbital ephemerides of the Sun, Moon, and planets is DE405/LE405 computed by the Jet Propulsion Laboratory (JPL) of the California Institute of Technology [16]. These ephemerides are obtained by precise numerical integration of the equations of motion of the bodies. This computation has four main ingredients: formulating the equations of motion, determining the initial conditions, performing the integrations, and making the results available in a useful form. It is believed that the difficulty of accurately determining the initial conditions is the limiting factor in the accuracy of the solutions.

The equations of motion are described in detail in [16]. They include "(a) point-mass interactions among the Moon, planets, and Sun; (b) general relativity (isotropic, parameterized post-Newtonian); (c) Newtonian perturbations of selected asteroids; (d) action upon the figure of the Earth from the Moon and Sun; (e) action upon the figure of the Moon from the Earth and Sun; (f) physical libration of the Moon, modeled as a solid body with tidal and rotational distortion, including both elastic and dissipational effects; (g) the effect upon the Moon's motion caused by the tides raised upon the Earth by the Moon and Sun; and (h) the perturbations of 300 asteroids upon the motions of Mars, the Earth, and the Moon." The reference frame for the ephemerides is the International Celestial Reference Frame [14]. The initial conditions are determined by fitting data from optical observations, meridian transits, photographic and CCD astrometry, occultation timings, astrolabe observations, radiometric emission measurements, ranging data, VLBI data, and lunar laser range data.

The planetary ephemerides are saved as files of Chebyshev polynomials for the Cartesian positions and velocities of the Sun, Moon, and planets, typically at 32-day intervals. They can be obtained from JPL's interactive "Horizons" website at http://ssd.jpl.nasa.gov/?horizons. The positional accuracy of the inner planet ephemerides is believed to be a few thousandths of an arcsecond, and a few hundredths of an arcsecond for the outer planets. Reference [16] also provides Keplerian elements for applications that do not require the full accuracy of an integrated ephemeris.

References

1. Amelina, T., Batyr, G., Dicky, V., Tumolskaya, N., Yurasov, V.S.: Comparison of atmospheric density models. In: 2nd U. S. - Russian Space Surveillance Workshop. Poznan, Poland (1996)
2. Bowman, B.R., Tobiska, W., Marcos, F.A., Huang, C.Y., Lin, C.S., Burke, W.J.: A new empirical thermospheric density model JB2008 using new solar and geomagnetic indices. In: AIAA/AAS Astrodynamics Specialist Conference. Honolulu (2008)
3. Bowman, B.R., Tobiska, W., Marcos, F.A., Valladares, C.: The JB2006 empirical thermospheric density model. J. Atmos. Sol. Terr. Phys. **70**(5), 774–793 (2008)
4. Finlay, C.C., Maus, S., Beggan, C.D., Bondar, T.N., Chambodut, A., Chernova, T.A., Chulliat, A., Golovkov, V.P., Hamilton, B., Hamoudi, M., Holme, R., Hulot, G., Kuang, W., Langlais, B., Lesur, V., Lowes, F.J., Lühr, H., Macmillan, S., Mandea, M., McLean, S., Manoj, C., Menvielle, M., Michaelis, I., Olsen, N., Rauberg, J., Rother, M., Sabaka, T.J., Tangborn, A., Tøffner-Clausen, L., Thébault, E., Thomson, A.W.P., Wardinski, I., Wei, Z., Zvereva, T.I.: International Geomagnetic Reference Field: the eleventh generation. Geophys. J. Int. **183**, 1216–1230 (2010)
5. Fröhlich, C., Lean, J.: Total solar irradiance (TSI) composite database (1978-present). http://www.ngdc.noaa.gov/stp/solar/solarirrad.html#composite (Accessed Sept. 20, 2013)
6. Harris, I., Priester, W.: Time-dependent structure of the upper atmosphere. J. Atmos. Sci. **19**, 286–301 (1962)
7. Jacchia, L.G.: New static models of the thermosphere and exosphere with empirical temperature profiles. Tech. Rep. SAO Special Report No. 313, Smithsonian Astrophysical Observatory, Cambridge. Smithsonian Institution Astrophysical Observatory (1970)
8. Jacchia, L.G.: Revised static models of the thermosphere and exosphere with empirical temperature profiles. SAO special report no. 332, Smithsonian Astrophysical Observatory, Cambridge. Smithsonian Institution Astrophysical Observatory (1971)
9. Jacchia, L.G.: Thermospheric temperature, density, and composition: New models. SAO special report no. 375, Smithsonian Astrophysical Observatory, Cambridge. Smithsonian Institution Astrophysical Observatory (1977)
10. Jacchia, L.G., Slowey, J., Verniani, F.: Geomagnetic pertubations and upper-atmosphere heating". J. Geophys. Res. **72**(5), 1423–1434 (1966)
11. Kopp, G., Lean, J.L.: A new, lower value of total solar irradiance: Evidence and climate significance. Geophys. Res. Lett. **38**(1), L01,706 (2011)
12. Langel, R.A.: The main field. In: Jacobs, J.A. (ed.) Geomagnetism, pp. 249–512. Academic Press, Orlando (1987)

13. Long, A.C., Cappellari Jr., J.O., Velez, C.E., Fuchs, A.J.: Goddard Trajectory Determination System (GTDS) mathematical theory (revision 1). FDD/522-89/001 and CSC/TR-89/6001, National Aeronautics and Space Administration/Goddard Space Flight Center, Greenbelt (1989)
14. Ma, C., Arias, E.F., Eubanks, T.M., Frey, A.L., Gontier, A.M., Jacobs, C.S., Sovers, O.J., Archinal, B.A., Charlot, P.: The International Celestial Reference Frame as realized by very long baseline interferometry. Astron. J. **116**, 516–546 (1998)
15. Roberts Jr., C.E.: An analytical model for upper atmosphere densities based upon Jacchia's 1970 models. Celestial Mech. **314**(4), 368–377 (1971)
16. Standish, E.M., Williams, J.G.: Orbital ephemerides of the Sun, Moon, and planets. In: Urban, S.E., Seidelmann, P.K. (eds.) Explanatory Supplement to the Astronomical Almanac, 3rd edn., chap. 8. University Science Books, Sausalito (2013)
17. Vallado, D.A.: Fundamentals of Astrodynamics and Applications, 3rd edn. Microcosm Press, Hawthorne and Springer, New York (2007)
18. Vallado, D.A., Finkleman, D.: A critical assessment of satellite drag and atmospheric density modelling. In: AIAA/AAS Astrodynamics Specialist Conference. Honolulu (2008)
19. Wertz, J.R. (ed.): Spacecraft Attitude Determination and Control. Kluwer Academic, The Netherlands (1978)
20. Winch, D.E., Ivers, D.J., Turner, J.P.R., Stening, R.J.: Geomagnetism and Schmidt quasi-normalization. Geophys. J. Int. **160**, 487–504 (2005)
21. Yurasov, V.S., Cefola, P.J.: Earth's upper atmosphere density model for ballistic suport of the flight of artificial Earth satellites. Tech. rep., National Standard of the Russian Federation (Gosstandart of Russia), Moscow, Russia (2004)
22. Yurasov, V.S., Nazarenko, A.I., Cefola, P.J., Alfriend, K.T.: Density corrections for the NRLMSIS-00 atmospheric model. Adv. Astronaut. Sci. **120**(2), 1079–1107 (2005)

Chapter 12
Review of Control and Estimation Theory

The purpose of this chapter is to provide a review of control and estimation theory. It is expected that the reader has some basic knowledge of dynamical systems and probability theory. Several of the concepts shown in this chapter are used throughout the text. First a basic review of system modeling using differential equations is shown. This is followed by linear and nonlinear control theory. Then estimation concepts, such as maximum likelihood and the Kalman filter, are reviewed. The reader is encouraged to read the several cited texts in this chapter for further information.

12.1 System Modeling

In this section system modeling is reviewed. This subject is introduced by first deriving the equations of motion for the classic inverted pendulum problem. Then, state and observation models for both nonlinear and linear systems are introduced. Finally, discrete-time systems are reviewed.

12.1.1 Inverted Pendulum Modeling

The classic inverted pendulum mounted to a motorized cart is shown in Fig. 12.1a, where x is the cart position, M is the mass of the cart, m is the mass of the pendulum, ℓ is the length to pendulum center of mass, $J = m\ell^2/3$ is the mass moment of inertia of the pendulum, modeling it as a uniform thin rod, θ is the pendulum angle from vertical, and u is the force applied to the cart. The associated free body diagram is shown in Fig. 12.1b, where N and P are reaction forces along the horizontal

F.L. Markley and J.L. Crassidis, *Fundamentals of Spacecraft Attitude Determination and Control*, Space Technology Library 33, DOI 10.1007/978-1-4939-0802-8_12,
© Springer Science+Business Media New York 2014

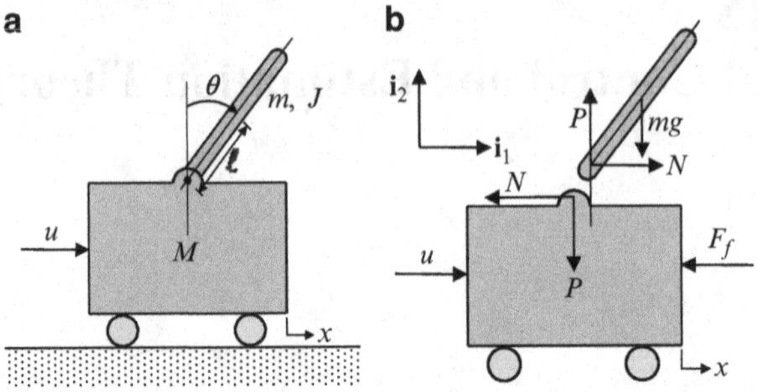

Fig. 12.1 Inverted pendulum (**a**) and associated free body diagram (**b**)

and vertical directions, respectively, and F_f is the friction force which is given by $F_f = c\dot{x}$, where c is a constant. The axes $\mathbf{i}_1$ and $\mathbf{i}_2$ are inertially fixed, which represent the horizontal and vertical directions, respectively.

The locations of the centroid of the pendulum, along the $\mathbf{i}_1$ and $\mathbf{i}_2$ axes, denoted by x_c and y_c, respectively, are given by

$$x_c = x + \ell \sin\theta \tag{12.1a}$$

$$y_c = \ell \cos\theta \tag{12.1b}$$

Taking two time derivatives of Eq. (12.1a) gives

$$\ddot{x}_c = \ddot{x} + \ell\ddot{\theta}\cos\theta - \ell\dot{\theta}^2\sin\theta \tag{12.2}$$

Therefore, summing the forces of the pendulum in the horizontal direction, $\mathbf{i}_1$, gives

$$N = m\ddot{x} + m\ell\ddot{\theta}\cos\theta - m\ell\dot{\theta}^2\sin\theta \tag{12.3}$$

Summing the forces of the cart in the horizontal direction gives

$$M\ddot{x} + c\dot{x} + N = u \tag{12.4}$$

Substituting Eq. (12.3) into Eq. (12.4) gives

$$(M + m)\ddot{x} + c\dot{x} + m\ell\ddot{\theta}\cos\theta - m\ell\dot{\theta}^2\sin\theta = u \tag{12.5}$$

Taking two time derivatives of Eq. (12.1b) gives

$$\ddot{y}_c = -(\ell\ddot{\theta}\sin\theta + \ell\dot{\theta}^2\cos\theta) \tag{12.6}$$

Summing the forces of the pendulum in the vertical direction, $\mathbf{i}_2$, gives

$$P = mg - m\ell\ddot{\theta}\sin\theta - m\ell\dot{\theta}^2\cos\theta \tag{12.7}$$

For the pendulum, summing the moments around its center of gravity gives the following rotational dynamic equations:

$$J\ddot{\theta} = (m\ell^2/3)\ddot{\theta} = P\ell\sin\theta - N\ell\cos\theta \tag{12.8}$$

Substituting Eqs. (12.3) and (12.7) into Eq. (12.8), and simplifying yields

$$(4\ell/3)\ddot{\theta} = g\sin\theta - \ddot{x}\cos\theta \tag{12.9}$$

Equations (12.5) and (12.9) provide the governing equations of motion.

We now linearize the governing equations about the vertical point, where $\theta = 0$. The first-order approximations are given by $\cos\theta \approx 1$, $\sin\theta \approx \theta$, and $\dot{\theta}^2 \approx 0$. Then the linearized governing equations become

$$(4\ell/3)\ddot{\theta} = g\theta - \ddot{x} \tag{12.10a}$$

$$(M + m)\ddot{x} + c\dot{x} + m\ell\ddot{\theta} = u \tag{12.10b}$$

Solving Eq. (12.10a) for $\ddot{\theta}$ and substituting the resultant into Eq. (12.10b) leads to

$$\ddot{x} = -\frac{c}{M'}\dot{x} - \frac{3mg}{4M'}\theta + \frac{1}{M'}u \tag{12.11}$$

where

$$M' \equiv M + m/4 \tag{12.12}$$

Substituting Eq. (12.11) into Eq. (12.10a) then leads to

$$\ddot{\theta} = \frac{3(M + m)g}{4M'\ell}\theta + \frac{3c}{4M'\ell}\dot{x} - \frac{3}{4M'\ell}u \tag{12.13}$$

We now define the following *state vector*:

$$\mathbf{x} = \begin{bmatrix} x \; \theta \; \dot{x} \; \dot{\theta} \end{bmatrix}^T \tag{12.14}$$

Taking the time derivative of Eq. (12.14), and using Eqs. (12.11) and (12.13) gives

$$\dot{\mathbf{x}} = \begin{bmatrix} 0 & 0 & 1 & 0 \\ 0 & 0 & 0 & 1 \\ 0 & -\dfrac{3mg}{4M'} & -\dfrac{c}{M'} & 0 \\ 0 & \dfrac{3(M+m)g}{4M'\ell} & \dfrac{3c}{4M'\ell} & 0 \end{bmatrix} \mathbf{x} + \begin{bmatrix} 0 \\ 0 \\ \dfrac{1}{M'} \\ -\dfrac{3}{4M'\ell} \end{bmatrix} u \tag{12.15a}$$

$$\mathbf{y} = \begin{bmatrix} 1\,0\,0\,0 \\ 0\,1\,0\,0 \end{bmatrix} \mathbf{x} + \begin{bmatrix} 0 \\ 0 \end{bmatrix} u \tag{12.15b}$$

where the outputs, given in the vector $\mathbf{y}$, are assumed to be x and θ.

12.1.2 State and Observation Models

Modeling is an important aspect for the analysis of system behavior and for the design of control and estimation algorithms. A model may either be static or dynamic. A static model's output at any given time only depends on the input at that time [19]. A dynamic model's present output depends on past inputs. Dynamic models frequently use differential or difference equations to describe physical systems. Both static and dynamic models can either be derived using a theoretical analysis or be identified using experimental data. Theoretical models help to provide a user with insight to any aspect which is modeled, but may not accurately represent the actual physical process. The major advantage of theoretical models is that the actual system need not be implemented to develop a model. Identified models usually describe actual physical processes more accurately, but may not provide all the information to describe all the desired aspects of the system. Choosing between a theoretical model or an identified model depends on a number of factors; including, the availability of obtaining experimental data, the desired accuracy of the model, the complexity of the physical system, etc.

12.1.2.1 State Models

An important quantity for modern modeling design and analysis is the state vector, such as the one given by Eq. (12.14). The components of a state vector are known as state variables. These variables describe a system's condition, and are directly related to the dissipation and storage of energy. The number of state variables is known as the order of the system, which describes how many energy storing elements are present in the system. The consequence of the order leads directly to the number of first-order differential or difference equations in the model.

An nth-order system can be written using n first-order equations (known as state equations), which are mathematically described in differential form by

$$\dot{\mathbf{x}}(t) = \mathbf{f}(\mathbf{x}(t), \mathbf{u}(t), t), \quad \mathbf{x}(t_0) = \mathbf{x}_0 \tag{12.16}$$

where $\mathbf{f}$ is an $n \times 1$ vector that is sufficiently differentiable, $\mathbf{x}$ is an $n \times 1$ state vector, and $\mathbf{u}$ is a $q \times 1$ vector that denotes an exogenous input representing any input to the system that does not depend on the state elements. An *autonomous* system is one that does not depend on time explicitly, so that $\dot{\mathbf{x}}(t) = \mathbf{f}(t)(\mathbf{x}(t), \mathbf{u}(t))$.

Equation (12.16) describes the general structure of a dynamic model. A subset of this general structure is a class of systems known as linear systems. A linear system follows the superposition principle [3], which states that a linear combination of inputs produces an output that is a linear combination of the outputs that would be produced if each input was applied separately [19]. Mathematically, this principle is described by

$$y = f(z) = f(\alpha_1 z_1 + \alpha_2 z_2) = \alpha_1 f(z_1) + \alpha_2 f(z_2) \tag{12.17}$$

where α_1 and α_2 are arbitrary constants, and z_1 and z_2 are arbitrary inputs. It is easy to show that the following system is linear:

$$\dot{\mathbf{x}}(t) = F(t)\mathbf{x}(t) + B(t)\mathbf{u}(t), \quad \mathbf{x}(t_0) = \mathbf{x}_0 \tag{12.18}$$

where the $n \times n$ matrix F and the $n \times q$ matrix B are known as the state matrix and input matrix, respectively. Equation (12.15a) shows the F and B matrices for the inverted pendulum system. The solution to Eq. (12.18) is given by

$$\mathbf{x}(t) = \Phi(t, t_0)\mathbf{x}_0 + \int_{t_0}^{t} \Phi(t, \tau)B(\tau)\mathbf{u}(\tau)\, d\tau \tag{12.19}$$

where $\Phi(t, t_0)$ is known as the *state transition matrix*, which has the following properties:

$$\Phi(t_0, t_0) = I \tag{12.20a}$$

$$\Phi(t_0, t) = \Phi^{-1}(t, t_0) \tag{12.20b}$$

$$\Phi(t_2, t_0) = \Phi(t_2, t_1)\Phi(t_1, t_0) \tag{12.20c}$$

$$\dot{\Phi}(t, t_0) = F(t)\Phi(t, t_0) \tag{12.20d}$$

Note that Eq. (12.20a) gives $\mathbf{x}(t_0) = \mathbf{x}_0$, which satisfies the initial condition given in Eq. (12.18). Differentiating Eq. (12.19) and using the properties in Eq. (12.20) shows that it is indeed the solution for Eq. (12.18):

$$\begin{aligned}\dot{\mathbf{x}}(t) &= \dot{\Phi}(t,t_0)\mathbf{x}_0 + \Phi(t,t)B(t)\mathbf{u}(t) + F(t)\int_{t_0}^{t}\Phi(t,\tau)B(\tau)\mathbf{u}(\tau)\,d\tau \\ &= F(t)\left[\Phi(t,t_0)\mathbf{x}_0 + \int_{t_0}^{t}\Phi(t,\tau)B(\tau)\mathbf{u}(\tau)\,d\tau\right] + B(t)\mathbf{u}(t) \\ &= F(t)\mathbf{x}(t) + B(t)\mathbf{u}(t)\end{aligned} \tag{12.21}$$

Note that since $\mathbf{x}(t)$ is the linear sum of a part due to the initial condition and a part due to the forcing input, it must follow the superposition principle.

Consider a simple single-input-single-output (SISO) nth-order linear and autonomous ordinary differential equation (ODE), given by

$$\frac{d^n y}{dt^n} + a_{n-1}\frac{d^{n-1}y}{dt^{n-1}} + \cdots + a_1\frac{dy}{dt} + a_0\,y = u \tag{12.22}$$

where y is the output variable and u is the input variable. In order to convert the ODE into first-order form, consider the following variable change:

$$\begin{aligned}x_1 &= y \\ x_2 &= \frac{dy}{dt} \\ &\vdots \\ x_n &= \frac{d^{n-1}y}{dt^{n-1}}\end{aligned} \tag{12.23}$$

This leads to the following equivalent system of n first-order equations:

$$\begin{aligned}\dot{x}_1 &= x_2 \\ \dot{x}_2 &= x_3 \\ &\vdots \\ \dot{x}_n &= -a_0x_1 - a_1x_2 - \cdots - a_{n-1}x_n + u\end{aligned} \tag{12.24}$$

which can be represented in matrix form by

$$\dot{\mathbf{x}} = F\mathbf{x} + Bu \tag{12.25}$$

where the vector $\mathbf{x}$ contains the state variables

$$\mathbf{x} = \begin{bmatrix}x_1 \; x_2 \cdots x_n\end{bmatrix}^T \tag{12.26}$$

and the matrices F and B are given by

$$F = \begin{bmatrix} 0 & 1 & 0 & \cdots & 0 \\ 0 & 0 & 1 & \cdots & 0 \\ \vdots & \vdots & \vdots & \ddots & \vdots \\ 0 & 0 & 0 & \cdots & 1 \\ -a_0 & -a_1 & -a_2 & \cdots & -a_{n-1} \end{bmatrix} \tag{12.27a}$$

$$B = \begin{bmatrix} 0 \, 0 \cdots 1 \end{bmatrix}^T \tag{12.27b}$$

If the linear system is autonomous, then the solution given by Eq. (12.19) simplifies to

$$\mathbf{x} = e^{F(t-t_0)}\mathbf{x}_0 + \int_{t_0}^{t} e^{F(t-\tau)} B\mathbf{u}(\tau)\,d\tau \tag{12.28}$$

where $e^{F(t-t_0)}$ is the matrix exponential.

A linear model can also be used to describe an autonomous nonlinear model. This is accomplished by performing a linearization about an equilibrium state vector and input, so that

$$\mathbf{0} = \mathbf{f}(\mathbf{x}_e, \mathbf{u}_e) \tag{12.29}$$

where $\mathbf{x}_e$ and $\mathbf{u}_e$ denote the known equilibrium quantities. Next, a multivariable Taylor series expansion about and $\mathbf{x}_e$ and $\mathbf{u}_e$ is performed, keeping only the first-order terms, so that

$$\dot{\mathbf{x}} \approx \mathbf{f}(\mathbf{x}_e, \mathbf{u}_e) + \left.\frac{\partial \mathbf{f}}{\partial \mathbf{x}}\right|_{\mathbf{x}_e} (\mathbf{x} - \mathbf{x}_e) + \left.\frac{\partial \mathbf{f}}{\partial \mathbf{u}}\right|_{\mathbf{u}_e} (\mathbf{u} - \mathbf{u}_e) \tag{12.30}$$

Now, the following linearized state and input vectors are defined as

$$\Delta\mathbf{x} = \mathbf{x} - \mathbf{x}_e \tag{12.31a}$$

$$\Delta\mathbf{u} = \mathbf{u} - \mathbf{u}_e \tag{12.31b}$$

Therefore, a linearized model of an autonomous nonlinear system is given by

$$\Delta\dot{\mathbf{x}} = F\Delta\mathbf{x} + B\Delta\mathbf{u} \tag{12.32}$$

where

$$F \equiv \left.\frac{\partial \mathbf{f}}{\partial \mathbf{x}}\right|_{\mathbf{x}_e} \tag{12.33a}$$

$$B \equiv \left.\frac{\partial \mathbf{f}}{\partial \mathbf{u}}\right|_{\mathbf{u}_e} \tag{12.33b}$$

This model can now be used to describe and analyze the characteristics of the nonlinear system using linear tools, such as superposition, linear stability, etc.

Choosing state variables usually depends on the complexity of the system. A complex system such as a spacecraft potentially has an extremely large number of state variables, such as orbital elements, kinematic elements, rotational dynamic elements, flexible dynamic elements, thermal effects, disturbance model elements, etc. Often, state variables do not themselves represent physical quantities, but are needed to form a relationship to physical quantities. Implementing all state variables into a full model can be a formidable task. Ideally, one chooses the least number of state variables to represent the characteristics of sought physical quantities. For example, for attitude estimation using rate-integrating gyros the state vector may only contain the attitude elements and gyro biases. Other effects, such as flexible dynamic effects, may often be neglected in the attitude estimation, or may not be required for on-orbit implementation. It is important to note that although this simple example describes the least number of state variables for attitude estimation, other state variables in other models may have already been used before or concurrently with the attitude estimator. For example, orbital element states may be required in the attitude estimator whenever the sensor supplying the data produces an observation which depends on spacecraft position in the orbit. An example of this scenario is a magnetometer-based attitude estimator, which requires knowledge of the orbital position state vector in order to develop the reference magnetic field vector. Therefore, other state variables can indirectly affect the state quantities used in the primary model.

12.1.2.2 Observation Models

State vectors are used to model the dynamic motion of a particular system. Observation vectors are used to show how a sensor relates to various state quantities. Observations are developed from sensor output quantities. The number of elements in an observation vector is usually determined by both the number of available sensors and the number of individual sensor quantities (e.g. a three-axis magnetometer provides observations of three quantities). The general form of an observation vector is given by

$$\mathbf{y}(t) = \mathbf{h}(\mathbf{x}(t), \mathbf{u}(t), t) \tag{12.34}$$

where $\mathbf{h}$ is a $p \times 1$ observation vector. This vector shows how each sensor is related to input and state quantities. The dimension of the observation vector (p) may be smaller or larger than the dimension of the state vector (n). As seen from Eq. (12.34), the observation vector may also be a function of the input $\mathbf{u}$ (it must always be a function of the state vector).

Linear observation models follow the same structure as linear state models. The standard form for the linear observation model is given by

$$\mathbf{y}(t) = H(t)\mathbf{x}(t) + D(t)\mathbf{u}(t) \tag{12.35}$$

where H is the $p \times n$ observation matrix (also referred to as the output or sensitivity matrix) and D is the $p \times q$ direct transmission matrix. Equation (12.15b) shows the H and D matrices for the inverted pendulum system.

The general SISO nth-order linear and autonomous ODE is given by

$$\frac{d^n y}{dt^n} + a_{n-1}\frac{d^{n-1}y}{dt^{n-1}} + \cdots + a_1\frac{dy}{dt} + a_0\, y$$
$$= b_n\frac{d^n u}{dt^n} + b_{n-1}\frac{d^{n-1}u}{dt^{n-1}} + \cdots + b_1\frac{du}{dt} + b_0\, u \tag{12.36}$$

In order to convert the ODE into first-order form we first rewrite Eq. (12.36) into an equivalent form involving two ODEs, given by

$$y = b_n\frac{d^n x}{dt^n} + b_{n-1}\frac{d^{n-1}x}{dt^{n-1}} + \cdots + b_1\frac{dx}{dt} + b_0\, x \tag{12.37a}$$
$$u = \frac{d^n x}{dt^n} + a_{n-1}\frac{d^{n-1}x}{dt^{n-1}} + \cdots + a_1\frac{dx}{dt} + a_0\, x \tag{12.37b}$$

where x is an intermediate variable. Now, consider the following variable change:

$$\begin{aligned} x_1 &= x \\ x_2 &= \frac{dx}{dt} \\ &\vdots \\ x_n &= \frac{d^{n-1}x}{dt^{n-1}} \end{aligned} \tag{12.38}$$

This leads to the following equivalent system of n first-order equations, given in matrix form by

$$\dot{\mathbf{x}} = F\mathbf{x} + Bu \tag{12.39a}$$
$$y = H\mathbf{x} + Du \tag{12.39b}$$

where the matrices F and B are given by Eq. (12.27), and H and D are given by

$$H = \left[(b_0 - b_n\, a_0)\ (b_1 - b_n\, a_1)\ \cdots\ (b_{n-1} - b_n\, a_{n-1})\right] \tag{12.40a}$$
$$D = b_n \tag{12.40b}$$

Clearly, if $b_0 = 1$ and the remaining coefficients $b_i = 0,\ i = 1, 2, \ldots, n$, then the intermediate variable $x = y$, which reduces the general case in Eq. (12.36) to the simple case in Eq. (12.22).

The *transfer function* for autonomous linear systems can be found by by taking the Laplace transform [12] of both sides of Eqs. (12.18) and (12.35) with zero initial conditions:

$$s\,\mathbf{X}(s) = F\,\mathbf{X}(s) + B\,\mathbf{U}(s) \tag{12.41a}$$

$$\mathbf{Y}(s) = H\,\mathbf{X}(s) + D\,\mathbf{U}(s) \tag{12.41b}$$

where s is the Laplace variable. Solving for $\mathbf{X}(s)$ in Eq. (12.41a) and substituting the resulting expression into Eq. (12.41b) yields

$$\mathbf{Y}(s) = \left[H\,(sI - F)^{-1}\,B + D\right]\mathbf{U}(s) \tag{12.42}$$

Since the inverse of $(sI - F)$ is given by its adjoint divided by its determinant, then the determinant of $(sI - F)$ gives the *poles* of the transfer function. Also, the eigenvalues of F are equivalent to the roots of the denominator of the transfer function. All of the eigenvalues must lie left of the imaginary axis, i.e. in the *left-hand plane*, for a stable response. Even if one eigenvalue lies to the right of the imaginary axis then the system is unstable.

Linearized observation models can also be developed, similar in fashion to the linearized state equations. This is accomplished by

$$H \equiv \left.\frac{\partial \mathbf{h}}{\partial \mathbf{x}}\right|_{\mathbf{x}_e} \tag{12.43a}$$

$$D \equiv \left.\frac{\partial \mathbf{h}}{\partial \mathbf{u}}\right|_{\mathbf{u}_e} \tag{12.43b}$$

which are again evaluated at the equilibrium conditions.

12.1.3 Discrete-Time Systems

All of the concepts shown previously extend to discrete-time systems. Discrete-time systems have now become standard in most dynamic applications with the advent of digital computers, which are used to process sampled-data systems for estimation and control purposes. The mechanism that acts on the sensor output and supplies numbers to the digital computer is the analog-to-digital (A/D) converter. Then, the numbers are processed through numerical subroutines and sent to the dynamic system input through the digital-to-analog (D/A) converter. This allows the use of software driven systems to accommodate the estimation/control aspect of a dynamic system, which can be modified simply by uploading new subroutines to the computer.

Fig. 12.2 Continuous signal and sampled zero-order hold

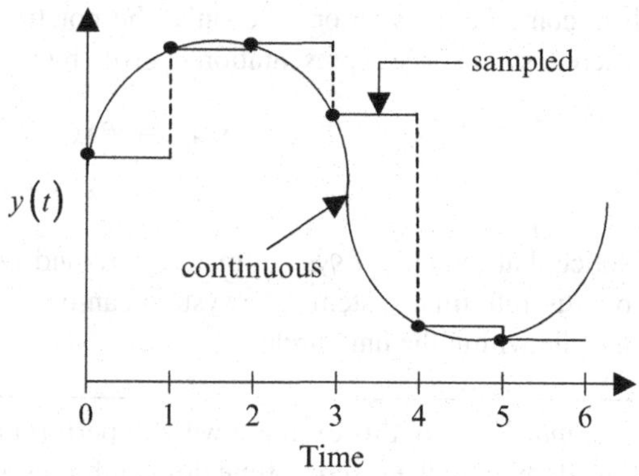

We shall only consider the most common sampled-type system given by a "zero-order hold" which holds the sampled point to a constant value throughout the interval. Figure 12.2 shows a sampled signal using a zero-order hold. Obviously, as the sample interval decreases the sampled signal more closely approximates the continuous signal. Consider the case where time is set to the first sample interval, denoted by Δt, and F and B are constants in Eq. (12.18). Then, Eq. (12.28) reduces to

$$\mathbf{x}(\Delta t) = e^{F\Delta t}\mathbf{x}(0) + \left[\int_0^{\Delta t} e^{F(\Delta t-\tau)}\,d\tau\right] B\,\mathbf{u}(0) \tag{12.44}$$

The integral can be simplified by defining $\zeta = \Delta t - \tau$, which leads to

$$\int_0^{\Delta t} e^{F(\Delta t-\tau)}\,d\tau = -\int_{\Delta t}^{0} e^{F\zeta}\,d\zeta = \int_0^{\Delta t} e^{F\zeta}\,d\zeta \tag{12.45}$$

Therefore, Eq. (12.44) becomes

$$\mathbf{x}(\Delta t) = \Phi\,\mathbf{x}(0) + \Gamma\,\mathbf{u}(0) \tag{12.46}$$

where

$$\Phi \equiv e^{F\Delta t} \tag{12.47a}$$

$$\Gamma \equiv \left[\int_0^{\Delta t} e^{Ft}\,dt\right] B \tag{12.47b}$$

Expanding Eq. (12.46) for $k+1$ samples gives

$$\mathbf{x}[(k+1)\Delta t] = \Phi\mathbf{x}(k\,\Delta t) + \Gamma\mathbf{u}(k\,\Delta t) \tag{12.48}$$

It is common convention to change the notation of Eq. (12.48) so that the entire discrete state-space representation is given by

$$\mathbf{x}_{k+1} = \Phi \mathbf{x}_k + \Gamma \mathbf{u}_k \tag{12.49a}$$

$$\mathbf{y}_k = H \mathbf{x}_k + D \mathbf{u}_k \tag{12.49b}$$

Notice that the output system matrices H and D are unaffected by the conversion to a discrete-time system. The system can be shown to be stable if all eigenvalues of Φ lie within the unit circle [7].

Example 12.1. In this example we will perform a conversion from the continuous-time domain to the discrete-time domain for a second-order system, given by

$$F = \begin{bmatrix} -1 & 0 \\ 1 & 0 \end{bmatrix}, \quad B = \begin{bmatrix} 1 \\ 0 \end{bmatrix}$$

To compute Φ we will enlist the help of Laplace transforms, with

$$\begin{aligned} \Phi = e^{F\Delta t} &= \left\{ \mathscr{L}^{-1}[sI - F]^{-1} \right\}\Big|_{\Delta t} = \left\{ \mathscr{L}^{-1} \begin{bmatrix} \dfrac{1}{s+1} & 0 \\ \dfrac{1}{s(s+1)} & \dfrac{1}{s} \end{bmatrix} \right\}\Bigg|_{\Delta t} \\ &= \begin{bmatrix} e^{-\Delta t} & 0 \\ 1 - e^{-\Delta t} & 1 \end{bmatrix} \end{aligned}$$

where $\mathscr{L}^{-1}$ denotes the inverse Laplace transform. The matrix Γ is computed using Eq. (12.47b):

$$\Gamma = \int_0^{\Delta t} \begin{bmatrix} e^{-t} \\ 1 - e^{-t} \end{bmatrix} dt = \begin{bmatrix} 1 - e^{-\Delta t} \\ \Delta t + e^{-\Delta t} - 1 \end{bmatrix}$$

If the sampling interval is chosen to be $\Delta t = 0.1$ s, then Φ and Γ become

$$\Phi = \begin{bmatrix} 0.9048 & 0 \\ 0.0952 & 1 \end{bmatrix}, \quad \Gamma = \begin{bmatrix} 0.0952 \\ 0.0048 \end{bmatrix}$$

Determining analytical expressions for Φ and Γ can be tedious and difficult for large-order systems. Fortunately, several numerical approaches exist for computing these matrices [18]. A computationally efficient and accurate approach involves a series expansion:

$$\Phi = I + F\Delta t + \frac{1}{2!} F^2 \Delta t^2 + \frac{1}{3!} F^3 \Delta t^3 + \cdots \tag{12.50}$$

and the matrix Γ is obtained from integration of Eq. (12.50):

$$\Gamma = \left[I\Delta t + \frac{1}{2!}F\Delta t^2 + \frac{1}{3!}F^2\Delta t^3 + \cdots \right] B \tag{12.51}$$

Adequate results can be obtained in most cases using only a few of the terms in the series expansion. For the matrices in Example 12.1, using three terms in the series expansion yields

$$\Phi = \begin{bmatrix} 0.9048 & 0 \\ 0.0952 & 1 \end{bmatrix}, \quad \Gamma = \begin{bmatrix} 0.0952 \\ 0.0048 \end{bmatrix} \tag{12.52}$$

The series results for Φ and Γ are accurate to within four significant digits. Results vary with sampling interval. As a general rule of thumb, if the sampling interval is below Nyquist's upper limit, then three to four terms in the series expansion gives accurate results [20].

12.2 Control Theory

In this section control theory is reviewed. First, the fundamentals of basic linear control are introduced. A single axis attitude control system design is shown next to highlight the various aspects of linear control. Then, stability of nonlinear systems is discussed. Finally, a popular nonlinear controller law, called sliding-mode control, is reviewed.

12.2.1 Basic Linear Control Design

A fundamental block diagram of a SISO negative feedback control system is shown in Fig. 12.3, where $R(s)$ is the reference, $Y(s)$ is the output signal, $D(s)$ is the disturbance, $U(s)$ is the control input, $G_p(s)$ is the plant to be controlled, $G_c(s)$ is the controller, and $H(s)$ is the sensor dynamics. The disturbance is added to the control input because for spacecraft the disturbance is an external torque that "turns" it just as a control input does. This is similar to a current-controlled motor system [5]. Assuming the disturbance is zero, the transfer function between the output and reference is given by

$$\frac{Y(s)}{R(s)} = \frac{G_c(s)\,G_p(s)}{1 + G_c(s)\,G_p(s)\,H(s)} \tag{12.53}$$

The denominator of Eq. (12.53) is known as the *characteristic equation* whose roots define the type of response. All roots of the associated polynomial must be to the

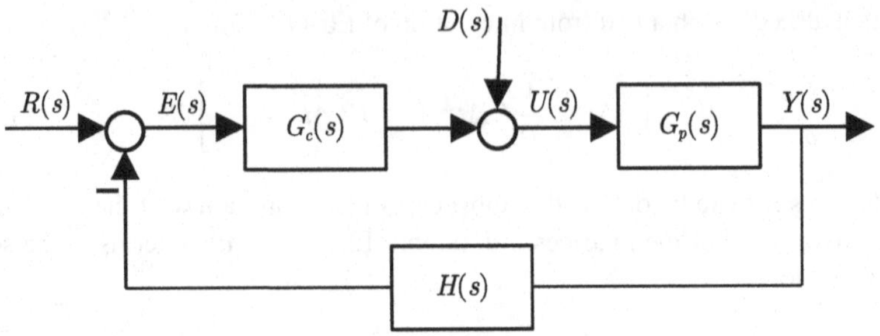

Fig. 12.3 Control block diagram

left of the imaginary axis to insure that the closed-loop system is asymptotically stable. Clearly, the poles of the open-loop system, $G_c(s)\,G_p(s)\,H(s)$, can now be manipulated by the controller, $G_c(s)$, to either stabilize an unstable plant or obtain some desired characteristics in the closed-loop system. The sensor dynamics can also affect the closed-loop response, although for most cases this is not a destabilizing issue. Assuming the reference is zero, the transfer function between the output and disturbance is given by

$$\frac{Y(s)}{D(s)} = \frac{G_p(s)}{1 + G_c(s)\,G_p(s)\,H(s)} \tag{12.54}$$

Note that the denominators of Eqs. (12.53) and (12.54) are equivalent, which means that the stability dynamics are the same for the reference and disturbance closed-loop responses. The numerators are different which means that the actual output responses to a reference input and a disturbance input are different. The total response is given by the sum of the two transfer functions.

12.2.1.1 Single Axis Attitude Control

In this section a simple single axis attitude control system is designed. For this case Euler's rotational equations of motion reduce down to simply

$$J\,\ddot{\theta} = u \tag{12.55}$$

where J is the inertia, θ is the angle, and u is the applied torque. The open-loop transfer function is given by

$$\frac{\Theta(s)}{U(s)} = \frac{1}{J\,s^2} \tag{12.56}$$

Clearly this system is unstable since it has a double pole at the origin. Several different types of controllers can be designed to stabilize this system. Here we focus on a "proportional-integral-derivative" (PID) controller. We first will attempt a simple proportional controller, where the error signal, $E(s)$, is multiplied simply by a scalar gain, k_p. Assuming that $H(s) = 1$ the closed-loop dynamics in Eq. (12.53) are given by

$$\frac{Y(s)}{U(s)} = \frac{k_p}{J\,s^2 + k_p} \tag{12.57}$$

Clearly, using only proportional control results in a closed-loop system that yields a sinusoidal response and thus the response is not asymptotically stable.

We now will attempt a PD controller, with $G_c(s) = k_p + k_d\,s$, where k_d is the scalar gain associated with the derivative portion of the controller. The closed-loop system now becomes

$$\frac{Y(s)}{U(s)} = \frac{k_p + k_d\,s}{J\,s^2 + k_d\,s + k_p} \tag{12.58}$$

Now the closed-loop dynamics can be made stable. The characteristic equation can be written in the familiar form of $s^2 + 2\zeta\,\omega_n\,s + \omega_n^2$, where ζ is the *damping ratio* and ω_n is the *natural frequency*, which is the frequency of the sinusoidal response for the undamped response. From Eq. (12.58) we have $\omega_n = \sqrt{k_p/J}$ and $\zeta = k_d/(2\sqrt{J\,k_p})$. Four types of responses can be given for various values of ζ:

- $\zeta < 0$. This results in an unstable system.
- $\zeta = 0$. This is the *undamped* response which yields a pure sinusoidal closed-loop response with frequency given by ω_n. The undamped response is given by $y(t) = A\,\sin(\omega_n t) + B\,\cos(\omega_n t)$, where A and B are constants determined from the initial conditions.
- $0 < \zeta < 1$. This results in exponentially decaying sinusoidal, known as the *under-damped* response, with frequency $\omega_d = \omega_n\sqrt{1-\zeta^2}$, called the *damped natural frequency*. The response is given by $y(t) = e^{-\zeta\omega_n t}[A\,\sin(\omega_d t) + B\,\cos(\omega_d t)]$, where A and B are constants determined from the initial conditions. Figure 12.4 gives plots for various values of ζ. Clearly, the oscillations damp out faster as ζ increases.
- $\zeta = 1$. This gives a *critically damped* response that converges as fast as possible without oscillating. The roots of the characteristic equation are repeated for this case with $s_{1,2} = -\omega_n$. The response is given by $y(t) = (A + B\,t)e^{-\zeta\omega_n t}$, where A and B are constants determined from the initial conditions. Note that the term t grows with time, but the overall system is still stable because the exponential "decays" the response faster than the t "grows" the response.
- $\zeta > 1$. This gives an *over-damped* response since all the roots, denoted by s_1 and s_2, are real and negative. The response is given by $y(t) = A\,e^{s_1 t} + B\,e^{s_2 t}$, where A and B are constants determined from the initial conditions.

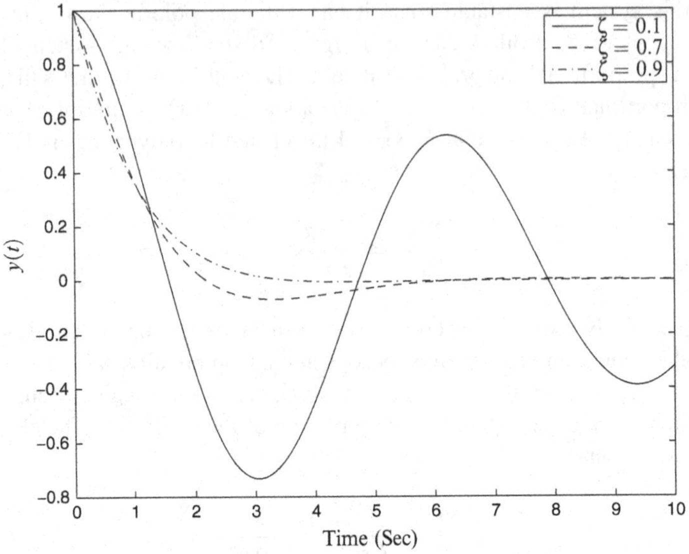

Fig. 12.4 Response for various damping ratio values

The single-axis PD control design explains a fundamental concept for basic spacecraft design. That is, spacecraft attitude control *must* incorporate derivative feedback in order to achieve an asymptotically stable response. This derivative information is achieved either through gyros or by taking a finite difference of the attitude. For the latter case, Eq. (3.174) can be used to determine the angular velocity. However, it is important to note that employing a finite difference amplifies noise since higher frequencies are amplified. This is clearly seen by the fact that a magnitude plot of s shows a 20 decibel per decade increase. Typically, a low-pass filter is employed to filter the noise. This approach works well when loose attitude and jitter requirements are given. Some researchers imply that spacecraft can be controlled without angular velocity measurements [14]. However, a derivative is always given in these works, which must be derived from a finite difference approach in practice. For tight pointing and/or jitter requirements the finite difference approach will typically not meet the requirements and high quality gyros must be employed.

From Eq. (12.58) it can be seen that a unit step input will result in a unit response as $t \to \infty$. Thus zero steady-state error is achieved which results in perfect tracking. The closed-loop disturbance response is given by

$$\frac{Y(s)}{D(s)} = \frac{1}{J\,s^2 + k_d\,s + k_p} \tag{12.59}$$

From the *final value theorem* [5] the steady-state response for a unit disturbance input is given by $1/k_p$. Ideally we want the steady-state disturbance response to

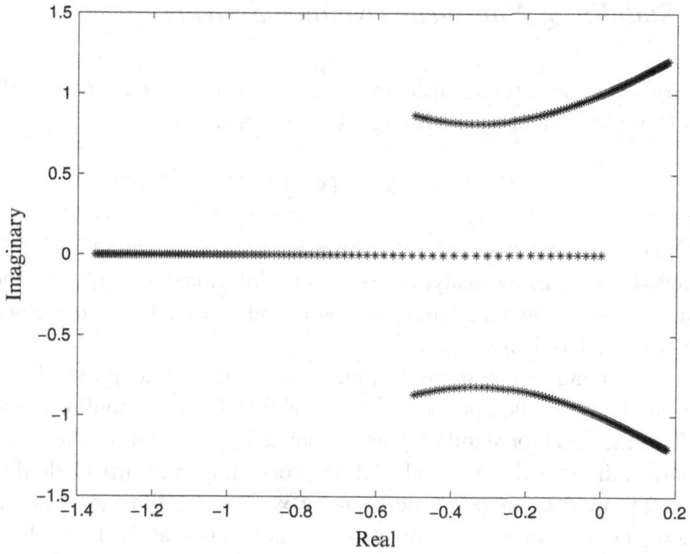

Fig. 12.5 Pole locations for increase k_i

be zero. We employ a full PID controller with $G_c(s) = k_p + k_d\,s + k_i/s$, where k_i is the integral gain, to make this happen. The closed-loop disturbance response is now given by

$$\frac{Y(s)}{D(s)} = \frac{s}{J\,s^3 + k_d\,s^2 + k_p\,s + k_i} \tag{12.60}$$

From the final value theorem the disturbance response at steady-state is now zero. However, one needs to be careful. From Eq. (12.59) a stable response is given when the PD gains are positive. However, the addition of integral control can actually destabilize the system. We show this by example. We set $J = k_p = k_d = 1$ and vary k_i from 0 to 2. A plot of the root locations, also known as a *root locus* plot, for varying k_i is shown in Fig. 12.5. When $k_i = 0$ the roots are given by 0 and $-1/2 \pm \sqrt{3}/2\,j$. As k_i increases the two complex poles move towards the imaginary axis and cross it when $k_i > 1$. While the 0 pole moves further down the left-hand plane's real axis, any integral gain that is greater than 1 causes the closed-loop system to become unstable. This is a typical characteristic of integral control and the reason why care must be taken when it is employed in a control design.

12.2.2 *Stability of Nonlinear Dynamic Systems*

We now consider the circumstance in which the original system of differential equations is nonlinear and can be brought to the standard form

$$\dot{\mathbf{x}} = \mathbf{f}(\mathbf{x}) \tag{12.61}$$

Some of the nonlinear systems of differential equations encountered in applications can be solved for an exact analytical solution. Unfortunately, only a minority of these systems have known analytical solutions and no standardized methods exist for finding exact analytical solutions.

Stability for nonlinear systems is much more difficult to prove. Fortunately, Lyapunov methods can be applied to show stability for both nonlinear and linear systems. Two methods for stability were introduced by Lyapunov. The first is given by Lyapunov's linearization method. Before proceeding with this method we must first define an equilibrium point, denoted by $\mathbf{x}_e$. An equilibrium is defined as a point where the system states remain indefinitely, so that $\dot{\mathbf{x}}(t) = \mathbf{0}$ for all t. For linear systems there is usually only one equilibrium point given at $\mathbf{x}_e = \mathbf{0}$, although there are exceptions. In Lyapunov's linearization method each equilibrium point is considered and evaluated in a linearized model. The equilibrium point is said to be Lyapunov stable if we can select a bound on initial conditions that results in trajectories that remain with a chosen finite limit. Furthermore, the equilibrium point is asymptotically stable if the linearized state also approaches zero as time approaches infinity. Denoting F as the Jacobian of $\mathbf{f}(\mathbf{x})$ evaluated at the equilibrium point, Lyapunov's linearization method gives the following stability conditions [25]:

- The equilibrium point is asymptotically stable for the actual nonlinear system if the linearized system is strictly stable, with all eigenvalues of F strictly in the left-hand plane.
- The equilibrium point is unstable for the actual nonlinear system if the linearized system is strictly unstable, with at least one eigenvalue strictly on the right-hand plane.
- Nothing can be concluded if the linearized system is marginally stable, with at least one eigenvalue of F on the imaginary axis and the remainder in the left-hand plane (the equilibrium point may be stable or unstable for the nonlinear system).

Lyapunov's linearization method provides a powerful approach to help qualify the stability of a system if a control (or estimation) scheme is designed to remain within a linear region, but does not give a thorough understanding of the nonlinear system in many cases.

Lyapunov's direct method gives a global stability condition for the general nonlinear system. This concept is closely related to the energy of a system, which is a *scalar* function. The scalar function must in general be continuous and have continuous derivatives with respect to all components of the state vector. Lyapunov showed that if the total energy of a system is dissipated, then the state is confined to a volume bounded by a surface of constant energy, so that the system must eventually settle to an equilibrium point. This concept is valid for both linear and nonlinear systems. Lyapunov stability is given if a chosen scalar function $V(\mathbf{x})$ satisfies the following conditions:

- $V(\mathbf{x}_e) = 0$
- $V(\mathbf{x}) > 0$ for $\mathbf{x} \neq \mathbf{x}_e$
- $\dot{V}(\mathbf{x}) \leq 0$

If these conditions are met, then $V(\mathbf{x})$ is a *Lyapunov function*. Furthermore, if $\dot{V}(\mathbf{x}) < 0$ for $\mathbf{x} \neq \mathbf{x}_e$, then the equilibrium point, $\mathbf{x}_e$, is asymptotically stable. For the case when the condition $\dot{V}(\mathbf{x}) \leq 0$ can only be shown, LaSalle's theorem [28] allows us to prove asymptotic stability. We first define an invariant set, as given in [25]: A set G is an *invariant set* for a dynamic system if every system trajectory which starts from a point G remains in G for all future time. Now assume that $\dot{V}(\mathbf{x}) \leq 0$ is true over the entire state space and that $V(\mathbf{x}) \to \infty$ as $\|\mathbf{x}\| \to \infty$. Let R be the set of all points $\dot{V}(\mathbf{x}) = 0$ and M be the largest invariant set in R. LaSalle's theorem states that all solutions globally asymptomatically converge to M as time approaches infinity. More details on Lyapunov methods for stability can be found in [25].

Example 12.2. Consider the following spring-mass-damper system with nonlinear spring and damper components:

$$m\ddot{x} + c\dot{x}|\dot{x}| + k_1 x + k_2 x^3 = 0$$

where m, c, k_1, and k_2 have positive values. The system can be represented in first-order form by defining the following state vector $\mathbf{x} = \begin{bmatrix} x \ \dot{x} \end{bmatrix}^T$:

$$\begin{aligned} \dot{x}_1 &= x_2 \\ \dot{x}_2 &= -(k_1/m)x_1 - (k_2/m)x_1^3 - (c/m)x_2|x_2| \end{aligned}$$

The system has only one equilibrium point at $\mathbf{x} = \begin{bmatrix} 0 \ 0 \end{bmatrix}^T$ that is physically correct (the others are complex). We wish to investigate the stability of this nonlinear system using Lyapunov's direct method. Intuitively, we choose a candidate Lyapunov function that is given by the total mechanical energy of the system, which is the sum of its kinetic and potential energies:

$$V(\mathbf{x}) = \frac{1}{2}m\dot{x}^2 + \int_0^x (k_1 x + k_2 x^3)\, dx$$

Evaluating this integral yields

$$V(\mathbf{x}) = \frac{1}{2}m\dot{x}^2 + \frac{1}{2}k_1 x^2 + \frac{1}{4}k_2 x^4$$

Note that zero energy corresponds to the equilibrium point ($\mathbf{x} = \mathbf{0}$), which satisfies the first condition for a valid Lyapunov function. Also, the second condition, $V(\mathbf{x}) > 0$ for $\mathbf{x} \neq \mathbf{0}$, is clearly satisfied. Taking the time derivative of $V(\mathbf{x})$ gives

$$\dot{V}(\mathbf{x}) = m\ddot{x}\dot{x} + (k_1 x + k_2 x^3)\dot{x}$$

Solving the original system equation for $m\ddot{x}$, and substituting the resulting expression into the equation for $\dot{V}(\mathbf{x})$ yields

$$\dot{V}(\mathbf{x}) = -c|\dot{x}|^3$$

Clearly, $\dot{V}(\mathbf{x}) \leq 0$ for all nonzero values of $\dot{x}$. Therefore, $V(\mathbf{x})$ is a Lyapunov function and shows that the system is stable. But $\dot{V}(\mathbf{x})$ does not depend on x. To prove asymptotic stability we consider the set R, which consists of $\mathbf{x}$ such that $x_2 = 0$ and x_1 is anything. Suppose $\mathbf{x}(t)$ is a trajectory starting in R. Then $\dot{x}_1 = x_2 = 0$, which implies that $x_1(t)$ is constant. But if this constant is not zero, then $\dot{x}_2$ is not necessarily zero and so x_2 will become nonzero. Therefore, any initial point in R must leave R. Since R is not attracting, then LaSalle's theorem implies the origin is asymptotically stable. This example shows how an "energy-like" function can be used to find a Lyapunov function, since the energy of this system is dissipated by the damper until the mass settles down.

Lyapunov's global method can be shown to be valid for linear autonomous systems with $\dot{\mathbf{x}} = F\mathbf{x}$. Consider the function $V(\mathbf{x}) = \mathbf{x}^T P\mathbf{x}$, where P is a positive definite symmetric matrix. Clearly, $V(\mathbf{x}) > 0$ for all $\mathbf{x} \neq \mathbf{0}$. The time derivative of $V(\mathbf{x})$ is given by

$$\dot{V}(\mathbf{x}) = \dot{\mathbf{x}}^T P\mathbf{x} + \mathbf{x}^T P\dot{\mathbf{x}} = \mathbf{x}^T (F^T P + PF)\mathbf{x} \tag{12.62}$$

Next, define the following matrix Lyapunov equation:

$$F^T P + PF = -Q \tag{12.63}$$

If Q is strictly positive definite then the system is asymptotically stable. Lyapunov showed that this condition is true if and only if all eigenvalues of F are strictly in the left-hand plane. See [4] for the proof.

12.2.3 *Sliding-Mode Control*[1]

In this section a popular control approach, called *sliding-mode control*, for nonlinear systems is briefly reviewed. More details can be found in [25]. Sliding-mode control is also referred to as *variable structure control* because the structure of the

[1]The authors would like to thank Agamemnon L. Crassidis for many of the contributions in this section.

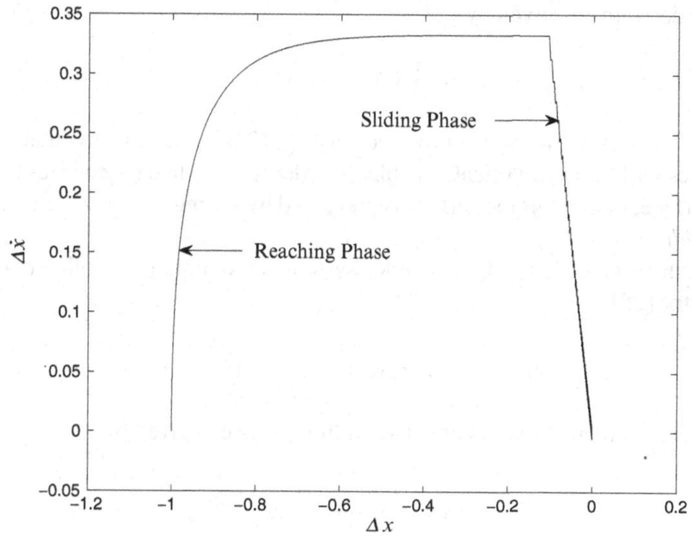

Fig. 12.6 Phase portrait

control law varies, i.e. switches, based on the position of the state trajectory. Before introducing the actual sliding control approach the basic concepts of switching theory must first be introduced. A candidate Lypaunov function is first chosen so that the system's state trajectories in the phase plane remain stable and point toward the origin. The switching law forces the state trajectories onto a surface in the phase plane, which is called the *sliding surface* or *switching surface*. The control law switches between two sets of control laws depending if the states are above or below the surface. Once the state trajectories reach the sliding surface, referred to as the *reaching phase*, the discontinuous controller forces the states to slide towards the origin, referred to as the *sliding phase*. A phase portrait showing these two phases for the system used in Example 12.3 is shown in Fig. 12.6.

We begin our introduction to sliding-mode control by considering the following second-order system:

$$\ddot{x} = f(x, \dot{x}) + u \tag{12.64}$$

where $f(x, \dot{x})$ may be a linear or nonlinear function. Let x_c be the desired state and $\Delta x = x - x_c$. The sliding surface, denoted by s, for a second-order system can be mathematically described by the following:

$$s = \Delta\dot{x} + \lambda\, \Delta x \tag{12.65}$$

for some scalar λ. Consider the following candidate Lyapunov function:

$$V(\Delta x) = \frac{1}{2}s^2 \tag{12.66}$$

The time derivative is given by

$$\dot{V}(\Delta x) = s\,\dot{s} \tag{12.67}$$

A control law can now be found based on Eq. (12.67) that ensures that the state trajectories will be asymptotically stable. In order for the state trajectories to remain on the surface, motion off the surface is prevented by setting $\dot{s} = 0$, which is referred to as a *sliding mode*.

This can be extended to higher-order systems by defining the following tracker error vector [25]:

$$\Delta \mathbf{x} = \mathbf{x} - \mathbf{x}_c = \left[\Delta x \;\; \Delta \dot{x} \;\; \cdots \;\; \Delta x^{(n-1)}\right]^T \tag{12.68}$$

where n is the order of the system. The sliding surface is given by

$$s = \left(\frac{d}{dt} + \lambda\right) \Delta x \tag{12.69}$$

For example, when $n = 3$ we have $s = \Delta \ddot{x} + 2\lambda\,\Delta \dot{x} + \lambda^2 \Delta \dot{x}$.

Sliding-mode control essentially reduces the order of the system to a first-order system by a form of feedback linearization. Direct feedback linearization is attractive for nonlinear systems since the nonlinear system dynamics are transformed into a system with linear dynamics. Typically, however, the dynamics of the system are not known exactly. This may lead to severe tracking errors when a direct feedback linearization control law is implemented on an actual system. In other words, robustness in the presence of parametric and un-modeled uncertainties cannot be ensured. Sliding-mode control solves the shortcomings and limitations of direct feedback linearization by forcing the state trajectories to track desired state trajectories in the phase portrait. Stability is ensured by Lyapunov's direct method and forms the basis for deriving the control law. As an example consider the following assumed second-order system:

$$\ddot{x} = \bar{f}(x, \dot{x}) + u \tag{12.70}$$

where the assumed model $\bar{f}(x, \dot{x})$ will be used to develop the control law. Taking the derivative of s in Eq. (12.65) and substituting $\ddot{x} = \bar{f}(x, \dot{x}) + u$ gives

$$\begin{aligned} \dot{s} &= \ddot{x} - \ddot{x}_c + \lambda\,\Delta \dot{x} \\ &= \bar{f}(x, \dot{x}) + u - \ddot{x}_c + \lambda\,\Delta \dot{x} \end{aligned} \tag{12.71}$$

Using the condition $\dot{s} = 0$ from Eq. (12.67) leads to the following control law:

$$u_e = -\bar{f}(x, \dot{x}) + \ddot{x}_c - \lambda\,\Delta \dot{x} \tag{12.72}$$

where u_e is interpreted as the "best" estimate of the equivalent control [25]. In order to account for model uncertainties a discontinuous term is added across the sliding surface:

$$u = u_e - k\,\mathrm{sign}(s) \tag{12.73}$$

for some scalar k. Substituting this control law into Eq. (12.71) and then substituting the resultant into Eq. (12.67) yields

$$\begin{aligned}\dot{V}(\Delta x) &= \left[f(x, \dot{x}) - \bar{f}(x, \dot{x}) - k\,\mathrm{sign}(s)\right] s \\ &= \left[f(x, \dot{x}) - \bar{f}(x, \dot{x})\right] s - k\,|s|\end{aligned} \tag{12.74}$$

Let us assume that the model error on $f(x, \dot{x})$ can be bounded by some known function $F(x, \dot{x})$, so that

$$|f(x, \dot{x}) - \bar{f}(x, \dot{x})| \leq F(x, \dot{x}) \tag{12.75}$$

Let the maximum value of $F(x, \dot{x})$ be denoted by $F_{\max}$. Then setting $k = F_{\max} + \eta$ for some positive scalar η gives $\dot{V}(\Delta x) \leq -\eta |s|$.

The resulting sliding-mode control law provides stability in the face of modeling uncertainties if k is chosen to be large enough, but it is important to note that as k increases the control discontinuity also increases. The control switching law will cause the closed-loop system to "chatter" due to the discontinuous term, which is undesirable for most applications. The chattering can be eliminated by smoothing the control signal in a thin boundary layer. To accomplish this task the signum function is replaced with a saturation function with a varying boundary layer thickness. The applied sliding-mode control is now given by

$$u = u_e - k\,\mathrm{sat}(s, \epsilon) \tag{12.76}$$

where ϵ is the boundary layer thickness and

$$\mathrm{sat}(s, \epsilon) \equiv \begin{cases} 1 & \text{for} \quad s > \epsilon \\ s/\epsilon & \text{for} \quad |s| \leq \epsilon \\ -1 & \text{for} \quad s < -\epsilon \end{cases} \tag{12.77}$$

The tracking performance is suboptimal using the saturation function, compared to using the signum function, but this provides a good tradeoff between robustness and practical implementation for a smoothed control input. The extension of sliding-mode control to multiple inputs is given in [25]. Other properties of this control approach can also be found in this reference.

Example 12.3. In this example the nonlinear system shown in Example 12.2 is controlled to a desired trajectory. The system with a control input is given by

$$m\ddot{x} + c\dot{x}|\dot{x}| + k_1 x + k_2 x^3 = u$$

where m, c, k_1, and k_2 have positive values. The system can be represented in first-order form by defining the following state vector $\mathbf{x} = \begin{bmatrix} x & \dot{x} \end{bmatrix}^T$:

$$\dot{x}_1 = x_2$$
$$\dot{x}_2 = -(k_1/m)x_1 - (k_2/m)x_1^3 - (c/m)x_2|x_2| + u$$

The desired trajectories are simulated using the model parameters: $m = 1$, $c = 2$, $k_1 = 1$, and $k_2 = 0.5$ with $u = 0$ for all time and an initial condition of $\mathbf{x}_0 = [1 \;\; 0]^T$. A time interval of 0.01 s is used for the integration process with a final time of 50 s.

The sliding-mode control law assumes the model parameters: $\bar{m} = 1$, $\bar{c} = 3$, $\bar{k}_1 = 1.5$, and $\bar{k}_2 = 0.75$. The control parameters are given by $\lambda = 3$ and $k = 1$. The boundary layer thickness when the saturation function is used in the control law is given by $\epsilon = 1$. The assumed initial condition is given by zero for both states. Simulation results using the signum and saturation functions are shown in Fig. 12.7. Comparing Fig. 12.7a and c shows that using the signum function drives the position error to zero faster than using the saturation function. But comparing Fig. 12.7b and d shows that using the signum function produces the classical chattering phenomenon for the control input, while the saturation function produces a much smoother control input.

12.3 Estimation Theory

In this section the basic concepts of estimation theory are reviewed. We begin with a discussion on the differences between static-based and filter-based estimation. Then, methods of least squares are presented. Formulations are shown for both batch and recursive least squares methods involving linear processes. Then, maximum likelihood estimation is reviewed, including a review of the Cramér-Rao lower bound. Next, nonlinear least squares is reviewed. An example is shown that involves estimating the attitude of a spacecraft from body and reference observations. This is followed by a brief discussion of the advantages and disadvantages of least squares methods. Then, a review of state estimation methods is provided, which includes the Kalman filter for linear systems, the extended Kalman filter for nonlinear systems, and a batch smoothing algorithm. This is followed by a discussion on the concept of linear covariance analysis. Finally, the separation theorem is reviewed.

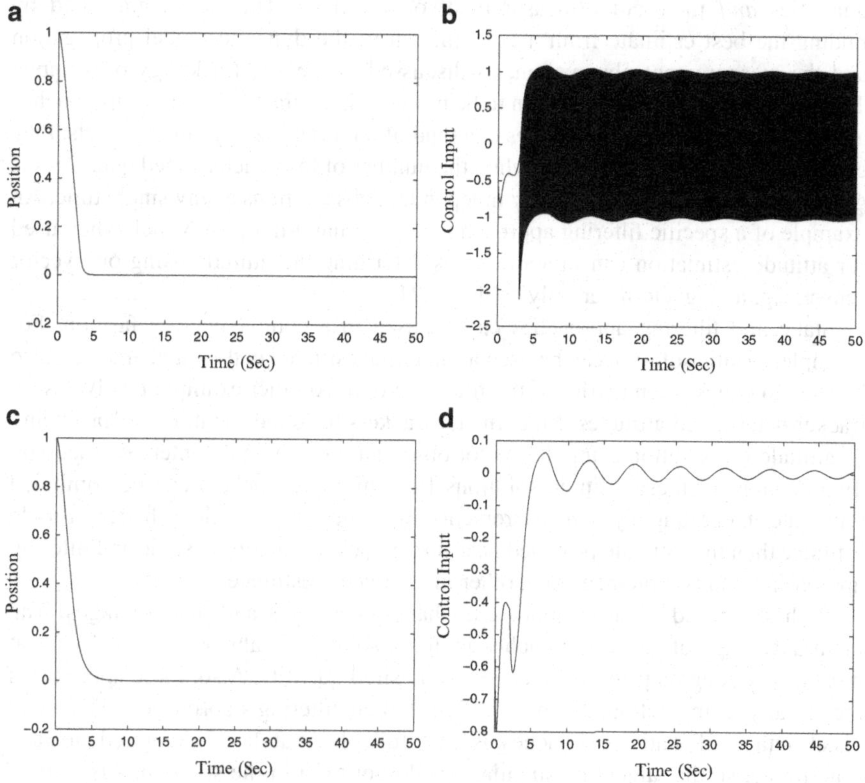

Fig. 12.7 Sliding-mode simulation results. (**a**) Position error—signum function. (**b**) Control input—signum function. (**c**) Position error—saturation function. (**d**) Control input—saturation function

12.3.1 Static-Based and Filter-Based Estimation

Filtering refers to the process of estimating the current state of a system from a set of measured observations using a priori information, typically given by the previous time state estimate. The filtering problem involves finding the best estimate of the true system state using an assumed model and measurements each corrupted with random noise of known statistics. The variables to be estimated are usually collected into a state vector, which typically include more variables than just the attitude. For example, star tracker measurements can be combined with a kinematic model, which is propagated using gyroscopic measurements. However, all gyros have inherent drift, which cause inaccuracies in the propagated model. A filter is used to estimate the attitude and gyro drift, oftentimes referred to as a bias because over short periods the drift occurs slowly, from the measurements. An important aspect of filtering approaches is that they can be used to simultaneously estimate quantities *and* filter noisy measurement observations. This is accomplished by finding the best estimate from a combination of the dynamic model propagation and the measurement observation, as discussed in the star tracker/gyro example. The attitude estimate can be shown to be more accurate than using static approaches alone. Because filtering approaches can (and often do) utilize dynamic models, they can in theory estimate quantities when the number of to-be-determined quantities is greater than the number of linearly independent observations at any single time. An example of a specific filtering approach is the Kalman filter [11], which when used for attitude estimation can in some cases determine the attitude using one vector sensor, e.g. a magnetometer-only solution [21].

Static and filtering approaches can be used in a complementary fashion. For example, a static solution can be used as an initial estimate in the filter, and may also be used to check the integrity of the filter solution. Another example involves star-tracker determined attitudes. Modern star trackers have the capability of providing an attitude-out solution using the vector observations only in the internal processor. Two or more of these attitude solutions from multiple trackers can be combined with rate-integrating gyro measurements in a filter to provide a better attitude estimate than the estimate provided be a single tracker. Therefore, static and filtering approaches can complement each other in the overall estimate.

Both static and filtering approaches have advantages and disadvantages. The main advantage of static approaches is that a solution is always provided with at most a very rough a priori estimate of the desired quantity. Also, these approaches are usually computationally more efficient than filtering approaches. The main disadvantage of static approaches is that full observability is required at each time frame, so that algebraic singularities do not exist in the solution. Also, some variables cannot be included or determined from a static solution. Finally, optimally combining determined quantities with the proper statistical balance may be difficult to do using static approaches.

The main advantage of filtering approaches is that estimates may still be found even when deterministic methods fail, such as in the algebraic singularity situation. Also, other variables, such as biases, can be appended into the state estimation variables, and can easily be combined with other variables with the proper statistical balance. Another advantage is that filters often provide statistical measures, such as estimate error covariances, as part of their solutions, while simultaneously filtering noisy observations. The main disadvantage of filtering approaches is that an a priori estimate is usually required for the solution. Also, filters may be prone to divergence problems. Finally, the computational load, coding size, and implementation effort are usually greater for filtering approaches than with static approaches.

This section provides a review of both static-based and filtering-based estimation approaches. It is expected that the reader has basic knowledge of probability and statistics, especially the theory behind expectations of a random variable, Gaussian distributions and the definition of the covariance matrix. First, an introduction to the notation used for the estimation concepts throughout the text is shown.

This is followed by a review of static-based methods, including linear and nonlinear least squares. Then, filter-based methods, such as the Kalman filter and extended Kalman filter are reviewed.

The foundation of all estimation algorithms involves taking sensor measurements to estimate unknown variables. For any variable or parameter in estimation, there are three quantities of interest: the true value, the measured value, and the estimated value. The true value (or "truth") is usually unknown in practice. This represents the actual value sought of the quantity being approximated by the estimator. In this text the true value contains the word "true" as a superscript, such as x^{true}. Measurements are never perfect, since they will always contain errors. Thus, measurements are usually modeled using a function of the true values plus some error. Other quantities used commonly in estimation are the measurement error (measurement value minus true value) and the residual error (measurement value minus estimated value). Thus, for a measurable quantity x, the following two equations hold:

$$\begin{array}{ccccc} \text{measured value} & = & \text{true value} & + & \text{measurement error} \\ x & = & x^{\text{true}} & + & v \end{array}$$

and

$$\begin{array}{ccccc} \text{measured value} & = & \text{estimated value} & + & \text{residual error} \\ x & = & \hat{x} & + & \epsilon \end{array}$$

The actual measurement error, v, like the true value, is never known in practice. However, the errors in the mechanism that physically generate this error are usually approximated by some known process (often by a zero-mean Gaussian noise process with known variance). These assumed known statistical properties of the measurement errors are often employed to weight the relative importance of various measurements used in the estimation scheme. Unlike the measurement error, the residual error is known explicitly and is easily computed once an estimated value has been found. The residual error is often used to drive the estimator itself. It should be evident that both measurement errors and residual errors play important roles in the theoretical and computational aspects of estimation.

12.3.2 Batch Least Squares Estimation

The principle of batch least squares, developed simultaneously by Karl Gauss and Adrien-Marie Legendre in the early Nineteenth Century, is used to estimate the elements of a constant vector from redundant observations. Therefore, least squares methods are known as "static" estimators. It is important to note that these methods may still involve dynamic models; however, the elements to be estimated are always constant. The simplest least squares problem has a linear observation model, given by

$$\mathbf{y} = H\mathbf{x}^{\text{true}} + \mathbf{v} \tag{12.78}$$

where $\mathbf{y}$ is an $m \times 1$ vector of observations, $\mathbf{x}^{\text{true}}$ is an $n \times 1$ vectors of to-be-estimated variables, H is an $m \times n$ constant observation matrix, and $\mathbf{v}$ is an $m \times 1$ vector of measurement noise which is usually represented by a zero-mean Gaussian white-noise process with covariance

$$R \equiv E\{\mathbf{v}\mathbf{v}^T\} \tag{12.79}$$

where $E\{\}$ denotes *expectation* [4]. Here it is assumed that R is a positive definite matrix. The optimum estimate for $\mathbf{x}^{\text{true}}$, denoted by $\hat{\mathbf{x}}$, is found by minimizing a loss function involving the weighted sum square residual errors, given by

$$J(\hat{\mathbf{x}}) = \frac{1}{2}(\mathbf{y} - H\hat{\mathbf{x}})^T R^{-1}(\mathbf{y} - H\hat{\mathbf{x}}) \tag{12.80}$$

Minimizing Eq. (12.80) can easily be shown to be equivalent to maximizing the likelihood function [26], which will be shown later. The best least squares estimate is obtained by taking the derivative of Eq. (12.80) with respect to $\hat{\mathbf{x}}$ and setting the resultant to zero, which yields

$$\hat{\mathbf{x}} = (H^T R^{-1} H)^{-1} H^T R^{-1}\mathbf{y} \tag{12.81}$$

A sufficiency test for locating the minimum is found by using the second derivative of $J(\hat{\mathbf{x}})$,

$$\frac{\partial^2 J(\hat{\mathbf{x}})}{\partial \hat{\mathbf{x}}^2} = H^T R^{-1} H \tag{12.82}$$

which is an $n \times n$ *Hessian* matrix. It is positive definite when H has rank n, thus providing a sufficient condition for a minimum. For the simple case of estimating a scalar variable, it is easy to show that Eq. (12.81) reduces to the average of a number of observations. Variants to the least squares problem, such as the constrained problem, can be found in [4].

The structure of Eq. (12.81) can also be used to prove that the estimator is "unbiased." An estimator $\hat{\mathbf{x}}(\mathbf{y})$ is said to be an "unbiased estimator" of $\mathbf{x}$ if $E\{\hat{\mathbf{x}}(\mathbf{y})\} = \mathbf{x}^{\text{true}}$ for every possible value of $\mathbf{x}^{\text{true}}$ [4].[2] If $\hat{\mathbf{x}}$ is biased, the difference $E\{\hat{\mathbf{x}}(\mathbf{y})\} - \mathbf{x}^{\text{true}}$ is called the "bias" of $\hat{\mathbf{x}}$. Substituting Eq. (12.78) into Eq. (12.81) gives

$$\hat{\mathbf{x}} = \mathbf{x}^{\text{true}} + (H^T R^{-1} H)^{-1} H^T R^{-1}\mathbf{v} \tag{12.83}$$

Taking the expectation of both sides gives $E\{\hat{\mathbf{x}}\} = \mathbf{x}^{\text{true}}$, since $\mathbf{v}$ has zero mean. Therefore, Eq. (12.81) is indeed an unbiased estimate.

[2]This implies that the estimate is a *function* of the measurements.

12.3.3 Sequential Least Squares Estimation

The batch least squares problem can be run as a sequential process. This is accomplished by adding an observation and deriving a sequential estimate based on the previous estimate [4]. The sequential estimate follows the following recursion equation:

$$\hat{\mathbf{x}}_{k+1} = \hat{\mathbf{x}}_k + K_{k+1}(\mathbf{y}_{k+1} - H_{k+1}\hat{\mathbf{x}}_k) \tag{12.84}$$

where

$$K_{k+1} = P_{k+1}H_{k+1}^T R_{k+1}^{-1} \tag{12.85}$$

$$P_{k+1}^{-1} = P_k^{-1} + H_{k+1}^T R_{k+1}^{-1} H_{k+1} \tag{12.86}$$

The matrix P is known as the *estimate error-covariance matrix*, which will be described shortly. Equation (12.84) modifies the previous best correction $\hat{\mathbf{x}}_k$ by an additional correction to account for the information contained in the $(k + 1)$th measurement subset. Another form for Eq. (12.85) is given by using the *matrix inversion lemma* or *Sherman-Morrison-Woodbury formula* [9]:

$$(A + U\,C\,V)^{-1} = A^{-1} - A^{-1}U\,(C^{-1} + V\,A^{-1}U)^{-1}V\,A^{-1} \tag{12.87}$$

This leads to the *covariance recursion form*, given by

$$\hat{\mathbf{x}}_{k+1} = \hat{\mathbf{x}}_k + K_{k+1}(\mathbf{y}_{k+1} - H_{k+1}\hat{\mathbf{x}}_k) \tag{12.88}$$

where

$$K_{k+1} = P_k H_{k+1}^T \left[H_{k+1} P_k H_{k+1}^T + R_{k+1}\right]^{-1} \tag{12.89}$$

$$P_{k+1} = [I - K_{k+1}H_{k+1}]\,P_k \tag{12.90}$$

An alternative form for the sequential covariance expression involves using the Joseph form [17], which has been shown to be more numerically stable. This is given by

$$P_{k+1} = [I - K_k H_{k+1}]P_k[I - K_k H_{k+1}]^T + K_k R_{k+1} K_k^T \tag{12.91}$$

Equation (12.91) requires more computations, but guarantees that the covariance matrix will remain positive definite.

In sequential least squares, the new estimate is the old estimate plus a linear correction term involving the residual between a new observation and the value predicted by the old estimate. Also, if the estimate covariance is "large," then more weight is given to the residual. This intuitively makes sense since a large covariance implies a large estimate error, so that more reliance should be placed on the residual than the previous estimate. As the covariance decreases more weight is placed on the previous estimate, so that the effect of residual correction is less. Thus, the addition of more observations does not significantly affect the new estimate.

12.3.4 Maximum Likelihood Estimation

Maximum likelihood estimation (MLE) was first introduced by R. A. Fisher, a geneticist and statistician in the 1920s. Maximum likelihood yields estimates for the unknown quantities which maximize the probability of obtaining the observed set of data. The conditional probability density function (pdf) for the measurement model in Eq. (12.78) is simply given by [4]

$$p(\mathbf{y}|\mathbf{x}^{\text{true}}) = \frac{1}{(2\pi)^{m/2}(\det R)^{1/2}} \exp\left[-\frac{1}{2}(\mathbf{y} - H\mathbf{x}^{\text{true}})^T R^{-1}(\mathbf{y} - H\mathbf{x}^{\text{true}})\right] \tag{12.92}$$

We wish to maximize the conditional pdf to determine an estimate for $\mathbf{x}^{\text{true}}$, denoted by $\hat{\mathbf{x}}$. Due to the monotonic aspect of this function, the maximization problem can be solved by maximizing the natural log of Eq. (12.92), which gives

$$\ln\left[p(\mathbf{y}|\mathbf{x}^{\text{true}})\right] = -\frac{1}{2}(\mathbf{y} - H\mathbf{x}^{\text{true}})^T R^{-1}(\mathbf{y} - H\mathbf{x}^{\text{true}}) - \frac{m}{2}\ln(2\pi) - \frac{1}{2}\ln(\det R) \tag{12.93}$$

The last two terms on the right hand side of Eq. (12.93) can be ignored because they do not depend on $\mathbf{x}^{\text{true}}$. Maximizing Eq. (12.93) leads to

$$\hat{\mathbf{x}} = (H^T R^{-1} H)^{-1} H^T R^{-1}\mathbf{y} \tag{12.94}$$

which is the familiar least squares batch estimator given by Eq. (12.81). In order to guarantee that the function is maximized the matrix $-(H^T R^{-1} H)^{-1}$ must be negative definite. Since R is positive definite, then this condition is true if $m \geq n$ and there are at least n independent rows in H.

The likelihood function, $\ell(\mathbf{y}|\mathbf{x}^{\text{true}})$, is also a pdf, given by

$$\ell(\mathbf{y}|\mathbf{x}^{\text{true}}) = \prod_{i=1}^{N} p(\mathbf{y}_i|\mathbf{x}^{\text{true}}) \tag{12.95}$$

where N is the total number of density functions (a product of a number of density functions is also a density function in itself). Note that the distributions used in Eq. (12.95) are the same, but the measurements belong to a different sample drawn from the conditional density. The goal of the method of maximum likelihood is to choose as our estimate of the unknown parameters $\mathbf{x}^{\text{true}}$ that value for which the probability of obtaining the observations $\mathbf{y}$ is maximized.

Many likelihood functions, such as the Gaussian pdf, contain exponential terms, which can complicate the mathematics involved in obtaining a solution. However, as seen previously, since $\ln[\ell(\mathbf{y}|\mathbf{x}^{\text{true}})]$ is a monotonic function of $\ell(\mathbf{y}|\mathbf{x}^{\text{true}})$, then maximizing $\ln[\ell(\mathbf{y}|\mathbf{x}^{\text{true}})]$ is equivalent to maximizing $\ell(\mathbf{y}|\mathbf{x}^{\text{true}})$.[3] It follows that for a maximum we have the following:

necessary condition

$$\left\{\frac{\partial}{\partial \mathbf{x}^{\text{true}}} \ln\left[\ell(\mathbf{y}|\mathbf{x}^{\text{true}})\right]\right\}\bigg|_{\hat{\mathbf{x}}} = \mathbf{0}_n \tag{12.96}$$

sufficient condition

$$\frac{\partial^2}{\partial \mathbf{x}^{\text{true}}\, \partial (\mathbf{x}^{\text{true}})^T} \ln\left[\ell(\mathbf{y}|\mathbf{x}^{\text{true}})\right] \text{ must be negative definite} \tag{12.97}$$

Equation (12.96) is often called the *likelihood equation.*

Maximum likelihood has many desirable properties. The first is the *invariance principle*, which is stated as follows: Let $\hat{\mathbf{x}}$ be the maximum likelihood estimate of $\mathbf{x}^{\text{true}}$. Then the maximum likelihood estimate of any function $g(\mathbf{x}^{\text{true}})$ of these parameters is the function $g(\hat{\mathbf{x}})$ of the maximum likelihood estimate. This is a powerful tool since we do not have to take more partial derivatives to determine the maximum likelihood estimate! Another property is that maximum likelihood is an *asymptotically efficient* estimator. This means that if the sample size is large, the maximum likelihood estimate is approximately unbiased and has a variance that approaches the smallest that can be achieved by any estimator. Finally, the estimation errors in the maximum likelihood estimate can be shown to be *asymptotically Gaussian* no matter what density function is used in the likelihood function. Proofs of these properties can be found in Sorenson [26].

The sufficient condition in Eq. (12.97) is useful to also compute the Cramér-Rao inequality which gives a lower bound on the expected errors between the estimated quantities and the values from the known statistical properties of the measurement errors. The Cramér-Rao inequality for an unbiased estimate $\hat{\mathbf{x}}$, i.e. $E\{\hat{\mathbf{x}}\} = \mathbf{x}^{\text{true}}$, is given by

$$P \equiv E\left\{\left(\hat{\mathbf{x}} - \mathbf{x}^{\text{true}}\right)\left(\hat{\mathbf{x}} - \mathbf{x}^{\text{true}}\right)^T\right\} \geq F^{-1} \tag{12.98}$$

where the *Fisher information matrix*, F, is given by

$$F = -E\left\{\frac{\partial^2}{\partial \mathbf{x}^{\text{true}}\, \partial (\mathbf{x}^{\text{true}})^T} \ln[p(\mathbf{y}|\mathbf{x}^{\text{true}})]\right\} \tag{12.99}$$

---

[3]Also, taking the natural logarithm changes a product to a sum which often simplifies the problem to be solved.

The first- and second-order partial derivatives are assumed to exist and to be absolutely integrable. A formal proof of the Cramér-Rao inequality can be found in [4]. The Cramér-Rao inequality gives a *lower* bound on the expected errors. When the equality in Eq. (12.98) is satisfied, then the estimator is said to be *efficient*. This can be useful for the investigation of the quality of a particular estimator. It should be stressed that the Cramér-Rao inequality gives a lower bound on the expected errors only for the case of unbiased estimates.

As an example of the Cramér-Rao inequality, consider the least squares estimator of Eq. (12.94). The Fisher information matrix using Eq. (12.99) is found to be given by

$$F = (H^T R^{-1} H) \tag{12.100}$$

Substituting Eq. (12.78) into Eq. (12.94) gives

$$\hat{\mathbf{x}} - \mathbf{x}^{\text{true}} = (H^T R^{-1} H)^{-1} H^T R^{-1} \mathbf{v} \tag{12.101}$$

Since $E\{\mathbf{v}\} = \mathbf{0}$ then clearly the least squares estimate is unbiased. Using $E\{\mathbf{v}\,\mathbf{v}^T\} = R$ leads to the following estimate covariance:

$$P = (H^T R^{-1} H)^{-1} \tag{12.102}$$

Therefore, the *equality* in Eq. (12.98) is satisfied, so the least squares estimate leads to an efficient estimator. Note that estimate covariance matrix does not depend on the observations, so that the performance of the estimator can be evaluated before any measurements are taken. This includes finding the 3σ bounds on the estimate error parameters, even though the truth, $\mathbf{x}^{\text{true}}$, is never known in practice!

12.3.5 Nonlinear Least Squares

Linear least squares provides a simple and efficient closed-form solution for linear models. Unfortunately, many systems in spacecraft orbit and attitude estimation involve nonlinear models. The *nonlinear least squares* problem is more complicated, since in general closed-form solutions are not possible. Consider the following nonlinear observation model:

$$\mathbf{y} = \mathbf{h}(\mathbf{x}^{\text{true}}) + \mathbf{v} \tag{12.103}$$

where $\mathbf{h}(\mathbf{x}^{\text{true}})$ represents some nonlinear, continuously differentiable function of $\mathbf{x}^{\text{true}}$. Also, it is assumed that the measurement noise is an additive function to the observation model $\mathbf{h}(\mathbf{x}^{\text{true}})$. This may not always be true, which further complicates the problem. The nonlinear least squares problem still minimizes the weighted sum square residual errors, given by

$$J(\hat{\mathbf{x}}) = \frac{1}{2}[\mathbf{y} - \mathbf{h}(\hat{\mathbf{x}})]^T R^{-1} [\mathbf{y} - \mathbf{h}(\hat{\mathbf{x}})] \tag{12.104}$$

In principle the solution follows exactly as in the linear case, where the derivative of Eq. (12.104) is set to zero. However, in practice the estimate cannot be found in closed-form due to the nonlinear nature of the observation vector $\mathbf{h}(\mathbf{x})$. We assume that an initial estimate of $\mathbf{x}$ is known, denoted by $\mathbf{x}_c$, so that the estimate is related by an unknown set of corrections, $\Delta\mathbf{x}$, as

$$\hat{\mathbf{x}} = \mathbf{x}_c + \Delta\mathbf{x} \tag{12.105}$$

If the components of $\Delta\mathbf{x}$ are sufficiently small, it may be possible to solve for approximations to them and thereby update $\mathbf{x}_c$ with an improved estimate of $\mathbf{x}$ from Eq. (12.105). With this assumption, we may linearize $\mathbf{h}(\hat{\mathbf{x}})$ about $\mathbf{x}_c$ using a first-order Taylor series expansion as

$$\mathbf{h}(\hat{\mathbf{x}}) \approx \mathbf{h}(\mathbf{x}_c) + H\Delta\mathbf{x} \tag{12.106}$$

where

$$H \equiv \left.\frac{\partial \mathbf{h}}{\partial \mathbf{x}}\right|_{\mathbf{x}_c} \tag{12.107}$$

The measurement residual "after the correction" can now be linearly approximated as

$$\Delta\mathbf{y} \equiv \mathbf{y} - \mathbf{h}(\hat{\mathbf{x}}) \approx \mathbf{y} - \mathbf{h}(\mathbf{x}_c) - H\Delta\mathbf{x} = \Delta\mathbf{y}_c - H\Delta\mathbf{x} \tag{12.108}$$

where the residual "before the correction" is

$$\Delta\mathbf{y}_c \equiv \mathbf{y} - \mathbf{h}(\mathbf{x}_c) \tag{12.109}$$

Recall that the objective is to minimize the weighted sum squares, J, given by Eq. (12.104). The local strategy for determining the approximate corrections ("differential corrections") in $\Delta\mathbf{x}$ is to select the particular corrections that lead to the *minimum sum of squares of the linearly predicted residuals* J_p:

$$J = \frac{1}{2}\Delta\mathbf{y}^T R^{-1}\Delta\mathbf{y} \approx J_p \equiv \frac{1}{2}(\Delta\mathbf{y}_c - H\Delta\mathbf{x})^T R^{-1}(\Delta\mathbf{y}_c - H\Delta\mathbf{x}) \tag{12.110}$$

Before carrying out the minimization, we note that the minimization of J_p in Eq. (12.110) is equivalent to the minimization of J in Eq. (12.104). If the process is convergent, then $\Delta\mathbf{x}$ determined by minimizing Eq. (12.110) would be expected to decrease on successive iterations until (on the final iteration) the linearization is an extremely good approximation.

Observe that the minimization of Eq. (12.110) is completely analogous to the previously minimized quadratic form. Thus, any algorithm for solving the least squares problem directly applies to solving for $\Delta\mathbf{x}$ in Eq. (12.110). Therefore, the corrections is given by

$$\Delta\mathbf{x} = (H^T R^{-1} H)^{-1} H^T R^{-1} \Delta\mathbf{y}_c \tag{12.111}$$

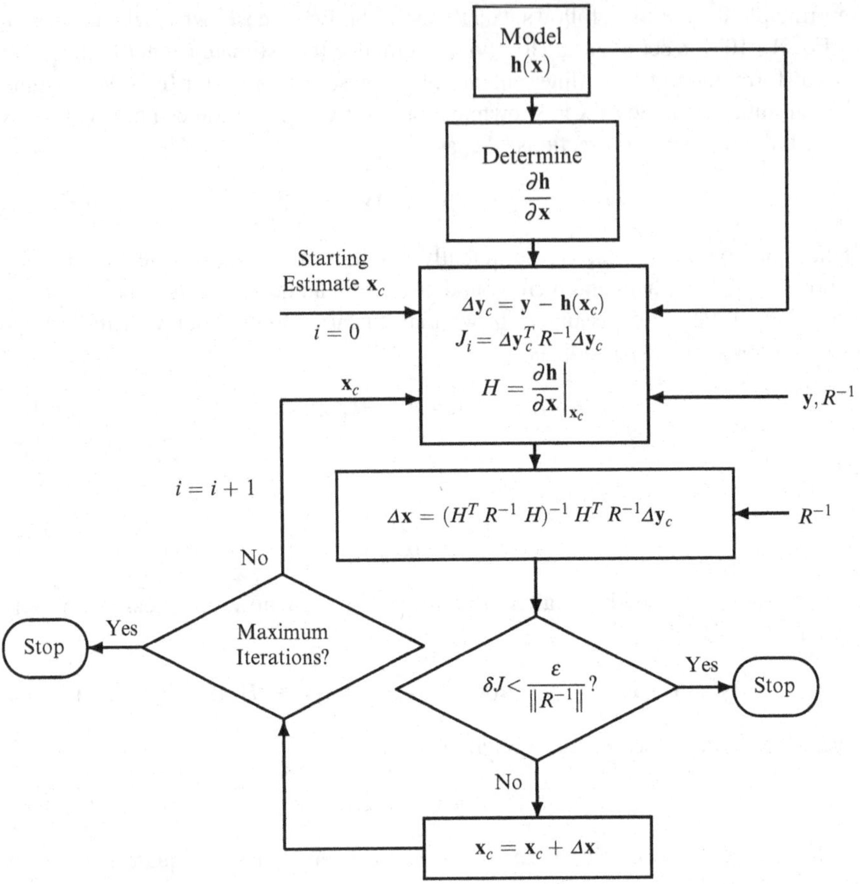

Fig. 12.8 Nonlinear least squares algorithm

The complete nonlinear least squares algorithm is summarized in Fig. 12.8. An initial guess $\mathbf{x}_c$ is required to begin the algorithm. Equation (12.111) is then calculated using the residual measurements ($\Delta\mathbf{y}_c$), Jacobian matrix (H), and measurement covariance matrix (R^{-1}), so that the current estimate can be updated. A stopping condition with an accuracy dependent tolerance for the minimization of J is given by

$$\delta J \equiv \frac{|J_i - J_{i-1}|}{J_i} < \frac{\varepsilon}{\|R^{-1}\|} \tag{12.112}$$

where ε is a prescribed small value. If Eq. (12.112) is not satisfied, then the update procedure is iterated with the new estimate as the current estimate until the process converges, or unsatisfactory convergence progress is evident (e.g. a maximum allowed number of iterations is exceeded, or J increases on successive iterations).

Therefore, a linear update is performed. Nonlinear least squares implies that an fairly accurate estimate of **x** must be known a priori, which may not always be possible. Also, a necessary condition for $\hat{\mathbf{x}}$ to be the least squares estimate is that the residual must be orthogonal to the columns of H. This is not true in general. However, in practice $\hat{\mathbf{x}}$ is used to invoke an iterative procedure with $\mathbf{x}_c$ replaced by $\hat{\mathbf{x}}$ so that the process continues until the estimate is no longer improved. This iterative search procedure is known as the Gauss method. The estimate covariance is still given by Eq. (12.102), but is theoretically evaluated at the truth. Since this is not known, the final estimate is used in practice.

Example 12.4. In this example nonlinear least squares is used to compute the modified Rodrigues parameters (MRPs) from two vector observations. The true body vectors are given by

$$\mathbf{b}_1^{\text{true}} = \begin{bmatrix} 0 \\ 0 \\ 1 \end{bmatrix}, \quad \mathbf{b}_2^{\text{true}} = \begin{bmatrix} 0 \\ 0.1 \\ \sqrt{1-0.1^2} \end{bmatrix}$$

Note that the angle between these vectors is 5.74°. The true MRP and associated attitude matrix are given by

$$\mathbf{p}^{\text{true}} = \begin{bmatrix} \dfrac{1}{1+\sqrt{2}} \\ 0 \\ 0 \end{bmatrix}, \quad A(\mathbf{p}^{\text{true}}) = \begin{bmatrix} 1 & 0 & 0 \\ 0 & 0 & 1 \\ 0 & -1 & 0 \end{bmatrix}$$

The reference vectors are computed using $\mathbf{r}_i = A^T \mathbf{b}_i^{\text{true}}$ for $i = 1, 2$. To generate body measurements the QUEST measurement model in Eq. (5.107b) is used with $\sigma_{b_i} = 0.001°$ for $i = 1, 2$. Note that the measurement covariance is fully populated. To overcome the issue of generating synthetic noise with correlations in any general semi-definite covariance matrix R, we first will diagonalize this matrix using an eigenvalue/eigenvector decomposition:

$$R = V \Lambda V^T$$

where V is a matrix of eigenvectors and Λ is a matrix of eigenvalues. Since R is positive semi-definite, then V is an orthogonal matrix and the elements of Λ are real and positive. A random sample, denoted by **z**, using scalar sampling can be generated using the matrix Λ, where the elements of Λ are the variances of the elements of **z**. To determine the correlated measurement noise, we simply rotate the vector **z** using V with

$$\mathbf{v} = V\,\mathbf{z}$$

where $\mathbf{v}$ is the correlated noise associated with R. To see that this is correct, we compute $E\left\{\mathbf{v}\,\mathbf{v}^T\right\} = VE\left\{\mathbf{z}\,\mathbf{z}^T\right\}V^T = V\,\Lambda\,V^T = R$. This process is done on both R_{b_1} and R_{b_2} to generate synthetic $\mathbf{v}_1$ and $\mathbf{v}_2$ vectors. The measured body vectors are then given by

$$\mathbf{b}_1 = \frac{\mathbf{b}_1^{\text{true}} + \mathbf{v}_1}{||\mathbf{b}_1^{\text{true}} + \mathbf{v}_1||}, \quad \mathbf{b}_2 = \frac{\mathbf{b}_2^{\text{true}} + \mathbf{v}_2}{||\mathbf{b}_2^{\text{true}} + \mathbf{v}_2||}$$

To compute the matrix H a first-order expansion of the attitude matrix in the error MRPs is used. First the true attitude is related to the estimated attitude, denoted by $A(\hat{\mathbf{p}})$, by using an error attitude, denoted $A(\boldsymbol{\delta}\mathbf{p})$, through

$$A(\mathbf{p}^{\text{true}}) = A(\boldsymbol{\delta}\mathbf{p})A(\hat{\mathbf{p}})$$

The first-order expansion of $A(\boldsymbol{\delta}\mathbf{p})$ is given by

$$A(\boldsymbol{\delta}\mathbf{p}) = I_3 - 4[\boldsymbol{\delta}\mathbf{p}\times]$$

The estimated ith body vector is given by $\hat{\mathbf{b}}_i = A(\hat{\mathbf{p}})\mathbf{r}_i$. Computing $\Delta\mathbf{b}_i \equiv \mathbf{b}_i^{\text{true}} - \hat{\mathbf{b}}_i$ gives

$$\begin{aligned}\Delta\mathbf{b}_i &= -4[\boldsymbol{\delta}\mathbf{p}\times]A(\hat{\mathbf{p}})\mathbf{r}_i \\ &= 4[A(\hat{\mathbf{p}})\mathbf{r}_i\times]\boldsymbol{\delta}\mathbf{p}\end{aligned}$$

Taking the derivative with respect to $\boldsymbol{\delta}\mathbf{p}$ gives the following H matrix:

$$H = \begin{bmatrix} 4[A(\hat{\mathbf{p}})\mathbf{r}_1\times] \\ \\ 4[A(\hat{\mathbf{p}})\mathbf{r}_2\times] \end{bmatrix}$$

The vector $\Delta\mathbf{y}$ and matrix R are given by

$$\Delta\mathbf{y} = \begin{bmatrix} \mathbf{b}_1 - \hat{\mathbf{b}}_1 \\ \mathbf{b}_2 - \hat{\mathbf{b}}_2 \end{bmatrix}, \quad R = \sigma^2 I_6$$

Equation (12.111) can now be used to perform the nonlinear least squares iterations. A multiplicative correction using Eq. (2.150) is used to update the MRP:

$$\hat{\mathbf{p}} = \frac{\left(1 - \|\mathbf{p}_c\|^2\right)\Delta\mathbf{x} + \left(1 - \|\Delta\mathbf{x}\|^2\right)\mathbf{p}_c - 2\,\Delta\mathbf{x}\times\mathbf{p}_c}{1 + \|\mathbf{p}_c\|^2\|\Delta\mathbf{p}_c\|^2 - 2\,\Delta\mathbf{x}\cdot\mathbf{p}_c} \tag{12.113}$$

where $\hat{\mathbf{p}}$ is the estimated MRP and $\mathbf{p}_c$ is the initial MRP.

The nonlinear least squares process is started with an initial MRP of $\mathbf{p}_c = \mathbf{0}_{3\times1}$. The stopping criterion is given when $\Delta\mathbf{x} < 1\times10^{-9}$. One thousand Monte Carlo

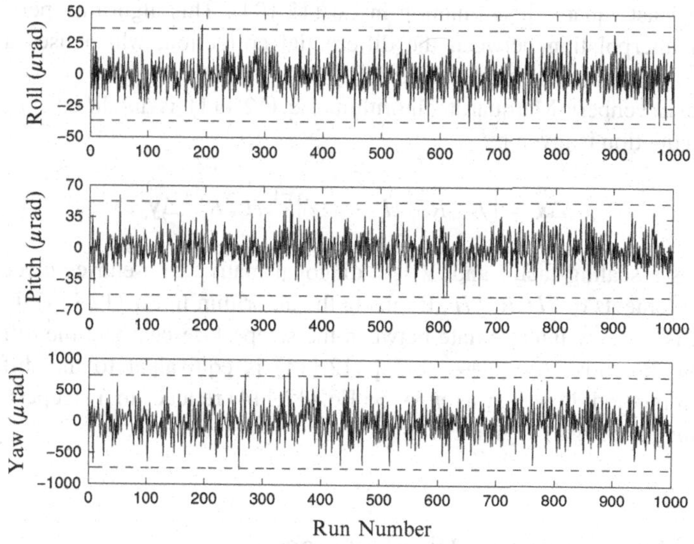

Fig. 12.9 Attitude errors

runs are executed. In each run the nonlinear least squares process converges in exactly 5 iterations. This is rather remarkable considering the large error in the initial estimate! The MRP errors are also computed using Eq. (2.150). The inverse MRP is given by its negative. Thus, $\boldsymbol{\delta}\mathbf{p}$ is given by

$$\boldsymbol{\delta}\mathbf{p} = \frac{\left(1 - \|\hat{\mathbf{p}}\|^2\right)\mathbf{p}^{\text{true}} - \left(1 - \|\mathbf{p}^{\text{true}}\|^2\right)\hat{\mathbf{p}} + 2\,\mathbf{p}^{\text{true}} \times \hat{\mathbf{p}}}{1 + \|\hat{\mathbf{p}}\|^2\|\mathbf{p}^{\text{true}}\|^2 + 2\,\mathbf{p}^{\text{true}} \cdot \hat{\mathbf{p}}}$$

The attitude errors are computed by multiplying $\boldsymbol{\delta}\mathbf{p}$ by 4. Note that covariance of the attitude errors, $16\sigma^2(H^T H)^{-1}$, evaluated at the true values, is equivalent to Eq. (5.113). A plot of the attitude errors along with their respective 3σ bounds is shown in Fig. 12.9. This indicates that the nonlinear least squares approach provides good estimates. Also, note that the yaw errors are an order of magnitude larger than the roll and pitch errors. This is due to the small angle between the two body vectors.

Nonlinear least squares may not be suitable to locate the optimum estimate since convergence is not guaranteed unless the a priori estimate is close to the optimum value. The gradient search method [22, 26] overcomes this difficulty by adjusting the estimate so that the search direction is always along the negative gradient of $J(\hat{\mathbf{x}})$. However, this method has been known to converge slowly as the solution approaches the minimum $J(\hat{\mathbf{x}})$. The Levenberg-Marquardt algorithm [16] overcomes both the difficulties of the standard differential correction approach when an accurate initial estimate is not given, and the slow convergence problems of the method of steepest descent when the solution is close to minimizing the nonlinear least squares loss function in Eq. (12.104). This algorithm performs an optimum interpolation between the differential correction, which uses a Taylor series expansion, and the gradient method.

In the Levenberg-Marquardt algorithm, Eq. (12.111) is modified so that the iteration equation is given by

$$\Delta \mathbf{x} = \left[H^T R^{-1} H + \eta \mathscr{R}\right]^{-1} H^T R^{-1} \Delta \mathbf{y}_c \tag{12.114}$$

where η is a scaling factor and $\mathscr{R}$ is a diagonal matrix with entries given by the diagonal elements of $H^T R^{-1} H$. By using the algorithm in Eq. (12.114) the search direction is actually intermediate between the steepest descent and the differential correction direction. As $\eta \to 0$, Eq. (12.114) is equivalent to the differential correction method; however, as $\eta \to \infty$ Eq. (12.114) reduces to a steepest descent search along the negative gradient of J.

12.3.6 Advantages and Disadvantages

Linear batch estimators are simple to design and implement provided that the system can be represented accurately using linear observation models. A major advantage of the linear batch estimator is that the error covariance can be found without using actual observations. A disadvantage of batch estimators is that the $n \times n$ matrix, which must be inverted, may be ill-conditioned. This problem occurs frequently when the observability of the system is low. This problem can be alleviated by using matrix square-root or factorization methods, such as a Cholesky decomposition which factors a symmetric matrix into the product of a lower triangular matrix and its transpose. Also, a large number of observations may be required to improve the accuracy of the estimate, which may substantially increase computer storage. Sequential estimators have certain advantages since the inverse of an $n \times n$ matrix can be reduced to a scalar inverse, and since the estimate is found sequentially computer storage can be kept to a minimum. However, convergence of sequential methods cannot always be guaranteed (especially for bad initialization parameters), and erroneous observations cannot easily be disregarded.

12.3.7 State Estimation Techniques

The least-square methods shown previously apply to "static" estimation problems. When an estimate of a varying state is required, "dynamic" methods must be used. The most common technique for state estimation using dynamic models is the Kalman filter [11]. The term "filter" is used since it not only estimates dynamic states, but filters noisy processes. The Kalman filter is a sequential process, but can also estimate for constant variables as least-squares methods do. However, the essential feature of the Kalman filter is the utilization of state models for dynamic propagation. In addition, the Kalman filter compensates for dynamic model inaccuracy by incorporating a noise term (commonly known as *process noise*), that gives the filter a fading memory. So, each observation has a gradually diminishing effect on future state estimates.

The Kalman filter satisfies an optimality criterion which minimizes the trace of the covariance of the estimate error between the state estimates and true state quantities. Statistical properties of the process noise and measurement error are used to design an optimal filter. This section introduces the fundamental equations used in a Kalman filter. The linear Kalman filter is first shown in order to introduce the fundamental concepts. The steady-state case is also shown for autonomous system models. The more useful formulation for attitude and orbit estimation is the nonlinear Kalman filter (also known as the *extended Kalman filter*), which incorporates nonlinear dynamic and/or observation models. Divergence and practical considerations are also discussed.

12.3.7.1 Linear Kalman Filter

The linear Kalman filter development begins by assuming a truth model using the linear dynamic model in Eq. (12.18) appended with a noise term, given by

$$\dot{\mathbf{x}}^{\text{true}} = F\mathbf{x}^{\text{true}} + B\mathbf{u} + G\mathbf{w} \tag{12.115}$$

where $\mathbf{w}$ is a zero-mean $p \times 1$ Gaussian white-noise process vector with spectral density Q, and G is an $n \times p$ process noise distribution matrix. If we ignore the measurement for now, then the "propagated" estimated state, $\hat{\mathbf{x}}$, follows

$$\dot{\hat{\mathbf{x}}} = F\hat{\mathbf{x}} + B\mathbf{u} \tag{12.116}$$

Next, a residual error is defined between the true state in Eq. (12.115) and the estimated state in Eq. (12.116):

$$\boldsymbol{\epsilon} \equiv \mathbf{x}^{\text{true}} - \hat{\mathbf{x}} \tag{12.117}$$

Taking the time derivative of Eq. (12.117) leads to

$$\dot{\boldsymbol{\epsilon}} = F\boldsymbol{\epsilon} + G\mathbf{w} \tag{12.118}$$

The covariance of the residual error:

$$P \equiv E\{\boldsymbol{\epsilon}\boldsymbol{\epsilon}^T\} \tag{12.119}$$

obeys the equation[4]

$$\dot{P} = E\{\epsilon\dot{\epsilon}^T\} + E\{\dot{\epsilon}\epsilon^T\} = F\,P + P\,F^T + G\,Q\,G^T \tag{12.120}$$

which is solved for $P(t)$ with initial condition $P(t_0) = P_0$. Equations (12.116) and (12.120) define the propagated system for the estimated dynamic model and associated error covariance.

In general the measurements are discrete, which are modeled by

$$\mathbf{y}_k = H_k \mathbf{x}_k^{\text{true}} + \mathbf{v}_k \tag{12.121}$$

where $\mathbf{v}_k$ is a $m \times 1$ zero-mean Gaussian white-noise process vector with covariance R_k. A discrete "update" equation is required to process the measurements. This is very similar to the sequential least-squares problem. The linear discrete-time update equation is given by

$$\hat{\mathbf{x}}_k^+ = \hat{\mathbf{x}}_k^- + K_k[\mathbf{y}_k - H_k\hat{\mathbf{x}}_k^-] \tag{12.122}$$

where the superscripts $-$ and $+$ denote the discrete times just before and after a discrete measurement update, respectively, i.e. $\hat{\mathbf{x}}_k^-$ comes from the propagated estimate using Eq. (12.116). Also, the subscript k denotes the variable at time t_k. The updated covariance expression is given by computing

$$P_k^+ \equiv E\left\{\left(\mathbf{x}_k^{\text{true}} - \hat{\mathbf{x}}_k^+\right)\left(\mathbf{x}_k^{\text{true}} - \hat{\mathbf{x}}_k^+\right)^T\right\} \tag{12.123}$$

Substituting Eqs. (12.121) and (12.122) into (12.123), and performing this expectation leads to

$$P_k^+ = [I - K_k H_k]P_k^-[I - K_k H_k]^T + K_k R_k K_k^T \tag{12.124}$$

where P_k^- is given from the propagated system by Eq. (12.120). Note that Eq. (12.124) is valid for any gain K_k. Also note that it is equivalent to Eq. (12.91), except that the propagated P_k^- is used. The optimal gain is determined by minimizing the trace of the updated covariance:

$$J_k = \operatorname{tr} P_k^+ \tag{12.125}$$

Performing this minimization leads to a gain that is identical to the sequential least-squares problem:

$$K_k = P_k^- H_k^T [H_k P_k^- H_k^T + R_k]^{-1} \tag{12.126}$$

[4]This follows formally from $E\{\epsilon\mathbf{w}^T\} = \frac{1}{2}GQ$, but there are more rigorous derivations [4].

Table 12.1 Continuous-discrete linear Kalman filter

Model	$\dot{\mathbf{x}}^{\text{true}} = F\mathbf{x}^{\text{true}} + B\mathbf{u} + G\mathbf{w}$, where $\mathbf{w} \sim N(\mathbf{0}, Q)$ $\mathbf{y}_k = H\mathbf{x}_k^{\text{true}} + \mathbf{v}_k$, where $\mathbf{v}_k \sim N(\mathbf{0}, R_k)$
Initialize	$\hat{\mathbf{x}}(t_0) = \hat{\mathbf{x}}_0$ $P(t_0) = P_0$
Propagation	$\dot{\hat{\mathbf{x}}} = F\hat{\mathbf{x}} + B\mathbf{u}$ $\dot{P} = F\,P + P\,F^T + G\,Q\,G^T$
Gain	$K_k = P_k^- H_k^T [H_k P_k^- H_k^T + R_k]^{-1}$
Update	$\hat{\mathbf{x}}_k^+ = \hat{\mathbf{x}}_k^- + K_k[\mathbf{y}_k - H_k\hat{\mathbf{x}}_k^-]$ $P_k^+ = [I - K_k H_k]P_k^-$

Substituting Eq. (12.126) into Eq. (12.124) gives

$$P_k^+ = [I - K_k H_k]P_k^- \tag{12.127}$$

This the preferred form for the covariance update, but the Joseph form in Eq. (12.124) can still be used with the gain in Eq. (12.126) when numerical instabilities are a concern.

A summary of the continuous-discrete linear Kalman filter is shown in Table 12.1, where $N(\mathbf{0}, Q)$ denotes a Gaussian distribution with zero mean and spectral density Q for continuous-time systems, and $N(\mathbf{0}, R_k)$ denotes a Gaussian distribution with zero mean and covariance R_k for discrete-time systems. The filter is first initialized using knowledge of the initial state, $\hat{\mathbf{x}}_0$, and error covariance, P_0. Then the state and error covariance are updated using the measured observation. Finally, the state dynamic model and associated error covariance are propagated using the updated values to the next discrete measurement time.

The Kalman filter shown in Table 12.1 can be used with time-varying model matrices. However, if the model is autonomous a steady-state expression for the error covariance can be used. This is due to the fact that the covariance usually converges rapidly for this case. Also, for a large class of problems the discrete-time version of the dynamic model in Eq. (12.115) is usually quite adequate, given by

$$\mathbf{x}_{k+1}^{\text{true}} = \Phi_k \mathbf{x}_k^{\text{true}} + \Gamma_k \mathbf{u}_k + \Upsilon_k \mathbf{w}_k \tag{12.128}$$

where $\mathbf{w}_k$ is a $p \times 1$ zero-mean Gaussian white-noise process vector with covariance Q_k. The propagation estimate is simply given by

$$\hat{\mathbf{x}}_{k+1}^- = \Phi_k \hat{\mathbf{x}}_k^+ + \Gamma_k \mathbf{u}_k \tag{12.129}$$

Combining (12.129) with (12.122) gives, for the post-update state estimates,

$$\hat{\mathbf{x}}_{k+1} = \Phi_k \hat{\mathbf{x}}_k + \Gamma_k \mathbf{u}_k + K_{k+1}[\mathbf{y}_{k+1} - H_{k+1}(\Phi_k \hat{\mathbf{x}}_k + \Gamma_k \mathbf{u}_k)] \tag{12.130}$$

where the superscript $+$ on the state estimates is understood. The pre-update covariance is needed to compute the Kalman gain. The discrete-time covariance propagation is given by [13]

$$P_{k+1}^{-} = \Phi_k P_k^{+} \Phi_k^T + \Upsilon_k Q_k \Upsilon_k^T \tag{12.131}$$

Combining the discrete error covariance propagation (12.131) and update equation (12.127) gives

$$P_{k+1} = \Phi_k P_k \Phi_k^T - \Phi_k P_k H_k^T [H_k P_k H_k^T + R]^{-1} H_k P_k \Phi_k^T + \Upsilon_k Q_k \Upsilon_k^T \tag{12.132}$$

where the superscript $-$ on the covariances is understood. If the system is autonomous, then the following steady-state equation can be used to find P:

$$P = \Phi P \Phi^T - \Phi P H^T [H P H^T + R]^{-1} H P \Phi^T + \Upsilon Q \Upsilon^T \tag{12.133}$$

which is known as the discrete *algebraic Riccati equation*. The solution to Eq. (12.133) is given by forming a $2n \times 2n$ Hamiltonian matrix

$$\mathscr{H} \equiv \begin{bmatrix} \Phi^{-T} & \Phi^{-T} H^T R^{-1} H \\ \Upsilon Q \Upsilon^T \Phi^{-T} & \Phi + \Upsilon Q \Upsilon^T \Phi^{-T} H^T R^{-1} H \end{bmatrix} \tag{12.134}$$

which can be shown to have n stable eigenvalues (inside the unit circle) and n reciprocal unstable eigenvalues [4]. The eigenvectors of Eq. (12.134) are then partitioned into $n \times n$ submatrices:

$$V = \begin{bmatrix} V_{11} & V_{12} \\ V_{21} & V_{22} \end{bmatrix} \tag{12.135}$$

where V_{11} and V_{21} contain the eigenvectors corresponding to the unstable eigenvalues, and V_{12} and V_{22} contain the eigenvectors corresponding to the stable eigenvalues. The steady-state solution for P is found by using V_{11} and V_{21}

$$P = V_{21} V_{11}^{-1} \tag{12.136}$$

Therefore, the gain in Eq. (12.126) is now constant. Other efficient methods for the solution of the Eq. (12.133) are shown in [1]. The steady-state discrete-time Kalman filter is summarized in Table 12.2.

12.3.7.2 Extended Kalman Filter

A large class of orbit and attitude estimation problems involve nonlinear models. For several reasons, state estimation for nonlinear systems is considerably more

Table 12.2 Discrete and autonomous linear Kalman filter

Model	$\mathbf{x}_{k+1}^{\text{true}} = \Phi \mathbf{x}_k^{\text{true}} + \Gamma \mathbf{u}_k + \Upsilon \mathbf{w}_k$, where $\mathbf{w}_k \sim N(\mathbf{0}, Q)$ $\mathbf{y}_k = H \mathbf{x}_k^{\text{true}} + \mathbf{v}_k$, where $\mathbf{v}_k \sim N(\mathbf{0}, R)$
Initialize	$\hat{\mathbf{x}}(t_0) = \hat{\mathbf{x}}_0$
Covariance	$P = \Phi P \Phi^T - \Phi P H^T [H P H^T + R]^{-1} H P \Phi^T + \Upsilon Q \Upsilon^T$
Gain	$K = P H^T [H P H^T + R]^{-1}$
Estimate	$\hat{\mathbf{x}}_{k+1} = \Phi \hat{\mathbf{x}}_k + \Gamma \mathbf{u}_k + K[\mathbf{y}_{k+1} - H(\Phi \hat{\mathbf{x}}_k + \Gamma \mathbf{u}_k)]$

Table 12.3 Continuous-discrete extended Kalman filter

Model	$\dot{\mathbf{x}}^{\text{true}} = \mathbf{f}(\mathbf{x}^{\text{true}}, \mathbf{u}, \mathbf{w}, t)$, where $\mathbf{w} \sim N(\mathbf{0}, Q)$ $\mathbf{y}_k = \mathbf{h}(\mathbf{x}_k^{\text{true}}) + \mathbf{v}_k$, where $\mathbf{v}_k \sim N(\mathbf{0}, R_k)$
Initialize	$\hat{\mathbf{x}}(t_0) = \hat{\mathbf{x}}_0$ $P(t_0) = P_0$
Propagation	$\dot{\hat{\mathbf{x}}} = \mathbf{f}(\hat{\mathbf{x}}, \mathbf{u}, t)$ $\dot{P} = F\,P + P\,F^T + G\,Q\,G^T$ $F \equiv \left.\dfrac{\partial \mathbf{f}}{\partial \mathbf{x}}\right\vert_{\hat{\mathbf{x}}}, \quad G \equiv \left.\dfrac{\partial \mathbf{f}}{\partial \mathbf{w}}\right\vert_{\hat{\mathbf{x}}}$
Gain	$K_k = P_k^- H_k^T [H_k P_k^- H_k^T + R_k]^{-1}$ $H_k \equiv \left.\dfrac{\partial \mathbf{h}}{\partial \mathbf{x}}\right\vert_{\hat{\mathbf{x}}_k^-}$
Update	$\hat{\mathbf{x}}_k^+ = \hat{\mathbf{x}}_k^- + K_k[\mathbf{y}_k - \mathbf{h}(\hat{\mathbf{x}}_k^-)]$ $P_k^+ = [I - K_k H_k] P_k^-$

difficult, and admits a wider variety of solutions than the linear problem [8]. Some of these problems are seen by considering the general nonlinear system model

$$\dot{\mathbf{x}}^{\text{true}} = \mathbf{f}(\mathbf{x}^{\text{true}}, \mathbf{u}, \mathbf{w}, t) \tag{12.137}$$

Clearly, the probability density function of $\mathbf{w}$ is altered as it is transmitted through the nonlinear elements. So, if the noise term cannot be linearly separated from this equation, then the simple covariance expression in Eq. (12.137) cannot be used, since a Gaussian input causes a non-Gaussian response. Fortunately, a Kalman filter can still be derived using nonlinear models. The extended Kalman filter, though not precisely "optimum," has been successfully applied to many nonlinear systems over the past many years. The fundamental concept of this filter involves the notion that the true state is sufficiently close to the estimated state. Therefore, the error dynamics can be represented fairly accurately by a linearized first-order Taylor series expansion, given by Eq. (12.30). The extended Kalman filter uses the full nonlinear dynamic equation for model propagation, but uses the same linear correction as the linear Kalman filter. Also, the nominal values are replaced by the current estimates in the filter propagation. This allows the extended Kalman filter to compute the gain matrix online, as opposed to pre-computing a gain matrix sequence. A summary of the extended Kalman filter algorithm is shown in Table 12.3.

12.3.7.3 Smoothing

A batch version of both the linear and extended Kalman filters is useful for significantly smoothing noisy processes. Smoothing problems general fall into three cases. The fixed-point smoother seeks an estimate at a single fixed point in time, while the observations continue ahead of the estimation. The fixed-lag smoother seeks an estimate at a fixed length of time back in the past. The fixed-interval smoother uses a fixed time interval of observations, and seeks estimates at some or all of the interior points. This is the most common smoother for batch estimation. An example of such a smoother is the Rauch-Tung-Striebel (RTS) optimal smoother [23]. This smoother essentially processes the observation first forward in time and then backwards in time. The forward estimates are given by the Kalman filter. The backwards process is initialized using the final time forward estimates:

$$\hat{\mathbf{x}}_{s_N} = \hat{\mathbf{x}}_N^+ \tag{12.138a}$$

$$P_{s_N} = P_N^+ \tag{12.138b}$$

where the subscript s denotes the smoothed quantity. The RTS smoothed estimates and covariance are computed using

$$\mathscr{K}_k \equiv P_k^+ \Phi_k^T (P_{k+1}^-)^{-1} \tag{12.139a}$$

$$\hat{\mathbf{x}}_{s_k} = \hat{\mathbf{x}}_k^+ + \mathscr{K}_k (\hat{\mathbf{x}}_{s_{k+1}} - \hat{\mathbf{x}}_{k+1}^-) \tag{12.139b}$$

$$P_{s_k} = P_k^+ - \mathscr{K}_k (P_{k+1}^- - P_{s_{k+1}}) \mathscr{K}_k^T \tag{12.139c}$$

Note that storage of the forward estimates, state matrices, and covariance is required, but storage of the measurements is not required. The advantages of smoothing algorithms is that the resulting error covariance is always less than either the forward or backward process alone. However, smoothing algorithms cannot be implemented for realtime application.

12.3.7.4 Stability and Performance

The linear and autonomous Kalman filter is known to be extremely stable (i.e. the estimates will not diverge from the true values), and provides accurate estimates under the properly defined conditions. However, the stability of the extended Kalman filter must be properly addressed before on-board implementation. Many factors affect filter stability for this case. One common problem is in the error covariance update and propagation, which may become non-positive definite chiefly due to numerical instabilities. A measure of the potential for difficulty in an ill-conditioned matrix can be found by using the *condition number* [27]. This problem may be overcome by using the Joseph form (shown previously) or by using matrix square root or factorization methods. Square root methods guarantee positive definiteness, but typically require 50–150 % more computational time than that required by the standard Kalman filter [27]. A more efficient algorithm is given by the *U-D* filter, which factors the covariance matrix using

$$P = U\,D\,U^T = \left[U\,D^{1/2}\right]\left[U\,D^{1/2}\right]^T \tag{12.140}$$

where U is an upper triangular matrix and D is diagonal. The gain matrix, covariance propagation and update are given in terms of these matrices [2].

For filter performance, non-Gaussian measurement and process noise errors are a major source of concern. Typically, in most real-world applications these error sources are indeed non-Gaussian. For example, consider the problem of estimating a spacecraft's attitude using magnetometer data. Flight data results have clearly shown that the magnetic field measurements contain errors which are a function of orbit rate and higher harmonics [24]. This can severely degrade the filter's performance, since this is typically assumed to be modeled by a Gaussian process. A possible solution to these type of problems involves using a *colored-noise filter* [13], which pre-filters (shapes) a Gaussian noise process to more accurately model actual measurement error processes. However, even if the Gaussian noise assumption is true, the proper values for the process and measurement covariances may not be straightforward to choose. This leads to a process known as filter tuning, which involves adjusting the filter parameters to achieve the best possible estimation performance, most often using actual measured observations. Filter tuning can be performed off-line using numerical optimization techniques [22] or online using adaptive methods [10]. However, in practice manual optimization is more prevalent. A common procedure involves choosing small values for the measurement error covariance (which puts more weight or reliance on the measurements) and then adjusting the process noise covariance and initial covariance to achieve a reasonable performance level. Then, the tuning process is refined until the desired performance is achieved.

There are many other concerns to consider in the design and implementation of a Kalman filter. For an extended Kalman filter the obvious question is: how accurate is the linearized model? This is an aspect which must be addressed in order to design an accurate filter. For example, the quaternion kinematic model loses normalization in a straightforward linearization. Therefore, the physical nature of the model can be lost. Also, the observability may vary from one time step to the next. This must be addressed in order to check that observability can be obtained throughout the orbit. If the Kalman filter is used in conjunction with a control system, then the filter's convergence rate (i.e. when transients decay) can significantly affect the control system's performance. This is usually accounted for during the initialization procedure; however, the filter may lose "lock" during mission mode and may need proper initialization before the control system is again invoked. All or part of these factors need to be accounted early in the design (before actual operation) in order to reduce risk of failures due to inadequate filter performance.

12.3.8 Linear Covariance Analysis

Soon after the development of the Kalman filter, it was recognized that the covariance matrix can be computed without processing actual data, and that the resulting covariance is very useful for assessing the expected estimation errors resulting from candidate dynamic models and measurement strategies [6, 10]. This is the basic idea of *linear covariance analysis*, which executes only the covariance initialization, covariance propagation, gain computation, and covariance update steps of Table 12.1. It has been widely employed during the development of attitude estimation systems, to determine the type of sensors required, their locations, accuracy, and required data frequency, as well as the refinement required of dynamic models.

The number of parameters describing a dynamical system can be quite large, and it is often impractical to solve for all of them, especially in real-time applications. Thus it can be important to consider the effect on estimation accuracy of errors in parameters that are unknown but not solved for. Reference [15] considers the generalization of linear covariance analysis to include such *consider parameters*, including references to the earlier literature.

If the models are state-dependent, as is usually the case, a nominal state trajectory is specified to evaluate the partial derivative matrices $F \equiv \partial\mathbf{f}/\partial\mathbf{x}$, $G \equiv \partial\mathbf{f}/\partial\mathbf{w}$, and $H_k \equiv \partial\mathbf{h}/\partial\mathbf{x}$. In most cases, the nominal trajectory is also required to account for sensor occultation, star availability for star cameras, spacecraft attitude maneuvers, and similar effects. Note that, unlike the EKF, the nominal state trajectory is specified at the outset of the analysis and is not affected by the filter equations.

12.3.9 Separation Theorem

Oftentimes a full-state controller is used in the controller. However, just as often all states in a system cannot be measured from sensors. Thus, an estimator is used in concert with a controller, which provides the full state information required in the controller. For example, in the inverted pendulum derived at the beginning of this chapter only measurements of position and angle may be provided, and these along with their derivatives are estimated by the Kalman filter to provide full state knowledge to a controller. A natural question arises: how does the dynamics of either the controller or the estimator affect the dynamics of the overall (combined) controller/estimator system?

First, a description of the closed-loop system is provided. Here only linear and autonomous systems are considered. Also, only the *regulation* case is considered where the goal is to drive the true states to zero. A controller with full state feedback is given by $\mathbf{u} = -L\mathbf{x}^{\text{true}}$, where L is a constant gain. Substituting this expression into Eq. (12.21) with the definition $\mathbf{x}^{\text{true}} \equiv \mathbf{x}$ gives

$$\dot{\mathbf{x}}^{\text{true}} = (F - BL)\mathbf{x}^{\text{true}} \tag{12.141}$$

The closed-loop controller dynamics are given by eigenvalues of $(F - BL)$. The control gain, L, is chosen to give some desired closed-loop response. Now consider the continuous-time version of the Kalman filter estimate equation, given by

$$\dot{\hat{\mathbf{x}}} = F\hat{\mathbf{x}} + K(\mathbf{y} - H\hat{\mathbf{x}}) \tag{12.142}$$

Assuming no noise on $\mathbf{y}$, so that $\mathbf{y} = H\mathbf{x}^{\text{true}}$, then the error dynamics of the estimator, defined by Eq. (12.117), are given by

$$\dot{\boldsymbol{\epsilon}} = (F - KH)\boldsymbol{\epsilon} \tag{12.143}$$

The error dynamics of the estimator are given by eigenvalues of $F - KH$. In practice the control input uses the estimated states, with

$$\mathbf{u} = -L\hat{\mathbf{x}} \tag{12.144}$$

Substituting $\hat{\mathbf{x}} = \mathbf{x}^{\text{true}} - \boldsymbol{\epsilon}$ into (12.144), and then substituting the resultant into Eq. (12.21) leads to

$$\dot{\mathbf{x}}^{\text{true}} = (F - BL)\mathbf{x}^{\text{true}} + BL\boldsymbol{\epsilon} \tag{12.145}$$

Hence, the closed-loop dynamics of the overall controller/estimator system are given by

$$\begin{bmatrix} \dot{\mathbf{x}}^{\text{true}} \\ \dot{\boldsymbol{\epsilon}} \end{bmatrix} = \begin{bmatrix} F - BL & BL \\ 0_{n \times n} & F - KH \end{bmatrix} \begin{bmatrix} \mathbf{x}^{\text{true}} \\ \boldsymbol{\epsilon} \end{bmatrix} \tag{12.146}$$

Since the matrix in Eq. (12.146) is block-triangular, its eigenvalues are given by the eigenvalues of $F - BL$ and $F - KH$. This shows that the overall controller/estimator system can be designed by separating the estimator from the controller, which is exactly the *separation theorem.* A block diagram of the overall controller/estimator system is shown in Fig. 12.10.

Example 12.5. In this example a linear Kalman filter is used to estimate for the states of the inverted pendulum. A full state controller is then used to bring the pendulum to its zero (vertical) position. The following parameters are used for the model: $M = 2$ kg, $m = 0.1$ kg, $\ell = 0.1$ m, and $c = 2$ Ns/m. Note that this system is unstable. The continuous-time system is converted to discrete-time using a sampling interval of 0.01 s. The discrete-time system matrices are given by

$$\Phi = \begin{bmatrix} 1 & -3.8687 \times 10^{-2} & 9.9485 \times 10^{-3} & -1.2597 \times 10^{-4} \\ 0 & 1.3910 & 3.9241 \times 10^{-4} & 1.1273 \times 10^{-2} \\ 0 & -8.1945 & 9.8924 \times 10^{-1} & -3.8687 \times 10^{-2} \\ 0 & 82.860 & 8.3117 \times 10^{-2} & 1.3910 \end{bmatrix}$$

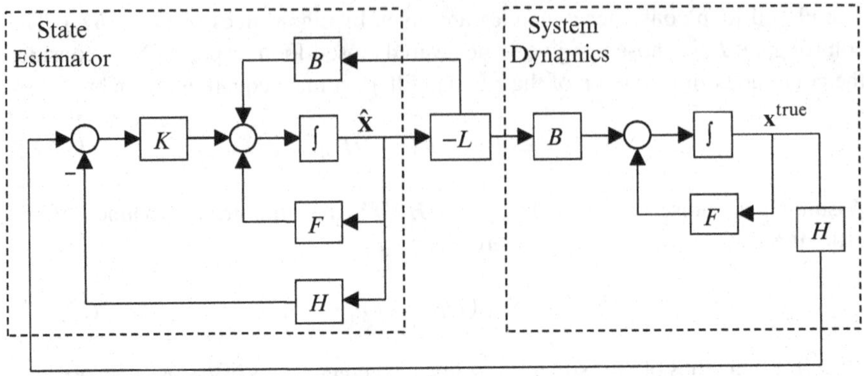

Fig. 12.10 Overall controller/estimator system

$$\Gamma = \begin{bmatrix} 2.5761 \times 10^{-5} \\ -1.9620 \times 10^{-4} \\ 5.3794 \times 10^{-3} \\ -4.1558 \times 10^{-2} \end{bmatrix}$$

The true and estimated initial states and the initial covariance are given by

$$\mathbf{x}^{\text{true}}(t_0) = \hat{\mathbf{x}}(t_0) = \begin{bmatrix} 0 \ (10\pi/180) \ 0 \ 0 \end{bmatrix}^T$$

$$P_0 = \begin{bmatrix} (1 \times 10^{-4})^2 & 0 & 0 & 0 \\ 0 & (0.15\pi/180)^2 & 0 & 0 \\ 0 & 0 & 0.1^2 & 0 \\ 0 & 0 & 0 & (0.1\pi/180)^2 \end{bmatrix}$$

The estimated initial and true states are equal, which is generally not true in practice. It is assumed here that the filter has converged before the controller is enabled so that the estimated states will be close to their respective true states and small values in P_0 are appropriate. The outputs are position and angle. Measurements are generated using a zero-mean Gaussian white-noise process with standard deviations of 0.001 m for position and $0.001\pi/180$ rad for the angle. The state matrices for the estimate are the same as the truth so no process noise is required. The Kalman filter provides estimates for position, velocity, angle, and angular rate for use in the controller.

The controller is given by $u_k = -L\hat{\mathbf{x}}_k$. The control gain, L, is chosen to stabilize the system within 1 s. This is achieved by:

$$L = \begin{bmatrix} -8.2701 \ -2.1249 \times 10^3 \ -5.3521 \ -18.386 \end{bmatrix}$$

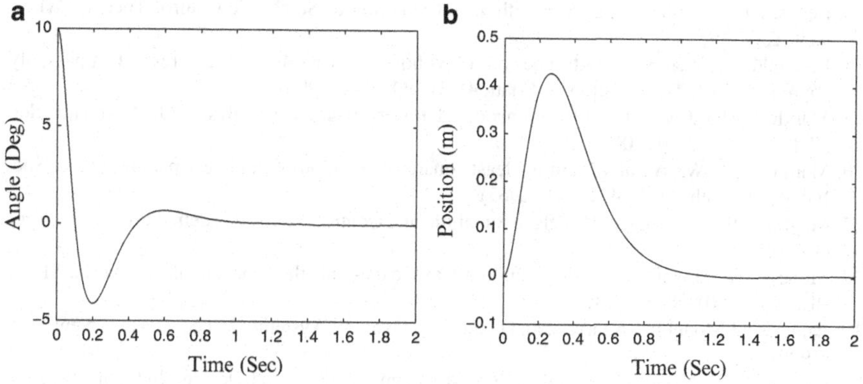

Fig. 12.11 Controlled response of the inverted pendulum. (**a**) Angle. (**b**) Position

Figure 12.11a shows a plot of the controlled pendulum angle response. The angle starts at 10° and then swings to the other side, finally coming to a near resting point of zero. Small oscillations are present at steady state, which are due to the noise in the state estimates. A plot of the controlled cart position response is shown in Fig. 12.11b. The cart starts at its initial resting point of zero and then returns to that same point once the pendulum is stabilized.

References

1. Arnold, W.F., Laub, A.J.: Generalized eigenproblem algorithms and software for algebraic Riccati equations. Proc. IEEE **72**(12), 1746–1754 (1984)
2. Bierman, G.J.: Factorization Methods for Discrete Sequential Estimation. Academic Press, Orlando (1977)
3. Chen, C.T.: Linear System Theory and Design. Holt, Rinehart and Winston, New York (1984)
4. Crassidis, J.L., Junkins, J.L.: Optimal Estimation of Dynamic Systems, 2nd edn. CRC Press, Boca Raton (2012)
5. Dorf, R.C., Bishop, R.H.: Modern Control Systems. Addison Wesley Longman, Menlo Park (1998)
6. Fagin, S.L.: Recursive linear regression theory, optimal filter theory, and error analysis of optimal systems. In: 1964 IEEE International Convention Record, pp. 216–240 (1964)
7. Franklin, G.F., Powell, J.D., Workman, M.: Digital Control of Dynamic Systems, 3rd edn. Addison Wesley Longman, Menlo Park (1998)
8. Gelb, A. (ed.): Applied Optimal Estimation. MIT Press, Cambridge (1974)
9. Golub, G.H., Van Loan, C.F.: Matrix Computations, 3rd edn. The Johns Hopkins University Press, Baltimore (1996)
10. Jazwinski, A.H.: Stochastic Processes and Filtering Theory. Academic Press, New York (1970)
11. Kalman, R.E., Bucy, R.S.: New results in linear filtering and prediction theory. J. Basic Eng. 95–108 (1961)
12. LePage, W.R.: Complex Variables and the Laplace Transform for Engineers. Dover Publications, New York (1980)

13. Lewis, F.L.: Optimal Estimation with an Introduction to Stochastic Control Theory. Wiley, New York (1986)
14. Lizarralde, F., Wen, J.T.Y.: Attitude control without angular velocity measurement: A passivity approach. IEEE Trans. Automat. Contr. **41**(3), 468–472 (1996)
15. Markley, F.L., Carpenter, J.R.: Generalized linear covariance analysis. J. Astronaut. Sci. **57**(1 & 2), 233–260 (2009)
16. Marquardt, D.W.: An algorithm for least-squares estimation of nonlinear parameters. J. Soc. Ind. Appl. Math. **11**(2), 431–441 (1963)
17. Maybeck, P.S.: Stochastic Models, Estimation, and Control, vol. 1. Academic Press, New York (1979)
18. Moler, C., van Loan, C.: Nineteen dubious ways to compute the exponential of a matrix. SIAM Rev. **20**(4), 801–836 (1978)
19. Palm, W.J.: Modeling, Analysis, and Control of Dynamic Systems, 2nd edn. Wiley, New York (1999)
20. Phillips, C.L., Nagle, H.T.: Digital Control System Analysis and Design, 2nd edn. Prentice Hall, Englewood Cliffs (1990)
21. Psiaki, M.L., Martel, F., Pal, P.K.: Three-axis attitude determination via Kalman filtering of magnetometer data. J. Guid. Contr. Dynam. **13**(3), 506–514 (1990)
22. Rao, S.S.: Engineering Optimization: Theory and Practice, 3rd edn. Wiley, New York (1996)
23. Rauch, H.E., Tung, F., Striebel, C.T.: Maximum likelihood estimates of linear dynamic systems. AIAA J. **3**(8), 1445–1450 (1965)
24. Sedlack, J., Hashmall, J.: Accurate magnetometer/gyroscope attitudes using a filter with correlated sensor noise. In: Proceedings of the Flight Mechanics/Estimation Theory Symposium, pp. 293–303. NASA-Goddard Space Flight Center, Greenbelt (1997)
25. Slotine, J.J.E., Li, W.: Applied Nonlinear Control. Prentice Hall, Englewood Cliffs (1991)
26. Sorenson, H.W.: Parameter Estimation, Principles and Problems. Marcel Dekker, New York (1980)
27. Stengel, R.F.: Optimal Control and Estimation. Dover Publications, New York (1994)
28. Żak, S.H.: Systems and Control. Oxford University press, New York (2003)

Appendix: Computer Software

All of the examples shown in the text have been programmed and simulated using MATLAB®. A website of these programs, listed by chapter, can be found at

http://www.buffalo.edu/~johnc/space_book.htm

For general information regarding MATLAB or related products, please consult MathWorks, Inc. at

http://www.mathworks.com

It has been our experience that to thoroughly understand the intricacies of a subject matter in this text, one must learn from basic fundamentals first. Although computer routines can provide some insights to the subject, we feel that they may hinder rigorous theoretical studies that are required to properly comprehend the material. Therefore, we strongly encourage students to program their own computer routines, using the codes provided from the website for verification purposes only. We have decided not to include a disk of programs with the text so that up-to-date versions of the computer programs can be maintained on the website. The programs have been written so that anyone with even a terse background in MATLAB should be able to comprehend the relationships between the examples in the text and the coded scripts. We hope that the reader will use these programs in the spirit that they are given; to supplement their reading and understanding of the material in printed text in order to bridge the gap between theoretical studies and practical applications.

Limit of Liability/Disclaimer of Warranty: The computer programs are provided as a service to readers. While the authors have used their best efforts in preparing these programs, they make no representation or warranties with respect to the accuracy or completeness of the programs. The book publisher (Springer), the authors, the authors' employers, or MathWorks, Inc. shall not be liable for any loss of profit or any other commercial or noncommercial damages, including, but not limited to, special, incidental, consequential, or other damages.

F.L. Markley and J.L. Crassidis, *Fundamentals of Spacecraft Attitude Determination and Control*, Space Technology Library 33, DOI 10.1007/978-1-4939-0802-8,
© Springer Science+Business Media New York 2014

Appendix. Computer Software

Index

F.L. Markley and J.L. Crassidis, *Fundamentals of Spacecraft Attitude Determination and Control*, Space Technology Library 33, DOI 10.1007/978-1-4939-0802-8,

© Springer Science+Business Media New York 2014

V

W

Printed in the United States
By Bookmasters